T0202946

# Lecture Notes in Artificial Intelligence 12986

Subseries of Lecture Notes in Computer Science

## Series Editors

Randy Goebel
*University of Alberta, Edmonton, Canada*

Yuzuru Tanaka
*Hokkaido University, Sapporo, Japan*

Wolfgang Wahlster
*DFKI and Saarland University, Saarbrücken, Germany*

## Founding Editor

Jörg Siekmann
*DFKI and Saarland University, Saarbrücken, Germany*

More information about this subseries at http://www.springer.com/series/1244

Carlos Soares · Luis Torgo (Eds.)

# Discovery Science

24th International Conference, DS 2021
Halifax, NS, Canada, October 11–13, 2021
Proceedings

*Editors*
Carlos Soares ⓘ
Universidade do Porto and Fraunhofer
Portugal AICOS
Porto, Portugal

Luis Torgo ⓘ
Dalhousie University
Halifax, NS, Canada

ISSN 0302-9743 ISSN 1611-3349 (electronic)
Lecture Notes in Artificial Intelligence
ISBN 978-3-030-88941-8 ISBN 978-3-030-88942-5 (eBook)
https://doi.org/10.1007/978-3-030-88942-5

LNCS Sublibrary: SL7 – Artificial Intelligence

This Springer imprint is published by the registered company Springer Nature Switzerland AG
The registered company address is: Gewerbestrasse 11, 6330 Cham, Switzerland

# Preface

This volume contains the papers selected for presentation at the 24th International Conference on Discovery Science (DS 2021), which was organized to be held in Halifax, Canada, during October 11–13, 2021. Due to the restrictions associated with the COVID-19 pandemic, the conference was moved online and held virtually over the same time period.

DS is a conference series that started in 1986. Held every year, DS continues its tradition as the unique venue for the latest advances in the development and analysis of methods for discovering scientific knowledge, coming from machine learning, data mining, and intelligent data analysis, along with their application in various scientific domains. In particular, major areas selected for DS 2021 included the folllowing: applications (including a relevant number of papers addressing the COVID-19 pandemic), classification, data streams, feature selection, graph and network mining, neural networks and deep learning, preferences, recommender systems, representation learning, responsible artificial intelligence, and spatial, temporal and spatiotemporal data.

DS 2021 received 76 international submissions that were carefully reviewed by three or more Program Committee (PC) members or external reviewers, with a few exceptions. After a rigorous reviewing process, 15 regular papers and 21 short papers were accepted for presentation at the conference and publication in the DS 2021 volume.

We would like to sincerely thank all people who helped this volume come into being and made DS 2021 a successful and exciting event. In particular, we would like to express our appreciation for the work of the DS 2021 PC members and external reviewers who helped assure the high standard of accepted papers. We would like to thank all authors of submissions to DS 2021, without whose work it would not have been possible to have such high-quality contributions in the conference.

We are grateful to the Steering Committee chair, Sašo Džeroski, for his extraordinary support in critical decisions concerning the event plan, particularly important in these challenging times. We are also very grateful to the Program Committee chairs of DS 2020, Annalisa Appice and Grigorios Tsoumakas, for all the information provided, which made the whole organization much easier. We wish to express our thanks to the local organization chairs, David Langstroth, Nuno Moniz, Paula Branco, Vitor Cerqueira, and Yassine Baghoussi, for their support and incredible work. We would like to express our deepest gratitude to all those who served as organizers, session chairs, and hosts, who made great efforts to meet the online challenge to make the virtual conference a real success. Finally, our thanks are due to Alfred Hofmann and Anna Kramer of Springer for their continuous support and work on the proceedings. We are grateful to Springer for a special issue on Discovery Science to be published in the Machine Learning journal. All authors were given the possibility to extend and rework versions of their papers presented at DS 2021 for a chance to be published in this prestigious journal. For DS 2021, Springer also supported a Best Student Paper Award, which was given to Bart Bussmann and his co-authors, Jannes Nys and Steven Latré, for their paper "Neural Additive Vector Autoregression Models for Causal Discovery in Time Series."

This paper presents high quality work on a very relevant topic and is complemented with the materials to reproduce it. We would like to congratulate the authors for this achievement.

September 2021                                              Carlos Soares
                                                           Luis Torgo

# Organization

## Program Committee Chairs

Carlos Soares          Universidade do Porto and Fraunhofer Portugal AICOS, Portugal
Luis Torgo             Dalhousie University, Canada

## Program Committee

Martin Atzmueller            Tilburg University, The Netherlands
Colin Bellinger              National Research Council of Canada, Canada
Paula Branco                 Ottawa University, Canada
Alberto Cano                 Virginia Commonwealth University, USA
Daniel Castro Silva          University of Porto, Portugal
Michelangelo Ceci            University of Bari Aldo Moro, Italy
Victor Cerqueira             Dalhousie University, Canada
Bruno Cremilleux             University of Caen Normandy, France
Nicola Di Mauro              University of Bari Aldo Moro, Italy
Ivica Dimitrovski            Ss. Cyril and Methodius University in Skopje, North Macedonia
Wouter Duivesteijn           Eindhoven University of Technology, The Netherlands
Sašo Džeroski                Jožef Stefan Institute, Slovenia
Johannes Fürnkranz           Johannes Kepler University Linz, Austria
Mohamed Gaber                Birmingham City University, UK
Dragan Gamberger             Rudjer Bošković Institute, Croatia
Manco Giuseppe               Italian National Research Council, Italy
Kouichi Hirata               Kyushu Institute of Technology, Japan
Dino Ienco                   Irstea, France
Nathalie Japkowicz           American University, USA
Stefan Kramer                Johannes Gutenberg University Mainz, Germany
Vincenzo Lagani              Ilia State University, Georgia
Pedro Larranaga              University of Madrid, Spain
Nada Lavrač                  Jožef Stefan Institute, Slovenia
Tomislav Lipic               Rudjer Bošković Institute, Croatia
Gjorgji Madjarov             Ss. Cyril and Methodius University in Skopje, North Macedonia
Rafael Gomes Mantovani       Federal University of Technology – Paraná, Brazil
Elio Masciari                Institute for High Performance Computing and Networking, Italy
Nuno Moniz                   University of Porto, Portugal
Anna Monreale                University of Pisa, Italy

| | |
|---|---|
| Catarina Oliveira | Portucalense University, Portugal |
| Sageev Oore | Dalhousie University and Vector Institute, Canada |
| Rita P. Ribeiro | University of Porto, Portugal |
| George Papakostas | International Hellenic University, Greece |
| Ruggero G. Pensa | University of Turin, Italy |
| Pedro Pereira Rodrigues | University of Porto, Portugal |
| Bernhard Pfahringer | University of Waikato, New Zealand |
| Gianvito Pio | University of Bari Aldo Moro, Italy |
| Pascal Poncelet | LIRMM Montpellier, France |
| Jan Ramon | Inria, France |
| André L. D. Rossi | Universidade Estadual Paulista, Brazil |
| Kazumi Saito | University of Shizuoka, Japan |
| Tomislav Smuc | Rudjer Bošković Institute, Croatia |
| Marina Sokolova | University of Ottawa and Dalhousie University, Canada |
| Jerzy Stefanowski | Poznan University of Technology, Poland |
| Luis Filipe Teixeira | University of Porto, Portugal |
| Herna Viktor | University of Ottawa, Canada |
| Baghoussi Yassine | University of Porto, Portugal |
| Albrecht Zimmermann | University of Caen Normandy, France |

## Additional Reviewers

Ana Mestrovic
Daniela Santos
Emanuel Guberovic
Fabrizio Lo Scudo
Francesca Pratesi
Francesco S. Pisani
Francesco Spinnato
Isabel Rio-Torto
Ivan Grubisic
João Almeida

João Alves
Leonardo Ferreira
Miha Keber
Narjes Davari
Nima Dehmamy
Nunziato Cassavia
Paolo Mignone
Rafaela Leal
Roberto Interdonato

# Contents

**Graph and Network Mining**

**Machine Learning for COVID-19**

**Neural Networks and Deep Learning**

**Preferences and Recommender Systems**

**Representation Learning and Feature Selection**

**Responsible Artificial Intelligence**

## Spatial, Temporal and Spatiotemporal Data

# Applications

# Automated Grading of Exam Responses: An Extensive Classification Benchmark

Jimmy Ljungman[✉], Vanessa Lislevand, John Pavlopoulos,
Alexandra Farazouli, Zed Lee, Panagiotis Papapetrou, and Uno Fors

Department of Computer and Systems Sciences,
Stockholm University, Stockholm, Sweden
{jimmy.ljungman,lislevand,ioannis,alexandra.farazouli,zed.lee,
panagiotis,uno}@dsv.su.se

**Abstract.** Automated grading of free-text exam responses is a very challenging task due to the complex nature of the problem, such as lack of training data and biased ground-truth of the graders. In this paper, we focus on the automated grading of free-text responses. We formulate the problem as a binary classification problem of two class labels: low- and high-grade. We present a benchmark on four machine learning methods using three experiment protocols on two real-world datasets, one from Cyber-crime exams in Arabic and one from Data Mining exams in English that is presented first time in this work. By providing various metrics for binary classification and answer ranking, we illustrate the benefits and drawbacks of the benchmarked methods. Our results suggest that standard models with individual word representations can in some cases achieve competitive predictive performance against deep neural language models using context-based representations on both binary classification and answer ranking for free-text response grading tasks. Lastly, we discuss the pedagogical implications of our findings by identifying potential pitfalls and challenges when building predictive models for such tasks.

**Keywords:** Automated grading · Answer grading · Natural language processing · Machine learning

## 1 Introduction

The assessment of students' knowledge and understanding in academic courses plays a crucial role in effective teaching and usually takes place in the form of distance or on-the-spot formal examinations. Exams are typically composed of different types of questions, such as multiple-choice, true/false, and free-text questions. Free-text questions, in particular, have benefits in terms of students' learning as students are required to recall external knowledge, and freely and concisely elaborate on a subject [14].

Supported by the AutoGrade project of Stockholm University.

C. Soares and L. Torgo (Eds.): DS 2021, LNAI 12986, pp. 3–18, 2021.
https://doi.org/10.1007/978-3-030-88942-5_1

**Table 1.** Example of an exam question, three responses, their grades (Scores), and their corresponding class labels (Converted; H for high, L for low).

| Question | Define the task of supervised learning and provide an example of a supervisied learning method. | Scores | Converted |
|---|---|---|---|
| Answer 1 | Supervised learning refers to artificial systems that learn to improve their performance by experience. In the case of supervised learning, experience corresponds to labeled objects, while performance refers to the ability to assign class labels to new (previously unseen) objects. An example of a supervised learning method is a decision tree | 10 | **H** |
| Answer 2 | Supervised learning refers to artificial systems that learn to improve their performance when given unseen data. Their experience is improved by the use of training examples that may contain class labels. An example of a supervised learning method is a decision tree built using training examples clustered into two groups corresponding to two class labels | 2.5 | **L** |
| Answer 3 | Supervised learning refers to clustering methods that group similar objects together using some similarity metric. An example of a supervised method is K-means | 0.0 | **L** |

Automated grading of free-text responses to exam questions is divided into two focus areas: Automated Essay Scoring (AES) and Automated Short-Answer Grading (ASAG). AES focuses on language style, coherence, and ideas organization, while ASAG focuses on assessing the correctness of the responses in terms of semantic content, completeness, and relevance to the questions. Moreover, open-ended responses, which include short-text answers (between a phrase and a paragraph [3]), may not be strictly limited in the word count, as students are usually allowed to write more than a paragraph. Although in this paper we use an ASAG approach in the focus of assessment and the use of computational methods, we are not limited to short-text answers, and we also include student answers up to 1,000 words.

Free-text responses are written in natural language and hence require more effort to grade. In Table 1 we see an example of an exam question (within the area of Data Mining), along with three graded responses. Observe that Answers 1 and 3 are easy to grade since the first corresponds to a model answer while the third one is clearly erroneous. However, Answer 2 contains several ambiguities that may require extra attention from the examiner. In this sense, automated grading of free-text questions can support academic staff by eliminating the burden of grading large amounts of exam papers. In particular, it can function as an effective means for filtering out easy to grade exam responses (e.g., Answers 1, 3) and focusing on those that require particular attention (e.g., Answer 2), while minimizing potential marking errors.

In this paper, we propose an experiment workflow for automated grading of exam responses and provide a benchmark on four machine learning methods for this task. Our main contributions are summarized as follows:

– We provide a workflow for automated grading of exams, which we refer to as `AutoGrade`, that includes an extensive benchmark of four machine-learning methods.
– We propose three evaluation protocols for the problem, which we refer to as FLATTEN, LOQO, and Q-BASED, depending on the objective of the experiment (see Sect. 4).
– We benchmark `AutoGrade` for the binary classification problem (high/low-grade) on two real-world datasets, one set of exams written in English and another set of exams in Arabic.
– We discuss the pedagogical implications of our findings and identify potential pitfalls and challenges when building predictive models for the task.

## 2 Related Work

Various approaches to address the automated grading task have been explored from traditional handcrafted features [15], similarity-based [5] and rule-based [28] formulations, to recent deep learning techniques [10,23].

To be more concrete, earlier studies involve manually defining features that try to capture the similarity of answers in multiple levels [23]. Those features include surface features, such as answer length ratios and lexical overlap; semantic similarity measures, which refer to sentence embedding distances or knowledge bases like WordNet [20]; and syntactic features like dependency graph alignment and measures based on the part-of-speech tag distribution in the student answers [18]. Leacock and Chodorow's [12] C-Rater is a scoring engine designed to grade short answers matching the teacher and students' answers based on the number of concepts (main key points from the curriculum) identified in students' answers matched with those concepts found in model answers. The matching is based on a set of rules and a canonical representation of the texts using syntactic variation, anaphora, morphological variation, synonyms, and spelling correction. Results show a 84% agreement with the human graders.

Likewise, Rodrigues and Oliveira [22] have matched student answers with referred correct answers by cosine similarity after preprocessing. Basu and Vanderwende [2] and Süzen et al. [25] have made use of clustering techniques, creating groups and subgroups of answers based on the similarity between the model and the student answer. Nandini and Maheswari [16] have proposed a system which compares the model with the students' answers and the grading depends on the keywords used and context of the phrases conveying meaning. This system employs question and answer classifications in the training phase according to the type and focus of the question and answer. Then, in the prediction phase the system examines if the question labels (factual, inductive, analytical) match the answer labels (chronological/sequence ordering, descriptive/enumerative, compare and contrast, problem/solution, cause and effect) and finally assigns grades accordingly and provides feedback.

Deep learning methods have also been adopted for the task of automated grading of free-text responses [10]. Sung et al. [24] have provided a solution based on BERT by adding a feed forward layer and fine-tuning the model to predict whether student answers are correct, incorrect, or contradictory compared to a reference answer. They have shown that fine-tuning BERT for the problem of short-answer grading increases its effectiveness and outperforms most of the existing techniques on a large number of tasks. However, little attention has been given to further pre-training the language model incorporating domain-specific knowledge. Similarly, in this work we fine tuned the state-of-the-art BERT-based transformer XLM-RoBERTa (XLM-R) [4], which is further described in Sect. 4.

Based on the above, it becomes apparent that similarity-based solutions, bag-of-words feature-based solutions, and deep learning based solutions can achieve promising results for the problem of automated free-text response grading. Nonetheless, there is still no extensive benchmark of these lines of solutions and no common experimental workflow for the problem.

## 3   Background

Let $\mathcal{D} = \{D_1, \ldots, D_n\}$ be a dataset of $n$ written exams. Each exam $D_i \in \mathcal{D}$ is defined over three sets of $k_i$ questions, answers, and grades. More concretely, each exam $D_i$ is a triplet $<\mathcal{Q}_i, \mathcal{A}_i, \mathcal{G}_i>$, with $\mathcal{Q}_i = \{Q_{(i,1)}, \ldots, Q_{(i,k_i)}\}$, $\mathcal{A}_i = \{A_{(i,1)}, \ldots, A_{(i,k_i)}\}$, and $\mathcal{G}_i = \{G_{(i,1)}, \ldots, G_{(i,k_i)}\}$. Each $Q_{(i,j)}$ and $A_{(i,j)}$ are texts and each $G_{(i,j)} \in \mathbb{R}$.

In our formulation we assume that the exams in $\mathcal{D}$ correspond to a single exam instance of a particular course. We additionally assume that each exam may not necessarily contain the same set and number of questions (e.g., in the case where exam instances are generated using a question bank). Figure 1 illustrates the structure of an exam $D_1$.

In this paper we assume a simplified formulation that considers the binary version of the problem, where $G_{(i,j)} \in \{L, H\}$, with $L, H$ corresponding to *low* and *high* grade, respectively.

*Problem 1.* **The AutoGrade Problem.** Given an answer $A_{(i,j)}$ to an exam question $Q_{i,j}$ with a corresponding ground-truth grade $G_{(i,j)}$, we want to learn a classification model $f_{\mathcal{D}}$ on a training dataset $\mathcal{D}_{train}$ with $f_{\mathcal{D}} : A_{(i,j)} \rightarrow \{L, H\}$, such that $||\tilde{G}_{(i,j)} - G_{(i,j)}||$ is minimized, with $\tilde{G}_{(i,j)}$ being the predicted value.

## 4   The AutoGrade Workflow

We propose a three-step workflow for solving Problem 1. The steps include (1) feature extraction and tokenization, (2) modeling, and (3) evaluation. Next we describe each step.

**Step I: Feature Extraction and Tokenization.** Depending on the ML model used in Step II we perform feature extraction as a pre-processing step. For the first three models we use TF-IDF, as it quantifies the relevance of a word

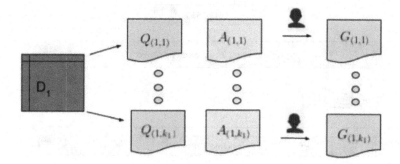

**Fig. 1.** Structure of the exam $D_1$. The exam contains a set of questions, a set of responses and a grade per response.

in a document with respect to the whole corpus [7].[1] The fourth model is a transformer-based method and uses contextualized sub-word embeddings based on SentencePiece [9] with a unigram language model [8].

**Step II: Modeling.** We employ four ML methods described below.

*Distance-Based Methods.* We use k-Nearest Neighbour (k-NN) that computes the distance similarity between a test instance and each training instance. For a new test instance, the output is the majority label of the k most similar training instances. For this paper, we use both 1-NN and 3-NN. The distance metric used is Euclidean distance.

*Statistical Methods.* We use Logistic Regression (LR) that applies the logistic (sigmoid) function on top of a linear regressor, yielding a score from 0 to 1, which can then be turned into a binary score with the use of a threshold. LR can estimate the likelihood that a new sample belongs to a class using the coefficients found in a classification environment.

*Ensemble-Based Methods.* Ensemble-based methods predict outcomes of new instances by either majority voting or averaging on multiple models. The algorithm used for this method is Random Forest (RF), where the ensemble models are decision trees. Each tree is trained on a subset of the training data, and RF classifies new instances using the majority vote of these trees. We set the number of decision trees in the forest to 100.

*Transformer-Based Methods.* Lastly, XLM-R is chosen as the transformer-based algorithm, which is a masked language model, fine-tuned to classify responses as either low- or high-grade. XLM-R is a multilingual Masked Language Model (MLM) that is based on Transformers [4,26], which are based on a multi-head self-attention mechanism. XLM-R is a combination of the Cross-lingual Language Model (XLM) and a Robustly Optimized BERT pretraining approach

---

[1] We use TfidfVectorizer for feature extraction with all parameters set to default.

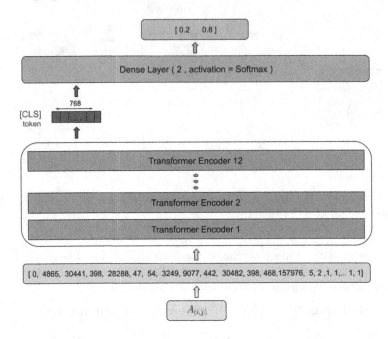

**Fig. 2.** Architecture of our XLM-R-based model.

(RoBERTa). XLM has achieved state-of-the-art results on cross-lingual classification, unsupervised and supervised machine translation [11]. It changes the MLM by using continuous streams of text instead of sentence pairs and demostrated that MLM can give strong cross-lingual features for pretraining. RoBERTa is an improved form of Bidirectional Encoder Representations from Transformer (BERT) [13], achieving state-of-the-art results on GLUE, SQuAD and RACE. This method proves that training BERT with alternative design choices and with more data, including the CommonCrawl News dataset, improves the performance on downstream tasks.

The main idea of XLM-R is to follow the XLM approach as closely as possible by training RoBERTa with the MLM objective on a huge multilingual dataset. The amount of training data is significantly increased, especially for low resource languages, by training XLM-R on one hundred languages using CommonCrawl data[2], in contrast to previous works such as BERT or XLM that are trained on Wikipedia. For our task, we build the model by adding a Feed Forward Neural Network (FFNN) on top of the pre-trained XLM-R. As Fig. 2 shows, we feed XLM-R with vectors (length of 200) that contain the input ids of the answers, and subsequently, XLM-R feeds the FFNN with its output, i.e., the context-aware embedding (length of 768) of the [CLS] token of each sentence. The number of nodes in the output layer is the same as the number of classes, i.e., 2, and each node in the output layer uses a softmax activation function, predicting a class probability between 0 and 1.

---

[2] https://commoncrawl.org/.

**Step III: Evaluation Protocol.** We experimented with three different strategies to create our training, development, and test sets. Initially, following an experimental strategy that we call FLATTEN, we trained and evaluated our models on responses from all questions, without discrimination. We then experimented with a leave-one-question-out strategy (dubbed LOQO), where all the responses of a particular question (one at a time) are considered as the test set. Thirdly, we trained and tested on the best and the balanced questions supported separately. We call this setting Q-BASED. The former two strategies, FLATTEN and LOQO, are based on the following in-domain transfer-learning assumption: *in-domain knowledge can be transferred among the responses to different questions*, i.e., a model trained to grade the responses of some questions will be able to grade the responses to other questions.

## 5 Empirical Evaluation

### 5.1 Datasets

We employ two real-world datasets in our experiments from two different subject areas: Data Mining (in English), referred to as DAMI and Cyber-crime (in Arabic), referred to as AR-ASAG. Since in this paper we pose automated grading as a binary classification problem, for both datasets the responses are labeled with two distinct classes as follows:

- **H:** If the grade of the response is greater than or equal to half of the maximum grade of the question.
- **L:** If the grade is lower than half of the maximum grade of the question.

In Fig. 3 we provide a summary of the two datasets regarding the length of the responses (in # of words) and the overall class label distribution. Moreover, Fig. 4 shows the class label distribution per question for each dataset. As observed, DAMI is more balanced (L:H of 41/59) compared to AR-ASAG (L:H of 33/67). Next, we provide more details about each dataset.

**DAMI.** This dataset is obtained from Stockholm University and contains a collection of three different sets of exams of a masters level course in Data Mining, written in English. The dataset carry a total of 31 questions and 1,131 responses. The grade of each response may range either from 0–10, from 0–8, or from 0–5.

As we observe in Fig. 3(a), high-grade responses (in green) are generally longer than low-grade responses while there are responses that are up to a thousand words long. Moreover, the number of responses and class distribution vary per question (see Fig. 4(a)) For example, Questions 1 and 2 (i.e., Q1 and Q2) have the highest support while low-grade is the most frequent class. By contrast, the vast majority of the rest of the questions comprised more high-grade responses. Q12 and Q16, for example, have only three low-grade responses, compared to 36 and 33 high-grade ones, respectively. Q1 has the best class balance while maintaining a respectable support with regards to other questions in the dataset. The vocabulary size of the dataset is 5,627.

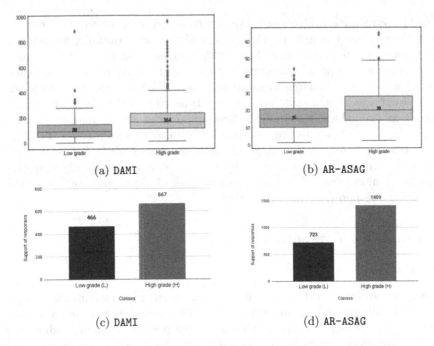

**Fig. 3.** Response length distribution per class, in words for both datasets (top row). Different scale of the vertical axis is used across the two, due to the much lengthier DAMI responses. Class distribution per dataset (bottom row). (Color figure online)

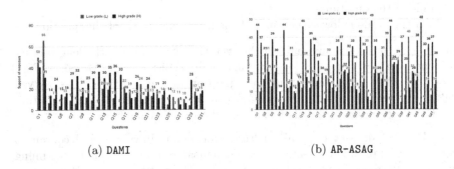

**Fig. 4.** Support of responses per question and per class for the DAMI (left) and the AR-ASAG (right) datasets. There are 37 responses on average per question on the DAMI dataset, and there are 44 responses on average on the AR-ASAG dataset.

**AR-ASAG.** This dataset has been collected by Ouahrani and Bennouar [17] and contains three different exams taken by three different classes of Masters students. Each exam contains 16 short answer questions with a total of 48 questions and 2,133 responses. The responses are independently graded by two human experts, using a scale from 0 (completely incorrect) to 5 (perfect answer). The average grade of the two annotators is treated as the gold standard.

Figure 3(b) shows that high-grade responses are generally longer than short-grade responses also for this dataset. However, both low- and high-grade AR-ASAG responses are shorter than the ones of DAMI, with the latter being up to two orders of magnitude longer and comprising more outliers compared to AR-ASAG. Just as for DAMI, the question-wise class balance and support differs between questions in AR-ASAG (Fig. 4(b)). For some questions, e.g., Question 1 (Q1), the class balance is heavily imbalanced towards high-grade responses, while it is heavily imbalanced towards low-grade responses on other questions like Question 30 (Q30). The vocabulary size of the dataset is 6,366.

## 5.2   Setup

Our experiments are based on the three evaluation strategies described in Sect. 4. All four ML methods are benchmarked alongside a random classifier that predicts a class randomly with equal probability. The datasets are partitioned in a stratified manner to maintain the class distribution. We keep 70% of the texts for training, 15% for development, and 15% for testing. In LOQO, we report the results directly on the test set, because we do not use any split. In FLATTEN and Q-BASED, the average score of five and ten repetitions of Monte Carlo Cross Validation (MCCV) is reported, respectively [21].[3]

All models except XLM-R uses the default parameters set by the sklearn library. For XLM-R, the implementation is based on Keras and Ktrain. We use XLM-R's tokenizer which is based on SentencePiece, and we define the maximum length of the text to be 200. We train our model for 15 epochs on LOQO and 30 epochs for the rest of the experiments, with a learning rate equal to 1e-5 and batch size equal to six. These parameters are selected after fine-tuning. The source code is publicly available in our GitHub repository.[4]

We report precision, recall, F1, AUC, AUPRC, and accuracy. We also report the Spearman's $\rho$ correlation coefficient between a model's (high-grade) probability estimates and the respective ground truth grades for all the responses in the test set. High correlation reflects the model's ability to effectively rank the responses correctly. The k-NN models, as well as the random classifier, have no notion of probability estimates, as the former are distance-based while the latter is random guessing. Therefore, Spearman's $\rho$ is not reported for these models.

## 5.3   Results

**FLATTEN.** As it can be observed in Table 2, XLM-R, RF, and LR are the best performing models in terms of accuracy, AUPRC, and AUC on DAMI. All three models also have equal performance in Spearman's rho. For precision, recall, and F1, the results do not lead to any clear winner. RF achieves a higher F1 on the high-grade responses due to a high recall, while XLM-R achieves the highest F1

---

[3] The size of FLATTEN restricts us from running the models for more repetitions.
[4] https://github.com/dsv-data-science/autograde_DS2021.

**Table 2.** Results of 5 repetitions of MCCV following the FLATTEN strategy, for both datasets (DAMI support 70/100; AR-ASAG support 109/211). RC is a random classifier.

| | | P | R | F1 | AUPRC | $\rho$ | AUC | Ac | P | R | F1 | AUPRC | $\rho$ | AUC | Ac |
|---|---|---|---|---|---|---|---|---|---|---|---|---|---|---|---|
| | | DAMI | | | | | | | AR-ASAG | | | | | | |
| XLM-R | L | 0.68 | 0.63 | **0.65** | 0.70 | 0.49 | 0.78 | **0.73** | **0.64** | 0.45 | **0.52** | **0.60** | 0.48 | **0.75** | **0.73** |
| | H | 0.76 | 0.79 | 0.77 | 0.82 | | | | **0.76** | 0.87 | **0.81** | **0.85** | | | |
| RF | L | **0.72** | 0.52 | 0.61 | **0.71** | **0.50** | **0.79** | 0.72 | 0.62 | 0.21 | 0.31 | 0.52 | 0.39 | 0.69 | 0.69 |
| | H | 0.73 | **0.86** | **0.79** | **0.84** | | | | 0.70 | 0.94 | 0.80 | 0.81 | | | |
| LR | L | 0.67 | 0.46 | 0.54 | 0.70 | 0.49 | 0.77 | 0.69 | 0.62 | 0.09 | 0.16 | 0.54 | 0.41 | 0.71 | 0.67 |
| | H | 0.69 | 0.85 | 0.76 | **0.84** | | | | 0.68 | **0.97** | 0.80 | 0.83 | | | |
| 3-NN | L | 0.52 | **0.80** | 0.63 | 0.52 | – | 0.67 | 0.61 | 0.53 | 0.32 | 0.40 | 0.45 | – | 0.66 | 0.67 |
| | H | **0.78** | 0.48 | 0.60 | 0.72 | | | | 0.71 | 0.85 | 0.78 | 0.75 | | | |
| 1-NN | L | 0.53 | 0.64 | 0.56 | 0.48 | – | 0.62 | 0.62 | 0.49 | 0.41 | 0.44 | 0.40 | – | 0.59 | 0.65 |
| | H | 0.72 | 0.60 | 0.64 | 0.66 | | | | 0.72 | 0.78 | 0.75 | 0.71 | | | |
| RC | L | 0.42 | 0.51 | 0.46 | 0.40 | – | 0.49 | 0.51 | 0.33 | **0.49** | 0.40 | 0.34 | – | 0.51 | 0.49 |
| | H | 0.60 | 0.50 | 0.55 | 0.59 | | | | 0.66 | 0.49 | 0.56 | 0.67 | | | |

on the low-grade ones due to equally high performance in precision and recall. LR underperforms both models in terms of F1.

For AR-ASAG, all models perform worse on the low-grade responses compared to DAMI. RF and LR even underperform RC on low-grade F1. This could be due to the fact that DAMI has a better class balance than AR-ASAG (see Fig. 3). Even so, XLM-R performs reasonably well, having the highest score on all metrics except for recall, and has similar accuracy, AUC, and Spearman's rho as in DAMI. XLM-R also performs better in terms of AUPRC, F1, and recall on the high-grade class, but drops in recall on the low-grade class which leads to an F1 reduction. The lower AUPRC on the low-grade responses also indicates that the above-mentioned weakness holds for all classification thresholds.

**LOQO.** Table 3 presents the performance of all models when we followed the LOQO experiment on both datasets. One can observe similar behaviors as for FLATTEN, namely that RF, LR, and XLM-R are the best performing models on DAMI, and that XLM-R is the best performing model on AR-ASAG. However, in this case, the performance of XLM-R deteriorates much more when we move from DAMI to AR-ASAG. Moreover, even though LR has the best performance on F1 on the high-grade class, it fails to classify the low-grade responses. When we focus on Spearman's rho, all models fail. Similarly to FLATTEN, when focusing on DAMI, XLM-R achieves the best F1 for low-grade, while RF achieves the best F1 for high-grade.

**QBASED.** We use the two questions with the highest support and equal balance to experiment with for DAMI (Q1 and Q2; see Fig. 4(a)). For AR-ASAG, we use Q13 and Q33. Q13 has the best support in the dataset while Q33 has the best class balance (see Fig. 4(b)).

Table 4 presents the results when following the Q-BASED approach on DAMI. For the most balanced question of the two (Q1; topmost), LR and RF are the

**Table 3.** Results following the LOQO strategy.

| | | P | R | F1 | AUPRC | $\rho$ | AUC | Ac | P | R | F1 | AUPRC | $\rho$ | AUC | Ac |
|---|---|---|---|---|---|---|---|---|---|---|---|---|---|---|---|
| | | DAMI | | | | | | | AR-ASAG | | | | | | |
| XLM-R | L | 0.66 | 0.59 | **0.62** | 0.64 | 0.43 | 0.74 | **0.71** | **0.40** | 0.44 | **0.42** | 0.39 | 0.17 | 0.56 | 0.58 |
| | H | **0.74** | 0.78 | 0.76 | 0.77 | | | | 0.69 | 0.65 | 0.67 | 0.70 | | | |
| RF | L | **0.69** | 0.52 | 0.59 | **0.70** | 0.53 | 0.77 | 0.71 | 0.35 | 0.26 | 0.30 | 0.34 | 0.03 | 0.50 | 0.58 |
| | H | 0.72 | **0.84** | **0.77** | **0.81** | | | | 0.65 | 0.75 | 0.70 | 0.70 | | | |
| LR | L | **0.60** | 0.56 | 0.58 | 0.65 | 0.45 | 0.71 | 0.67 | 0.28 | 0.13 | 0.18 | 0.33 | 0 | 0.49 | 0.58 |
| | H | 0.71 | 0.74 | 0.72 | 0.77 | | | | 0.64 | **0.82** | **0.72** | 0.66 | | | |
| 3-NN | L | 0.42 | **0.87** | 0.57 | 0.44 | – | 0.55 | 0.46 | 0.31 | 0.36 | 0.33 | 0.33 | – | 0.47 | 0.50 |
| | H | 0.66 | 0.17 | 0.27 | 0.62 | | | | 0.63 | 0.58 | 0.60 | 0.64 | | | |
| 1-NN | L | 0.44 | 0.85 | 0.58 | 0.44 | – | 0.55 | 0.49 | 0.35 | 0.46 | 0.39 | 0.35 | – | 0.50 | 0.52 |
| | H | 0.70 | 0.25 | 0.37 | 0.62 | | | | 0.65 | 0.55 | 0.60 | 0.65 | | | |
| RC | L | 0.41 | 0.47 | 0.44 | 0.41 | – | 0.49 | 0.50 | 0.35 | **0.49** | 0.41 | 0.35 | – | 0.50 | 0.51 |
| | H | 0.59 | 0.52 | 0.55 | 0.59 | | | | 0.66 | 0.52 | 0.58 | 0.65 | | | |

**Table 4.** Results of 10 repetitions of MCCV on Q1 and Q2 per class for DAMI.

| | | P | R | F1 | AUPRC | $\rho$ | AUC | Ac | P | R | F1 | AUPRC | $\rho$ | AUC | Ac |
|---|---|---|---|---|---|---|---|---|---|---|---|---|---|---|---|
| | | Question 1 | | | | | | | Question 2 | | | | | | |
| XLM-R | L | 0.77 | 0.76 | 0.74 | 0.88 | 0.58 | 0.82 | 0.74 | 0.70 | 0.86 | 0.76 | 0.82 | 0.30 | 0.66 | 0.65 |
| | H | 0.75 | 0.71 | 0.71 | 0.81 | | | | 0.27 | 0.24 | 0.25 | 0.58 | | | |
| RF | L | 0.80 | **0.88** | **0.83** | **0.91** | **0.73** | **0.89** | **0.81** | 0.68 | 0.93 | 0.78 | 0.82 | 0.30 | 0.63 | 0.65 |
| | H | **0.88** | 0.75 | **0.79** | 0.88 | | | | 0.20 | 0.10 | 0.13 | 0.52 | | | |
| LR | L | 0.81 | 0.84 | 0.82 | **0.91** | 0.67 | 0.88 | **0.81** | 0.67 | **1.00** | **0.80** | **0.86** | **0.40** | **0.71** | **0.67** |
| | H | 0.83 | 0.78 | **0.79** | **0.89** | | | | 0.00 | 0.00 | 0.00 | **0.59** | | | |
| 3-NN | L | 0.84 | 0.69 | 0.75 | 0.80 | – | 0.82 | 0.76 | 0.67 | 0.64 | 0.65 | 0.70 | – | 0.50 | 0.44 |
| | H | 0.72 | **0.85** | 0.77 | 0.69 | | | | 0.28 | 0.34 | 0.30 | 0.40 | | | |
| 1-NN | L | **0.86** | 0.69 | 0.74 | 0.75 | – | 0.77 | 0.76 | 0.67 | 0.56 | 0.60 | 0.68 | – | 0.50 | 0.52 |
| | H | 0.73 | **0.85** | 0.77 | 0.69 | | | | 0.33 | 0.44 | 0.37 | 0.36 | | | |
| RC | L | 0.53 | 0.59 | 0.50 | 0.53 | – | 0.48 | 0.50 | **0.72** | 0.45 | 0.55 | 0.67 | – | 0.47 | 0.51 |
| | H | 0.47 | 0.51 | 0.48 | 0.50 | | | | **0.36** | **0.62** | **0.45** | 0.35 | | | |

**Table 5.** Results of 10 repetitions of MCCV on Q13 and Q33 per class AR-ASAG.

| | | P | R | F1 | AUPRC | $\rho$ | AUC | Ac | P | R | F1 | AUPRC | $\rho$ | AUC | Ac |
|---|---|---|---|---|---|---|---|---|---|---|---|---|---|---|---|
| | | Question 13 | | | | | | | Question 33 | | | | | | |
| XLM-R | L | **0.85** | **0.75** | **0.78** | 0.95 | **0.65** | **0.98** | **0.93** | 0.78 | 0.75 | 0.73 | **0.94** | **0.81** | **0.93** | 0.79 |
| | H | **0.94** | 0.99 | **0.96** | 0.99 | | | | **0.86** | 0.81 | **0.77** | 0.95 | | | |
| RF | L | 0.00 | 0.00 | 0.00 | 0.77 | 0.54 | 0.90 | 0.78 | 0.77 | **1.00** | **0.85** | 0.91 | 0.78 | 0.91 | **0.80** |
| | H | 0.78 | **1.00** | 0.88 | 0.97 | | | | 0.80 | 0.62 | 0.68 | 0.94 | | | |
| LR | L | 0.00 | 0.00 | 0.00 | 0.84 | 0.53 | 0.92 | 0.78 | **0.82** | 0.59 | 0.63 | 0.81 | 0.55 | 0.77 | 0.69 |
| | H | 0.78 | **1.00** | 0.88 | 0.98 | | | | 0.61 | 0.78 | 0.68 | 0.82 | | | |
| 3-NN | L | 0.50 | 0.30 | 0.37 | 0.62 | – | 0.76 | 0.84 | 0.56 | 0.30 | 0.35 | 0.63 | – | 0.66 | 0.59 |
| | H | 0.84 | 1.00 | 0.91 | 0.89 | | | | 0.56 | **0.83** | 0.65 | 0.66 | | | |
| 1-NN | L | 0.67 | 0.45 | 0.51 | 0.54 | – | 0.71 | 0.86 | 0.44 | 0.44 | 0.43 | 0.52 | – | 0.50 | 0.51 |
| | H | 0.87 | 0.97 | 0.91 | 0.87 | | | | 0.49 | 0.55 | 0.51 | 0.55 | | | |
| RC | L | 0.20 | 0.45 | 0.27 | 0.26 | – | 0.45 | 0.48 | 0.48 | 0.44 | 0.43 | 0.51 | – | 0.50 | 0.54 |
| | H | 0.75 | 0.49 | 0.57 | 0.78 | | | | 0.59 | 0.66 | 0.59 | 0.57 | | | |

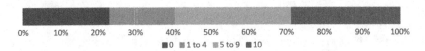

**Fig. 5.** Grades of the DAMI dataset in %.

best performing models with top-performance in F1, AUPRC, AUC, Spearman's rho, and accuracy. Moreover, the k-NN models perform reasonably well where 1-NN has a better F1 score on the high-grade class than XLM-R.

On the more supported but imbalanced question (Q2), the performance on all models deteriorates. This is especially true for predicting the minority class of the question (H), where RC outperforms in precision, recall, and F1, which means that all models fail to distinguish high-grade responses from low-grade ones. Focusing on specific evaluation measures, LR holds the best Spearman's rho, and AUPRC for both classes, with XLM-R and RF close behind. However, looking at precision, recall, and F1 of RF and LR, the accuracy stems from the high performance for the frequent low-grade class and is outweighed by the very low performance on the high-grade class. The high AUPRC and low precision, recall, and F1 scores of the models indicate that tuning the classification threshold would lead to better performance.

In AR-ASAG, the results follow similar patterns as was observed in DAMI. That is, the models deteriorate on the more imbalanced but higher supported question (Q13) in contrast to the more balanced but less supported question (Q33). Interestingly, while all other models fail to distinguish the two classes, XLM-R actually produces better scores on all metrics except Spearman's rho for Q13 than for Q33. XLM-R is also the best performing model overall for both questions (Table 5).

### 5.4   Lesson Learned

The experimental results indicate that there is not a single best-performing method for our classification task. When training and evaluating the responses of a single question (Q-BASED), the class balance generally seems more significant than the overall support. For the more supported but imbalanced questions, the transformer-based method (XLM-R) produces better results than any other method (DAMI Q2; AR-ASAG Q13). For the more balanced but less supported questions, the overall best method is, however, unclear. XLM-R, RF, and LR have similar performances on Q1, while XLM-R performs best on Q33. When training on all the responses except for those of a particular question (LOQO), the best performing XLM-R and RF achieve relatively high scores. As the models are not trained on any of the responses to the tested question, this indicates that the models could be used to assist the grader in evaluating responses to a new question. For AR-ASAG, however, results deteriorate compared to Q-BASED. The reason behind this is probably due to a lower semantic coverage between the responses of different questions. When training on all the responses without discriminating between the questions (FLATTEN), similar scores are achieved for DAMI, but are much higher for AR-ASAG. This is probably because the responses to all questions are used to train the models. Hence, problems such as a limited coverage between the questions can be bypassed (e.g., XLM-R achieves a correlation of 0.48, compared to 0.17 in LOQO).

Creating a system for automated grading of free-text responses is a challenging task due to complexity issues related to the data quality [27] and faith in the process [3]. More specifically these challenges regard 1) the limited availability of data; 2) the quality of such data; and 3) the complex process of analysis and understanding of natural language. The lack of publicly available test datasets for training and evaluating automated grading systems is a crucial challenge. Most datasets containing students' responses in exams are not easily shared due to privacy reasons, leading researchers to be limited in the use of their own institutions' data. Consequently, generalised conclusions cannot be made from the developments of systems based on limited data as this could lead to bias in terms of a specific domain, student group and teacher approach. The assessment of free-text responses requires the grader to extract the exact meaning of the students' answers and appropriately evaluate the knowledge obtained by the students. This type of question requires a deep understanding of a topic from students in order to be able to recall knowledge and synthesise their response freely. Furthermore, they are widely used in assessments because they can showcase complex learning goals more holistically and effectively [1,6]. This type of questions is very beneficial in terms of developing students' cognitive skills and demonstrating knowledge in short texts [14]. Through the assessment of such questions, the teacher may have a clearer view of the level of students' understanding of the subject in question and their knowledge gaps.

Taking a closer look at DAMI we observe that over 50% of the grades assigned to students' answers from the examiner fall into either the min or max grade (e.g., they are 0s and 10s; see Fig. 5). This observation led us further analyse the key requirements of such type of questions which are to test whether a student 1) knows/understands a concept and 2) can elaborate, explain, justify or provide examples in a specific topic or concept. Requirement 1 is a prerequisite to meet requirement 2 and grading is employed accordingly, meaning that the examiner awards maximum points when both requirements are met (correct answers), minimum points when none of them is met (incorrect answers) and in-between points when there are contradictory or mislead answers in the responses. Drawing inspiration from this, our future goal is to create an automated grading system which will award maximum and minimum scores when high-precision threshold is exceeded. In this way, the volume of the answers that the grader should pay special attention could be significantly decreased. In addition, we also aim to highlight the problematic areas of the grey answers and suggest a grade. This would help the graders in terms of revealing possible patterns of mistakes and providing formative feedback to the students.

Developing automated grading systems is expected to reduce teachers' burden and augment their judgment, by providing useful insights for their students and enhance the reliability of grading. However, we note that human graders will still be required to supervise and calibrate such systems as the main assessors in grading. Another challenge is that human graders may judge free-text responses differently, making it hard to find a 100% agreed "golden standard". The exact details of how human graders will employ such systems are not yet known. A possible solution could be borrowed from user-generated comment moderation, where toxicity detection systems are meant to assist and not substitute the expert [19]. In grading, a similar approach could be adopted, with systems being used with two thresholds, one to filter clearly low and one to filter clearly high graded responses. Any remaining responses, which systems were less confident to classify, are probably the ones that human graders would like to assess with full attention. This scenario, however, remains to be investigated in future work.

## 6    Conclusions

This work provided a classification benchmark on automated grading of student responses into high- and low-grade classes. Machine learning methods based on distance, regression, ensemble, and Transformers were applied on two real-world datasets, one from Cyber-crime exams in Arabic and one from Data Mining exams in English that was introduced in this work. The experimental results indicate that there is no single best-performing method for our classification task. When training and assessing per question, class balance seem more important than overall support. When the questions have many responses, XLM-R performs the best overall. When training on all responses except for some, XLM-R and RF perform considerably well in Arabic and in English. However, when training on all responses except for a particular question, XLM-R and RF perform considerably well only in English. In future work, we aim to study the collaboration between automated grading systems and human evaluators by investigating semi-automated grading and the employment of rationale generation mechanisms that could assist the grader decide between a low- and a high-grade.

**Acknowledgements.** This work was supported by the AutoGrade project (https://datascience.dsv.su.se/projects/autograding.html) of the Dept. of Computer and Systems Sciences at Stockholm University.

## References

1. Anderson, R.C., Biddle, W.B.: On asking people questions about what they are reading. In: Psychology of Learning and Motivation, vol. 9, pp. 89–132. Elsevier (1975)
2. Basu, S., Jacobs, C., Vanderwende, L.: Powergrading: a clustering approach to amplify human effort for short answer grading. TACL 1, 391–402 (2013)

3. Burrows, S., Gurevych, I., Stein, B.: The eras and trends of automatic short answer grading. IJAIED **25**(1), 60–117 (2014). https://doi.org/10.1007/s40593-014-0026-8

4. Conneau, A., et al.: Unsupervised cross-lingual representation learning at scale. In: ACL, pp. 8440–8451 (2020)

5. Horbach, A., Pinkal, M.: Semi-supervised clustering for short answer scoring. In: International Conference on Language Resources and Evaluation (2018)

6. Karpicke, J., Roediger, H.: The critical importance of retrieval for learning. Science **319**, 966–968 (2008)

7. Kim, S.W., Gil, J.M.: Research paper classification systems based on TF-IDF and LDA schemes. Hum. Cent. Comput. Inf. Sci. **9**, 30 (2019). https://doi.org/10.1186/s13673-019-0192-7

8. Kudo, T.: Subword regularization: improving neural network translation models with multiple subword candidates (2018)

9. Kudo, T., Richardson, J.: SentencePiece: a simple and language independent subword tokenizer and detokenizer for neural text processing (2018)

10. Kumar, S., Chakrabarti, S., Roy, S.: Earth mover's distance pooling over Siamese LSTMs for automatic short answer grading. In: IJCAI, pp. 2046–2052 (2017)

11. Lample, G., Conneau, A.: Cross-lingual language model pretraining. arXiv preprint arXiv:1901.07291 (2019)

12. Leacock, C., Chodorow, M.: C-rater: Automated scoring of short-answer questions. Comput. Humanit. **37**(4), 389–405 (2003)

13. Liu, Y., et al.: RoBERTa: a robustly optimized BERT pretraining approach. arXiv preprint arXiv:1907.11692 (2019)

14. McDaniel, M., Anderson, J.L., Derbish, M.H., Morrisette, N.: Testing the testing effect in the classroom. Eur. J. Cogn. Psychol. **19**, 494–513 (2007)

15. Mohler, M., Bunescu, R., Mihalcea, R.: Learning to grade short answer questions using semantic similarity measures and dependency graph alignments. In: ACL, pp. 752–762 (2011)

16. Nandini, V., Uma Maheswari, P.: Automatic assessment of descriptive answers in online examination system using semantic relational features. J. Supercomput. **76**(6), 4430–4448 (2018). https://doi.org/10.1007/s11227-018-2381-y

17. Ouahrani, L., Bennouar, D.: AR-ASAG an Arabic dataset for automatic short answer grading evaluation. In: LREC, pp. 2634–2643 (2020)

18. Padó, U.: Get semantic with me! The usefulness of different feature types for short-answer grading. In: COLING, pp. 2186–2195 (2016)

19. Pavlopoulos, J., Malakasiotis, P., Androutsopoulos, I.: Deep learning for user comment moderation. In: WOAH, pp. 25–35. ACL (2017)

20. Pedersen, T., Patwardhan, S., Michelizzi, J., et al.: WordNet: similarity-measuring the relatedness of concepts. In: AAAI, vol. 4, pp. 25–29 (2004)

21. Picard, R.R., Cook, R.D.: Cross-validation of regression models. J. Am. Stat. Assoc. **79**, 575–583 (1984)

22. Rodrigues, F., Oliveira, P.: A system for formative assessment and monitoring of students' progress. Comput. Educ. **76**, 30–41 (2014)

23. Saha, S., Dhamecha, T.I., Marvaniya, S., Sindhgatta, R., Sengupta, B.: Sentence level or token level features for automatic short answer grading?: use both. In: AIED, pp. 503–517 (2018)

24. Sung, C., Dhamecha, T.I., Mukhi, N.: Improving short answer grading using transformer-based pre-training. In: AIED, pp. 469–481 (2019)

25. Süzen, N., Gorban, A.N., Levesley, J., Mirkes, E.M.: Automatic short answer grading and feedback using text mining methods. Procedia Comput. Sci. **169**, 726–743 (2020)
26. Vaswani, A., et al.: Attention is all you need. In: NeurIPS, pp. 6000–6010 (2017)
27. Williamson, D., Xi, X., Breyer, F.: A framework for evaluation and use of automated scoring. Educa. Meas. Issues Pract. **31**, 2–13 (2012)
28. Willis, A.: Using NLP to support scalable assessment of short free text responses. In: BEA, pp. 243–253 (2015)

# Automatic Human-Like Detection
# of Code Smells

Chitsutha Soomlek[1]($^{(\boxtimes)}$)(iD), Jan N. van Rijn[2](iD), and Marcello M. Bonsangue[2](iD)

[1] Department of Computer Science, Khon Kaen University, Khon Kaen, Thailand
chitsutha@kku.ac.th
[2] Leiden Institute of Advanced Computer Science, Leiden University,
Leiden, The Netherlands
{j.n.van.rijn,m.m.bonsangue}@liacs.leidenuniv.nl

**Abstract.** Many code smell detection techniques and tools have been proposed, mainly aiming to eliminate design flaws and improve software quality. Most of them are based on heuristics which rely on a set of software metrics and corresponding threshold values. Those techniques and tools suffer from subjectivity issues, discordant results among the tools, and the reliability of the thresholds. To mitigate these problems, we used machine learning to automate developers' perception in code smells detection. Different from other existing machine learning used in code smell detection we trained our models with an extensive dataset based on more than 3000 professional reviews on 518 open source projects. We conclude by an empirical evaluation of the performance of the machine learning approach against PMD, a widely used metric-based code smell detection tool for Java. The experimental results show that the machine learning approach outperforms the PMD classifier in all evaluations.

**Keywords:** Code smells · Machine learning · Software engineering

## 1 Introduction

Code smells are properties of the source code that may indicate either flaws in its design or some poor implementation choices. Differently from a bug, a code smell does not necessarily affect the technical correctness of a program, but rather it may be a symptom of a bad design pattern affecting the quality of a software system. Also, the experimental evaluation shows a direct correlation between code smells and software evolution issues, design vulnerabilities, and software failure in the long run [6,25,30]. Even in well-managed and designed projects, code smells could be inadvertently added into the code by inexperienced developers, and as such it is very important to detect them early in the design process [18,28].

Typically, code refactoring is a solution to the design problem coming from code smells [10,11]. Due to the subjectivity of their definition, detection of code smells, and the associated refactoring, are non-trivial tasks. The manual detection process requires tremendous efforts and is infeasible for large-scale software.

© Springer Nature Switzerland AG 2021
C. Soares and L. Torgo (Eds.): DS 2021, LNAI 12986, pp. 19–28, 2021.
https://doi.org/10.1007/978-3-030-88942-5_2

Commonly used automated approaches in tools and academic prototypes are search-based, metric-based, symptom-based, visualization-based, probabilistic-based, and machine learning [2,13,24]. The metric-based approach defines code smells systematically using a fixed set of metrics and corresponding threshold values. It is the most commonly used approach in both open-source and commercial tools and the idea has been adopted for more than a decade. The major problems of the metric-based approach are: (1) matching subjective perception of developers who often perceive code smells differently than the metrics classification and (2) reliability of the threshold values. Currently, many well-known code smell detectors adopt the metrics and their threshold values from Lanza and Marinescu's work [15] in 2006 as reference points. However, finding the best-fit threshold values for a certain type of code smell requires significant efforts on data collection and calibration. For example, Lanza and Marinescu's analysis [15] is based on their manual review of few dozen mid-size projects. Moreover, the concept of code smells was introduced and cataloged more than 20 years ago. During this period, programming languages have been evolving to today's modern programming language which comprises both functional and advanced object-oriented features. To obtain more reliable code smell detection results, human perceptions on design issues should be integrated into an automated analysis. Machine learning is one of the promising solutions for this case because it enables a machine to mimic the intelligence and capabilities of humans to perform many functions.

Following this direction, we define the following research questions:

- RQ1: Can we mimic a developer's perception of a code smell?
- RQ2: How does machine learning perform when comparing to existing tools?

We use a large dataset of industry projects reviewed by developers [17], we clean and prepare the data so to be utilized in training a machine learning model, and finally, we compare the results with those coming from a modern tool using a metric-based approach. This includes the validation of the two approaches concerning the perception by human experts. We make the dataset publicly available on OpenML [29]. Each of the above steps can be considered as a contribution to our work on its own. From the experimental results, we can conclude a better performance of the machine learning approach for code smells detection, compared to the tools based on static rules.

## 2   Related Work

While there is no general agreement on the definition of code smells or of their symptoms, many approaches have been introduced in the literature to automate code smell identification. There exist both commercial and open-source tools. Detection approaches can be classified from guided manual inspection to fully automated: manual, symptom-based, metric-based, probabilistic, visualization-based, search-based, and cooperative-based [12]. The metric-based approach is the most used technique in both research and tools. In this case, the generic code smell identification process involves source code analysis and matching the examining source code to the code smell definition and specification by using specific software metrics [24]. Object-oriented metrics suite [4] and their threshold values are commonly used in the detection process. The accuracy of the metric-based approach depends on (i) the metric selection, (ii) choosing the right threshold values, and (iii) on their interpretation.

In recent years, many studies adopted artificial intelligence and machine learning algorithms for code smell identification [27], classification [5,8,13,16], and prioritization [9,20]. Machine learning techniques provide a suitable alternative to detect code smells without human intervention and errors, solving the difficulty in finding threshold values of metrics for identification of bad smells, lack of consistency between different identification techniques, and developers' subjectiveness in identifying code smells. Those techniques differ in the type of code smell detected, the algorithms used, the size of the dataset used for training, and the evaluation approaches and performance metrics.

Originally, Kreimer [14] proposed a prediction model based on decision trees and software metrics to detect blobs and long methods in Java code. The results were evaluated experimentally with regard to accuracy. The efficiency in using decision trees has been confirmed in mid-size open-source software projects [1]. For an extensive review of the existing approaches and their comparison, we refer the reader to [13].

Two closest works to ours are [8] and [5]. They both applied machine learning techniques to detect data class, blob, feature envy, and long method. In [8] the results of 16 supervised machine learning algorithms were evaluated and compared using 74 software systems from Qualitas Corpus. A large set of object-oriented metrics were extracted and computed from this training data by using deterministic rules in the sampling and manual labeling process. Labeling results were confirmed by master students. Due to their limited software engineering experience, relying on them can be considered a limitation of this work. The authors of [5] repeated the work of the authors of [8] to reveal critical limitations. In contrast, our work is based on a dataset constructed by more than 3000 reviews by human experts on more than 500 mid- to large-size software projects.

## 3    Dataset Construction and Pre-processing

To conduct our empirical study, we need to collect (1) data or a reliable and up-to-date dataset reporting human perceptions on a set of code smells that is large enough to train a machine learning model, and (2) software metrics of the software projects matching to (1). To achieve this, we first surveyed existing datasets and discussed with their contributors the scientific definitions and data collection process. From this study, we selected the dataset that fitted our requirements best. Finally, we used the selected dataset to define (1) a set of software projects, (2) the types of code smells we were interested in and their corresponding detectors, and (3) a set of software metrics to be extracted from the set of software projects.

**The MLCQ Dataset.** Madeyski and Lewowski contributed the MLCQ data set which contains industry-relevant code smells with severity levels, links to code samples, the location of each code smell in the Java project, and background information of the experts [17]. More specifically, they collaborated with 26 professional software developers to review the code samples with regard to four types of code smells both at class and function levels: blob, data class, feature envy, and long method. The reviews are based on four severity levels, i.e., *critical, major, minor*, and *none*. The *none* severity level is assigned to a code sample when the expert does not consider it as a code smell, i.e., a negative result. If however the sample is marked by any of the other severity levels then the sample should be considered as a positive result and thus as a code smell. In summary, the samples contain 984, 1057, 454, and 806 positive cases of blob, data class, feature envy, and long method, respectively. For negative results, 3092, 3012, 2883, and 2556 samples are available. The MLCQ dataset captures the contemporary understanding of professional developers towards code smells from 524 active Java open-source projects. This improves on other existing datasets that either rely on graduate and undergraduate students to collect and review software projects or use automatic code smell detectors tools that impose threshold values from legacy literature, to identify certain types of code smells. However, MLCQ is not ready to be used for our research as it does not provide any software metric of the code samples and software projects. Therefore, we expanded the dataset accordingly. Furthermore, since there are 14,853 reviews on 4,770 code samples, it is often more than one expert review on the same code samples. We thus needed to pre-process the dataset. Next, we describe this step.

**Pre-processing and Code Smell Selection.** Expert reviewers can disagree on the interpretation of a code smell on a given code sample, in particular to the severity levels assigned to it. To combine the multiple reviews on a code sample to a single result, we need to ensure the validity of the results. In other words, the combined result must stay positive when the majority of the reviewers did not evaluate the severity level of the code sample as *none*. Likewise, the combined results must be negative if the majority assigned *none* to the sample. Thus, we mapped the severity level of a review to a corresponding numerical severity score ($critical = 3$, $major = 2$, $minor = 1$, $none = 0$), and calculated the average severity score for each code sample. The result of the last step can be considered as the *average review score*. If the experts agree on the definition of a code smell but they have different opinions for the severity level, the approach still can identify which sample is a certain type of code smells and which is not.

Table 1 presents the distribution of the number of reviews together with the average of the average review score and standard deviation calculated for each group of reviews separately. Surprisingly, for blob and data class, the review results have no significant difference. However, when considering long method and feature envy, we noticed a considerable disagreement on the reviews. For blob, the variation is mostly within one category, either in *critical* to *major* (occurring in one sample with 6 reviews) or in *minor* to *none* (as we see in the cases of 3 to 5 reviews per sample). A similar situation happens when we consider the data class. In case of feature envy, however, when there are four

**Table 1.** Distribution of the number of reviews (first column). Per code smell, we show for each encountered number of reviewers, the average of the average review score (ARS) that was granted, as well as the average standard deviation (calculated across standard deviations per code sample).

| Rev. | Blob | | Data class | | Feature envy | | Long method | |
|---|---|---|---|---|---|---|---|---|
| | No. of Samp. | Average ARS ± Std | No. of Samp. | Average ARS ± Std | No. of Samp. | Average ARS ± Std | No. of Samp. | Average ARS ± Std |
| 1 | 1562 | 0.00 ± 0.00 | 1566 | 0.00 ± 0.00 | 1954 | 0.00 ± 0.00 | 1967 | 0.00 ± 0.02 |
| 2 | 147 | 0.88 ± 0.12 | 147 | 0.49 ± 0.00 | 82 | 0.20 ± 0.16 | 162 | 0.68 ± 1.16 |
| 3 | 409 | 0.50 ± 0.70 | 411 | 0.66 ± 0.69 | 303 | 0.43 ± 0.73 | 302 | 0.00 ± 0.31 |
| 4 | 198 | 0.66 ± 0.84 | 196 | 0.79 ± 0.79 | 70 | 0.73 ± 0.97 | 73 | 0.96 ± 1.96 |
| 5 | 39 | 0.73 ± 0.88 | 39 | 0.87 ± 0.88 | 6 | 0.53 ± 0.87 | 7 | 0.97 ± 2.20 |
| 6 | 1 | 2.33 ± 0.52 | 1 | 0.00 ± 0.00 | 0 | N/A | 0 | N/A |

reviews for a sample, the combined severity score has a variation from *none* to *major*, which indicates a diversity of reviewers' opinions. This is not an incident, as the numbers in the table refers to 70 samples. The situation for the long method is even worse. There are 73 code samples with four reviews, leading to an average of the average review score of 0.96 with a standard deviation of 1.96. There are 7 code samples with 5 reviewers, having an average of the average review score of 0.97 and standard deviation of 2.20. The high average standard deviation in these two cases reveal a more spread out disagreement among the human experts. Therefore, we decided to omit feature envy and long method from our experiments.

**Selecting Code Smell Detectors.** Among all the popular tools in the literature [7,19,24], we selected PMD [22] because it is an active source code analyzer that can automatically analyze a Java project to identify our targeted code smells and also long method, by using metrics and threshold values. By exploring PMD's documentation, relevant sets of Java rules, and PMD source code, we found that the latest version of PMD (6.35.0) available at the time of writing this paper still adopts metrics and threshold values from [15]. Therefore, we decided to use PMD as a representative of metric-based code smell detectors based on threshold values from legacy literature to describe the characteristics of a class containing a particular code smell.

PMD detects blobs by using the following metrics: weighted method count (WMC), access to foreign data (AFTD), and tight class cohesion (TCC) [22]. PMD detects data classes by using the following metrics: weight of class (WOC), number of public attributes (NOPA), number of accessor methods (NOAM), and WMC are employed to identify a sign of encapsulation violation, poor data behavior proximity, and strong coupling [22]. For long method, PMD uses 100 lines of code as a default threshold value to indicate excessive method length [22]. We could not find out on what study this value is based on. However, to the best of our knowledge, there seems not to be a common agreement on threshold values for long methods. According to the MLCQ dataset, the average length of code samples identified as a long method is 20.7 lines, a threshold much lower than the one used in PMD.

**Table 2.** Performance of PMD methods against code smells determined by human experts. The assessment of the human expert was labelled as code smell if the average review was higher than 0.75.

| Code smells | TP | TN | FP | FN | Precision | Recall | F1-score |
|---|---|---|---|---|---|---|---|
| Blob | 111 | 1822 | 207 | 186 | 0.349 | 0.374 | 0.361 |
| Data class | 80 | 1987 | 46 | 217 | 0.635 | 0.269 | 0.378 |
| Long method | 80 | 1830 | 326 | 166 | 0.197 | 0.325 | 0.245 |

As a preliminary experiment, we deploy the results from PMD on the classes evaluated in the MLCQ dataset. As such, we compared the detection results against the average review score we calculated from the MLCQ dataset. Because there are a few archives of Java projects in the MLCQ datasets that are either corrupted or no longer exist, we could only analyze and compare 518 projects in total. These can be considered as a baseline of the contemporary understanding of professional developers. Table 2 presents the comparison results in terms of the number of true positive (TP), true negative (TN), false positive (FP), false negative (FN), as well as precision, recall, and F1-Score.

From the comparison results, we can see that the metric-based approach is far from being accurate when considering human experts' perceptions of code smells. The only exception is perhaps the precision of data class (with a poor recall), but we will see later that even in this case, the machine learning approach will perform better. Note that the precision scores are the lowest for long method, indicating the discrepancy between the threshold value set by PMD and the much lower perceived average value calculated from the expert reviews.

**Collecting the Software Metrics.** In order to deploy machine learning models to detect code smells, we need to extract metrics from each code file. PMD is a metric-based code smell detector, and the API allows us to extract the metrics it calculates, which turns out to be a good starting point. When we employed PMD to analyze the 518 Java open-source projects, PMD presented the code smell identification results with the corresponding metrics. However, PMD does not provide any metric information for the healthy classes and methods as well as their locations in the project. In other words, PMD only provides positive cases with a set of metrics and identified locations. To obtain the negative cases, we customized PMD to present relevant metric information for every path PMD traversed. For instance, when the customized PMD is run and the *GodClass* rule (for detecting the code smell blob) for a Java program is called, WMC, AFTD, and TCC are calculated for the examining class. Note that there are cases when TCC cannot be calculated, e.g., there is no violation of a certain rule. In which case, PMD presents *NaN* as the metric value.

Additionally, we employed the Understand tool by SciTools [26] version 6.0 (build 1055) to analyzed the 518 Java projects. Understand is static code analysis and code visualization tool that can analyze 60 metrics for Java, project structure, and relationships among files, classes, functions, and variables. The metrics calculated by Understand also cover the CK metrics suite [4]. As a result, we constructed another set of data containing a wide variety of metrics and a very detailed program structure.

## 4 Empirical Study Definition and Evaluation

Generally, these machine learning models are induced based on a dataset $\mathcal{D} = \{(x_j, y_j) \mid j = 1, \ldots, n\}$ with $n$ datapoints to map an input $x$ to output $f(x)$, which closely represents $y$. In this setting, $x_i$ is typically the vector of numerical features $\mathcal{F}(\mathcal{C})$ from some code block $\mathcal{C}$. To deploy a machine learning classifier on the task of predicting whether a certain piece of code is considered code smell, we need to have a notion of ground truth (labels) and some features that describe labeled pieces of code. The human expert described per class that was inspected whether it is considered to be a certain type of code smell or not, representing our $n$ data points, and for each of these we now have label $y_j$ (human expert assessment). The main challenge is representing a piece of code $\mathcal{C}$ as feature vector $\mathcal{F}(\mathcal{C})$. For this, we will use two sources of features: features extracted by PMD and features extracted by the Understand tool.

Although both Understand and the human experts report a fully-qualified Java name (which can also be at a subclass in a specific file), PMD uses a different convention. Although PMD is also capable of reporting at (sub-)class level, it does not report the fully-qualified Java name, meaning that ambiguities can arise with duplicated class names, when automatically making the mapping with the class labels. We solve this by predicting code smells at the file level, rather than at the class level. As such, we have more observations in our dataset than actual observations from the human expert dataset. For each class in which a code smell was detected, we now need to study all subclasses as well. If in either of these a code smell was detected, we consider this file a positive case.

We are confronted with the following design choice. The human experts have graded the code smell severity with four levels, i.e., *none, minor, major* and *critical*. Additionally, since some code pieces were judged by multiple reviewers, we have a broad range of severity levels. If we were to employ a binary classifier, we have to decide from which severity level we consider a piece of code a positive class (code smell). As such, we have to determine a *severity threshold*. To avoid subjectivity, we run the experiment with various ranges of severity. The categorical assessments of the human experts are averaged as described in Sect. 3, such that severity levels around 0 correspond to a negative class, severity levels around 1 correspond to minor code smell, severity levels around 2 correspond to major code smell, and severity levels around 3 correspond to critical code smell. As such, the assumption is that when increasing the severity threshold, detecting the code smell should become easier for the machine learning approach. We ran the experiment several times, with each severity threshold ranging from 0.25 until 2.50 with intervals of 0.25.

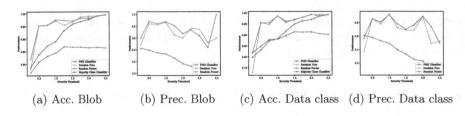

|          (a) Acc. Blob          |          (b) Prec. Blob          |          (c) Acc. Data class          |          (d) Prec. Data class          |

**Fig. 1.** Results of various machine learning classifiers on prediction whether a certain file has a code smell of the indicated type, where Acc. is accuracy and Prec. is precision.

## 5 Results and Discussion

We compared the performance of the machine learning approach in identifying the code smell based on the ground truth decided by human experts. Although we could use any model, we show decision tree [23] and random forest [3], as these both have a good trade-off between performance and interpretability. Both are used as implemented in Scikit-learn [21]. Additionally, we also show the majority class classifier. As most important baseline, we employ the PMD classifier. Indeed, the PMD classifier can also identify code smell based on its own decision rules, and this can be evaluated against the ground truth set by human experts. Note that the PMD classifier is a set of static decision rules, whereas the machine learning models learn these patterns based on the data. Per figure, we show a different performance measure: accuracy and precision. For the majority class classifier we only show accuracy, as it fails to identify any positive class.

Figure 1a and b show the results for blob. The $x$-axis shows the severity threshold at which a certain experiment was run, and the $y$-axis shows the performance of the given experiment. As can be seen, the machine learning approaches outperform both baselines in terms of accuracy and precision. Also in terms of recall, the machine learning models are better than the PMD classifier for most of the severity thresholds (figures omitted). It seems that the random forest has a slight edge over the decision tree classifier, and also focuses slightly more on precision.

For data class, the results in Fig. 1c and d confirm that the machine learning techniques have superior accuracy and precision. Altogether, the results seem to indicate that the machine learning techniques are capable of better identifying blob and data class than the static PMD rules.

## 6 Conclusion and Future Directions

This research intends to mimic contemporary developer's perception of code smell to machine learning and support automated analysis. More specifically, our first research question was 'Can we mimic a developer's perception of a code smell?'

To this aim, we investigated which data we could leverage for building a machine learning classifier. A recent and reliable dataset containing code smells and developers' perceptions of the design flaws are crucial for this. MLCQ contains four types of code smell, due to constraints with other tools we could use two (data class and blob) for this research. This provides for a wide number of classes information whether a developer considers it a code smell or not. As such, we can employ a binary classifier. We enriched this dataset by automatically extracted metrics from two common tools, i.e., Understand and PMD. These features enable us to train machine learning models on the data and make the machine learning model detect code smells. The machine learning models were able to outperform a majority class baseline on all settings.

The second research question was 'How does machine learning perform when comparing to existing tools?' We compared this machine learning model to PMD, a static metric-based code smell detection tool. We employed both the random forest and decision tree classifier, in settings that had to classify code smells from various severity levels. We measured both accuracy and precision. The results indicate that the machine learning-based models outperform the metric-based tool for both code smells.

Finally, we also make the dataset derived from MLCQ and developed in this research publicly available on OpenML [29]. The dataset would elevate and support advanced studies in the research areas.

**Acknowledgements.** We would like to thank Khon Kaen University for awarding KKU Outbound Visiting Scholarship to the first author.

# References

1. Amorim, L., Costa, E., Antunes, N., et al.: Experience report: evaluating the effectiveness of decision trees for detecting code smells. In: 2015 IEEE 26th International Symposium on ISSRE, pp. 261–269. IEEE (2015)
2. Azeem, M.I., Palomba, F., Shi, L., et al.: Machine learning techniques for code smell detection: a systematic literature review and meta-analysis. Inf. Softw. Technol. **108**, 115–138 (2019)
3. Breiman, L.: Random forests. Mach. Learn. **45**(1), 5–32 (2001)
4. Chidamber, S.R., Kemerer, C.F.: Towards a metrics suite for object oriented design. In: Conference Proceedings on Object-Oriented Programming Systems, Languages, and Applications, pp. 197–211 (1991)
5. Di Nucci, D., Palomba, F., Tamburri, D.A., et al.: Detecting code smells using machine learning techniques: are we there yet? In: 2018 IEEE 25th International Conference on SANER, pp. 612–621. IEEE (2018)
6. Elkhail, A.A., Cerny, T.: On relating code smells to security vulnerabilities. In: 2019 IEEE 5th International Conference on BigDataSecurity, IEEE International Conference on HPSC, and IEEE International Conference on IDS, pp. 7–12. IEEE (2019)
7. Fernandes, E., Oliveira, J., Vale, G., et al.: A review-based comparative study of bad smell detection tools. In: Proceedings of the 20th International Conference on Evaluation and Assessment in Software Engineering, pp. 1–12 (2016)
8. Arcelli Fontana, F., Mäntylä, M.V., Zanoni, M., et al.: Comparing and experimenting machine learning techniques for code smell detection. Empir. Softw. Eng. **21**, 1143–1191 (2016). https://doi.org/10.1007/s10664-015-9378-4

9. Fontana, F.A., Zanoni, M.: Code smell severity classification using machine learning techniques. Knowl. Based Syst. **128**, 43–58 (2017)
10. Fowler, M., Beck, K., Brant, W., et al.: Refactoring: Improving the Design of Existing Code. Addison-Wesley Longman Publishing Co. Inc., Boston, USA (1999)
11. Fowler, M.: Refactoring: Improving the Design of Existing Code. Addison-Wesley Professional (2018)
12. Kamaraj, N., Ramani, A.: Search-based software engineering approach for detecting code-smells with development of unified model for test prioritization strategies. Int. J. Appl. Eng. Res. **14**(7), 1599–1603 (2019)
13. Kaur, A., Jain, S., Goel, S., et al.: A review on machine-learning based code smell detection techniques in object-oriented software system(s). Recent Adv. Electr. Electron. Eng. **2021**(14), 290–303 (2021)
14. Kreimer, J.: Adaptive detection of design flaws. Electron. Notes Theoret. Comput. Sci. **141**(4), 117–136 (2005)
15. Lanza, M., Marinescu, R.: Object-Oriented Metrics in Practice: Using Software Metrics to Characterize, Evaluate, and Improve the Design of Object-Oriented Systems. Springer, Heidelberg (2006). https://doi.org/10.1007/3-540-39538-5
16. Liu, H., Jin, J., Xu, Z., et al.: Deep learning based code smell detection. IEEE Trans. Softw. Eng. **47**, 1811–1837 (2019)
17. Madeyski, L., Lewowski, T.: MLCQ: industry-relevant code smell data set. In: Proceedings of the Evaluation and Assessment in Software Engineering, pp. 342–347 (2020)
18. Martin, R.C.: Clean Code: A Handbook of Agile Software Craftsmanship, 1st edn. Prentice Hall, USA (2008)
19. Paiva, T., Damasceno, A., Figueiredo, E., Sant'Anna, C.: On the evaluation of code smells and detection tools. J. Softw. Eng. Res. Develop. **5**(1), 1–28 (2017). https://doi.org/10.1186/s40411-017-0041-1
20. Pecorelli, F., Palomba, F., Khomh, F., et al.: Developer-driven code smell prioritization. In: Proceedings of the the 17th International Conference on MSR, pp. 220–231 (2020)
21. Pedregosa, F., Varoquaux, G., Gramfort, A., et al.: Scikit-learn: machine learning in Python. J. Mach. Learn. Res. **12**, 2825–2830 (2011)
22. PMD: an extensible cross-language static code analyzer. https://pmd.github.io. Accessed 31 May 2021
23. Quinlan, J.R.: Learning decision tree classifiers. ACM Comput. Surv. (CSUR) **28**(1), 71–72 (1996)
24. Rasool, G., Arshad, Z.: A review of code smell mining techniques. J. Softw. Evol. Process **27**(11), 867–895 (2015)
25. Santos, J.A.M., Rocha-Junior, J.B., Prates, L.C.L., et al.: A systematic review on the code smell effect. J. Syst. Softw. **144**, 450–477 (2018)
26. Understand by SciTools. https://www.scitools.com/. Accessed 31 May 2021
27. Sharma, T., Efstathiou, V., Louridas, P., et al.: Code smell detection by deep direct-learning and transfer-learning. J. Syst. Softw. **176**, 110936 (2021)
28. Sirikul, K., Soomlek, C.: Automated detection of code smells caused by null checking conditions in Java programs. In: 2016 13th International Joint Conference on Computer Science and Software Engineering (JCSSE), pp. 1–7. IEEE (2016)
29. Vanschoren, J., Van Rijn, J.N., Bischl, B., Torgo, L.: OpenML: networked science in machine learning. ACM SIGKDD Expl. Newsl. **15**(2), 49–60 (2014)
30. Yamashita, A., Moonen, L.: To what extent can maintenance problems be predicted by code smell detection? An empirical study. Inf. Softw. Technol. **55**(12), 2223–2242 (2013)

# HTML-LSTM: Information Extraction from HTML Tables in Web Pages Using Tree-Structured LSTM

Kazuki Kawamura[✉] and Akihiro Yamamoto

Graduate School of Informatics, Kyoto University, Yoshida-Honmachi,
Sakyo-ku, Kyoto 606-8501, Japan
Kazuki.Kawamura@sony.com, akihiro@i.kyoto-u.ac.jp

**Abstract.** In this paper, we propose a novel method for extracting information from HTML tables with similar contents but with a different structure. We aim to integrate multiple HTML tables into a single table for retrieval of information containing in various Web pages. The method is designed by extending tree-structured LSTM, the neural network for tree-structured data, in order to extract information that is both linguistic and structural information of HTML data. We evaluate the proposed method through experiments using real data published on the WWW.

## 1 Introduction

Tables in Web pages are useful for displaying data representing relationships. We can find them on Web pages showing, for example, syllabus in universities, product information in companies, and flight information in airlines. Our research aims at integrating tables from various pages but representing the same type of relational data into a single table for retrieval of information. In this paper, we propose a novel method, called HTML-LSTM, for extracting information from tables in HTML with the same type of contents but with a different structure.

When we browse Web pages, tables representing the same type of relationships look to have similar visual structures but may not be matched completely. In some tables, every tuple is represented in a row, and in other pages, it is in a column. The ordering features (or attributes) may be different. Moreover, they are not always presented as similar HTML source code because different pages are usually designed by different organizations. The source code may contain noises such as codes for visual decorations and additional information. Therefore in order to extract and amalgamate relations from tables in different Web pages, we need to unify them in a common set of features (that is, relational schema), as well as in the structure in the level of HTML source codes. For this purpose, many methods have been proposed [2], but they work well only for HTML tables having almost the same structure in source codes or for the

---

K. Kawamura—Now at Sony Group Corporation.

© Springer Nature Switzerland AG 2021
C. Soares and L. Torgo (Eds.): DS 2021, LNAI 12986, pp. 29–43, 2021.
https://doi.org/10.1007/978-3-030-88942-5_3

case where each feature to be extracted clearly differs from each other. Therefore, extracting and amalgamating relations from tables of similar content but of non-uniform structure is still a major challenge. We solve this problem by using neural networks developed recently and present our solution as HTML-LSTM.

Some neural networks for extracting effectively features from tree and graph structures have been proposed [5,7,13,25]. The Tree-LSTM [25] neural network is a generalization of LSTM to handle tree-structured data, and it is shown that the network effectively works mainly as a feature extractor for parse trees in the field of natural language processing. Since the source codes of HTML are also parsed into tree structures, we extend Tree-LSTM into HTML-LSTM for extracting features in relational data and tree structure from HTML data simultaneously.

We cannot apply Tree-LSTM to HTML data for the following reason: In parse trees of texts in natural language, linguistic information is given only in the leaves, while the other nodes are given information about the relationship between words. Therefore, Tree-LSTM transfers information in the direction from the leaves to the root and is often has been applied to tasks such as machine translation and sentiment analysis [6,25]. On the other hand, when an HTML source code is parsed into a tree, information is attached not only leaves but internal nodes and the root. In addition, when extracting the features of each element in HTML source codes for tables representing relational data, the path from the root tag <table> to the element is quite essential. This means that for extracting information from table data, manipulating parsing trees in the direction from the root to leaves as well as in the direction from leaves to the root. Therefore HTML-LSTM is designed so that information can be transferred in both directions.

In applying HTML-LSTM to information extraction from real HTML data, we first extract the substructure of a table in the data and convert it into a tree. Next, we extract features from the obtained tree structure using HTML-LSTM and classify the features to be extracted for each node. Finally, we integrate the extracted information into a new table. We also introduce a novel data augmentation method for HTML data in order to improve the generalization performance of information extraction.

We evaluate and confirm the effectiveness of the HTML-LSTM method for integrating HTML data formats by applying the method explained above to tables of preschools of local governments and tables of syllabuses published by universities, which are published on the Web. As the results, we succeeded in extracting the required information with an accuracy of an $F_1$-measure of 0.96 for the data of preschools and an $F_1$-measure of 0.86 for the syllabus data. Furthermore, our experimental results show that HTML-LSTM outperforms Tree-LSTM.

This paper is organized as follows. In Sect. 2, we describe previous methods for extracting information from Web pages. In Sect. 3, we introduce the architecture of our proposed method, HTML-LSTM, and how to use HTML-LSTM for information extraction. In Sect. 4, we summarize the results of experiments using HTML data on the Web. Finally, in Sect. 5, we provide our conclusion.

## 2   Related Work

Extracting information from documents and websites and organizing it into a user-friendly form is called information extraction and is widely studied. The research field originates with the Message Understanding Conference (MUC) [8,24], which started in the 1980s. At this conference, every year, a competition is held to extract information from newspapers on, for example, terrorist activities, product development, personnel changes, corporate mergers, rocket launches, and participants competed for some scores evaluating their technique for information extraction.

In the early years of the research area, rule-based methods were widely used [4,22], where rules are defined based on features such as the representation of the characters of the tokens in the sentence, the notation of the tokens (uppercase, lowercase, mixed case of uppercase and lowercase, spaces, punctuation, *etc.*), and the parts of speech of the tokens. Such rule-based methods require experts who manually create rules depending on the types of objects they want to extract. Since it is very time-consuming, algorithms have been developed to automatically create rules using labeled data [1,3,23]. In recent years, statistical methods have also been used in order to treat documents which may have many noises. Example of methods are Support Vector Machine (SVM) [26], Hidden Markov Model (HMM) [21], Maximum Entropy Markov Model (MEMM) [17], Conditional Markov Model (CMM) [16], and Conditional Random Fields (CRF) [19].

Our key idea is to introduce natural language processing methods for information extraction and simultaneously handle structural and linguistic information. Every Web page is written as a source code in HTML, with a clear tree structure after parsing it. The method to extract a specific part from a Web page using the structural information is called Web wrapper [14]. Some of the methods extract information by regarding a specific part in a Web page as data in a tree structure [11,18]. These methods work for Web pages of almost similar structure, and it is difficult to apply them to pages whose structure is completely different, but the meaning of them is the same. This situation often appears in tables representing relational data.

Other types of methods are treating linguistic features of Web pages based on natural language processing, in other words, treating the meaning of the texts on each page. These methods have the disadvantage that they cannot capture the structure of pages. However, natural processing has greatly advanced thanks to the introduction of greatly improved neural network techniques. Some researchers propose new types of neural networks which treat the parsing tree of texts in natural languages. This motivates us to apply such neural networks to extracting information taking into account structure and meaning simultaneously.

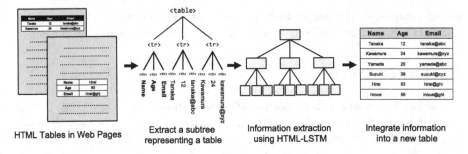

Fig. 1. The **HTML-LSTM** framework for information extraction and integration from HTML tables in web pages

## 3    HTML-LSTM

The overview of our proposed method is shown in Fig. 1. First, we extract the substructure of the table from the entire HTML data. Since Table tags (`<table>` `</table>`) are usually used to represent tables in HTML, we can extract the table by focusing on the region surrounded by the table. tags. Next, we convert the HTML data into a tree structure called DOM tree so that HTML-LSTM can take it as its input. Then, the obtained tree structure data is input to HTML-LSTM for feature extraction, and the obtained features of nodes are used to classify which attribute values each node belongs to. Finally, we pick up the nodes' information classified into the extraction target's attribute values and integrate the information in a new single table.

### 3.1    Extracting Information

In this subsection, we explain the details of HTML-TLSM and the extraction of information from HTML data using it. The workflow of information extraction is shown in Fig. 2. First, each element of the HTML data is encoded using Bi-LSTM (Bidirectional LSTM) [10,20] in order to obtain the language representation of each element. Next, HTML-LSTM is applied to obtain the features of the HTML data, considering the relationship between the positions of the elements in the parsed tree. In order to use the information of the whole tree structure of HTML data effectively, HTML-LSTM extends Tree-LSTM, in which the information flows only from leaf to root, to enable the flow of information from root to leaf as well as from leaf to root. Finally, the features of each node obtained by the HTML-LSTM are passed through the fully connected layer, and the softmax classifier is applied to determining which attribute value each node is classified as.

**Encoding of HTML Data:** The DOM tree that is fed into HTML-LSTM is obtained by parsing the HTML source code. In general, when parsing HTML data to a tree, the values of the nodes in the tree structure are HTML tags. In our method, in order to extract information by effective use of the linguistic and

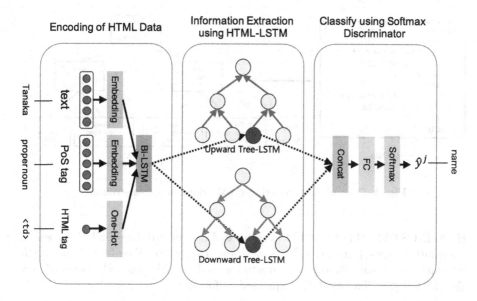

**Fig. 2.** The workflow of information extraction using HTML-LSTM

structural information of HTML, each node of the tree has three types of values: the HTML tag, the text between the start and end tags, and the PoS (part-of-speech) tags of the text, as shown in Fig. 3. The extracted text is treated as a sequence of words. The PoS tag is the sequence of parts-of-speech data corresponding to the sequence of words of the text. If the attribute names to be extracted differ from a Web page to a Web page, a dictionary is used to unify the attribute names. Finally, the obtained tree is converted into a binary tree because HTML-LSTM accepts only binary trees as input.

After converting the HTML data into a binary tree, the text, the sequence of PoS tags, and the HTML tag in each node are converted using a neural network to a representation for input to the HTML-LSTM. In particular, we combine the one-hot encoding of the tag $t^j$ of a node $j$ and the text $w^j$ and PoS tags $p^j$ converted by the embedding matrices $E_{\text{content}}$ and $E_{\text{pos}}$, and feed them to Bi-LSTM:

$$e_t^j = \left[\text{onehot}(t^j)\|E_{\text{content}}(w_t^j)\|E_{\text{pos}}(p_t^j)\right],$$
$$\overrightarrow{h_t^j} = \overrightarrow{\text{LSTM}}(e_t^j, \overrightarrow{h_{t-1}^j}),$$
$$\overleftarrow{h_t^j} = \overleftarrow{\text{LSTM}}(e_t^j, \overleftarrow{h_{t-1}^j}),$$

where onehot is a function that converts a tensor to one-hot a representation and $\|$ is the concatenation of two tensors. The function $\overrightarrow{\text{LSTM}}$ is the forward LSTM and $\overleftarrow{\text{LSTM}}$ is the backward LSTM of the Bi-LSTM. The outputs at the last time $T$ of the forward and backward LSTMs are combined to obtain the representation $x^j = \left[\overrightarrow{h_T}\|\overleftarrow{h_T}\right]$ for each node.

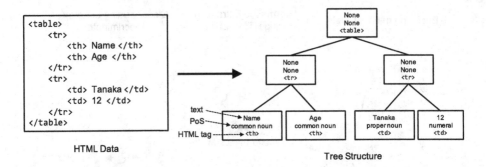

**Fig. 3.** Example of converting HTML data to a tree structure

**HTML-LSTM:** HTML-LSTM is composed of Upward Tree-LSTM, in which information flows from leaves to roots, and Downward Tree-LSTM, in which information is transmitted from roots to leaves. Upward Tree-LSTM uses Binary Tree-LSTM [25]. The model is expressed as follows:

$$i^j = \sigma\left(W^{(i)}x^j + \sum_{k \in \{L,R\}} U_k^{(i)}h_k^j + b^{(i)}\right),$$

$$f_{[L,R]}^j = \sigma\left(W^{(f)}x^j + \sum_{k \in \{L,R\}} U_{[L,R]k}^{(f)}h_k^j + b^{(f)}\right),$$

$$o^j = \sigma\left(W^{(o)}x^j + \sum_{k \in \{L,R\}} U_k^{(o)}h_k^j + b^{(o)}\right),$$

$$u^j = \tanh\left(W^{(u)}x^j + \sum_{k \in \{L,R\}} U_k^{(u)}h_k^j + b^{(u)}\right),$$

$$c^j = i^j \odot u^j + \sum_{k \in \{L,R\}} f_k^j \odot c_k^j,$$

$$h^j = o^j \odot \tanh\left(c^j\right).$$

Upward Tree-LSTM has a forget gate $f^j$, an input gate $i^j$, an output gate $o^j$, a memory cell $c^j$, and a hidden state $h^j$, just like a simple LSTM. In the expressions, $\sigma$ denotes the sigmoid function and $\odot$ denotes the element-wise product. Both of the parameters $W$ and $U$ are weights, and $b$ is the bias. All of these parameters are learnable. As shown in Fig. 4a, Upward Tree-LSTM is a mechanism that takes two inputs and gives one output. In this case, the forget gate uses its own parameters $U_L$ and $U_R$ to select the left and right children's information $c_L^j, c_R^j$.

On the other hand, Downward Tree-LSTM, in which information flows from roots to leaves, is as follows:

$$i^j = \sigma\left(W^{(i)}x^j + U^{(i)}h^j + b^{(i)}\right),$$

$$f^j_{[L,R]} = \sigma\left(W^{(f)}x^j + U^{(f)}_{[L,R]}h^j + b^{(f)}\right),$$

$$o^j_{[L,R]} = \sigma\left(W^{(o)}x^j + U^{(o)}_{[L,R]}h^j + b^{(o)}\right),$$

$$u^j = \tanh\left(W^{(u)}x^j + U^{(u)}h^j + b^{(u)}\right),$$

$$c^j_{[L,R]} = i^j \odot u^j + \sum_{k \in \{L,R\}} f^j_k \odot c^j,$$

$$h^j_{[L,R]} = o^j_{[L,R]} \odot \tanh(c^j_{[L,R]}).$$

As shown in Fig. 4b, this mechanism takes one input and gives two outputs. In this case, the forgetting gate generates two outputs by operating on the input $c^j$ with different parameters $U_L$ and $U_R$, so that the model can choose information to transmit to the left and right children.

Finally, we combine the Upward Tree-LSTM hidden state $h^j_\uparrow$ and the Downward Tree-LSTM hidden states $h^j_{l\downarrow}$, $h^j_{jr\downarrow}$ to obtain $h^j = [h_{j\uparrow}\|h_{jl\downarrow}\|h_{jr\downarrow}]$ as the feature of each node. A softmax classifier predicts the label $\hat{y}^j$ from among the $N$ classes,

$$p = \text{softmax}\left(W^{(s)}h^j + b^{(s)}\right),$$

$$\hat{y}^j = \underset{i \in \{1,\dots,N\}}{\arg\max}\, p_i,$$

where $W^{(s)}, b^{(s)}$ are the learnable parameters. The set of labels consists of the set of attributes to be extracted and a special label *Other* that does not belong to any of the attributes to be extracted. For example, when extracting the attributes *Name* and *Age* from the table shown in Fig. 1 (left), there are three types of classes: *Name*, *Age*, and *Other*. The attribute values "Hirai" and "8" in the table belong to the class of *Name* and *Age*, respectively, while the attribute value "hirai@ghi" and the attribute names "Name", "Age", and "Email" belong to the class of *Other*. We classify all the nodes in the HTML tree to determine what attribute each node is (or is not included in any of the attributes to be extracted).

Every HTML source code may contain much information that is not required to be extracted and noise for decoration. In treating real data, most nodes are not the target of extraction, and therefore the trees as the inputs of HTML-LSTM tend to be imbalanced. In order to treat such trees, we use Focal Loss [15] which is an extension of Cross-entropy Loss to deal with class imbalance. Focal Loss is defined as follows with the correct label $N$ and one-hot vector $t$:

$$\mathcal{L}_{\text{focal}} = -\alpha_i \sum_{i=1}^{N} (1 - p_i)^\gamma\, t_i \log(p_i),$$

where $\alpha_i$ is the frequency inverse of each class and $\gamma$ is the hyperparameter.

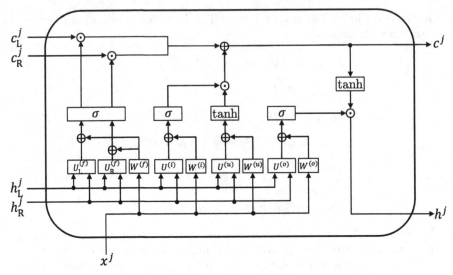

(a) Upward Tree-LSTM

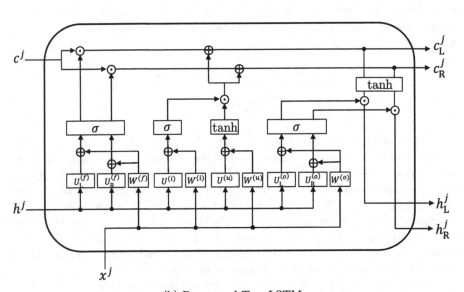

(b) Downward Tree-LSTM

**Fig. 4.** HTML-LSTM architecture

Also, in order to improve scores of the model's recall and precision in a well-balanced way, F1 loss is used jointly. The F1 Loss is given by $1 - F_1$, where $F_1$ is the average of the $F_1$ measures of each class. This is denoted as $\mathcal{L}_{\text{f1}}$, and the final loss function $\mathcal{L}$ is given as

$$\mathcal{L} = \mathcal{L}_{\text{focal}} + \mathcal{L}_{\text{f1}}.$$

Furthermore, the table data on the Web is equivalent to the original data even if the order of the rows and columns of the table is changed. Therefore, we introduce a data augmentation technique that randomly changes the order of rows and columns, thereby increasing the number of HTML data used for training.

## 3.2   Integrating Information

After classifying the class of each node in the HTML tree using HTML-LSTM, we extract the required information from the tree and integrate it into a new table. For each class (attribute), the node with the highest classification score in the HTML tree is selected, and the text of that node is extracted and put into the table. Here, the classification score is the maximum value of the output of the softmax classifier for each class, *i.e.*, $\max_i p_i$. The left side of Fig. 5 shows an example of the tree after classifying the class of each node. Each node contains the text of the original element (top row in the box), the class with the highest classification score (bottom row left in the box), and the classification score (bottom row right in the box). For example, to extract the information of *Name* class from this tree, extract the text of the node elements classified as *Name* class. In this case, the only node classified in the *Name* class is "Tanaka", so we extract "Tanaka" as the *Name* information of this HTML tree and put it in the table. In the same way, when we extract the *Age* class information, we find that there are two nodes classified as *Age* class: "12" and "None". Comparing the classification scores, the classification score of "12" is 0.87, and "None" is 0.23. Since the score of "12" is higher, we extract "12" as the *Age* information of this HTML tree and put it in the table. However, if multiple values should be classified into a certain class, there will be omissions in the extraction. Therefore, when it is expected that there are multiple values classified into a certain class in the web table, a threshold value is set, and all the texts of the nodes that exceed the threshold value are extracted. For example, suppose that in a given HTML tree, there are three nodes classified into the *Name* class, "Tanaka", "Suzuki", and "Apple" with classification scores of 0.97, 0.89, and 0.23, respectively. If the threshold is set to 0.5, then "Tanaka" and "Suzuki" will be extracted as the *Name* class information for this tree.

## 3.3   Implementation Details

The dimensions of the hidden layer of the model are 128 and 5 for the embedding layer of the text and part-of-speech tags, respectively, in the information

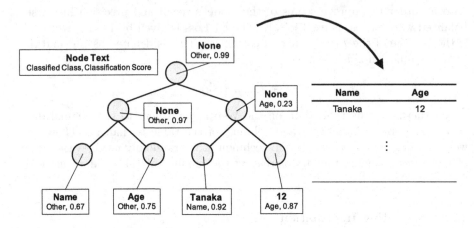

**Fig. 5.** Example of information integration

embedding part of the HTML data. The dimensions of the hidden layer of the HTML-LSTM are 64, and the dimensions of the linear layer used for classification are 64.

We use Adam [12] as the optimization algorithm, with a minibach size of 128. The learning rate starts from $10^{-2}$ and is divided 2 every 15 epochs, and the model is trained for 50 epochs. The parameters of Adam are $\alpha$ of $10^{-2}$, $\beta_1$ of 0.9 and $\beta_2$ of 0.999. We use Dropout [9] with a probability of 0.5 is used to prevent overfitting.

## 4    Experiments

We evaluate our method on tables of preschools published by local governments and tables of syllabus published by universities. These data are published on the Web, and the information is presented using table structures represented by HTML. For evaluation, we use Recall, Precision, and their harmonic mean, $F_1$ measure. $F_1$ measure and Precision and Recall are defined by

$$F_1 = 2 \cdot \frac{\text{Precision} \cdot \text{Recall}}{\text{Precision} + \text{Recall}},$$

$$\text{Precision} = \frac{TP}{TP + FP}, \quad \text{Recall} = \frac{TP}{TP + FN},$$

where $TP$, $TN$, $FP$, and $FN$ are the true positives, true negatives, false positives, false negatives, respectively.

### 4.1    Experiments on Preschool Data

Many local governments publish a list of preschools in their localities in a table on the Web. Those pages have common information, such as the preschools' name,

**Table 1.** Information extraction result for the preschool data

| Attribute | Precision | Recall | $F_1$ measure |
|---|---|---|---|
| name | 0.94 | 1 | 0.97 |
| address | 0.92 | 1 | 0.96 |
| telephone number | 0.87 | 1 | 0.92 |
| other | 1 | 0.98 | 0.99 |
| mean | 0.93 | 0.99 | 0.96 |

**Table 2.** Information integration result for the preschool data

| Name | Address | Telephone number |
|---|---|---|
| とうりん幼稚園小規模保育こみち<br>Tourin Kindergarten | 北区紫竹高縄町 43<br>43 Shichiku Takanawacho Kita-ku | 492-4717 |
| かも保育ルーム<br>Kamo Nursery Room | 北区上賀茂池殿町 59<br>59 Kamigamo Ikedonocho Kita-ku | 585-5958 |
| ののはな保育園<br>Nonohana Preschool | 北区小山西大野町 37<br>37 Koyama Nishionocho Kita-ku | 354-6927 |
| 西藤保育所<br>Nisihuji Preschool | 西藤町 1584-1<br>1584-1 Nishifujicho | 0848-55-6920 |
| 御調中央保育所<br>Mitsugi Chuo Preschool | 御調町花尻 94<br>94 Mitsugicho Hanajiri | 0848-76-0044 |

address, and phone number, but each page has a different HTML structure. In this experiment, we extracted and integrated the information of *name, address,* and *phone number* from these pages. We collected a total of 47 HTML data from 47 local governments that contain information on preschools for the experiment. The HTML data were converted into a tree with a text, PoS tags, and HTML tags at each node. The obtained ordered trees had 22–249 nodes (107 nodes on average) and contained 16 types of PoS tags. Since the data collected for the experiment was in Japanese, the word boundaries are not obvious. Therefore, the series of text and PoS tags were obtained by morphological analysis using Janome[1]. We also unified attribute names that have the same meaning but different notation, and labeled each node manually. The classes were four types of labels: *name, address, phone number,* and *other.*

Table 1 shows the results of information extraction for each attribute in the case of 10 fold cross-validation of preschool data. Table 2 shows the integrated table of the information extracted from the preschool data. We can see that our model does good work on information extraction and integration from these results.

---

[1] https://mocobeta.github.io/janome/.

**Table 3.** Information extraction result for the syllabus data

| Attribute | Precision | Recall | $F_1$ measure |
|---|---|---|---|
| course title | 0.76 | 0.82 | 0.77 |
| instructor name | 0.81 | 0.79 | 0.80 |
| target student | 0.90 | 0.76 | 0.82 |
| target year | 0.94 | 0.75 | 0.83 |
| year/term | 0.97 | 0.83 | 0.89 |
| day/period | 0.89 | 0.87 | 0.88 |
| number of credits | 0.83 | 0.94 | 0.88 |
| other | 0.99 | 0.99 | 0.99 |
| mean | 0.89 | 0.84 | 0.86 |

## 4.2 Experiments on Syllabus Data

The syllabus is the data which shows the contents of lectures in universities and is published on the Web by many universities, mainly using a table format. We collected the syllabus from 22 different universities on the Web and used the HTML data of 20,257 pages for the experiment. The syllabus data was converted into a tree structure in the same way as the data of preschools, and the attribute names with different notations were unified and labeled. The obtained ordered tree has 19 to 1,591 nodes (109 on average) and contains 25 kinds of tags. Since some of the obtained trees contain much noise other than necessary information, we clipped the nodes after the 100th node in the post-order in the syllabus data. The extracted attributes are *course title, instructor name, target student, target Year, year/term, day/period,* and *number of credits.* Therefore, there are eight types of labels in the syllabus data, including *other* in addition to these seven classes.

Table 3 shows the results of information extraction for each attribute in the case of 5 fold cross-validation of syllabus data. In each split, 18 (or 17) of the 22 universities were used for training, and 4 (or 5) were used for testing. Table 4 shows the integrated table of the information extracted from the syllabus data. The $F_1$ measure of the information extraction in the syllabus data is less than that in the preschool data. We believe this is because more attributes are extracted than in the preschool data, and more noise is included in the syllabus data. The result of the information integration shows that information can be extracted even in the blank areas (*i.e.,* areas that originally had no information). This shows that our model can use not only linguistic information but also structural information.

**Table 4.** Information integration result for the syllabus data

| Course Title | Instructor Name | Target Student | Target Year | Year/Term | Day/Period | Number of Credits |
|---|---|---|---|---|---|---|
| 先端科学機器分析及び実習 II<br>Instrumental Analysis, Adv. II | 大江浩一（工学研究科教授）<br>OOE KOUICHI<br>(Graduate School of Engineering Professor) | 大学院生<br>Graduate Student | 修士・博士<br>Master's student<br>Doctoral student | 2020・後期<br>Second semester | 木 4・5<br>Thu.4・5 | 1 |
| 臨床実践指導学特論 II<br>Advanced Studies : Educational of Clinical Psychologist II | 西見奈子（教育学研究科准教授）<br>NISHI MINAKO<br>(Graduate School of Education Associate Professor) | 大学院生<br>Graduate student | 博士<br>Doctoral student | 2020・後期<br>Second semester | 火 5<br>Tue.5 | 1 |
| 租税法 2 [TaxLawII]<br>Tax Law II | 岡村恵生<br>OKAMURA TADAO | 大学院生<br>Graduate student | 2・3 | 2020・後期<br>Second semester | 金 5<br>Fri.5 | 2 |
| 社会基盤材料特論 II<br>Advanced Materials Science & Engineering in Industries II | 辻伸泰（工学研究科教授）<br>TSUJI NOBUHIRO<br>(Graduate School of Engineering Professor) | 大学院生<br>Graduate student | 修士・博士<br>Master's student<br>Doctoral student | 2020・後期<br>Second semester | 火 4<br>Tue.4 | 2 |
| 臨床心理学講読演習 I<br>Advanced Reading on Clinical Psychology I | 名取琢自（非常勤講師）<br>NATORI TAKUJI<br>(Part-time Lecturer) | 大学院生<br>Graduate student | 修士<br>Master's student | 2020・前期<br>First semester | 火 1<br>Tue.1 | 2 |
| 食品有機化学 III<br>Organic Chemistry in Food Science III | 入江一浩（農学研究科教授）<br>IRIE KAZUHIRO<br>(Graduate School of Agriculture Professor) | 学部生<br>Undergraduate student | 2 回生<br>2nd year students | 2020・後期<br>Second semester | 火 3<br>Tue.3 | 2 |
| 生物環境物理学特論<br>Advanced Environmental Biophysics | 坂部綾香（白眉センター特定助教）<br>SAKABE AYAKA<br>(Hakubi Center Assistant Professor)<br>小杉緑子（農学研究科教授）<br>KOSUGI YOSHIKO<br>(Graduate School of Agriculture Professor) | 大学院生<br>Graduate student |  | 2020・前期集中<br>Intensive First semester | 集中前期集中：7月3日、<br>27日、28日、<br>29日を予定<br>Intensive, First semester<br>Scheduled for July 3, 27, 28, and 29 | 2 |
| 居住圏環境共生学<br>Innovative Humano-habitability | 柳川綾（生存圏研究所助教）<br>YANAGAWA AYA<br>(Research Institute for Sustainable Humanosphere Assistant Professor)<br>畑俊充（生存圏研究所講師）<br>HATA TOSHIMITSU<br>(Research Institute for Sustainable Humanosphere Lecturer)<br>吉村剛（生存圏研究所教授）<br>YOSHIMURA TSUYOSHI<br>(Research Institute for Sustainable Humanosphere Professor) | 大学院生<br>Graduate student |  | 2020・前期集中<br>Intensive, First semester | 集中 5/8（金）、<br>5/15（金）、5/22 日（金）<br>Intensive, 5/8 (Fri.),<br>5/15 (Fri.), 5/22 (Fri.) | 2 |
| 書論・書写演習 B<br>Exercises in Calligraphy and Copying B | 長谷川千尋（人間・環境学研究科准教授）<br>HASEGAWA CHIHIRO<br>(Graduate School of Human and Environmental Studies Associate Professor) | 学部生<br>Undergraduate student | 2-4 回生<br>2nd-4th year students | 2020・後期<br>Second semester | 水 2<br>Wed.2 | 2 |
| エネルギー情報学特論<br>Energy and Information, Adv. | 大林実明（非常勤講師）<br>OOBAYASHI FUMIAKI<br>(Part-time Lecturer) | 大学院生<br>Graduate student | 博士<br>Doctoral students | 2020・前期集中<br>Intensive, First semester | 集中<br>Intensive | 2 |
| ドイツ語 I A（文powered）D1120<br>Primary German A | 駒田奈美（非常勤講師）<br>KOMODA NAMI<br>(Part-time Lecturer) | 学部生<br>Undergraduate student | 全回生<br>All students | 2020・前期<br>First semester | 火 3<br>Tue.3 | 2 |

**Table 5.** Ablation study result

| Method | $F_1$ measure |
|---|---|
| Tree-LSTM [25]<br>(Upward Tree-LSTM) | 0.8285 |
| HTML-LSTM<br>(Upward Tree-LSTM + Downward Tree-LSTM) | 0.8414 |
| HTML-LSTM<br>w/ HTML data augmentaion | 0.8575 |

## 4.3  Ablation Experiments

We conducted ablation studies to investigate the effect of adding root-to-leaf information transfer, which is the opposite direction of the traditional Tree-LSTM, and the effect of HTML data augmentation introduced in this study. In the HTML data augmentation, the order of any pair of rows and any pair of columns in the table was switched with a probability of 0.5. The setting of the experiment is the same as the previous experiment on syllabus data, and we compare the average values of all classes of $F_1$ measure of the traditional Tree-LSTM, our HTML-LSTM, and the HTML-LSTM with data augmentation.

The results are shown in the Table 5. This result shows that the ability of information extraction can be improved by using not only the root-to-leaf direction but also the leaf-to-root direction. We can also see that the data augmentation of HTML can further improve the accuracy of information extraction.

## 5 Conclusion

In this paper, we proposed HTML-LSTM, a method for extracting and integrating required information from tables contained in multiple Web pages. The method is an extension of Tree-LSTM, which is mainly used in the field of natural language processing and extracts words in texts attached to the leaves of DOM trees of HTML data in a bottom-up manner. Our method treats DOM trees in a bottom-up manner and then a top-down manner to extract sequences of part-of-speeches and tags attached to nodes in the DOM trees. We applied HTML-LSTM to a list of childcare facilities and syllabus data that are opened on the Web and confirmed that HTML-LSTM could extract information with $F_1$ measures of 0.96 and 0.86, respectively.

In the future, we would improve HTML-LSTM to extract information from fragments of HTML data other than tables. Such fragments are also transformed into DOM trees. For tables or lists, some special tags are prepared in HTML, but other fragments may not have such tags. In order to overcome the problem, choosing good positive and negative examples would be important. Also, modifying the HTML-LSTM algorithm would be needed.

## References

1. Aitken, J.S.: Learning information extraction rules: an inductive logic programming approach. In: ECAI, pp. 355–359 (2002)
2. Chang, C.H., Kayed, M., Girgis, M., Shaalan, K.: A survey of web information extraction systems. IEEE Trans. Knowl. Data Eng. **18**(10), 1411–1428 (2006)
3. Ciravegna, F.: Adaptive information extraction from text by rule induction and generalisation. IJCAI **2**, 1251–1256 (2001)
4. Cunningham, H., Maynard, D., Bontcheva, K., Tablan, V.: GATE: a framework and graphical development environment for robust NLP tools and applications. In: ACL, pp. 168–175 (2002)
5. Defferrard, M., Bresson, X., Vandergheynst, P.: Convolutional neural networks on graphs with fast localized spectral filtering. In: NeurIPS, pp. 3844–3852 (2016)
6. Eriguchi, A., Hashimoto, K., Tsuruoka, Y.: Tree-to-sequence attentional neural machine translation. In: ACL, vol. 2, pp. 823–833 (2016)
7. Goller, C., Kuechler, A.: Learning task-dependent distributed representations by backpropagation through structure. Neural Netw. **1**, 347–352 (1996)
8. Grishman, R.: Message understanding conference-6: a brief history. In: COLING, pp. 466–471 (1996)
9. Hinton, G.: Dropout: a simple way to prevent neural networks from overfitting. JMLR **15**, 1929–1958 (2014)
10. Hochreiter, S., Urgen Schmidhuber, J.: Long short-term memory. Neural Comput. **9**(8), 1735–1780 (1997)
11. Kashima, H., Koyanagi, T.: Kernels for semi-structured data. In: ICML, pp. 291–298 (2002)
12. Kingma, D.P., Ba, J.L.: Adam: a method for stochastic optimization. In: ICLR (2015)
13. Kipf, T.N., Welling, M.: Semi-supervised classification with graph convolutional networks. In: ICLR (2017)

14. Kushmerick, N.: Wrapper induction: efficiency and expressiveness. Artif. Intell. **118**(1–2), 15–68 (2000)
15. Lin, T.Y., Goyal, P., Girshick, R., He, K., Dollar, P.: Focal loss for dense object detection. In: ICCV, pp. 3844–3852 (2017)
16. Malouf, R.: Markov models for language-independent named entity recognition. In: CoNLL, pp. 187–190 (2002)
17. Michael, A.: Maximum entropy Markov models for information extraction and segmentation Andrew. In: ICML, pp. 591–598 (2000)
18. Muslea, I., Minton, S., Knoblock, C.: Active learning for hierarchical wrapper induction. In: AAAI, p. 975 (1999)
19. Peng, F., McCallum, A.: Information extraction from research papers using conditional random fields. Inf. Process. Manage. **42**(4), 963–979 (2006)
20. Schuster, M., Paliwal, K.K.: Bidirectional recurrent neural networks. IEEE Trans. Sig. Process. **45**(11), 2673–2681 (1997)
21. Seymore, K., Mccallum, A., Rosenfeld, R.: Learning hidden Markov model structure. In: AAAI Workshop, pp. 37–42 (1999)
22. Shaalan, K., Raza, H.: Arabic named entity recognition from diverse text types. In: Nordström, B., Ranta, A. (eds.) GoTAL 2008. LNCS (LNAI), vol. 5221, pp. 440–451. Springer, Heidelberg (2008). https://doi.org/10.1007/978-3-540-85287-2_42
23. Soderland, S.: Learning information extraction rules for semi-structured and free text. Mach. Learn. **34**(1), 233–272 (1999)
24. Sundheim, B.M.: Overview of the fourth message understanding evaluation and conference. In: 4th Message Understanding Conference, pp. 3–22 (1992)
25. Tai, K.S., Socher, R., Manning, C.D.: Improved semantic representations from tree-structured long short-term memory networks. In: ACL-IJCNLP, vol. 1, pp. 1556–1566 (2015)
26. Takeuchi, K., Collier, N.: Use of support vector machines in extended named entity recognition. In: COLING, pp. 1–7 (2002)

# Predicting Reach to Find Persuadable Customers: Improving Uplift Models for Churn Prevention

Théo Verhelst[1]([envelope]) [ORCID], Jeevan Shrestha[2], Denis Mercier[2],
Jean-Christophe Dewitte[2], and Gianluca Bontempi[1] [ORCID]

[1] Machine Learning Group, Université Libre de Bruxelles, Brussels, Belgium
{Theo.Verhelst,Gianluca.Bontempi}@ulb.be
[2] Data Science Team, Orange Belgium, Brussels, Belgium
{Jeevan.Shrestha,Denis1.Mercier,Jean-Christophe.Dewitte}@orange.com

**Abstract.** Customer churn is a major concern for large companies (notably telcos), even in a big data world. Customer retention campaigns are routinely used to prevent churn, but targeting the right customers on the basis of their historical profile is a difficult task. Companies usually have recourse to two data-driven approaches: churn prediction and uplift modeling. In churn prediction, customers are selected on the basis of their propensity to churn in a near future. In uplift modeling, only customers reacting positively to the campaign are considered. Though uplift is better suited to maximize the efficiency of the retention campaign because of its causal aspect, it suffers from several estimation issues. To improve the uplift accuracy, this paper proposes to leverage historical data about the reachability of customers during a campaign. We suggest several strategies to incorporate reach information in uplift models, and we show that most of them outperform the classical churn and uplift models. This is a promising perspective for churn prevention in the telecommunication sector, where uplift modeling has failed so far to provide a significant advantage over non-causal approaches.

**Keywords:** Causal inference · Churn prediction · Uplift modeling

## 1 Introduction

The telecommunication market is saturated, and companies need to invest in customer relationship management to keep their competitive edge. It is common knowledge that preventing churn is less expensive than attracting new customers [11]. The classical strategy for churn prevention consists in ranking customers according to their churn risk and offering the most probable to leave an incentive to remain (e.g. a promotional offer). Predicting churn is a difficult problem, involving large class imbalance, high dimension, latent information, low class separability, and large quantities of data. A wide variety of machine learning models have been applied to this problem in the literature [13,19,21,22,26,30].

© Springer Nature Switzerland AG 2021
C. Soares and L. Torgo (Eds.): DS 2021, LNAI 12986, pp. 44–54, 2021.
https://doi.org/10.1007/978-3-030-88942-5_4

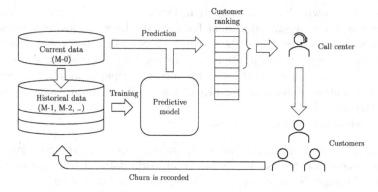

**Fig. 1.** Overview of the pipeline for customer retention.

The pipeline for a typical customer retention campaign is outlined in Fig. 1. First, a predictive model is trained on historical data from past campaigns. Then, this model predicts a score for each customer and ranks them accordingly. The list of customers with the highest scores is randomly split in a target and a control group, and the target group is sent to a call center. The call center contacts each of them individually, and the reaction of the customer is recorded and added to the historical data set for training future models.

The customers' ranking is provided by a predictive model estimating the probability of churn. This approach, however, disregards the causal aspect of the problem. Targeting high-risk customers is not necessarily the best strategy: for instance, some customers slightly less inclined to churn could be far more receptive to retention offers, and focusing the campaign on these customers could be more effective. This idea is exploited by uplift models, which estimate the causal effect of the campaign on an individual customer, rather than the risk of churn [10]. A wide variety of uplift models has been developed in the literature [1, 8,14,17,29].

However, the added value of uplift modeling over churn prediction has been seldom assessed empirically. While it is clear that uplift is less biased than churn for estimating causal effects, the gain in performance is debated and context-dependent [5,7,27]. In settings such as customer retention, characterized by non-linearity, low class separability, and high dimensionality, the theoretical advantages of uplift might be insufficient to outweigh its drawbacks with respect to the usual strategy of churn prediction.

In this article, we suggest leveraging information about the reaction of the customer to the campaign to improve uplift estimation. In the marketing domain, *reach* denotes the proportion of the population exposed to the campaign, more specifically for advertisement campaigns [6]. In this article, we define *reach* as the reaction of the customer to the attempted call, that is, whether or not the customer picked up the phone and had a conversation with the phone operator. This variable is potentially informative about customer behavior, and, as a result, could improve the estimation of customer uplift. It is important to note that reach is only known after the campaign. Thus, it cannot be simply added as input to the model as an additional feature. We have to devise a dedicated approach to incorporate it into the learning process. In this sense, reach serves as an inductive bias for the uplift model, rather than an additional predictive feature. This paper shows that an uplift model, properly adapted to account for this new source of information, provides a significant improvement over the state-of-the-art.

The main contributions of this paper are:

- The proposal of 4 original strategies to incorporate reach in uplift models.
- An assessment of these strategies on a real-world data set from our industrial partner Orange Belgium, a major telecom company in Belgium.
- A significant improvement of uplift estimation, clearly outperforming state-of-the-art uplift models and the classical churn prediction approach.

The rest of this paper is divided as follows. In Sect. 2, we define basic notions in churn prediction and uplift modeling. In Sect. 3, we present reach modeling and various strategies to improve uplift estimation. In Sect. 4, we evaluate these strategies against several baselines, and we present our results in Sect. 5. We discuss our findings and suggest future work in Sect. 6.

## 2   Churn Prediction and Uplift Modeling

In what follows, uppercase letters denote random variables, bold font denotes sets, and lowercase letters denote realizations of random variables. Causal inference notions are formalized using Pearl's notation [23]: an intervention fixing a variable $T$ to a value $t$ is noted $do(T = t)$, and a random variable $Y$ in a system under such an intervention is noted $Y_t$. For example, $Y_0$ is the churn indicator when the customer is in the control group ($T = 0$), whereas $Y_1$ is the churn indicator for the target group. We also denote customer features by a set of variables $X$, with a realisation $x$. Finally, $R$ is the reach indicator ($R = 1$ for reached customer, $R = 0$ otherwise).

Let us first formalize in probabilistic terms the two main approaches for selecting customers in a retention campaign: churn prediction and uplift modeling. Churn prediction estimates the probability $P(Y = 1 \mid X = x)$ that a customer churns ($Y = 1$) given the customer descriptive features $x$. Typical examples of descriptive features are tariff plan, metadata on calls and messages, mobile data usage, invoice amount, customer hardware, etc. Conventional supervised learning models can be used to predict churn [16,20,24,25]. An extensive review of machine learning for churn prediction is given in [15]. The main drawback of this approach is the absence of causal insight: in fact, there is no indication that the campaign will be most effective on customers with a high probability of churn. The causal perspective is instead adopted by uplift modeling.

Uplift modeling estimates the causal effect of the campaign on the customer's probability of churn. To estimate this effect, it considers two scenarios: the intervention case $do(T = 1)$ (i.e. the customer is offered an incentive) vs the control case do($T = 0$) (i.e. the customer is not contacted). The uplift is the difference in the probability of churn between these two scenarios. For a set of descriptive features $\boldsymbol{X} = \boldsymbol{x}$, it is

$$U(\boldsymbol{x}) = P(Y_0 = 1 \mid \boldsymbol{X} = \boldsymbol{x}) - P(Y_1 = 1 \mid \boldsymbol{X} = \boldsymbol{x}). \tag{1}$$

Note that, unlike probabilities, uplift can be negative. A negative uplift indicates that the customer is more likely to churn when contacted by the call center. An uplift model is trained on historical data from one or more past campaigns with a randomized group assignment (target or control). The reaction of the customer (e.g. stay or churn) is then monitored for a fixed period of time, typically some months. The group assignment and customer churn records can then be used to update the historical data set, and subsequently train a new uplift model. Several approaches exist to estimate uplift, either using one or more predictive models [14,17] or estimating uplift directly [1,8,29]. For a review of state-of-the-art uplift models, we refer the reader to [10].

## 3   Reach Modeling

While uplift modeling is theoretically unbiased for maximizing campaign efficiency, there is some evidence in the literature that it suffers from estimation issues [7,27]. This aspect can be so relevant as to cancel the benefits related to its causal design. Nevertheless, there is an additional piece of information that can be used to improve uplift estimation: the reaction of the customer to the call. More specifically, some customers will not pick up the phone, will hang up immediately, or more generally will not respond positively to the call. This information, automatically recorded by the call center, is a strong marker of customer receptivity. In email and online advertisement, a similar notion exists, under the name of click-through-rate [28] or response rate [12]. Although response models have been developed to improve direct marketing [2,9,12], current literature on uplift modeling ignores this information during the learning process. Expert knowledge in the telecom sector indicates that customers who do not pick up the phone or hang up immediately should be avoided because targeting them can increase their propensity to churn. We denote with $R = 1$ *reached* customers, i.e. customers who picked up the phone and had a dialogue with the phone operator. Otherwise, the customer is deemed *unreached* ($R = 0$). We present three ways to integrate reach information to improve uplift estimation. The four resulting equations are summarized in Table 1.

*Reach Probability as a Feature.* The first approach (called R-feature) consists in building a predictive model of reach from historical data, and integrating the reach probability $\hat{r}$ among the input features of the uplift model. Note that we cannot directly plug the reach indicator as an input feature, since such information is not available before the campaign. This approach consists in learning the function $U(\boldsymbol{x}) = P(Y_0 = 1 \mid \boldsymbol{x}, \hat{r}) - P(Y_1 = 1 \mid \boldsymbol{x}, \hat{r})$.

*Decomposition of Probability.* The second approach (R-decomp) is based on the decomposition of the probability of churn with respect to the reach:

$$U(x) = P(Y_0 = 1 \mid x) - P(Y_1 = 0 \mid x) \tag{2}$$
$$= P(Y_0 = 1 \mid x) - P(R_1 = 0 \mid x)P(Y_1 = 1 \mid x, R_1 = 0)$$
$$- P(R_1 = 1 \mid x)P(Y_1 = 1 \mid x, R_1 = 1) \tag{3}$$
$$= P(Y_0 = 1 \mid x) - P(R_1 = 0 \mid x)P(Y_1 = 1 \mid x, R_1 = 0)$$
$$- [1 - P(R_1 = 0 \mid x)]P(Y_1 = 1 \mid x, R_1 = 1) \tag{4}$$
$$= P(Y_0 = 1 \mid x) - P(Y_1 = 1 \mid x, R_1 = 1)$$
$$+ P(R_1 = 0 \mid x)[P(Y_1 = 1 \mid x, R_1 = 1) - P(Y_1 = 1 \mid x, R_1 = 0)]. \tag{5}$$

The last equation contains 5 terms but can be estimated with two uplift models and a simple classifier. The first two terms, $P(Y_0 = 1 \mid x) - P(Y_1 = 1 \mid x, R_1 = 1)$, can be estimated with a uplift model by restricting the target group to reached customers. The third term, $P(R_1 = 1 \mid x)$, can be estimated by a predictive model of reach. The last two terms between brackets, $P(Y_1 = 1 \mid x, R_1 = 1) - P(Y_1 = 1 \mid x, R_1 = 0)$, can also be returned by an uplift model, but using the reach indicator $R$ instead of $T$ as the treatment indicator for the model.

*Bounds on Uplift.* In marketing, there is empirical evidence that non-reached customers tend to have a negative uplift. Not reaching a customer has thus a doubly detrimental effect: the resources of the call center are wasted, and the customer is more likely to churn than if no call had been made. This domain knowledge may be translated into an inequality $P(Y_1 = 1 \mid x, R_1 = 0) \geq P(Y_0 = 1 \mid x)$. We derive the third approach (R-upper) using this assumption and the decomposition in Eq. (3):

$$U(x) = P(Y_0 = 1 \mid x) - P(Y_1 = 1 \mid x)$$
$$\leq (1 - P(R_1 = 0 \mid x))P(Y_0 = 1 \mid x)$$
$$- P(Y_1 = 1 \mid x, R_1 = 1)P(R_1 = 1 \mid x)$$
$$= P(R_1 = 1 \mid x)[P(Y_0 = 1 \mid x) - P(Y_1 = 1 \mid x, R_1 = 1)]. \tag{6}$$

Equation (6) requires two models: a simple predictive model of the reach variable (using only the target group), and an uplift model where the target group has been restricted to reached customers.

A symmetrical reasoning may lead to the hypothesis that a reached customer is less likely to churn than if not contacted: $P(Y_1 = 1 \mid x, R_1 = 1) \leq P(Y_0 = 1 \mid x)$. From such assumption and (3), we derive a lower bound:

$$U(x) \geq P(R_1 = 0 \mid x)[P(Y_0 = 1 \mid x) - P(Y_1 = 1 \mid x, R_1 = 0)]. \tag{7}$$

Equation (7) is similar to Eq. (6) but it requires the probability of not being reached, and the target group of the uplift model's training set is restricted to non-reached customers. This approach is named R-lower. Note that, among all methods presented in this section, R-upper and R-lower are the only biased estimators of uplift (since they estimate a bound instead). R-feature and R-decomp both estimate uplift, although they differ in the way they incorporate reach information.

## 4    Experiment

This experimental session benchmarks the approaches of Sect. 3 against several baselines:

**Table 1.** Summary of the approaches used to integrate reach in uplift modeling. The conditioning on $x$ is implicit in every term.

| Approach | Equation |
|---|---|
| R-feature | $P(Y_0 = 1 \mid \hat{r}) - P(Y_1 = 1 \mid \hat{r})$ |
| R-decomp | $P(Y_0 = 1) - P(Y_1 = 1 \mid R_1 = 1)$ |
|  | $+ P(R_1 = 0) \cdot [P(Y_1 = 1 \mid R_1 = 1) - P(Y_1 = 1 \mid R_1 = 0)]$ |
| R-upper | $P(R_1 = 1) [P(Y_0 = 1) - P(Y_1 = 1 \mid R_1 = 1)]$ |
| R-lower | $P(R_1 = 0) [P(Y_0 = 1) - P(Y_1 = 1 \mid R_1 = 0)]$ |

- **Uplift:** An uplift model with no information about reach.
- **ML approach:** A classical churn prediction model[1] returning $P(Y = 1 \mid x)$.
- **R-target:** Using the estimated probability of reach as a score, that is, $P(R = 1 \mid x)$.

Since the first two baselines are state-of-the-art strategies, it is important to check whether incorporating reach information outperforms those approaches. The baseline R-target is introduced to check whether the reach alone may be used to find persuadable customers. Based on previous experiments [27], we used the X-learner algorithm [17] to build uplift models, and random forests [3] to learn churn and reach predictive models. The unbalancedness between churners and non-churners is addressed with the EasyEnsemble strategy [18], averaging models trained on positive instances (churners) with models trained on equally-sized sampled subsets of negative instances (non-churners).

The dataset is provided by our industrial partner Orange Belgium and relates to a series of customer retention campaigns in 2020, spanning over 3 months. A monthly dataset concerns about 4000 customers, for a total of 11896 samples. Each campaign includes a control group of about 1000 customers (for a total of 2886 control samples, 24.3% of the total), and a target group whose size depends on the load of the call center. Customer churn is monitored up to two months following the call. The churn rate in the control group is 3.6%, and 3.4% in the

---

[1] Note that ML stands for *maximum likelihood* of churn.

target group. The reach rate is 44.1% in the target group. Additional details cannot be disclosed for evident confidentiality reasons.

Results are evaluated in terms of uplift curve [10], which estimates the causal effect of the campaign for different numbers of customers. The uplift curve measures the difference in probability of churn between customers in the target and control groups. For a given predictive model $f$, and a threshold $\tau$ over the score provided by $f$, the uplift curve is defined as

$$\text{Uplift}(\tau) = P(Y_0 = 1 \mid f(\boldsymbol{X}) > \tau) - P(Y_1 = 1 \mid f(\boldsymbol{X}) > \tau). \tag{8}$$

This quantity is estimated empirically by subtracting the proportion of churners in the control and target groups, restricted to the customers with a score above the threshold. The uplift curve then is obtained by varying the threshold over all possible values.

In order to obtain a measure of the performance variability, we created 50 independent random splits of the data set into training and test sets, in proportion 80%/20%. Each of these splits is used to train each model, and we report the area under the uplift curve on the test set, averaged over the 50 runs.

We also evaluated several variations of the 4 approaches listed in Table 1. But, since they did not provide any significant improvement, we did not include them in the results. These variations are: i) the average of `R-lower` and `R-upper`, ii) the product of the reach and uplift model predictions, and iii) the average of the reach and uplift models prediction.

## 5   Results

**Table 2.** Area under the uplift curve (AUUC), averaged over 50 runs. The confidence interval is one standard deviation. The best approach is underlined.

| Approach | AUUC |
|---|---|
| `R-feature` | 0.894 ($\pm$0.419) |
| `R-decomp` | 0.767 ($\pm$0.605) |
| `R-upper` | 0.608 ($\pm$0.525) |
| `R-lower` | 0.747 ($\pm$0.519) |
| `Uplift` | 0.663 ($\pm$0.537) |
| `ML approach` | 0.665 ($\pm$0.568) |
| `R-target` | 0.28 ($\pm$0.417) |

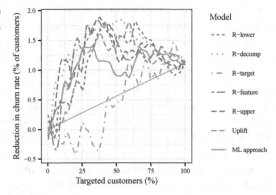

**Fig. 2.** Uplift curves for the first run of the experiment.

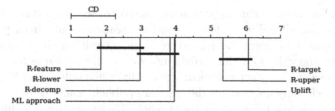

**Fig. 3.** Average ranking of the different approaches, with a line grouping approaches which do not have a significant rank difference. The critical mean rank difference is CD = 1.24, based on a Friedman-Nemenyi test with $p = 0.05$.

Table 2 reports the average area under the uplift curve (AUUC) over 50 runs while the uplift curves of the first run are in Fig. 2. A Friedman-Nemenyi test of rank [4] is reported on Fig. 3, which indicates the mean rank of each approach over the 50 runs. A method is considered significantly better if the mean rank difference is larger than CD = 1.24, based on a p-value of $p = 0.05$. The best performing model, in terms of area under the uplift curve and standard deviation, is R-feature. It is significantly better than all other models, except for R-lower. Among the approaches integrating reach, R-decomp and R-lower perform similarly, while R-upper is not able to outperform the baselines. The two baselines Uplift and ML approach have similar performances, and, as expected, R-target performs quite poorly.

Note that, due to the small size of the dataset, the standard deviation of the AUUC is quite high. The data set contains only 11896 samples, 20% of these samples are used in the test set, and the churn rate is only a few percent. This leaves a very limited number of churners in the test set, and thus induces a high variability in the uplift curve between the different runs of the experiment.

## 6   Conclusion and Future Work

This paper shows the potential of reach information to improve the estimation of uplift. The superiority of reach models (such as R-feature) over conventional churn or uplift models is not surprising, since the information provided by the reach indicator is not available to the baseline methods. However, since reach information is not directly available before the campaign, specific strategies must be used. In these strategies, reach plays more the role of inductive bias than the one of churn predictor.

A potential advantage of this approach is that reach models are relevant to a wider range of use cases than churn prevention. It is common for telecom companies to perform different campaigns using the voice call channel, such as *up-sell* (to propose a better product to the customer), or *cross-sell* (to present additional products). A model of reach can be used in these contexts as well while using the same training data. This is a significant advantage, both in terms of computation time and data volume.

The applicability of this approach is limited by several factors. Firstly, as it is the case for all uplift models, it requires historical data from past retention campaigns. Our approach further requires records on the reaction of the customers to the call. This data might not be readily available for companies with no experience in direct marketing. Secondly, since uplift modeling is a new area of research, only a few uplift datasets are publicly available online. None of these datasets include information about reach. Therefore, it is difficult to assess new approaches exploiting reach information outside the scope of a collaboration with a private company.

We plan to evaluate our approach in future live retention campaigns. Currently, customer retention campaigns are still based on the churn prediction approach, since uplift models have failed so far to provide a significant improvement. This is a unique opportunity to evaluate the added value of our improved uplift model over the classical approach, and going beyond the use of historical data sets. From the perspective of a practitioner, several improvements of the approach can be devised: for example, we considered only the random forest model to predict reach. Other machine learning models might provide better performances. Also, our pipeline addresses class unbalancedness with the Easy Ensemble strategy, and the reach model is included during this step. Since the reach indicator is not as heavily imbalanced as the churn indicator, it might be beneficial to train the reach model separately. Finally, we did not investigate the use of more fine-grained reach information, such as the time of call, or a more detailed description of the customer's reaction. This could potentially further improve uplift estimation. Such detailed information can also be exploited proactively, by calling the customer at a time and a day which maximizes the probability of reach.

# References

1. Athey, S., Imbens, G.: Recursive partitioning for heterogeneous causal effects. Proc. Nat. Acad. Sci. **113**(27), 7353–7360 (2016). https://doi.org/10.1073/PNAS.1510489113, https://www.pnas.org/content/113/27/7353
2. Bose, I., Chen, X.: Quantitative models for direct marketing: a review from systems perspective. Eur. J. Oper. Res. **195**(1), 1–16 (2009)
3. Breiman, L.: Random forests. Mach. Learn. **45**(1), 5–32 (2001)
4. Demšar, J.: Statistical comparisons of classifiers over multiple data sets. J. Mach. Learn. Res. **7**, 1–30 (2006)
5. Devriendt, F., Berrevoets, J., Verbeke, W.: Why you should stop predicting customer churn and start using uplift models. Inf. Sci. **584**, 497–515 (2019)
6. Farris, P.W., Bendle, N., Pfeifer, P.E., Reibstein, D.: Marketing Metrics: The Definitive Guide to Measuring Marketing Performance. Pearson Education (2010)
7. Fernández, C., Provost, F.: Causal Classification: Treatment Effect vs. Outcome Prediction. NYU Stern School of Business (2019)
8. Guelman, L., Guillén, M., Pérez-Marín, A.M.: Uplift random forests. Cybern. Syst. **46**(3–4), 230–248 (2015). https://doi.org/10.1080/01969722.2015.1012892
9. Guido, G., Prete, M.I., Miraglia, S., De Mare, I.: Targeting direct marketing campaigns by neural networks. J. Mark. Manag. **27**(9–10), 992–1006 (2011)

10. Gutierrez, P., Gérardy, J.Y.: Causal inference and uplift modelling: a review of the literature. In: Hardgrove, C., Dorard, L., Thompson, K., Douetteau, F. (eds.) Proceedings of the 3rd International Conference on Predictive Applications and APIs. Proceedings of Machine Learning Research, Microsoft NERD, Boston, USA, vol. 67, pp. 1–13. PMLR (January 2016). http://proceedings.mlr.press/v67/gutierrez17a.html

11. Hadden, J., Tiwari, A., Roy, R., Ruta, D.: Computer assisted customer churn management: state-of-the-art and future trends. Comput. Oper. Res. **34**(10), 2902–2917 (2007)

12. Hansotia, B.J., Rukstales, B.: Direct marketing for multichannel retailers: issues, challenges and solutions. J. Database Mark. Customer Strategy Manage. **9**(3), 259–266 (2002)

13. Idris, A., Khan, A.: Ensemble based efficient churn prediction model for telecom. In: 2014 12th International Conference on Frontiers of Information Technology (FIT), pp. 238–244 (2014). https://doi.org/10.1109/fit.2014.52

14. Jaskowski, M., Jaroszewicz, S.: Uplift modeling for clinical trial data. In: ICML Workshop on Clinical Data Analysis (2012)

15. Kayaalp, F.: Review of customer churn analysis studies in telecommunications industry. Karaelmas Fen ve Mühendislik Dergisi **7**(2), 696–705 (2017). https://doi.org/10.7212/zkufbd.v7i2.875

16. Keramati, A., Jafari-Marandi, R., Aliannejadi, M., Ahmadian, I., Mozaffari, M., Abbasi, U.: Improved churn prediction in telecommunication industry using data mining techniques. Appl. Soft Comput. **24**, 994–1012 (2014). https://doi.org/10.1016/j.asoc.2014.08.041

17. Künzel, S.R., Sekhon, J.S., Bickel, P.J., Yu, B.: Metalearners for estimating heterogeneous treatment effects using machine learning. Proc. Natl. Acad. Sci. U.S.A. **116**(10), 4156–4165 (2019). https://doi.org/10.1073/pnas.1804597116

18. Liu, X.Y., Wu, J., Zhou, Z.H.: Exploratory undersampling for class-imbalance learning. IEEE Trans. Syst. Man Cybern. Part B (Cybern.) **39**(2), 539–550 (2009). https://doi.org/10.1109/tsmcb.2008.2007853

19. Mitrović, S., Baesens, B., Lemahieu, W., De Weerdt, J.: On the operational efficiency of different feature types for telco Churn prediction. Eur. J. Oper. Res. **267**(3), 1141–1155 (2018). https://doi.org/10.1016/j.ejor.2017.12.015

20. Olle, G.D.O., Cai, S.: A hybrid churn prediction model in mobile telecommunication industry. Int. J. e-Educ. e-Bus. e-Manage. e-Learn. **4**(1), 55 (2014)

21. Óskarsdóttir, M., Bravo, C., Verbeke, W., Sarraute, C., Baesens, B., Vanthienen, J.: Social network analytics for churn prediction in telco: Model building, evaluation and network architecture. Exp. Syst. Appl. **85**, 204–220 (2017). https://doi.org/10.1016/j.eswa.2017.05.028

22. Óskarsdóttir, M., Van Calster, T., Baesens, B., Lemahieu, W., Vanthienen, J.: Time series for early churn detection: using similarity based classification for dynamic networks. Exp. Syst. Appl. **106**, 55–65 (2018). https://doi.org/10.1016/j.eswa.2018.04.003

23. Pearl, J.: Causality: Models, Reasoning, and Inference, vol. 6. Cambridge University Press (2009)

24. Umayaparvathi, V., Iyakutti, K.: Attribute selection and customer churn prediction in telecom industry. In: 2016 International Conference on Data Mining and Advanced Computing (SAPIENCE), pp. 84–90. IEEE (2016)

25. Vafeiadis, T., Diamantaras, K.I., Sarigiannidis, G., Chatzisavvas, K.C.: A comparison of machine learning techniques for customer churn prediction. Simul. Model. Pract. Theor. **55**, 1–9 (2015). https://doi.org/10.1016/j.simpat.2015.03.003

26. Verbeke, W., Martens, D., Baesens, B.: Social network analysis for customer churn prediction. Appl. Soft Comput. **14**, 431–446 (2014). https://doi.org/10.1016/j.asoc.2013.09.017
27. Verhelst, T., Caelen, O., Dewitte, J.C., Bontempi, G.: Does causal reasoning help preventing churn? (2021, under submission)
28. Winer, R.S.: A framework for customer relationship management. Calif. Manage. Rev. **43**(4), 89–105 (2001)
29. Zaniewicz, L., Jaroszewicz, S.: Support vector machines for uplift modeling. In: 2013 IEEE 13th International Conference on Data Mining Workshops, pp. 131–138. IEEE (2013)
30. Zhu, B., Baesens, B., vanden Broucke, S.K.L.M.: An empirical comparison of techniques for the class imbalance problem in churn prediction. Inf. Sci. **408**, 84–99 (2017). https://doi.org/10.1016/j.ins.2017.04.015

# Classification

# A Semi-supervised Framework
# for Misinformation Detection

Yueyang Liu[(✉)], Zois Boukouvalas, and Nathalie Japkowicz

American University, Washington D.C. 20016, USA
yueyang.liu@student.american.edu, {boukouva,japkowic}@american.edu

**Abstract.** The spread of misinformation in social media outlets has become a prevalent societal problem and is the cause of many kinds of social unrest. Curtailing its prevalence is of great importance and machine learning has shown significant promise. However, there are two main challenges when applying machine learning to this problem. First, while much too prevalent in one respect, misinformation, actually, represents only a minor proportion of all the postings seen on social media. Second, labeling the massive amount of data necessary to train a useful classifier becomes impractical. Considering these challenges, we propose a simple semi-supervised learning framework in order to deal with extreme class imbalances that has the advantage, over other approaches, of using actual rather than simulated data to inflate the minority class. We tested our framework on two sets of Covid-related Twitter data and obtained significant improvement in F1-measure on extremely imbalanced scenarios, as compared to simple classical and deep-learning data generation methods such as SMOTE, ADASYN, or GAN-based data generation.

**Keywords:** Semi-supervised learning · Class imbalance · Misinformation detection

## 1 Introduction

The spread of misinformation in social media outlets has become a prevalent societal problem and is the cause of many kinds of social unrest. Curtailing its prevalence is of great importance and machine learning advances have shown significant promise for the detection of misinformation [11]. However, to build a reliable model a large data set of reliable posts as well as posts containing misinformation is needed. In practice, this is not feasible since detecting posts containing misinformation is inherently a class imbalanced problem: the majority of posts are reliable whereas a very small minority contains misinformation. For instance, according to The Verge, Twitter removed 2,230 misleading tweets between March 16 and April 18, 2020[1]. Given that, on average, 6,000 tweets

---

[1] https://www.theverge.com/2020/4/22/21231956/twitter-remove-covid-19-tweets-call-to-action-harm-5g.

© Springer Nature Switzerland AG 2021
C. Soares and L. Torgo (Eds.): DS 2021, LNAI 12986, pp. 57–66, 2021.
https://doi.org/10.1007/978-3-030-88942-5_5

are tweeted every second[2], the class imbalance ratio is around 0.000014% for that month, or 1 unreliable Tweet for every 71,428 reliable ones, an extreme imbalance ratio.

The class imbalance problem has been pervasive in the Machine Learning field for over two decades [3]. Over the years, many techniques for dealing with class imbalances have been proposed including classical methods for inflating the minority class such as SMOTE [4] and ADASYN [8] and Deep-Learning based methods such as DEAGO [1] and GAMO [15], which use an autoencoder and a Generative Adversarial Network, respectively. One of the issues with previously proposed minority-class oversampling methods for the class imbalance problem is that either the data used to inflate the minority class is real but simply repeated from the existing minority class, as in random oversampling, or it is artificial as in SMOTE, ADASYN, DEAGO and GAMO Random oversampling is not an acceptable solution given that it is known to cause overfitting [6]. Artificial oversampling, while not overfitting as much as random oversampling, generates artificial data. While this kind of data approximates real data fairly well in continuous domains such as computer vision, it is not as representative in non-continuous domains such as text [9].

Semi-Supervised Learning for text data is not new and was first proposed in the context of class imbalance in [13]. However, we are dealing with such an extremely imbalanced data set that solutions of the type proposed in [13] are not expected to work. Semi-supervised learning in the class-imbalanced setting is also not new. Authors in [10] review existing approaches and propose their own. However, they focus on algorithmic modifications rather than the simpler and more practical re-sampling strategy.

Our framework is similar to standard approaches previously designed to tackle the class imbalance problem, but it differs from them in one important way. On the one hand, like methods such as SMOTE, GAMO and so on, it proposes to oversample the minority class, but on the other hand, unlike these approaches, instead of using generated samples it identifies candidates from the unlabeled data set to inflate the minority class with. Although the search for such candidates could be extremely costly, we show how the use of a K-D Tree makes it tractable.

We evaluate our framework on two data sets related to Covid-19 misinformation in social media, the one collected and curated in-house, early in the pandemic [2], and a data set obtained from English COVID-19 Fake News and Hindi Hostile Posts data set [17]. Our framework takes two forms: the direct approach in which the labeled minority samples alone are used to search the unlabeled data set; and the indirect approach, designed to increase the diversity of the search, where artificial data are first generated from the minority class and these samples, along with the original minority samples, are used to search the unlabeled set. Different instantiations of these approaches are compared to traditional ways of overcoming the class imbalance problem and to the results obtained on the original imbalanced data set. The results show that the direct

---

[2] https://www.internetlivestats.com/twitter-statistics/.

implementation of our framework is superior to the indirect approach, which in turn, is superior to the traditional approaches. All of them improve upon not attempting to counter the class imbalance problem.

The remainder of the paper is organized as follows. In Sect. 2, we discuss previous work on oversampling methods for class imbalances, semi-supervised learning, and discuss the functionality of K-D Trees. Section 3 introduces our framework and discusses its direct and indirect instantiations. The experimental set-up is discussed in Sect. 4, and the results of our experiments are presented in Sect. 5. Section 6 concludes the paper.

## 2   Related Work

This section reviews previous work related to this study. We first discuss the methods for inflating the minority class that were previously proposed in the context of the class imbalance problem, and we then move to a discussion of previous work in semi-supervised learning, especially for class imbalanced data. We then describe the K-D Tree data structure along with the Nearest Neighbor Search algorithm associated with it, and used in this paper.

### 2.1   Methods for Oversampling the Minority Class

We focus on four methods previously proposed for oversampling the minority class.

**SMOTE and ADASYN.** The Synthetic Minority Oversampling Technique (SMOTE) [4], is an oversampling approach that generates minority class instances to balance data sets. It searches for the $K$ closest minority neighbors of each sample point in the minority class using the Euclidean distance. For each minority class sample $x_i$, the algorithm randomly chooses a number of samples from its $K$ closet minority neighbors denoted as $x_i(nn)$. For each $x_i$, we generate new samples using the following formula

$$x_i^{new} = x_i + \alpha \left( x_i(nn) - x_i \right),$$

where $\alpha$ is a random number from 0 to 1. For the purpose of this work, we use the implementation found in the imbalanced-learn python library, with $K = 2$. In Adaptive Synthetic Sampling (ADASYN) [8], a mechanism is used to automatically determine how many synthetic samples need to be generated for each minority sample. For each minority class sample $x_i$, with its $K$ nearest neighbors $x_i(nn)$, it is possible to calculate the ratio $r_i = \frac{x_i(nn)}{K}$, and then normalize this ratio to obtain the density distribution $\Gamma_i = \frac{r_i}{\sum r_i}$. The calculation of a synthetic sample is obtained by $g_i = \Gamma_i \times G$, where $G$ is the discrepancy between 2 classes. For the purpose of this work, we use the ADASYN package from the imbalanced-learn python library, with $K = 2$.

**Generating Adversarial Networks (GANs).** A generative adversarial network (GAN) [7] consists of two neural networks: a generator $G$ and a discriminator $D$. These two networks are trained in opposition to one another. The generator $G$ takes as input a random noise vector $z \sim p(z)$ and outputs $m$ sample $\widetilde{x}^i = G\left(z^i\right)$. The discriminator $D$ receives as input the training sample $x^i$ and $\widetilde{x}^i$ and uses the loss function

$$\check{V}_{max} = \frac{1}{m}\sum_{i=1}^{m}\log D\left(x^i\right) + \frac{1}{m}\sum_{i=1}^{m}\log\left(1 - D\left(\check{x}^i\right)\right)$$

to update the discriminator $D$ 's parameters $\theta_d$;
Then it uses another random noise vector $z \sim p(z)$ and loss function:

$$\check{V}_{min} = \frac{1}{m}\sum_{i=1}^{m}\log\left(1 - \left(D\left(G\left(z^i\right)\right)\right)\right)$$

to update the generator $G$'s parameters $\theta_g$.

A VAE-GAN is a Variational Autoencoder combined with a Generative Adversarial Network [12]. It uses a GAN discriminator that can be used in place of a Variational Autoencoder (VAE) decoder to learn the loss function. The VAE loss function equals the negative sum of the expected log-likelihood (the reconstruction error) and a prior regularization term as well as a binary cross-entropy in the discriminator. This is what was used in this work.

## 2.2 Semi-supervised Learning

Semi-supervised learning is highly practical since labeling work is usually costly in terms of manpower and material resources [21]. There are two common methods used in semi-supervised learning [20]. The first one relies on the "clustering assumption" which assumes that the data follows a cluster structure and that samples in the same cluster belong to the same category. Another method follows the "manifold assumption" which assumes that the data is distributed on a manifold structure and that adjacent samples on that structure should output similar values. In such methods, the degree of proximity is often used to described the degree of similarity. The manifold hypothesis can be viewed as a generalization of the clustering hypothesis. Since we are working in the context of detection, a special case of classification, the "clustering assumption" is sufficient for our purposes.

Semi-supervised learning in the context of the class imbalance problem was considered in [10,13,19]. These works, however, do not consider the approach that consists of inflating the minority class nor do they look at the extremely imbalanced context.

## 2.3 K-Dimensional Tree and Nearest Neighbor Search

The search for nearest neighbors that we propose to undertake to identify data close to the labeled minority class data is computationally expensive. K-D Trees

(K-dimension trees) are a kind of binary trees, which divide the k-dimensional data space hierarchically, and stores the points in the k-dimension space in order to query its tree-shaped data structure afterwards [16]. K-D Trees are built with a recursive rule that splits the data according to the dimension/feature with highest variance. The dimension selected for splitting is set as the root node of the K-D Tree or subtree under consideration. This is done by finding the median for this dimension and using that value as a segmentation hyperplane, i.e., all the points whose value in that dimension is smaller than the median value are placed in the left child, and all the points with a greater value are placed in the right child. This procedure is followed recursively until nodes cannot be split anymore. A query search using a K-D Tree starts from the root node and moves down the tree recursively. It goes to the left or right child of the node it is currently visiting depending on its relation to the node value. Once the search reaches a leaf node, the algorithm sets it as "best current result". It then searches the other side of the parent to find out whether a better solution is available there. If so, it continues its search there, looking for a closer point. If such a point does not exist, it moves up the tree by one level and repeats the process. The search is completed when the root node is reached.

Using K-D Trees can reduce the search space compared to other clustering algorithm such as K-Nearest Neighbors which have a time complexity of $O(n \times m)$, where $n$ is the size of the data set and $m$ is its dimension. Commonly, the K-D Tree can be constructed in $O(n \log n)$, and the query algorithm has a running time $O(\sqrt{n} + k)$ where $k$ is the number of nearest points reported.

## 3   Our Framework

We propose a data augmentation method which, instead of randomly sampling from the minority class or generating synthetic minority samples based on the existing minority samples, leverages the unlabeled data set. The method is geared at non-continuous feature spaces such as those emanating from text applications, which present particular difficulty for data generation processes. Our approach works on binary data and takes as input a labeled imbalanced data set $LI$ and an unlabeled data set $U$, drawn from the same population. It outputs a labeled balanced data set $LB$ that is then used for classification. We consider two instantiations of our framework: the direct approach and the indirect approach. The direct approach constructs a K-D Tree from the labeled minority instances present in $LI$. Because the minority class can contain a very small number of samples, we also propose the indirect approach which implements Step 2. The rationale for the indirect approach is that the minority data set may be very small and not diverse enough to help direct the search for appropriate additional instances from $U$. We now describe each of the steps of our algorithm in detail:

**Step 1: Pre-processing:** We conduct the corpus cleaning work first. Since we are working with Twitter textual data, we remove all special symbols, white spaces, emoticon icons and stop words from the tweets. We used the pre-trained checkpoints provided by the Digital Epidemiology Lab [14] as a starting checkpoint and trained a BERT model [5], to compute an embedding for our tweet corpora.

**Step 2: Synthetic Sample Generation:** In this step, used by the indirect approach, we generate synthetic samples by using both classical and deep-learning means. In particular, we use: SMOTE, ADASYN and a VAE-GAN. The samples are generated according to the processes described in Sect. 2 for each of the approaches.

**Step 3: K-D Tree Construction and Nearest Neighbor Search:** In this step, we construct a K-D tree and search through it for the nearest neighbors, using the procedures described in Sect. 2.3.

**Step 4: Balanced Data Set Formation and Classification:** We first assume there are $n_{max}$ instances of the majority class and $n_{min}$ instances of the minority class in the data set. For the K-D Tree method, we augment the data using the following rule: For each minority data $x_i$, we traverse the tree composed of unlabeled data and find the $n_{aug_i} = \frac{n_{max} - n_{min}}{n_{min}}$. We add $n_{aug} = \sum_{i=1}^{n_{min}} n_{aug_i}$ to the data set after assigning them to the minority class. For SMOTE, ADASYN and VAE-GAN, we generate $n_{aug} = (n_{max} - n_{min})$ artificial samples, set them as minority class and add to the data set. After the data set is balanced, a logistic regression classifier is trained.

## 4  Experimental Evaluation

### 4.1  Data Sets

*Data Set 1:* The first data set was collected for the study by [2] which initially randomly collected a sample of 282,201 Twitter users from Canada by using the Conditional Independence Coupling (CIC) method [18]. A carefully curated and labeled sub data set was carved out from the random subset and includes 280 reliable and 280 unreliable Tweets and represent data set $LI$. The remaining 1,040 samples are unlabeled and correspond to data set $U_1$. We created a testing set $Test_1$ by randomly selecting 40 reliable and 40 unreliable tweets from $LI$. From the rest of the labeled data, we created several $LI_{1_n}$ data sets with all the 240 reliable tweets and different numbers, $n$, of unreliable tweets.

*Data Set 2:* The second data set is the COVID-19 Fake News Data set from [17] which includes a manually annotated data set of 10,700 social media posts and articles of real and fake news on Covid-19. We randomly selected 6,000 of them (3,000 true news and 3,000 fake news) for $LI$ and randomly selected 100 reliable and 100 unreliable tweets from $LI$ to create our testing set, $Test_2$. To create training sets $LI_{2_n}$, we randomly selected 900 samples from the true news subset and different numbers, $n$, from the fake news subset. The samples that were not selected were stripped of their labels and constitute the unlabeled data set $U_2$.

## 4.2  Training and Testing Method

*Training:* In our experiments, we trained the logistic regression classifier on the two data sets (Data Set 1 and Data Set 2), using the different data augmentation methods previously discussed to balance the training set. In more detail, we ran the following series of experiments on both Data Sets 1 and 2. Each experiment was repeated for a minority class of size $n$ where $n$ belongs to $\{5, 6, 7, 8, 9, 10, 20, 30, 40, 50, 100, 150\}$. Each of the 150 generated data sets are called $LI_n$.

- Train Logistic Regression on $LI_n$. The results for this series of experiments are seen on the curve called "Original".Which is also regard as baseline.
- Train Logistic Regression on $LI_n$ augmented by: SMOTE, ADASYN, VAE-GAN. The SMOTE and ADASYN functions we used in our task come from the python package "imblearn". The results for this series of experiments are reported on the curves called "SMOTE", ADASYN" and "VAE-GAN" respectively.
- Train Logistic Regression on $LI_n$ augmented using the K-D Tree and Nearest Neighbor Search technique on the n instances of the minority class present in $LI_n$. We recall that technique selects data from $U$, the unlabeled set, that most closely resembles the $n$ samples of the minority class. This is the Direct implementation of our framework that skips Step 2 in the Algorithm of Sect. 3. The results for this series of experiments are reported on the curve called "K-D Tree".
- Train Logistic Regression on $LI_n$ augmented using the K-D Tree and nearest neighbor search technique on the $n$ instances of the minority class and their augmentations through: SMOTE, ADASYN and VAE-GAN. We recall that this technique selects data from $U$, the unlabeled set, that most closely resembles the n samples of the minority class and the synthetic samples generated from them using one of the generation method shown above. This is the indirect implementation of our framework that uses Step 2 in the Algorithm of Sect. 3. The results for this series of experiments are reported on the curves called "SMOTE-KD, "ADASYN-KD", and "Replace-GAN".

*Testing Regimen:* In total, we conducted 192 tests on 24 $LI$ data sets with different numbers of minority class samples from the 2 data sets. The final result for each of these 192 experiments are reported based on the testing sets $Test_1$ and $Test_2$, respectively. We use the Bootstrap error estimation technique to evaluate the performance of our Method[3]. We report the F1, Precision, and Recall values of all the classifiers tested on the test set.

## 5  Results

The results of our experiments appear in Fig. 1. In all the graphs, the horizontal axis represents the number of labeled minority instances used to train the

---

[3] Bootstrap technique is implemented by repeating the sampling and testing procedure previously described 100 times and using the results of these experiments to estimate the real F1, Precision and Recall values and evaluate their standard deviation.

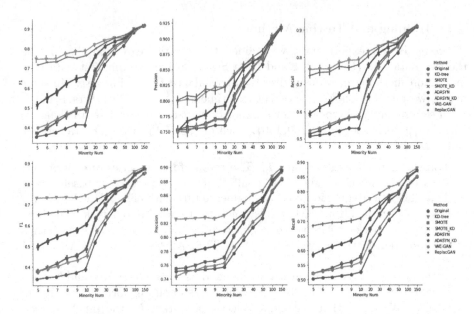

**Fig. 1.** The F1, Precision and Recall values of Data sets 1 (up) and 2 (down)

Logistic Regression classifier (with or without the different kinds of minority class inflation considered in this work). The graphs show that distinct differences in the results really happen when $n$, the number of minority instances initially present, is small. As $n$ increases, the differences between the methods becomes less and less visible. We also find that the results are similar for dataset 1 and dataset 2.

In general, we find that "Original", where no correction for the class imbalance domain is made, obtains the worst performance. This is followed closely by the three synthetic approaches (SMOTE, ADASYN and VAE-GAN) with a slight advantage for VAE-GAN in Data Set 1 and a slight advantage for SMOTE and ADASYN (which show identical performance in all experiments) in Data Set 2. As shown in all graphs, the advantage gained by these synthetic resampling methods is modest. Next, in terms of performance, come the three indirect methods of our framework, SMOTE-KD, ADASYN-KD, ReplaceGAN. We recall that these are the methods that generate synthetic data but do not use them directly. Instead, they are used to identify appropriate unlabeled samples to add to the minority class. The results show that these approaches obtain noticeably higher F1, Precision and Recall results, with a distinct advantage for Replace-GAN. This is true for both data sets, and suggests that the addition of real data through our semi-supervised scheme rather than synthetically generated data is a superior proposition. Finally, the results show, that in both domains, using the Direct implementation of our framework yields a better performance than the ReplaceGAN strategy. That difference is slight in Data Set 1, where the standard decision bars indicate that ReplaceGAN, while slightly less accurate, is more stable than K-D Tree, but it is unmistakable in Data Set 2 where ReplaceGAN is noticeably less accurate than K-D Tree. This suggests that our hypothesis regarding the advantage that a greater diversity to start off our unlabeled data set search for minority sample candidates did not pan out and the indirect implementation of our framework is less desirable than its Direct implementation.

In terms of run time, we tested the K-D tree and ReplaceGAN's running time with data set 2 and found, as expected, that the K-D Tree has a much lower running time than ReplaceGAN. The results are available on GitHub[4].

## 6  Discussion

In this paper, we presented a semi-supervised framework for identifying appropriate unlabeled samples to inflate the minority class and create a more balanced data set to learn from. The framework was designed specifically for non-continuous domains such as text, and tested on two misinformation/Fake news detection data sets where it obtained remarkable results, especially in cases of extreme class imbalance. Two categories of approaches of the framework were tested: the direct and indirect approach. The direct approach (K-D Tree) performed better than the indirect approach using a GAN (ReplaceGAN) but was not as stable in the smaller dataset (dataset 1). The direct approach is also more efficient than the indirect one, but the disparity is less noticeable in smaller data sets. The results obtained with our framework were significantly better than those obtained by methods that augment the data by synthetic generation, thus supporting the assumption that synthetic generation in non-continuous domains such as Text is not particularly useful and that semi-supervised methods such as ours fare much better.

In the future, we propose to investigate the utility of the ReplaceGAN indirect approach more carefully. We will also extend our framework to different domains (e.g., genetic and image domains) including continuous and discrete ones where an unlabeled data set exists, and test other classifiers on our resulting augmented data sets. This will allow us to test whether the advantage we noticed in text data and with logistic regression carries over to other types of domains and classifiers as well. More generally, we will also attempt to use our framework in the context of a data labeling tool having only a few seed labels to start from.

## References

1. Bellinger, C., Drummond, C., Japkowicz, N.: Manifold-based synthetic oversampling with manifold conformance estimation. Mach. Learn. **107**(3), 605–637 (2017). https://doi.org/10.1007/s10994-017-5670-4
2. Boukouvalas, Z., et al.: Independent component analysis for trustworthy cyberspace during high impact events: an application to Covid-19. arXiv:2006.01284 [cs, stat] (June 2020)

---

[4] https://github.com/Alex-NKG/Semi-Supervised-Framework-Misinfo-Detection.

3. Branco, P., Torgo, L., Ribeiro, R.P.: A survey of predictive modeling on imbalanced domains. ACM Comput. Surv. (CSUR) **49**, 1–50 (2016)
4. Chawla, N.V., Bowyer, K.W., Hall, L.O., Kegelmeyer, W.P.: SMOTE: synthetic minority over-sampling technique. J. Artif. Intell. Res. **16**, 321–357 (2002)
5. Devlin, J., Chang, M.-W., Lee, K., Toutanova, K.: BERT: pre-training of deep bidirectional transformers for language understanding (2019)
6. Drummond, C., Holte, R.C., et al.: C4. 5, class imbalance, and cost sensitivity: why under-sampling beats over-sampling. In: Workshop on Learning from Imbalanced Datasets II, vol. 11, pp. 1–8. Citeseer (2003)
7. Goodfellow, I.J., et al. Generative adversarial networks. arXiv preprint arXiv:1406.2661 (2014)
8. He, H., Bai, Y., Garcia, E.A., Li, S.: ADASYN: adaptive synthetic sampling approach for imbalanced learning. In: 2008 IEEE International Joint Conference on Neural Networks (IEEE World Congress on Computational Intelligence), pp. 1322–1328 (2008)
9. Hu, Z., Yang, Z., Liang, X., Salakhutdinov, R., Xing, E.: Toward controlled generation of text. In: ICML (2017)
10. Hyun, M., Jeong, J., Kwak, N.: Class-imbalanced semi-supervised learning. CoRR abs/2002.06815 (2020)
11. Islam, M.R., Liu, S., Wang, X., Xu, G.: Deep learning for misinformation detection on online social networks: a survey and new perspectives. Soc. Netw. Anal. Min. **10**(1), 1–20 (2020). https://doi.org/10.1007/s13278-020-00696-x
12. Larsen, A.B.L., Sønderby, S.K., Larochelle, H., Winther, O.: Autoencoding beyond pixels using a learned similarity metric. arXiv:1512.09300 [cs, stat] (February 2016)
13. Li, S., Wang, Z., Zhou, G., Lee, S.: Semi-supervised learning for imbalanced sentiment classification. In IJCAI (2011)
14. Müller, M., Salathé, M., Kummervold, P.E.: COVID-Twitter-BERT: a natural language processing model to analyse COVID-19 content on Twitter. arXiv preprint arXiv:2005.07503 (2020)
15. Mullick, S.S., Datta, S., Das, S.: Generative adversarial minority oversampling. In: 2019 IEEE/CVF International Conference on Computer Vision (ICCV), pp. 1695–1704 (2019)
16. Otair, D.M.: Approximate k-nearest neighbour based spatial clustering using KD-tree (2013)
17. Chakraborty, T., Shu, K., Bernard, H.R., Liu, H., Akhtar, M.S. (eds.): CONSTRAINT 2021. CCIS, vol. 1402. Springer, Cham (2021). https://doi.org/10.1007/978-3-030-73696-5
18. White, K., Li, G., Japkowicz, N.: Sampling online social networks using coupling from the past. In: 2012 IEEE 12th International Conference on Data Mining Workshops, pp. 266–272 (2012)
19. Yang, Y., Xu, Z.: Rethinking the value of labels for improving class-imbalanced learning. arXiv:abs/2006.07529 (2020)
20. Zhou, Z.-H.: Machine Learning. Springer, Singapore (2021). https://doi.org/10.1007/978-981-15-1967-3
21. Zhu, X.J.: Semi-supervised Learning Literature Survey (2008)

# An Analysis of Performance Metrics for Imbalanced Classification

Jean-Gabriel Gaudreault[1]([🖂]) [ID], Paula Branco[1] [ID], and João Gama[2] [ID]

[1] University of Ottawa, Ottawa, ON K1N6N5, Canada
{j.gaudreault,pbranco}@uottawa.ca
[2] University of Porto, Porto, Portugal
jgama@fep.up.pt

**Abstract.** Numerous machine learning applications involve dealing with imbalanced domains, where the learning focus is on the least frequent classes. This imbalance introduces new challenges for both the performance assessment of these models and their predictive modeling. While several performance metrics have been established as baselines in balanced domains, some cannot be applied to the imbalanced case since the use of the majority class in the metric could lead to a misleading evaluation of performance. Other metrics, such as the area under the precision-recall curve, have been demonstrated to be more appropriate for imbalance domains due to their focus on class-specific performance. There are, however, many proposed implementations for this particular metric, which could potentially lead to different conclusions depending on the one used. In this research, we carry out an experimental study to better understand these issues and aim at providing a set of recommendations by studying the impact of using different metrics and different implementations of the same metric under multiple imbalance settings.

**Keywords:** Imbalanced domains · Performance metrics · Performance evaluation · Precision-recall curve

## 1 Introduction

The choice of performance metrics constitutes a crucial part of assessing the effectiveness of a classification model. In imbalanced domains, due to the commonly used accuracy and error rate not being suitable metrics for performance evaluation [2], the research community has turned towards other metrics, such as the well-known $F_\beta$-measure and Geometric Mean (G-Mean).

While these are generally regarded as appropriate to assess the performance of classifiers on imbalanced data, they do present some drawbacks. For instance, they cannot provide a general picture of the classifier's performance at varying thresholds [13] and assume knowledge of the context where the classifier is being deployed [9]. Therefore, other metrics, namely the Receiver Operating Characteristic (ROC) curve, the precision-recall (PR) curve, and their corresponding

© Springer Nature Switzerland AG 2021
C. Soares and L. Torgo (Eds.): DS 2021, LNAI 12986, pp. 67–77, 2021.
https://doi.org/10.1007/978-3-030-88942-5_6

area under the curve (AUC), have risen in popularity. However, many have criticized the usage of the former [3,4,13], and the latter present some issues for which multiple implementations have been proposed [3,7,12,14].

With the advent of new, more complex algorithms, it becomes imperative to have standardized metrics to evaluate and compare their performance [8]. Given this wide array of performance metrics available, we wanted to explore the impact of using the different metrics proposed for imbalanced domains as well as study alternative implementations of the same metric.

Our main contributions are: (i) provide a synthesis and analysis of performance assessment metrics typically used under an imbalanced setting and their different estimation approaches; (ii) propose a novel rank-based measure to assess the divergence of the results when using different performance metrics; (iii) study the impact of using different metrics and different implementations of the same metric under multiple imbalance settings; and (iv) provide a set of recommendations regarding the evaluated metrics. The rest of this paper is structured as follows: Sect. 2 reviews some commonly used metrics to evaluate classification performance in imbalanced domains as well as different implementations/interpolations of the PR curve; Sect. 3 compares and discusses the experimental results obtained; and Sect. 4 concludes this paper.

## 2   Review of Performance Metrics and Estimation Approaches for Imbalanced Binary Classification

The performance evaluation of imbalanced classification problems is a common challenge for which multiple performance metrics have been defined. Using the classification proposed by Ferri et al. [6], these metrics can be clustered into three categories: threshold metrics, ranking metrics, and probabilistic metrics. While probabilistic metrics sometimes do behave well in imbalanced domains, it is not their probabilistic quality that makes them do so [9]. Hence, in this paper, we will focus on the most used threshold and ranking metrics.

These metrics will be defined in the case of a binary classification problem, using the prediction values of a confusion matrix and the inferred recall, specificity, and precision as demonstrated in Table 1, where we consider the negative class as the majority class and the positive class as the minority.

### 2.1   Threshold Metrics

Threshold-based metrics are single scalar values that aim at quantifying the model's prediction errors.

$F_\beta$-**Measure.** Metrics based on precision, recall, and specificity are often favored in imbalanced domains as they consider class-specific performance. The $F_\beta$-Measure aims to represent the trade-off between precision and recall in a single value. The value of $\beta$ represents the weight assigned to either metric. If $\beta$ is

**Table 1.** Confusion matrix of a binary classification problem.

|  |  | **Predicted** | | |
|---|---|---|---|---|
|  |  | Positive | Negative | |
| **Actual Values** | Positive | **True Positive** (TP) | **False Negative** (FN) | **Recall (Sensitivity)** $TP/(TP+FN)$ |
|  | Negative | **False Positive** (FP) | **True Negative** (TN) | **Specificity** $TN/(TN+FP)$ |
|  |  | **Precision** $TP/(TP+FP)$ | **Negative Predictive Value** $TN/(TN+FN)$ | |

1, precision and recall have the same importance, while for values of $\beta$ higher (lower) than 1, higher importance is given on recall (precision).

$$F_\beta = (1 + \beta^2) \cdot \frac{precision \cdot recall}{(\beta^2 \cdot precision) + recall} \tag{1}$$

**G-Mean.** The G-Mean is defined as the square root of the recall of each class [11]. This metric aims at maximizing the accuracies of both classes. However, it assigns equal importance to both classes. To address this issue, other formulations, such as one that replaces specificity with precision to focus only on the positive class, have also been proposed [9].

$$G_{Mean} = \sqrt{sensitivity \cdot specificity} \tag{2}$$

## 2.2   Ranking Metrics

Ranking metrics are based on how well a model ranks or separates the samples in different classes [6]. They use the probabilities of a sample belonging to a class outputted by the model and apply different thresholds to test the model's performance across the whole range of thresholds, thus identifying the model's ability to separate (rank) the classes.

These metrics are often represented in graph form, which allows the user to visualize the model's performance across the different thresholds and give more insight into the trade-off between the measurements. However, it can often be hard to compare multiple graphics between models and evaluate the best one, especially in cases where they intersect. Therefore, the AUC is often computed and used as it provides a single-value metric that allows for a more concise and straightforward comparison between models.

**Receiver Operating Characteristic Curve.** The ROC curve, as demonstrated in Fig. 1a, plots the true positive rate (recall) against the false positive rate: $FPR = \frac{FP}{FP+TN}$, with each point representing a threshold value.

While the AUC of the ROC curve, often referred to as the ROC AUC, can be defined as the definite integral over the whole range of values, there exist multiple ways of practically interpolating it, such as the trapezoidal method or the Mann–Whitney–Wilcoxon statistic interpretation [5].

Finally, it is worth noting that while the ROC curve is widely used in imbalanced domains, it has been criticized as it can present an overly optimistic evaluation of the model's performance when the imbalance is pronounced [3,5] and can lead to incorrect interpretations [13]. Other alternatives, such as the PR curve, can provide a different perspective on the evaluation.

**Precision-Recall Curve.** Similar to the ROC curve, the PR curve uses different thresholds on the model's predictions to compute different scores of two metrics: precision and recall. However, whereas the ROC curve had a fixed baseline, this is not the case for the PR curve. In fact, a random classifier would be displayed as a horizontal line at the $y$ value equal to the minority class proportion of the dataset, as it can be seen in Fig. 1b.

To obtain the AUC of the PR curve, one could consider linearly interpolating the points in the curve as it is carried out in the ROC AUC computation. However, the typical saw-tooth shape of the PR curve, which is explained by the fact that the precision value does not vary linearly with the recall value, makes it incorrect to use linear interpolation to compute the AUC under the PR curve since it would lead to an overly optimistic estimation of the performance [3]. Multiple solutions have been proposed to address this issue, and we will review the main ones in this paper.

*Interpolated Precision.* The interpolated precision, as described by Manning et al. [12], interpolates the precision of any recall point $r$ as the highest precision value found for any recall point greater than $r$. This process allows getting rid of the undesirable saw-tooth effect of the PR curve, and the AUC can then be computed using the usual methods.

*Average Precision.* The average precision (AP) is defined as the sum of precision at each point weighted by the difference in recall with the previous point, as described in Eq. 3 [14]. Therefore, its value is different from computing the AUC under the PR curve as it is not dependent on any curve and consequently does not require any interpolation strategy.

$$AP = \sum_n (Recall_n - Recall_{n-1})Precision_n \tag{3}$$

*Davis' Interpolation.* Davis and Goadrich proposed an interpolation of the PR AUC [3] that we will refer to as Davis' interpolation. Their proposition is based on the fact that every point of the PR curve is composed of their True Positive (TP) and False Positive (FP) underlying values. Keeping that in mind, we can interpolate the TP and FP values between two points. Considering A and B, two consecutive points, we can first evaluate how many negative examples it takes

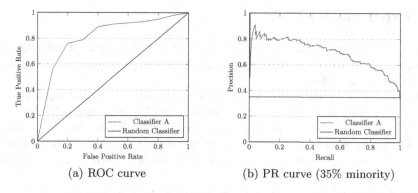

(a) ROC curve        (b) PR curve (35% minority)

**Fig. 1.** ROC and PR curves of a random and a typical classifier.

to equal one positive, referred to as the local skew: $\frac{FP_B - FP_A}{TP_B - TP_A}$. Then, new points are created such that $TP = TP_A + x$, where $x$ represent all integer values such that $1 \leq x \leq TP_B - TP_A$. Finally, we can interpolate FP using the local skew, and the resulting interpolated points can be described by Eq. 4.

$$(\frac{TP_A + x}{TotalPos.}, \frac{TP_A + x}{TP_A + x + FP_A + \frac{FP_B - FP_A}{TP_B - TP_A}x}) \tag{4}$$

**Precision-Recall-Gain Curve.** In their 2015 contribution [7], Flach and Kull demonstrate why the use of an arithmetic average over metrics that are represented on a harmonic scale, such as precision, recall, and F-Measure is fundamentally wrong and how it affects the PR curve. More specifically, they introduce a novel approach to the PR analysis by proposing the Precision-Recall-Gain curve (PRG curve) and its corresponding AUC. This new approach aims to address some of the drawbacks of the PR curve compared to the ROC curve, such as the lack of universal baseline and linear interpolation. In this context, two new measurements are defined as valid alternatives to the precision and recall metrics: the precision gain (Eq. 5) and recall gain (Eq. 6).

$$precG = \frac{prec - \pi}{(1 - \pi)prec} \tag{5}$$

$$recG = \frac{rec - \pi}{(1 - \pi)rec} \tag{6}$$

where $\pi$ represents the precision of the baseline, i.e., the always-positive classifier. By plotting the precision gain against the recall gain, one can then obtain the PRG curve and compute its corresponding AUC using linear interpolation.

# 3   Experimental Data

This section describes the experiments conducted to determine if different conclusions can be drawn from the same model while using different metrics or interpolation strategies.

In these experiments, we used 4 different classifiers: decision tree, random forest, XGBoost, and support vector machine (SVM). We used a 5-fold cross-validation procedure and evaluated the performance in 15 datasets obtained from the KEEL repository [1]. The datasets were selected based on their imbalance ratio (IR), which is defined as the number of majority class samples over the number of minority class samples.

We selected 5 datasets from each one of 3 different groups exhibiting different IR levels, as follows: less than 5 (slightly imbalanced), between 9 and 13 (moderately imbalanced), and higher than 16.9 (heavily imbalanced). We recorded the 8 previously discussed performance assessment metrics for each of these datasets and classifiers.

## 3.1   Assessing the Impact of the Performance Metrics

Our main goals are: (i) to determine if there is a significant disparity between the use of one metric compared to the others; and (ii) to determine the impact of using different interpolation methods to compute the same metric (PR curve AUC). For this reason, we will present the majority of the results using rankings. We will compute the score of each classifier using a set of performance metrics. Then, for each performance metric considered, each model will be ranked using the score obtained. To evaluate the differences in the rankings obtained, we use the following two measures: the well-known Krippendorff's Alpha score [10] and Mode Difference, a novel score specifically proposed to assess the impact in the performance in our particular setting.

In order to optimally observe which metric results differ from the others, we propose the use of a novel metric named Mode Difference (MD) (cf. Eq. 8). Suppose we have the results of a set of $n$ different performance metrics when using $k$ different classifiers. We start by obtaining the rankings for a given metric $m \in \{1, \cdots, n\}$ for all tested classifiers, $r_{m,1}, r_{m,2}, \ldots, r_{m,k}$. Then, we calculate the mode of all the rankings obtained for each classifier (cf. Eq. 7). This will provide the most frequent rank value (or the average of the most frequent values when there are ties) for a classifier. Thus, $Mo_j$ will show the most commonly occurring ranking across all metrics tested for classifier $j$.

$$Mo_j = Mode(r_{1,j}, r_{2,j}, \ldots, r_{n,j}) \tag{7}$$

$$MD_{i,j} = |r_{i,j} - Mo_j| \tag{8}$$

where $r_{i,j}$ represents the rank obtained for performance metric $i$ on classifier $j$. The value of $MD_{i,j}$ measures the deviation from the mode of the rank of a classifier in a metric.

Finally, to have a global overview of variation of the MD for a specific performance metric $i$, we calculate the $MD\_AVG_i$, as shown in Eq. 9. This average will quantify the observed variability in the rankings for each tested metric across all classifiers used.

Considering a total number of $k$ classifiers tested, we compute the average of the mode differences over a performance metric $i$ as follows:

$$MD\_AVG_i = Avg(MD_{i,1}, MD_{i,2}, \ldots, MD_{i,k}) \qquad (9)$$

We use Krippendorff's Alpha and Mode Difference to assess the divergence in the results obtained by the 8 performance metrics selected in our study.

## 3.2  Results and Discussion

The initial results of our experiments include the score of each of the 8 performance metrics evaluated for each one of the 15 datasets using 4 different classifiers. We transformed the scores into rankings and calculated the Mode Difference. Table 2 present the obtained rankings and mode difference scores for each classifier in each performance metric in a selected dataset. We chose to include only one dataset due to space constraints and for illustration purposes. The results on all the remaining datasets, as well as all the necessary code to reproduce our experiments, are freely available at https://github.com/jgaud/ PerformanceMetricsImbalancedDomains.

To present the results more concisely, we aggregated the results obtained per class imbalance level, i.e., we considered the 3 following IR levels: slightly $(IR < 5)$, moderately $(9 < IR < 13)$, and heavily $(16.9 < IR)$ imbalanced. Table 3 presents the average MD of all metrics for each class imbalance level.

Two main conclusions can be drawn by observing the first results presented in Table 3. The G-Mean, $F_1$ measure, and ROC AUC are the worst-performing metrics in terms of disagreement level across all imbalance ratios. In effect, these 3 metrics present the most disagreement with the rank given by the other alternative metrics. This behavior meets our initial expectations, as the remaining metrics are different interpolations of a same metric. It also confirms that these metrics (PR AUC with Linear Interpolation, Interpolated Precision AUC, AP, PR-Gain AUC, and PR AUC Davis) generally provide a similar ranking. A second interesting observation is related to the agreement level verified for the AUC computed from Davis' interpolation of the PR curve. This metric achieves the best scores, in terms of MD average, across all imbalance ratios, which means that it consistently agreed the most with the remaining metrics tested. We further establish this last conclusion by restricting our comparison to the different implementations of the PR curve AUC in Table 4.

**Table 2.** Ranks and mode difference of each performance metrics on the pima dataset (best performing model for each metric is underlined).

| Metric | Decision Tree | | Random Forest | | XGBoost | | SVM | | $MD_{AVG}$ |
|---|---|---|---|---|---|---|---|---|---|
| | Rank | MD | Rank | MD | Rank | MD | Rank | MD | |
| $F_1$ | 4 | 0 | <u>1</u> | 2 | 2 | 0 | 3 | 2 | 1 |
| $G_{Mean}$ | 3 | 1 | 2 | 1 | <u>1</u> | 1 | 4 | 3 | 1.5 |
| ROC AUC | 4 | 0 | 2 | 1 | <u>1</u> | 1 | 3 | 2 | 1 |
| PR AUC (Linear Interpolation) | 4 | 0 | 3 | 0 | 2 | 0 | <u>1</u> | 0 | 0 |
| Interpolated Precision AUC | 4 | 0 | 3 | 0 | 2 | 0 | <u>1</u> | 0 | 0 |
| AP | 4 | 0 | 3 | 0 | 2 | 0 | <u>1</u> | 0 | 0 |
| PR-Gain AUC | 4 | 0 | <u>1</u> | 2 | 2 | 0 | 3 | 2 | 1 |
| PR AUC Davis | 4 | 0 | 3 | 0 | 2 | 0 | <u>1</u> | 0 | 0 |

**Table 3.** MD of all metrics averaged by imbalance class (best score by imbalance level is underlined; best overall score is in boldface).

| Metric | IR < 9 | 9 < IR < 13 | 16.9 < IR |
|---|---|---|---|
| $F_1$ | 0.825 | 0.675 | 0.8 |
| $G_{Mean}$ | 0.925 | 0.775 | 0.9 |
| ROC AUC | 0.275 | 0.675 | 0.55 |
| PR AUC (Linear Interpolation) | 0.125 | **<u>0.025</u>** | 0.25 |
| Interpolated Precision AUC | 0.125 | 0.175 | 0.15 |
| AP | <u>0.075</u> | 0.125 | 0.45 |
| PR-Gain AUC | 0.375 | 0.225 | 0.35 |
| PR AUC Davis | <u>0.075</u> | **<u>0.025</u>** | <u>0.05</u> |

**Table 4.** MD of all PR AUC definitions averaged by imbalance class.

| Metric | IR < 9 | 9 < IR < 13 | 16.9 < IR |
|---|---|---|---|
| PR AUC (Linear Interpolation) | <u>0.1</u> | **<u>0</u>** | 0.125 |
| Interpolated Precision AUC | <u>0.1</u> | 0.2 | <u>0.025</u> |
| AP | <u>0.1</u> | 0.1 | 0.575 |
| PR AUC Davis | <u>0.1</u> | **<u>0</u>** | 0.075 |
| Krippendorff's Alpha | 0.95 | 0.94375 | 0.79375 |

With more detail, we observe from Table 4 that Davis' interpolation method performs the best in terms of agreement in two out of 3 class imbalance levels. For the case where it is not the best one, which corresponds to the datasets with a higher imbalance ratio, the PR AUC with Davis' interpolation is the second best. Moreover, we also observe a general trend where the metrics are more stable on less imbalanced data but display a higher divergence when moving on to highly imbalanced data, i.e., the MD average has higher discrepancies for higher class imbalance levels. This conclusion is also confirmed by the Krippendorff's alpha, which declines with the increase of the imbalance ratio, which means that the rankings produced by the scores of the tested metrics agree less with each other for datasets with a higher class imbalance level.

We continued to inspect further the results obtained, by computing the MD and Krippendorff's Alpha results, considering only the most consistent interpolation method tested. This means that we observed the results of the following metrics: $F_1$, G-Mean, ROC AUC, PR-Gain AUC, and PR AUC Davis. We selected Davis' interpolation as the metric representing the PR AUC because of our previous observations where it proved to be the most consistent metric among the tested interpolation methods. Table 5 shows the MD average of these metrics. We observe that G-Mean is the metric that deviates the most from the others across all datasets. It is important to note the discrepancy between threshold ($F_1$, G-Mean) and rank-based metrics (ROC AUC, PR-Gain AUC, and PR AUC Davis). We also observe a lower Krippendorff's Alpha score for datasets with higher imbalance, confirming that the agreement between the rankings obtained with these metrics decreases as the imbalance ratio increases.

**Table 5.** MD of each metric and the best performing PR AUC interpolation averaged by class imbalance level.

| Metric | IR < 9 | 9 < IR < 13 | 16.9 < IR |
|---|---|---|---|
| $F_1$ | 0.725 | 0.4 | 0.625 |
| $G_{Mean}$ | 0.875 | 0.5 | 0.725 |
| ROC AUC | **0.075** | 0.4 | 0.575 |
| PR-Gain AUC | 0.125 | 0.5 | 0.375 |
| PR AUC Davis | 0.175 | <u>0.35</u> | <u>0.325</u> |
| Krippendorff's Alpha | 0.639 | 0.6124 | 0.5098 |

**Table 6.** MD of threshold-based metrics against a rank-based metric averaged by class imbalance level.

| | Metric | IR < 9 | 9 < IR < 13 | 16.9 < IR |
|---|---|---|---|---|
| Test 1 | $F_1$ | <u>0.133</u> | <u>0.05</u> | <u>0</u> |
| | $G_{Mean}$ | 0.233 | 0.15 | 0.1 |
| | PR AUC Davis | 0.766 | 0.65 | 0.8 |
| Test 2 | $F_1$ | <u>0.166</u> | **0.016** | **0.016** |
| | $G_{Mean}$ | <u>0.166</u> | 0.183 | 0.116 |
| | ROC AUC | 0.633 | 0.616 | 1.183 |

Given the results described, we compared the agreement between threshold metrics and a rank-based metric. To investigate the presence of disparities in the agreement, we carried out two tests shown in Table 6. To this end, we observed the impact in MD average score when comparing: (i) the threshold metrics with Davis' interpolation of the PR AUC (Test 1 in the table); and (ii) the threshold metrics with the ROC AUC (Test 2 in the table). Both experiments show that the ranking metrics differ widely from the threshold metrics. In effect, we observe MD average results between 0 and 0.233 for the two threshold metrics while the ranking metrics exhibit much higher MD average values between 0.616 and 1.183. Moreover, we observe again that this divergence is more marked for datasets with a higher imbalance ratio.

## 4 Conclusion

This paper explores the impact of using different evaluation metrics and interpolation strategies for imbalanced classification tasks. We review the most frequently used metrics, presented some flaws when carrying out the interpolation of the PR curve, and presented methods to solve this issue. We propose MD, a novel metric for evaluating the level of agreement between the rankings of multiple raters.

The different interpolations of the PR AUC tested proved to give fairly similar rankings, especially when comparing them to other metrics. However, we observed that Davis' interpolation method performed the best, as it was the most consistent across all experiments, which means it agreed the most with the other interpolations tested. Another important observation was the noticeable discrepancy between ranking and threshold metrics, which shows the importance of using multiple types of metrics when evaluating the performance under imbalanced domains. Finally, a general trend was also observed throughout these experiments: as the imbalance increases, the metrics agreement tends to decrease. As future work, we plan to extend this study to the multi-class imbalance problem, where multiple ways of computing the ROC and PR curves have been defined.

## References

1. Alcalá-Fdez, J., et al.: Keel data-mining software tool: data set repository, integration of algorithms and experimental analysis framework. J. Multi-valued Logic Soft Comput. **17**, 255–287 (2011)
2. Branco, P., Torgo, L., Ribeiro, R.P.: A survey of predictive modeling on imbalanced domains. ACM Comput. Surv. **49**(2), 1–50 (2016)
3. Davis, J., Goadrich, M.: The relationship between precision-recall and ROC curves. In: Proceedings of the 23rd International Conference on Machine Learning, ICML 2006, pp. 233–240. Association for Computing Machinery, New York, NY, USA (2006)
4. Fawcett, T.: ROC graphs: notes and practical considerations for researchers. Mach. Learn. **31**, 1–38 (2004)
5. Fawcett, T.: An introduction to roc analysis. Pattern Recogn. Lett. **27**(8), 861–874 (2006)

6. Ferri, C., Hernández-Orallo, J., Modroiu, R.: An experimental comparison of performance measures for classification. Pattern Recogn. Lett. **30**(1), 27–38 (2009)

7. Flach, P.A., Kull, M.: Precision-recall-gain curves: PR analysis done right. In: NIPS, vol. 15 (2015)

8. He, H., Garcia, E.A.: Learning from imbalanced data. IEEE Trans. Knowl. Data Eng. **21**(9), 1263–1284 (2009)

9. Japkowicz, N.: Assessment metrics for imbalanced learning, chap. 8, pp. 187–206. Wiley (2013)

10. Krippendorff, K.: Computing Krippendorff's Alpha-Reliability (January 2011)

11. Kubat, M., Matwin, S., et al.: Addressing the curse of imbalanced training sets: one-sided selection. In: ICML, vol. 97, pp. 179–186. Citeseer (1997)

12. Manning, C.D., Raghavan, P., Schütze, H.: Introduction to Information Retrieval. Cambridge University Press, USA (2008)

13. Saito, T., Rehmsmeier, M.: The precision-recall plot is more informative than the ROC plot when evaluating binary classifiers on imbalanced datasets. PLOS ONE **10**(3), 1–21 (2015)

14. Su, W., Yuan, Y., Zhu, M.: A relationship between the average precision and the area under the ROC curve. In: Proceedings of the 2015 International Conference on the Theory of Information Retrieval, pp. 349–352 (2015)

# Combining Predictions Under Uncertainty: The Case of Random Decision Trees

Florian Busch[1]([✉]), Moritz Kulessa[1], Eneldo Loza Mencía[1],
and Hendrik Blockeel[2]

[1] Technische Universität Darmstadt, Darmstadt, Germany
florianpeter.busch@stud.tu-darmstadt.de,
{mkulessa,eneldo}@ke.tu-darmstadt.de
[2] KU Leuven, Leuven, Belgium
hendrik.blockeel@kuleuven.be

**Abstract.** A common approach to aggregate classification estimates in an ensemble of decision trees is to either use voting or to average the probabilities for each class. The latter takes uncertainty into account, but not the reliability of the uncertainty estimates (so to say, the "uncertainty about the uncertainty"). More generally, much remains unknown about how to best combine probabilistic estimates from multiple sources. In this paper, we investigate a number of alternative prediction methods. Our methods are inspired by the theories of probability, belief functions and reliable classification, as well as a principle that we call *evidence accumulation*. Our experiments on a variety of data sets are based on random decision trees which guarantees a high diversity in the predictions to be combined. Somewhat unexpectedly, we found that taking the average over the probabilities is actually hard to beat. However, evidence accumulation showed consistently better results on all but very small leafs.

**Keywords:** Random decision trees · Ensembles of trees · Aggregation · Uncertainty

## 1 Introduction

Ensemble techniques, such as bagging or boosting, are popular tools to improve the predictive performance of classification algorithms due to diversification and stabilization of predictions [5]. Beside learning the ensemble, a particular challenge is to combine the estimates of the single learners during prediction. The conventional approaches for aggregation are to follow the majority vote or to compute the average probability for each class. More sophisticated approaches are to use *classifier fusion* [14], *stacking* [18] or *mixture of experts* [19]. Most of these approaches mainly aim to identify strong learners in the ensemble, to give them a higher weight during aggregation. However, this ignores the possibility that the predictive performance of each learner in the ensemble may vary

© Springer Nature Switzerland AG 2021
C. Soares and L. Torgo (Eds.): DS 2021, LNAI 12986, pp. 78–93, 2021.
https://doi.org/10.1007/978-3-030-88942-5_7

depending on the instance to be classified at hand. To take these differences into account, we argue that considering the certainty and uncertainty of each individual prediction might be a better strategy to appropriately combine the estimates. This should be particularly the case for ensembles generated using randomization techniques since this potentially leads to many unstable estimates.

A popular and effective strategy is to randomize decision trees. The prediction of a single decision tree is usually based on statistics associated with the leaf node to which a test instance is forwarded. A straight-forward way of assessing the reliability of a leaf node's estimate is to consider the number of instances which have been assigned to the node during learning. For example, imagine a leaf containing 4 instances of class $A$ and 0 of $B$, and another leaf with 10 instances in class $A$ and 40 in class $B$. The first source of prediction has considerably less evidence to *support* its prediction in contrast to the second leaf. If both leafs need to be combined during prediction, it seems natural to put more trust on the second leaf and, therefore, one may argue that class $B$ should be preferred though the average probability of 0.6 for $A$ tells us the opposite. However, one may argue for $A$ also for another reason, namely the very high confidence in the prediction for $A$, or put differently, the absence of uncertainty in the rejection of $B$. The certainty of the prediction is also known as *commitment* in the theory of belief functions. In this work, we will review and propose techniques which are grounded in different theories on modelling uncertainty and hence take the *support and strength of evidence* differently into account.

More specifically, Shaker and Hüllermeier [16] introduced a formal framework which translates observed counts as encountered in decision trees into a vector of plausibility scores for the binary classes, which we use for the prediction, and scores for aleatoric and epistemic uncertainty. While aleatoric uncertainty captures the randomness inherent in the sample, the latter captures the uncertainty with regard to the lack of evidence. The theory of belief functions [15] contributes two well-founded approaches to our study in order to combine these uncertainties. In contrast to the theory of belief functions, in probability theory the uncertainty is usually expressed by the dispersion of the fitted distribution. We introduce a technique which takes advantage of the change of the shape of the distribution in dependence of the sample size. And finally, apart from combining the probabilities themselves, one can argue that *evidence* for a particular class label should be properly *accumulated*, rather than averaged across the many members of the ensemble. In this technique, we measure the strength of the evidence by how much the probabilities deviate from the prior probability.

As an appropriate test bed for combining uncertain predictions, we choose to analyse these methods based on random decision trees (RDT) [6]. In contrast to other decision tree ensemble learners, such as random forest, RDT do not optimize any objective function during learning, which results in a large diversity of class distributions in the leafs. This puts the proposed methods specifically to the test in view of the goal of the investigation and makes the experimental study independent of further decisions such as the proper selection of the splitting criterion. We have compared the proposed techniques on 21 standard binary classification data sets. Surprisingly, our experimental evaluation showed that methods, which take the amount of evidence into consideration, did only improve over the simple averaging baseline in some specific cases. However, we could observe an advantage for the proposed approach of *evidence accumulation* which takes the strength of the evidences into account.

## 2    Preliminaries

This section briefly introduces RDT, followed by a short discussion on previous work which is relevant to us. Throughout the paper, we put our focus on binary classification which is the task of learning a mapping $f : X^m \rightarrow y$ between an instance $x \in X^m$ with $m$ numerical and/or nominal features and a binary class label $y \in \{\ominus, \oplus\}$ through a finite set of observations $\{(x_1, y_1), \ldots, (x_n, y_n)\}$.

### 2.1    Random Decision Trees

Introduced by Fan et al. [6], the approach of RDT is an ensemble of randomly created decision trees which, in contrast to classical decision tree learners and random forest [2], do not optimize a objective function during training. More precisely, the inner tests in the trees are chosen randomly which reduces the computational complexity but still achieves competitive and robust performance.

**Construction.** Starting from the root node, inner nodes of a single random tree are constructed recursively by distributing the training instances according to the randomly chosen test at the inner node as long as the stopping criterion of a minimum number of instances for a leaf is not fulfilled. Discrete features are chosen without replacement for the tests in contrast to continuous features, for which additionally a randomly picked instance determines the threshold. In case that no further tests can be created, a leaf will be constructed in which information about the assigned instances will be collected. For binary classification, the number of positive $w^\oplus$ and negative $w^\ominus$ class labels are extracted. Hence, a particular leaf can be denoted as $\mathbf{w} = [w^\oplus, w^\ominus]$ where $\mathbf{w} \in \mathcal{W}$ and $\mathcal{W} = \mathbb{N}^2$.

**Prediction.** For each of the $K$ random trees in the ensemble, the instance to be classified is forwarded from the root to a leaf node passing the respective tests in the inner nodes. The standard approach mainly used in the literature to aggregate the assigned leaf nodes $\mathbf{w}_1, \ldots, \mathbf{w}_K$ is to first compute the probability for the positive class on each leaf which is then averaged across the ensemble.

### 2.2    Related Work

A very well known approach used for improving the probability estimates of decision trees is Laplace smoothing, which essentially incorporates epistemic uncertainty through a prior. However, whereas Provost and Domingos [12] showed a clear advantage over using the raw estimates for single decision trees, Bostrom [1] observed that for random forest better estimates are achieved without smoothing. Apart from that correction, the reliability of the individual predictions in decision trees is usually controlled via pruning or imposing leaf sizes [12,20].

While the common recommendation is to grow trees in an ensemble to their fullest extent, Zhou and Mentch [20] argue that depending on the properties of the underlying data greater leafs and hence more stable predictions are preferable. With respect to RDT, Loza Mencía [8] reward predictions with higher confidence in the ensemble using the inverted Gini-index. However, their weighting approach has not been directly evaluated.

The Dempster-Shafer theory of evidence, or theory of belief functions, [15] is also concerned with the strength of the confidences. The general framework for combining prediction of Lu [9] is based on this theory and the study showed that the proposed technique can improve upon the Bayesian approach. Raza et al. [13] similarly combined outputs of support vector machines but only compared to a single classifier in their evaluation. Nguyen et al. [10] take uncertainties in an alternative way into account by computing interval-based information granules for each classifier. They could improve over common ensemble techniques, however, their approach requires additional optimization during learning and during classification. As already discussed in Sect. 1, meta and fusion approaches also often require additional learning steps and ignore the uncertainty of individual predictions. Costa et al. [3] also ignore the uncertainties but propose interesting alternative aggregations for ensembles based on generalized mixture functions.

## 3   Aggregation of Scores from Leafs

Based on the leaf nodes $\mathbf{w}_1, \ldots, \mathbf{w}_K$ to which an instance has been assigned to during prediction, the concept explained in this section is to first convert the leafs into scores which are then combined using aggregation functions.

### 3.1   Scoring Methods

In this section, the scoring methods are introduced of which most are designed to take the uncertainty of the leafs into account. We denote $p : \mathcal{W} \to \mathbb{R}$ as the function which assigns a score $v = p(\mathbf{w})$ to a leaf $\mathbf{w} \in \mathcal{W}$ where $\mathbf{w} = (w^\oplus, w^\ominus)$. All scoring methods proposed in this work are designed such that the sign of the resulting score indicates whether the positive or the negative class would be predicted. In order to understand how the scoring methods deal with uncertainty, we introduce an approach to visualize and compare them in Fig. 1.

**Probability and Laplace.** Computing the probability $p_{prob}(\mathbf{w})$ is the conventional approach to obtain a score from a leaf. Hence, it will serve as a reference point in this work. In addition, we include the *Laplace* smoothing $p_{lap}(\mathbf{w})$ which corrects the estimate for uncertainty due to a lack of samples.

$$p_{prob}(\mathbf{w}) = \frac{w^\oplus}{w^\oplus + w^\ominus} - 0.5 \qquad p_{lap}(\mathbf{w}) = \frac{w^\oplus + 1}{w^\oplus + w^\ominus + 2} - 0.5$$

In Fig. 1, we can see how both $p_{prob}$ and $p_{lap}$ converge to the same score for larger leafs but that $p_{lap}$ has much less extreme predictions for small leafs.

**Plausibility.** The idea of this approach is to measure the uncertainty in terms of aleatoric and epistemic uncertainty. In order to define the uncertainty, Shaker and Hüllermeier [16] introduces the degree of support for the positive class $\pi(\oplus|\mathbf{w})$ and the negative class $\pi(\ominus|\mathbf{w})$ which can be calculated for a leaf $\mathbf{w}$ as follows:

$$\pi(\oplus|\mathbf{w}) = \sup_{\theta \in [0,1]} \min\left(\left(\frac{\theta}{\left(\frac{w^\oplus}{w^\oplus + w^\ominus}\right)}\right)^{w^\oplus}\left(\frac{1-\theta}{\left(\frac{w^\ominus}{w^\oplus + w^\ominus}\right)}\right)^{w^\ominus}, 2\theta - 1\right),$$

$$\pi(\ominus|\mathbf{w}) = \sup_{\theta \in [0,1]} \min\left(\left(\frac{\theta}{\left(\frac{w^\oplus}{w^\oplus + w^\ominus}\right)}\right)^{w^\oplus}\left(\frac{1-\theta}{\left(\frac{w^\ominus}{w^\oplus + w^\ominus}\right)}\right)^{w^\ominus}, 1 - 2\theta\right).$$

Based on this *plausibility* the epistemic uncertainty $u_e(\mathbf{w})$ and the aleatoric uncertainty $u_a(\mathbf{w})$ can be defined as:

$$u_e(\mathbf{w}) = \min\left[\pi(\oplus|\mathbf{w}), \pi(\ominus|\mathbf{w})\right] \qquad u_a(\mathbf{w}) = 1 - \max\left[\pi(\oplus|\mathbf{w}), \pi(\ominus|\mathbf{w})\right]$$

Following Nguyen et al. [11], the degree of preference for the positive class $s_\oplus(\mathbf{w})$ can be calculated as

$$s_\oplus(\mathbf{w}) = \begin{cases} 1 - (u_a(\mathbf{w}) + u_e(\mathbf{w})) & \text{if } \pi(\oplus|\mathbf{w}) > \pi(\ominus|\mathbf{w}), \\ \frac{1-(u_a(\mathbf{w})+u_e(\mathbf{w}))}{2} & \text{if } \pi(\oplus|\mathbf{w}) = \pi(\ominus|\mathbf{w}), \\ 0 & \text{if } \pi(\oplus|\mathbf{w}) < \pi(\ominus|\mathbf{w}) \end{cases}$$

and analogously for the negative class. We use the trade-off between the degrees of preference as our score:

$$p_{pls}(\mathbf{w}) = s_\oplus(\mathbf{w}) - s_\ominus(\mathbf{w}).$$

With respect to Fig. 1, we can observe that the plausibility approach models the uncertainty similarly to the Laplace method but also preserves a high

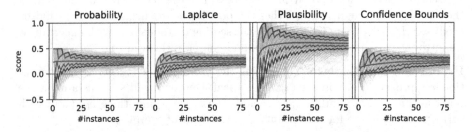

**Fig. 1.** Evolution of the scores ($y$-axis) for each scoring method based on simulating a leaf to which random samples are added ($x$-axis) with a probability of 75% for the positive class. This Bernoulli trial is repeated 100 times (cyan lines). The average over the trials scores is depicted by the red line and the 10%, 25%, 75% and 90% quantiles by the green and purple line, respectively. (Color figure online)

variability for small leaf sizes. From that point of view, it could be seen as a compromise between the Probability and the Laplace method.

**Confidence Bounds.** Another possibility to consider uncertainty is to model the probability of the positive and the negative class each with a separate probability distribution and to compare these. In this work, we use a beta-binomial distribution parameterized with $w^\oplus + w^\ominus$ number of tries, $\alpha = w^\oplus + 1$, and $\beta = w^\ominus + 1$ to model the probability of the positive class, analogously for the negative class. To get a measure $c(\mathbf{w})$ of how well both classes are separated, we take the intersection point in the middle of both distributions, normalized by the maximum height of the distribution, and use this as in the following:

$$p_{cb}(\mathbf{w}) = (1 - c(\mathbf{w})) \left( \frac{w^\oplus}{w^\oplus + w^\ominus} - 0.5 \right)$$

Hence, $c(\mathbf{w})$ generally decreases with 1) increasing class ratio in the leaf, or 2) with more examples, since this leads to more peaky distributions. In contrast to Laplace and Plausibility in Fig. 1, the scores of confidence bounds are increasing more steadily which indicates a weaker consideration of the size of the leafs than the aforementioned methods.

## 3.2   Aggregation Functions

In a final step the scores $v_1, \ldots, v_K$, where $v_i = p(\mathbf{w}_i)$, need to be combined using an aggregation function $h : \mathbb{R}^K \to \mathbb{R}$. Due to space restrictions, we limit our analysis in this work to the arithmetic mean and 0-1-voting

$$h_{avg}(\mathbf{v}) = \frac{1}{K} \sum_{i=1}^{K} v_i \qquad h_{vote}(\mathbf{v}) = \frac{1}{K} \sum_{i=1}^{K} sgn(v_i > 0)$$

with $sgn(\cdot)$ as the sign function. We consider $h_{vote}$ as a separate method in the following, since all presented scoring methods change their sign equally. Note that $p_{prob}$ in combination with $h_{avg}$ corresponds to what is known as *weighted voting*, and $p_{cb}$ an instantiation of *weighted averaging* over $p_{prob}$.

Similarly to Costa et al. [3], we found in preliminary experiments that alternative approaches, including the median, maximum and a variety of generalized mixture functions [7], could be beneficial in some cases, but did not provide meaningful new insights in combination with the explored scoring functions.

## 4   Integrated Combination of Leafs

In contrast to the aggregation of scores, the integrated combination of leafs skips the intermediate step of score generation and directly combines the statistics of the leafs $g : \mathcal{W}^K \to \mathbb{R}$ to form a prediction for the ensemble. Through this approach the exact statistics of the leafs can be considered for the aggregation which would otherwise be inaccessible using scores. In contrast to Fig. 1, Fig. 2 depicts the outputs of 100 simulated ensembles.

**Pooling.** The basic idea of pooling is to reduce the influence of leafs with a low number of instances which otherwise would obtain a high score either for the positive or the negative class. Therefore, this approach first sums up the leaf statistics and then computes the probability which can be defined as:

$$g_{pool}(\mathbf{w}_1, \mathbf{w}_2, \ldots, \mathbf{w}_K) = \frac{\sum_{i=1}^{K} w_i^{\oplus}}{\sum_{i=1}^{K} \left(w_i^{\oplus} + w_i^{\ominus}\right)} - 0.5$$

With respect to Fig. 2, we can observe that pooling behaves similarly to the probability approach. The reduced variability is mainly caused due to the simulation of more leafs.

**Dempster.** Based on the *Dempster-Shafer* framework, leaf aggregation can also be performed using the theory of belief functions [15]. For our purpose the states of belief can be defined as $\Omega = \{\oplus, \ominus\}$, where $\oplus$ represents the belief for the positive and $\ominus$ represents the belief for the negative class. A particular mass function $m : 2^{\Omega} \to [0, 1]$ assigns a mass to every subset of $\Omega$ such that $\sum_{A \subseteq \Omega} m(A) = 1$. Based on the plausibility (c.f. Sect. 3.1), we define the mass function for a particular leaf $\mathbf{w}$ as follows:

$$m_{\mathbf{w}}(\emptyset) = 0 \qquad\qquad m_{\mathbf{w}}(\{\oplus\}) = s_{\oplus}(\mathbf{w})$$
$$m_{\mathbf{w}}(\{\ominus\}) = s_{\ominus}(\mathbf{w}) \qquad m_{\mathbf{w}}(\{\oplus, \ominus\}) = u_e(\mathbf{w}) + u_a(\mathbf{w}).$$

In order to combine the predictions of multiple leafs, the mass functions can be aggregated using the (unnormalized) Dempster's rule of combination:

$$(m_{\mathbf{w}_1} \bigcirc m_{\mathbf{w}_2})(A) = \sum_{B \cap C = A} m_{\mathbf{w}_1}(B) m_{\mathbf{w}_2}(C) \qquad \forall A \subseteq \Omega$$

Hence, for the ensemble we can define the following belief function:

$$m_{dempster}(A) = (m_{\mathbf{w}_1} \bigcirc (m_{\mathbf{w}_2} \bigcirc (\ldots \bigcirc m_{\mathbf{w}_K})))(A)$$

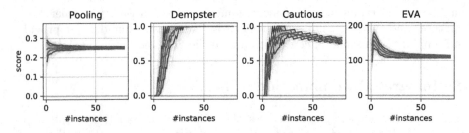

**Fig. 2.** Evolution of the scores ($y$-axis) for each combination method based on simulating 100 ensemble's final predictions (cyan lines) based each on 100 leafs sampled like in Fig. 1.

which can be used to form a score for the positive class as follows:

$$g_{dempster}(\mathbf{w}_1, \mathbf{w}_2, \ldots, \mathbf{w}_K) = m_{dempster}(\{\oplus\}) - m_{dempster}(\{\ominus\})$$

**Cautious.** A particular drawback of Dempster's rule of combination is that it requires independence among the combined mass functions which is usually not true for classifiers in an ensemble which have been trained over the same data. The independence assumption can be omitted by using the cautious rule of combinations [4]. The idea behind the cautious rule is based on the *Least Commitment Principle* [17] which states that, when considering two belief function, the least committed one should be preferred. Following this principle, the cautious rule $\bigwedge$ can be used instead of Dempster's rule of combination $\bigcap$. For more information we refer to Denœux [4]. Hence, for the ensemble we can define the following belief function

$$m_{cautious}(A) = (m_{\mathbf{w}_1} \bigwedge (m_{\mathbf{w}_2} \bigwedge (\ldots \bigwedge m_{\mathbf{w}_K})))(A)$$

which can be transformed to a score for the positive class as follows:

$$g_{cautios}(\mathbf{w}_1, \mathbf{w}_2, \ldots, \mathbf{w}_K) = m_{cautious}(\{\oplus\}) - m_{cautious}(\{\ominus\})$$

We use a small value $(10^{-5})$ as the minimum value for certain steps during the calculation to deal with numerical instabilities for both Dempster and Cautious.

In order to provide a more intuitive explanation, consider the following three estimates to combine: $s_\oplus(\mathbf{w}_1) = s_\oplus(\mathbf{w}_2) = 0.4$ and $s_\ominus(\mathbf{w}_3) = 0.4$. Dempster would consider that there are two estimates in favor of the positive class and hence predict it, whereas Cautious ignores several preferences for the same class, since they could result from dependent sources, and would just take the stronger one. Hence, Cautious would produce a tie (but high weights for $\emptyset$ and $\{\oplus, \ominus\}$). This behaviour can also be observed in Fig. 2 where the plot for Cautious resulted very similar to that of the maximum operator (not shown due to space restrictions). For Dempster, in contrast, the majority of leafs confirming the positive class very quickly push the prediction towards the extreme. Consider for that the following scenario, which also visualizes the behaviour under conflict. For $s_\oplus(\mathbf{w}_1) = s_\oplus(\mathbf{w}_2) = 0.8$ and $s_\ominus(\mathbf{w}_3) = 0.98$ Dempster's rule would prefer the negative over the positive class, but already give a weight of 0.94 to neither $\oplus$ nor $\ominus$, whereas Cautious would prefer the negative class with a score of 0.20 and 0.78 for $\emptyset$.

**Evidence Accumulation.** In the presence of uncertainty, it may make sense to reach a decision not by combining individual recommendations themselves, but by accumulating the *evidence* underlying each recommendation. This motivates a new combining rule, here called EVA (Evidence Accumulation).

To get the intuition behind the idea, consider a police inspector investigating a crime. It is perfectly possible that no clue by itself is convincing enough to arrest a particular person, but jointly, they do. This effect can never be achieved by expressing the evidence as probabilities and then averaging. E.g., assume there are 5 suspects, A, B, C, D, E; information from Witness $\mathbf{w}_1$ rules out D and E, and information from Witness $\mathbf{w}_2$ rules out A and B. Logical deduction leaves C as the only possible solution. Now assume $\mathbf{w}_1$ and $\mathbf{w}_2$ express their knowledge as probabilities: $\mathbf{w}_1$ returns $(0.33, 0.33, 0.33, 0, 0)$ (assigning equal probabilities to A, B and C and ruling out D and E), and $\mathbf{w}_2$ returns $(0, 0, 0.33, 0.33, 0.33)$. Averaging these probabilities gives $(0.17, 0.17, 0.33, 0.17, 0.17)$, when logic tells us it should be $(0, 0, 1, 0, 0)$. Note that 1 is out of the range of probabilities observed for C: no weighted average can yield 1. Somehow the evidence from different sources needs to be *accumulated*.

In probabilistic terms, the question boils down to: How can we express $P(y \mid \bigwedge \mathbf{w}_i)$ in terms of $P(y \mid \mathbf{w}_i)$? Translated to our example, how does the probability of guilt, given the joint evidence, relate to the probability of guilt given each individual piece? There is not one way to correctly compute this: assumptions need to be made about the independence of these sources. Under the assumption of class-conditional independence (as made by, e.g., Naive Bayes), we can easily derive a simple and interpretable formula:

$$P(y \mid \bigwedge_i \mathbf{w}_i) \sim P(\bigwedge_i \mathbf{w}_i \mid y) \cdot P(y) = P(y) \prod_i P(\mathbf{w}_i \mid y) \sim P(y) \prod_i \frac{P(y \mid \mathbf{w}_i)}{P(y)}$$

Thus, under class-conditional independence, evidence from different sources should be accumulated by multiplying the prior $P(Y)$ with a factor $P(y \mid \mathbf{w}_i)/P(y)$ for each source $\mathbf{w}_i$. That is, if a new piece of information, on its own, would make $y$ twice as likely, it also does so when combined with other evidence. In our example, starting from $(0.2, 0.2, 0.2, 0.2, 0.2)$, $\mathbf{w}_1$ multiplies the probabilities of A, B, C by $0.33/0.2$ and those of D, E by 0, and $\mathbf{w}_2$ lifts C, D, E by $0.33/0.2$ and A, B by 0; this gives $(0, 0, 0.54, 0, 0)$ which after normalization becomes $(0, 0, 1, 0, 0)$.

Coming back to the binary classification setting, we compute

$$g_{eva}(\mathbf{w}_1, \mathbf{w}_2, \ldots, \mathbf{w}_K) = P(\oplus) \prod_i \frac{P(\oplus \mid \mathbf{w}_i)}{P(\oplus)} - P(\ominus) \prod_i \frac{P(\ominus \mid \mathbf{w}_i)}{P(\ominus)}$$

where $P(\oplus)$ and $P(\ominus)$ is estimated by the class distribution on the training data and $P(y \mid \mathbf{w}_i)$ essentially comes down to $p_{prob}(\mathbf{w}_i)$. We apply similar tricks as in Naive Bayes for ensuring numerical stability such as a slight Laplace correction of 0.1 in the computation of $P(y \mid \mathbf{w}_i)$.

With respect to Fig. 2, we can observe an even stronger decrease of the absolute score with increasing leaf size than for Cautious. This indicates that the output of EVA can be highly influenced by smaller leafs which are more likely to carry more evidence than larger leafs, in the sense that their estimates are more committed.

## 5   Evaluation

A key aspect of our experimental evaluation is to compare our proposed methods for combining predictions with respect to the conventional approaches of voting and averaging probabilities. In order to put the combination strategies to the test, our experiments are based on random decision trees which, in contrast to other tree ensembles, guarantee a high diversity of estimates in the ensemble.

**Table 1.** Binary classification datasets and statistics.

| name | #instances | #features | class ratio | name | #instances | #features | class ratio |
|------|-----------|-----------|-------------|------|-----------|-----------|-------------|
| scene | 2407 | 299 | 0.18 | sonar | 208 | 60 | 0.47 |
| webdata | 36974 | 123 | 0.24 | mushroom | 8124 | 22 | 0.48 |
| transfusion | 748 | 4 | 0.24 | vehicle | 98528 | 100 | 0.50 |
| biodeg | 1055 | 41 | 0.34 | phishing | 11055 | 30 | 0.56 |
| telescope | 19020 | 10 | 0.35 | breast-cancer | 569 | 30 | 0.63 |
| diabetes | 768 | 8 | 0.35 | ionosphere | 351 | 34 | 0.64 |
| voting | 435 | 16 | 0.39 | tic-tac-toe | 958 | 9 | 0.65 |
| spambase | 4601 | 57 | 0.39 | particle | 130064 | 50 | 0.72 |
| electricity | 45312 | 8 | 0.42 | skin | 245057 | 3 | 0.79 |
| banknote | 1372 | 4 | 0.44 | climate | 540 | 20 | 0.91 |
| airlines | 539383 | 7 | 0.45 | | | | |

## 5.1   Experimental Setup

Our evaluation is based on 21 datasets for binary classification[1], shown in Table 1, which we have chosen in order to obtain diversity w.r.t. the size of dataset, the number features and the balance of the class distribution. To consider a variety of scenarios for combining unstable predictions, we have performed all of our evaluations with respect to an ensemble of 100 trees and minimum leaf sizes 1, 2, 3, 4, 8 or 32. Note that a single RDT ensemble for each leaf size configuration is enough in order to obtain the raw counts $\mathbf{w}_i$ and hence produce the final predictions for all methods. We used 5 times two fold cross validation in order to decrease the dependence of randomness on the comparability of the results.[2]

For measuring the performance, we have computed the area under receiver operating characteristic curve (AUC) and the accuracy. For a better comparison we have computed average ranks across all datasets. As an additional reference points to our comparison, we added the results of a single decision tree and a random forest ensemble. Note that though imposing an equal tree structure, these trees were trained with the objective of obtaining a high purity of class distributions in their leaves. Therefore, we did not expect to reach the performance w.r.t. classification but it was intended to show to what degree advanced combination strategies are able to close the gap between using tree models resulting merely from the data distribution and tree models specifically optimized for a specific task.

---

[1] Downloaded from the UCI Machine Learning Repository http://archive.ics.uci.edu and OpenML https://www.openml.org/.

[2] Our code is publicly available at https://github.com/olfub/RDT-Uncertainty.

## 5.2 Results

Figure 3 shows a comparison between all methods with respect to different tree structures. The first observation is that most of the methods perform around the baseline of taking the average probability. Exceptions are evidence accumulation, which is on top of the remaining approaches especially in terms of accuracy, and the approaches based on the theory of belief functions, which especially exhibit problems in producing rankings (AUC). Nonetheless, Dempster rule of combination has an advantage over the other methods for middle to large sized leafs in terms of accuracy. Figure 4 shows again how close the methods are together (Dempster, Cautious, Pooling and Voting were left out). We can see that $p_{lap}$ always is worse than $p_{prob}$, regardless of the leaf size. While $p_{pls}$ is not much worse than the best method ($p_{prob}$) on small leafs, it falls behind with increasing leaf size. Without $g_{eva}$, $p_{cb}$ would be the best method on medium sized or larger

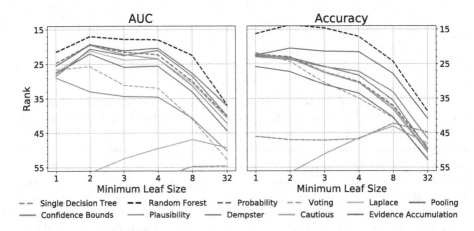

**Fig. 3.** Comparison of the average ranks with respect to the 11 methods and the 6 leaf configurations (worst rank is $6 \cdot 11 = 66$). Left: AUC. Right: Accuracy.

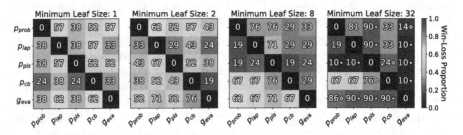

**Fig. 4.** Heatmap of pairwise comparisons. Each row and each column belongs to a method in the order $p_{prob}$, $p_{lap}$, $p_{pls}$, $p_{cb}$, and $g_{eva}$ and the number and color indicates how often the method in the row had a better AUC score than the method in the column. •, ◇ and ⋆ indicate a significant difference according to the Bonferroni corrected Wilcoxon signed-rank test with $\alpha = 0.05$, 0.01 and 0.001, respectively.

leafs but $g_{eva}$ has even better results there. Overall, $p_{prob}$ most often has the best results on very small leafs, but for not much larger leafs, $p_{prob}$ starts falling behind $g_{eva}$.

Figure 5 provides further insights on four selected datasets (single random 50% train/test split). It becomes apparent that the advantage of EVA is also often substantial in absolute terms. On the other hand, we can also observe cases where EVA falls behind the other approaches (*airlines*). Interestingly, the accuracy of EVA on *webdata* increases with increasing leaf size long after the other methods reach their peak, contrary to the general trend.

We can further observe that, as expected, the performances of the proposed methods lay between those of random forests and the single decision tree. However, Fig. 6 reveals that an important factor is the class ratio of the classification task. While random forest clearly outperform all other methods for highly imbalanced tasks, the advance is negligible for the balanced problems. This indicates a general problem of RDT with imbalanced data, since it gets less likely on such data to obtain leafs with (high) counts for the minority class.

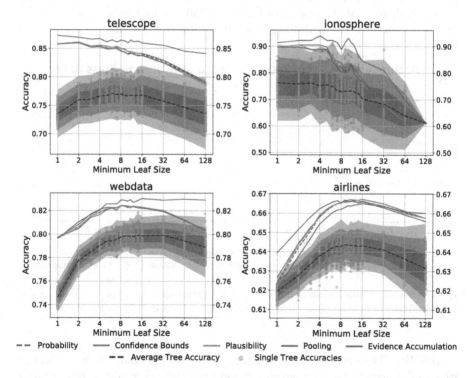

**Fig. 5.** Accuracy of selection of five combination strategies and distribution with mean (dashed blue line), 1, 2 and 3 standard deviations (blue shaded areas) over accuracies of the underlying individual trees in the ensemble (turquoise dots). (Color figure online)

## 5.3    Discussion

In the following, we discuss the results of each individual method in more detail.

**Probability, Laplace, Voting and Pooling.** Computing the expectation over $p_{prob}$ straightforwardly showed to be very effective compared to most of the alternative method ideas. Especially when small leafs are involved this behaviour was not expected since we believed that small leafs with overly certain but likely wrong estimates (probabilities 0 and 1) would out-weight the larger leafs which provide more evidence for their smoother estimates. Instead, the results indicate that a large enough ensemble (we used 100 trees) makes up for what we considered too optimistic $p_{prob}$ leaf predictions [1]. In fact, applying Laplace correction towards uniform distribution was hindering especially for small leaf sizes in our experiments. Voting is able to catch-up with the other methods only for the configuration of trees with very small leafs, where, indeed, the scores to be combined are identical to those of $p_{prob}$ except for leafs with size greater one. Interestingly, Pooling improves over the other methods on a few datasets as it can be seen for *airlines* in Fig. 5. Nevertheless, the results on most of the datasets indicate that the consideration of the leaf size by this method might be too extreme in most cases, leading generally to unstable results.

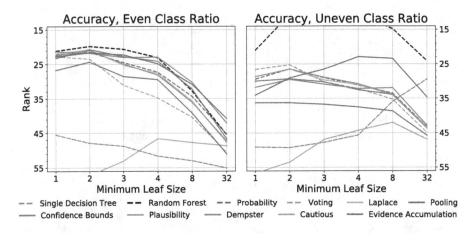

**Fig. 6.** Accuracy ranks on dataset with relatively even class ratio (0.4 to 0.6) on the left and relatively uneven class ratio (less than 0.3 or greater 0.7) on the right.

**Plausibility, Dempster and Cautious.** We expected that a precise computation and differentiation of plausibility and uncertainties would be beneficial in particular for small leaf sizes. However, we were not able to observe a systematic advantage of $p_{pls}$ over the $p_{prob}$ baseline in our experiments. Nonetheless, the usage of the uncertainty estimates in Dempster's rule of combinations shows the potential of having access to such scores. When the data was balanced, Dempster classified as good or better than EVA or even the random forest. The gap w.r.t. AUC demonstrates that the proposed translation of the components of the mass function into one score is not yet optimal. Moreover, the results suggests that the lack of independence is not problematic when combining predictions in RDT, but on the contrary, the way the Cautious rule addressed it caused serious problems.

**Confidence Bounds.** Among the methods with similar performance to $p_{prob}$, the approach using confidence bounds exhibits the greatest (but still small) advantage over the baseline for trees with medium to large leaf sizes. For very small leafs, however, the chosen statistical modeling is obviously not ideal, at least for ranking. Also, it might be possible to achieve even better results with the same idea by using distributions or more elaborated strategies which can achieve a better fit.

**Evidence Accumulation.** Using a small value for Laplace smoothing, $g_{eva}$ proved to be a very effective prediction strategy across all but the very smallest leafs. This suggests that the assumptions underlying EVA are well-met by RDTs. The randomness of the leaves fits EVA's independence assumption.

The stronger consideration of the class prior probability than in other methods may also play a role on why EVA works better on unbalanced data, particularly for larger leaf sizes (Fig. 6). Remind that EVA relates the estimates for each source to the prior probability and judges the evidence according to how much the leaf distribution deviates from it. Leafs for the minority class might hence have a higher impact on the final prediction than for other methods, where such leafs are more likely to be out-ruled by the average operation and the leafs following the class prior.

# 6   Conclusions

In this work, we proposed methods to combine predictions under uncertainty in an ensemble of decision trees. *Uncertainty* here refers not only to how uncertain a prediction is, but also to uncertainty about this uncertainty estimate. Our methods include Laplace smoothing, distinguishing between aleatoric and epistemic uncertainty, making use of the dispersion of probability distributions, combining uncertainties under the theory of belief functions, and accumulating evidence. Random decision trees ensured a high diversity of the predictions to combine and a controlled environment for our experimental evaluation. We found that including the support for the available evidence in the combination improves performance over averaging probabilities only in some specific cases. However, we could observe a consistent advantage for the proposed approach of evidence accumulation.

This result suggests that the strength of the evidence might be a better factor than the support for the evidence when combining unstable and diverse predictions from an ensemble. However, the principle of evidence accumulation revealed additional aspects, like the rigorous integration of the prior probability or the multiplicative accumulation of evidence, which might also be relevant for the observed improvement. Moreover, Dempster's rule showed that combining support (through the plausibilities) and commitment of evidence produces good results when certain conditions, such as even class ratio, are met. We plan to investigate these aspects in future work, e.g. by integrating a prior into Dempster's rule or aspects of reliable classification into EVA. Further research questions include which insights can be carried over to random forest ensembles, which are more homogeneous in their predictions, or if the presented strategies have a special advantage under certain conditions, such as small ensembles.

# References

1. Bostrom, H.: Estimating class probabilities in random forests. In: 6th International Conference on Machine Learning and Applications, pp. 211–216 (2007)
2. Breiman, L.: Random forests. Mach. Learn. **45**(1), 5–32 (2001)
3. Costa, V.S., Farias, A.D.S., Bedregal, B., Santiago, R.H., de P. Canuto, A.M.: Combining multiple algorithms in classifier ensembles using generalized mixture functions. Neurocomputing **313**, 402–414 (2018)
4. Denœux, T.: The cautious rule of combination for belief functions and some extensions. In: 9th International Conference on Information Fusion, pp. 1–8 (2006)
5. Dietterich, T.G.: Ensemble methods in machine learning. In: Kittler, J., Roli, F. (eds.) MCS 2000. LNCS, vol. 1857, pp. 1–15. Springer, Heidelberg (2000). https://doi.org/10.1007/3-540-45014-9_1
6. Fan, W., Wang, H., Yu, P.S., Ma, S.: Is random model better? On its accuracy and efficiency. In: 3rd IEEE International Conference on Data Mining (2003)
7. Farias, A.D.S., Santiago, R.H.N., Bedregal, B.: Some properties of generalized mixture functions. In: IEEE International Conference on Fuzzy Systems, pp. 288–293 (2016)
8. Kulessa, M., Loza Mencía, E.: Dynamic classifier chain with random decision trees. In: Soldatova, L., Vanschoren, J., Papadopoulos, G., Ceci, M. (eds.) DS 2018. LNCS (LNAI), vol. 11198, pp. 33–50. Springer, Cham (2018). https://doi.org/10.1007/978-3-030-01771-2_3
9. Lu, Y.: Knowledge integration in a multiple classifier system. Appl. Intell. **6**(2), 75–86 (1996)
10. Nguyen, T.T., Pham, X.C., Liew, A.W.C., Pedrycz, W.: Aggregation of classifiers: a justifiable information granularity approach. IEEE Trans. Cybern. **49**(6), 2168–2177 (2018)
11. Nguyen, V.L., Destercke, S., Masson, M.H., Hüllermeier, E.: Reliable multi-class classification based on pairwise epistemic and aleatoric uncertainty. In: International Joint Conference on Artificial Intelligence, pp. 5089–5095 (2018)
12. Provost, F., Domingos, P.: Tree induction for probability-based ranking. Mach. Learn. **52**(3), 199–215 (2003)
13. Raza, M., Gondal, I., Green, D., Coppel, R.L.: Classifier fusion using dempster-shafer theory of evidence to predict breast cancer tumors. In: IEEE Region 10 International Conference TENCON, pp. 1–4 (2006)
14. Ruta, D., Gabrys, B.: An overview of classifier fusion methods. Comput. Inf. Syst. **7**, 1–10 (2000)
15. Shafer, G.: A Mathematical Theory of Evidence, vol. 42. Princeton University Press (1976)

16. Shaker, M.H., Hüllermeier, E.: Aleatoric and epistemic uncertainty with random forests. In: Berthold, M.R., Feelders, A., Krempl, G. (eds.) IDA 2020. LNCS, vol. 12080, pp. 444–456. Springer, Cham (2020). https://doi.org/10.1007/978-3-030-44584-3_35

17. Smets, P.: Belief functions: the disjunctive rule of combination and the generalized Bayesian theorem. Int. J. Approximate Reasoning 9(1), 1–35 (1993)

18. Wolpert, D.: Stacked generalization. Neural Netw. 5(2), 241–259 (1992)

19. Yuksel, S.E., Wilson, J.N., Gader, P.D.: Twenty years of mixture of experts. IEEE Trans. Neural Netw. Learn. Syst. 23(8), 1177–1193 (2012)

20. Zhou, S., Mentch, L.: Trees, forests, chickens, and eggs: when and why to prune trees in a random forest. arXiv preprint arXiv:2103.16700 (2021)

# Shapley-Value Data Valuation
# for Semi-supervised Learning

Christie Courtnage and Evgueni Smirnov[✉]

Department of Data Science and Knowledge Engineering, Maastricht University,
P.O. BOX 616, 6200 MD Maastricht, The Netherlands
smirnov@maastrichtuniversity.nl

**Abstract.** Semi-supervised learning aims at training accurate prediction models on labeled and unlabeled data. Its realization strongly depends on selecting pseudo-labeled data. The standard approach is to select instances based on the pseudo-label confidence values that they receive from the prediction models. In this paper we argue that this is an indirect approach w.r.t. the main goal of semi-supervised learning. Instead, we propose a direct approach that selects the pseudo-labeled instances based on their individual contributions for the performance of the prediction models. The individual instance contributions are computed as Shapley values w.r.t. characteristic functions related to the model performance. Experiments show that our approach outperforms the standard one when used in semi-supervised wrappers.

**Keywords:** Semi-supervised learning · Data valuation · Shapley value

## 1 Introduction

Semi-supervised learning is an effective paradigm for training prediction models on labeled and unlabeled training data [18]. It can improve the model performance if the union of these data form meaningful clusters. Successful examples of semi-supervised learning include applications in web, drug discovery, part-of-speech tagging etc. for problems with small labeled and large unlabeled data.

There are several methods for semi-supervised learning [18]. Among them we focus on inductive wrappers due to their simplicity and widespread use. A wrapper is a meta method that can be iteratively applied to any prediction model following two consecutive steps. First, a prediction model is trained on the labeled data and then used to label the unlabeled data. Second, the pseudo-labeled data is added to the labeled data. The wrapper repeats the steps until a stopping condition is met. Hence, in the next iterations (a) the prediction model is retrained on the initial labeled data and selected pseudo-labeled data, and (b) labeling the unlabeled data is realized by a continuously updated prediction model. Once the stopping condition holds, the wrapper outputs the prediction model trained on the initial labeled data and final selected pseudo-labeled data.

C. Soares and L. Torgo (Eds.): DS 2021, LNAI 12986, pp. 94–108, 2021.
https://doi.org/10.1007/978-3-030-88942-5_8

Central to the success of inductive wrappers is the step of selecting pseudo-labeled data. If this step is accurate, the prediction models can be accurate as well, and, thus labeling the unlabeled data. All the existing wrapper methods select pseudo-labeled instances individually using confidence values that the prediction models employed output for the labels of those instances[1]. If a label confidence value is high enough for a new-labeled instance, then the instance is selected; otherwise, it is rejected. Thus, the main assumption is that the distribution of highly-confident pseudo-labeled instances is close to the distribution of the labeled data. This implies that adding those instances to training process can boost the performance of the prediction models in inductive wrappers and, thus, the overall process of semi-supervised learning. Although it is a plausible assumption, it may not always hold in practice, for example, when the prediction models are not calibrated (see Sect. 2 for a detailed explanation).

In this paper we propose an alternative approach to selecting pseudo-labeled instances that is not based on the above assumption. The approach selects the instances based on their individual contributions to the performance of the prediction models used in inductive wrappers; i.e. it selects only those pseudo-labeled instances that boost the performance of the models. In this respect our approach is *simpler and more direct* than the one based on confidence values.

To estimate the individual contribution of pseudo-labeled instances, we propose to use data-valuation techniques that are based on the Shapley value [6,13]. The Shapley value of a labeled instance is defined as the average marginal contribution of the instance for the generalization performance of a prediction model estimated by a characteristic function (related to metrics such as accuracy, ROC AUC, F1 score etc.). It is computed w.r.t. the available data and thus instance dependencies are taken into account.

We employ an algorithm for computing exact Shapley values based on nearest neighbor classification [13]. The characteristic function outputs the average probability of correct labels of validation instances. We incorporate the algorithm into the simplest inductive wrapper, the self-trainer wrapper [21]. Depending on the validation procedure of the characteristic function we propose two wrapper versions: hold-out Shapley-value self trainer and cross-validation Shapley-value self trainer. Both wrappers are experimentally compared to existing semi-supervised methods from a survey provided in [17]. The experiments show that despite their simplicity, the proposed wrappers outperform most of the existing methods.

The paper is organized as follows: Sect. 2 provides related work. Section 3 introduces semi-supervised classification and wrappers. Section 4 discusses data valuation and its algorithms. Section 5 introduces the two new semi-supervised wrappers. Sections 6 and 7 provide experiments and conclusions.

---

[1] For example, if the prediction model is a probabilistic classifier, the confidence value is the posterior probability of the label for an unlabeled instance.

## 2   Related Work

The existing semi-supervised methods select pseudo-labeled instances using confidence values that the employed prediction models output for the labels of those instances. There are many such models [17,18] from purely probabilistic classifiers [3,8,14,19,21,22,24] to scoring ensembles [4,7,23]. The probabilistic classifiers output directly posterior class probability estimates. The non-probabilistic scoring classifiers usually transform the posterior class scores to probability estimates, i.e. through score normalization, function transformation etc.

To secure robustly selecting pseudo-labeled instances, the probability estimates need to be well calibrated [25]. A prediction model is well calibrated if among the instances $x$ that receive the same probability estimates $\hat{p}(y|x)$ for class $y$, the proportion of those that belong to $y$ equals $\hat{p}(y|x)$. This implies that calibrated prediction models allow us to control the accuracy of selecting pseudo-labeled instances above a user-defined lower bound and, thus, to improve semi-supervised learning. Unfortunately, most of the existing inductive wrapper methods do not employ calibrated prediction models and, recently, this problem was recognized in [12,25]. The author in [12] proposed to calibrate the prediction models by training them on mixed training and pseudo-labeled instances. The authors in [25] used the class label scores of the prediction models to compute a $p$-value for each class label (using the conformal framework). Then the acceptance of any class label is realized on a user-predefined significance level; i.e. the pseudo-labeled instances are selected below a user-predefined error.

Selecting pseudo-labeled instances is a multi-test process that sequentially decides whether to accept/reject any successive pseudo-labeled instance $(x, y)$ based on the probability estimate $\hat{p}(y|x)$ for the assigned class label $y$. Due to the probabilistic instance selection, this process accumulates an error after handling $n$ pseudo-labeled instances even if the employed prediction model is calibrated. Depending on the assumption for the generation of the training data and unlabeled data (i.i.d./non-i.i.d), there exist several methods developed that try to keep this error bounded [1,10], however, without any statistical guarantees.

In this paper we argue that selecting pseudo-labeled instances based on the probability estimates of the assigned classes is an indirect approach: we first label the unlabeled data and assume that the distribution of highly-confident pseudo-labeled instances is close to the distribution of the labeled data. Then we employ the pseudo-labeled instances to boost the performance of the prediction models in inductive wrappers. Although our main assumption is plausible, it may not always hold in practice, for example, when the prediction models are not calibrated and, thus, the overall error of selecting pseudo-labeled instances is uncontrollable. Thus, we need a direct approach that is not dependent on the properties of the prediction models such as calibration.

In this paper we propose selecting pseudo-labeled instances to be based on their contribution for the performance of the predictions models used in inductive wrappers; i.e. we relate instance selection directly to the final goal - boosting the performance of the final prediction models. In addition, it is worth noting that our approach does not suffer from problems of the existing approaches to selecting pseudo-labeled instances such as calibrating the class probability estimates and controlling the overall error of sequentially labeling.

# 3   Semi-supervised Classification and Inductive Wrappers

Semi-supervised classification assumes the presence of instance space $X$, discrete class variable $Y$, and unknown probability distribution $P_{X \times Y}$ over $X \times Y$. The labeled training data set $L$ is a set of $N$ instances $(x_n, y_n) \in X \times Y$ i.i.d. drawn from $P_{X \times Y}$. The unlabeled training data set $U$ is a set of $M$ instances $x_m \in X$ i.i.d. drawn from another unknown probability distribution $Q_X$n. The semi-supervised classification task is to find a class estimate $\hat{y}$ for a test instance $x \in X$ according to $P_{X \times Y}$ using information from the training data sets $L$ and $U$ assuming that $P_X$ and $Q_X$ are close enough. To solve the task a classifier $h : X \to Y$ is identified in a hypothesis space $H$ using the sets $L$ and $U$. For the rest of the paper we assume that for any test instance $x \in X$ classifier $h$ outputs a class estimate $\hat{y}$ plus a confidence value.

Inductive wrappers are classifier-agnostic methods for semi-supervised classification. The standard self-trainer wrapper [9] is presented in Algorithm 1.

---

**Algorithm 1:** Standard Self Trainer Wrapper

**Input**   : Labeled training data set $L$ and Unlabeled training data set $U$.
1  Train a base classifier $h$ on data $L$;
2  Set the set $L'$ of selected pseudo-labeled data equal to $\emptyset$;
3  **repeat**
4  |   $UL \leftarrow \{(x, h(x)) \mid x \in U\}$;
5  |   Select subset $S \subseteq UL$ of the most relevant pseudo-labeled instances in a function of $L$, $L'$, and $UL$;
6  |   **if** $S \neq \emptyset$ **then**
7  |   |   $U \leftarrow \{x \in U \mid (x, y) \notin S\}$;
8  |   |   $L' \leftarrow L' \cup S$;
9  |   |   Train $h$ on $L \cup L'$;
10 |   **end**
11 **until** $S = \emptyset$ or maximum iterations reached;
12 **Output** set $L'$ and classifier $h$.

---

The standard self-trainer wrapper starts by training classifier $h$ on the labeled training data $L$ (step 1). It uses classifier $h$ to pseudo-label all the instances in the unlabeled training data $U$ and adds them to set $UL$ (step 4). The most relevant pseudo-labeled instances are assigned to set $S$ (step 5) while the remaining are assigned to set $U$ without the pseudo-labels (step 7). If $S$ is nonempty, set $S$ is added to the set $L'$ of the selected pseudo-labeled instances so far (step 8) and classifier $h$ is retrained on the union of $L$ and $L'$ (step 9). This process is repeated until convergence when $S = \emptyset$ or a maximum iteration number is reached.

The standard approach to select pseudo-labeled instances is to use confidence values provided by classifier $h$. In this paper we propose to select those instances based on their contributions for the performance of classifier $h$. Computing these contributions can be realized with data valuation introduced in the next section.

## 4   Data Valuation

This paper employs data valuation based on Shapley values [6,13]. The Shapley values were first proposed in cooperative game theory [16]. A Shapley value of a player in a game is the player's averaged marginal contribution for all coalitions of players, where each coalition has a known value. In Shapley-value data valuation, any labeled instance $(x, y)$ in a training data set $T$ is considered as a player and any coalition is defined as a subset $S \subseteq T$, where $T$ is considered as the "grand" coalition. The value of any training data subset (coalition) $S$ is defined w.r.t. a characteristic function $v(h, S, V)$. The function outputs the value of the chosen performance metric for a classifier $h$ when trained on $S$ and tested on a validation set $V$. In this context, the Shapley value $SV(x, y)$ for any training instance $(x, y) \in T$ can be computed using Eq. 1.

$$SV(x, y) = \frac{1}{|T|} \sum_{S \subseteq T \setminus (x, y)} \frac{1}{\binom{|T|-1}{|S|}} (v(h, S \cup \{(x, y)\}, V) - v(h, S, V)) \quad (1)$$

The Shapley value is the average marginal contribution of the training instance $(x, y) \in T$ for the performance of classifier $h$ estimated by the characteristic function $v(h, S, V)$ when trained over all possible subsets $S$ of the "grand" coalition $T$ and tested on $V$. This implies that the Shapley value of $(x, y)$ is w.r.t. all the instances (players) of the "grand" coalition $T$.

Shapley values are used in data valuation due to the following properties:

- Group Rationality: $v(h, T, V) = \sum_{(x, y) \in T} SV(x, y)$.
- Fairness: for any subset $S \subseteq T$ and any two instances $(x_i, y_i), (x_j, y_j) \in T$ if $v(h, S \cup \{(x_i, y_i)\}, V) = v(h, S \cup \{(x_j, y_j)\}, V)$ then $SV(x_i, y_i) = SV(x_j, y_j)$.
- Additivity: for any instance $(x, y) \in T$ we have $SV_1(x, y) + SV_2(x, y) = SV_{12}(x, y)$, where $SV_1(x, y)$, $SV_2(x, y)$, and $SV_{12}(x, y)$ are Shapley values of $(x, y) \in T$ for characteristic functions $v_1$, $v_2$, and $v_1 + v_2$, respectively.

The brute-force Shapley-value algorithm based on Eq. 1 is exponential in the size of $T$. There exist several approximation algorithms based on Monte-Carlo simulation [6]. Although they are faster, they do not find exact Shapley values. Recently, authors in [13] proposed an algorithm for computing exact Shapley values for $K$-nearest neighbor classification ($K$-ESVNN). It was shown to be more computationally efficient than any predecessor.

To introduce $K$-ESVNN let $(x_j, y_j) \in V$ be a validation instance. We order training instances $(x_i, y_i) \in T$ in an increasing order of their distance to $(x_j, y_j)$ and consider the first $K$ of them: $(x_{\pi(1)}, y_{\pi(1)}), (x_{\pi(2)}, y_{\pi(2)}), \ldots, (x_{\pi(K)}, y_{\pi(K)})$. They can be used to estimate the probability $p(y_j | x_j)$ of validation instance

$(x_j, y_j)$ as $\frac{1}{K} \sum_{k=1}^{K} \mathbb{1}[y_{\pi(k)} = y_n]$, where $\mathbb{1}$ is the indicator function. $p(y_j|x_j)$ can be viewed as the characteristic value $v(T, h, \{(x_j, y_j)\})$ of the "grand" coalition $T$ for the validation instance $(x_j, y_j)$ through $K$-nearest-neighbor classifier $h$:

$$v(T, h, \{(x_j, y_j)\}) = \frac{1}{K} \sum_{k=1}^{min(K,|V|)} \mathbb{1}[y_{\pi(k)} = y_j]. \tag{2}$$

Following the additivity property characteristic values $v(T, h, \{(x_j, y_j)\})$ are summed over all the instances in the validation set $V$ to receive characteristic value $v(T, h, V)$ of the "grand" coalition $T$ for the whole $V$:

$$v(T, h, V) = \frac{1}{|V|} \sum_{j=1}^{|V|} v(T, h, \{(x_j, y_j)\}). \tag{3}$$

K-ESVNN computes exact Shapley values using the characteristic function from Eq. (3). For this purpose, it first computes Shapley value $SV_j(x_i, y_i)$ of every training instance $(x_i, y_i) \in T$ for every validation instance $(x_j, y_j) \in V$. This is realized as follows: for any validation instance $(x_j, y_j)$ the algorithm sorts the training instances $(x_i, y_i) \in T$ in an increasing order of their distance to $(x_j, y_j)$. Then it visits each instance in the sorted sequence $(x_{\pi(1)}, y_{\pi(1)})$, $(x_{\pi(2)}, y_{\pi(2)}), \ldots, (x_{\pi(|T|)}, y_{\pi(|T|)})$ in reverse order and assigns the Shapley values $SV_j(x_i, y_i)$ according to a recursive rule:

$$s_j(\pi(|T|)) = \frac{\mathbb{1}[y_{\pi(|T|)} = y_j]}{|T|}, \tag{4}$$

$$s_j(\pi(i)) = s_n(\pi(i+1)) + \frac{\mathbb{1}[y_{\pi(i)} = y_j] - \mathbb{1}[y_{\pi(i+1)} = y_j]}{K} \frac{min(K, i)}{i}. \tag{5}$$

Once Shapley values $SV_j(x_i, y_i)$ of all the training instances $(x_i, y_i) \in T$ have been computed for all validation instances $(x_j, y_j) \in V$, the algorithm computes the final Shapley value $SV(x_i, y_i)$ of any training instance $(x_i, y_i) \in T$. The latter is computed following the additive property as the average of the Shapley values $SV_j(x_i, y_i)$ over all the validation instances $(x_j, y_j) \in V$.

Figure 1 presents distributions of the exact Shapley values of correctly and incorrectly labeled training instances. The right shift of the distribution of the correctly labeled training instances shows that Shapley values can be used to discriminate correctly from incorrectly labeled training instances. This is a motivation for the semi-supervised wrappers presented in the next section.

## 5   Shapley-Value Self-trainer Wrappers

In this section we propose a new approach to selecting pseudo-labeled instances. The approach selects the instances based on their individual contributions to the performance of the classifiers employed in inductive wrappers. The individual instance contributions are computed as Shapley values w.r.t. characteristic functions $v$ related to the classifiers' performance.

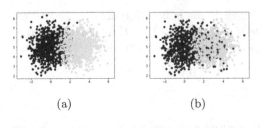
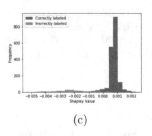

(a)                          (b)                                (c)

**Fig. 1.** (a) The original labels of a two class problem. (b) The same two class problem with class labels randomly flipped with probability of $\frac{1}{10}$. The randomly flipped instances are considered as incorrectly labeled instances. (c) The Shapley value distribution of the correctly and incorrectly labeled instances.

Our approach can be directly used in the standard self-trainer wrapper. Consider step 5 in Algorithm 1. In this step we select the most relevant pseudo-labeled instances from the pseudo-labeled training data set $UL$ produced in the current iteration. The instance selection has to be performed in a function of the labeled training data $L$, already-selected pseudo-labeled training data $L'$, and $UL$ itself. Following our approach we need to compute the Shapley values for all the instances in $UL$ using a characteristic function $v$ related to the performance of classifier $h$ in the wrapper. However, since $h$ has been trained on fixed $L \cup L'$ and on a to-be-selected subset of $UL$, we compute the Shapley values for all the instances in $L \cup L' \cup UL$ for a "grand" data coalition of $L \cup L' \cup UL$.[2]

Given the "grand" data coalition of $L \cup L' \cup UL$ and classifier $h$, to complete our design of the Shapley-value computations we need to determine the performance metric for $h$ and validation data set $V$ for the characteristic function $v$. Depending on the way we acquire set $V$, we employ different validation methods for classifier $h$ when implementing function $v$. This results in different characteristic functions $v$ and, thus, in different self-trainer wrappers introduced in the next two sub-sections.

### 5.1   Hold-Out Shapley Value Self-trainer Wrapper

If we acquire the validation set $V$ as a randomly selected subset of the labeled training data $L$, then we can use $L \setminus V$ for training the wrapper classifier $h$. This implies that we can compute the Shapley values only for the instances in the union $(L \setminus V) \cup L' \cup UL$ for a reduced "grand" data coalition of $(L \setminus V) \cup L' \cup UL$. Hence, the characteristic function $v$ in this case has to output performance-metric values for the wrapper classifier $h$ trained on subsets $S \subseteq (L \setminus V) \cup L' \cup UL$ and tested on $V$; i.e. $v$ has to be based on hold-out validation.

If we employ the characteristic function $v$ based on hold-out validation for computing Shapley values and employ these values for selecting pseudo-labeled instances, we will receive the Hold-Out Validation Shapley Value Self-Trainer

---

[2] Note that $T$ equals $L \cup L' \cup UL$ in formula 1 in our case.

Wrapper (HSVSTW). This wrapper is similar to the standard self-training wrapper and is present in Algorithm 2. Its execution steps can be described analogously (see Sect. 3). The main difference is that the wrapper classifier $h$ is trained on $(L\backslash V)\cup L'$ each time due to the hold-out validation and of course that pseudo-labeled instances are selected based on their Shapley values.

---

**Algorithm 2:** Hold-Out Validation Shapley Value Self-Trainer Wrapper

**Input** : Labeled training data set $L$,
Unlabeled training data set $U$,
Validation labeled data set $V$,
Acceptance threshold $\gamma$.

1  Randomly select validation set $V$ as a subset of $L$;
2  Set the set $L'$ of selected pseudo-labeled data equal to $\emptyset$;
3  Train a base classifier $h$ on $(L \setminus V) \cup L'$;
4  **repeat**
5      $UL \leftarrow \{(x, h(x)) \mid x \in U\}$;
6      Calculate Shapley value $SV(x, y)$ for any instance
       $(x, y) \in (L \setminus V) \cup L' \cup UL$ w.r.t. $V$ using characteristic function $v$ based
       on the hold-out validation of the wrapper classifier $h$;
7      $S \leftarrow \{(x, y) \in UL \mid SV(x, y) > \gamma\}$;
8      **if** $S \neq \emptyset$ **then**
9         $U \leftarrow \{x \in U \mid (x, y) \notin S\}$;
10        $L' \leftarrow L' \cup S$;
11        Train $h$ on $(L \setminus V) \cup L'$;
12     **end**
13 **until** $S = \emptyset$ or maximum iterations reached;
14 **Output** set $L'$ and classifier $h$.

---

HSVSTW computes the Shapley values for all the instances in $(L\backslash V)\cup L'\cup UL$ in each iteration. It is possible that after some iterations the labeled instances in $L \cup L'$ receive low Shapley values; i.e. their contributions for the performance of the wrapper classifier $h$ diminish. To remove these instances for next iterations we introduce option *Remove-L*. When the option is on, the labeled instances in $L \cup L'$ with Shapley values below a threshold parameter $\beta$ are removed.

The accuracy of HSVSTW is related to the trade-off between the data sets $L\backslash V$ and $V$. On one hand: the larger the set $L\backslash V$, the more accurate the wrapper classifier $h$ and the smaller the set $V$. Hence, we can label more correctly but we can estimate Shapley values less precisely. On the other hand: the smaller the set $L \setminus V$, the less accurate the wrapper classifier $h$ and the bigger the set $V$. Hence, we can label less correctly but we can estimate Shapley values more precisely. To avoid the trade-off between the data sets $L \setminus V$ and $V$, we introduce another wrapper in the next section that is based on cross validation.

## 5.2  Cross-Validation Shapley Value Self-trainer Wrapper

Assume that we randomly divide the labeled training data $L$ into $K$ equally-sized folds $L_k$ in a stratified manner. Then for any $k \leq K$ we can compute the Shapley values of the instances in $(L \setminus L_k) \cup L' \cup UL$ using hold-out validation (as described in the previous section). These values indicate the individual instance contributions for the performance of the wrapper classifier $h$ on the validation folder $L_k$. If we perform the aforementioned operation for all $k \leq K$, then:

- any instance $(x, y) \in L$ will receive $K - 1$ Shapley values $SV_k(x, y)$, and
- any instance $(x, y) \in L' \cup UL$ will receive $K$ Shapley values $SV_k(x, y)$.

The Shapley values $SV_{k_1}(x, y)$ and $SV_{k_2}(x, y)$ that any instance $(x, y)$ receives for $k_1 \neq k_2$ belong to different data "grand" coalitions $(L \setminus L_{k_1}) \cup L' \cup UL$ and $(L \setminus L_{k_2}) \cup L' \cup UL$. Hence, the additivity property does not hold. However, if we assume that $K$ is big enough, then:

- the Shapley value for any instance $(x, y) \in L$ can be approximated by $\widehat{SV}(x, y) = \sum_{k=1}^{K} SV_k(x, y)/(K - 1)$ since $SV_k(x, y) = 0$ for $(x, y) \in L_k$,
- the Shapley value for any instance $(x, y) \in L' \cup UL$ can be approximated by $\widehat{SV}(x, y) = \sum_{k=1}^{K} SV_k(x, y)/K$.

If we employ this cross-validation manner for computing approximated Shapley values $\widehat{SV}(x, y)$ and employ these values for selecting pseudo-labeled instances, we get the Cross-Validation Shapley Value Self-Trainer (CVSVSTW). This wrapper is also similar to the standard self-training wrapper and is present in Algorithm 3. Its execution steps can be described analogously (see Sect. 3).

The main advantages of CVSVSTW are as follows. First, there is no trade-off between the training and validation data as in HSVSTW. Second, CVSVSTW trains the wrapper classifier $h$ on the union $L \cup L'$ in each iteration. This means that all the available labeled data $L$ is used which benefits accurately labeling the unlabeled data. Third, any instance in $L \cup L' \cup UL$ receives an estimation of the true Shapley value with reduced variance due to averaging (see steps 11 and 13 in Algorithm 3). This improves selecting pseudo-labeled instances and, thus, semi-supervised learning organized by CVSVSTW.

An interesting property of CVSVSTW is that it computes an approximated Shapley value $\widehat{SV}(x, y)$ for any instance $(x, y) \in L$ in each iteration. This implies that we have $\widehat{SV}(x, y)$ even when $(x, y)$ is used for validation in some folder $L_k$. Hence, if $\widehat{SV}(x, y)$ is low, we can exclude $(x, y)$ from valuating instances in $(L \setminus L_k) \cup L' \cup UL$ which can benefit estimating instance Shapley values. This functionality is activated by option *Exclude-V*. When it is on, all the labeled instances in $L$ with Shapley values below a threshold parameter $\beta$ are not used for instance valuation but not removed from $L$. In addition, we note that for the same reason as in HSVSTW option *Remove-L* is present in CVSVSTW.

We illustrate CVSVSTW on two semi-supervised binary classification data sets when it employs the $K$-ESVNN algorithm. The first (second) set is class (non-) linearly-separable set given in Fig. 2a (Fig. 3a). The labeled and unlabeled

---

**Algorithm 3:** Cross-Validation Shapley-Value Self-Trainer Wrapper

---

**Input** : Labeled training data set $L$,
            Unlabeled training data set $U$,
            Number $K$ of folds,
            Acceptance threshold $\gamma$.

1   Split labeled training data set $L$ into $K$ folds $L_k$
2   Set the set $L'$ of selected pseudo-labeled data equal to $\emptyset$;
3   Train a base classifier $h$ on $L \cup L'$
4   **repeat**
5      $UL \leftarrow \{(x, h(x)) \mid x \in U\}$
6      **for** $k \leftarrow 1$ **to** $K$ **do**
7          Calculate Shapley value $SV_k(x, y)$ for any instance
                 $(x, y) \in (L \setminus L_k) \cup L' \cup UL$ w.r.t. $L_k$ using characteristic function $v$
                 based on the hold-out validation of the wrapper classifier $h$;
8          Set Shapley value $SV_k(x, y)$ for any instance $(x, y) \in L_k$ equal to 0;
9      **end**
10     **for** $(x, y) \in UL \cup L \cup L'$ **do**
11        **if** $(x, y) \in L$ **then**
12           $\widehat{SV}(x, y) \leftarrow \sum_{k=1}^{K} SV_k(x, y)/(K - 1)$
13        **else**
14           $\widehat{SV}(x, y) \leftarrow \sum_{k=1}^{K} SV_k(x, y)/(K)$
15        **end**
16      **end**
17      $S \leftarrow \{(x, y) \in UL \mid \widehat{SV}(x, y) > \gamma\}$
18      **if** $S \neq \emptyset$ **then**
19        $U \leftarrow \{x \in U \mid (x, y) \notin S\}$
20        $L' \leftarrow L' \cup S$
21        Train $h$ on $T \cup L'$
22      **end**
23   **until** $S = \emptyset$ *or maximum iterations reached*;
24   **Output** set $L'$ and classifier $h$.

---

data are generated by the same distributions but the labels of the unlabeled data are omitted.

Figures 2b, 2c, 3b, and 3c show pseudo-labeled instances added in the first iteration of CVSVSTW. Figures 2b and 3b show that when the pseudo-labeled instances with the highest 90% of Shapley values are added, they appear to be evenly spread over the distributions of the training labeled instances. In contrast, Figs. 2c and 3c show that when the pseudo-labeled instances with the highest 10% of Shapley values are added, they tend to be closer to the centers of the distributions of the training labeled instances. This means that adding high Shapley-value pseudo-labeled instances can be safe, and, thus, useful for semi-supervised learning.

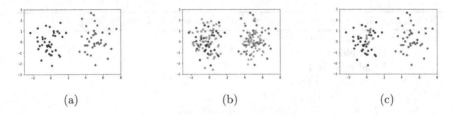

(a)                     (b)                     (c)

**Fig. 2.** (a) Linearly-separable labeled data, (b) the same linearly-separable labeled data with pseudo-labeled instances with the highest 90% of Shapley values, and (c) the same linearly-separable labeled data with pseudo-labeled instances with the highest 10% of Shapley values. **Note:** the darker colors represent the originally labeled instances, the lighter colors represent the pseudo-labeled instances.

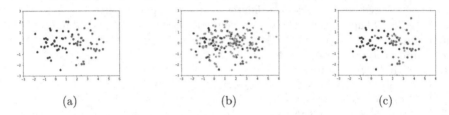

(a)                     (b)                     (c)

**Fig. 3.** (a) Non-linearly-separable labeled data, (b) the same non-linearly-separable labeled data with pseudo-labeled instances with the highest 90% of Shapley values, and (c) the same non-linearly-separable labeled data with pseudo-labeled instances with the highest 10% of Shapley values. **Note:** the darker colors represent the originally labeled instances, the lighter colors represent the pseudo-labeled instances.

## 6    Experiments

In this section we experimentally compare HSVSTW and CVSVSTW with 14 semi-supervised methods from an extensive empirical study presented in [17].

### 6.1    Methods' Setup

HSVSTW and CVSVSTW are set as follows. The wrapper classifier is 3-nearest neighbor classifier. The Shapley value algorithm is the 3-ESVNN algorithm (see Sect. 4). The internal hold-out validation of HSVSTW is stratified with 66.67% of the data reserved for training and 33.33% for testing. The internal cross-validation of CVSVSTW is stratified for 10 folds. The remaining parameters are set to maximize the performance of both wrappers.

The 14 semi-supervised methods from [17] are as follows: Self-Training Wrapper [21], Ant Based Semi-supervised Classification (APSSC) [8], Self-Training with Editing (SETRED) [14], Semi-Supervised learning based on Nearest Neighbor rule and Cut Edges (SNNRCE) [20], Co-training [2], CoTraining by Committee: Bagging (CoBagging) [7], Co-trained Random Forest (CoForest) [15], Adaptive Data Editing based CoForest (ADE-CoForest) [4], TriTraining [24], Tri-Training with Data Editing (DE-TriTraining) [3], Local Cluster Centers (CLCC)

[11], Democratic Co-Learning (Democratic-Co) [23], Random Subspace Method for Co-Training (RASCO) [19], and Co-Training with Relevant Random Sub-Spaces (Rel-RASCO) [22]. The classifiers setups are those from [17]. All the methods select pseudo-labeled data using pseudo-label confidence values. All classifier-agnostic methods employ a nearest-neighbor classifier as a base one to ensure a fare comparison with HSVSTW and CVSVSTW.

## 6.2  Data Sets

The experiments are performed on ten UCI data sets from [5] employed in [17]. The features of the data sets are summarized in Table 1.

Table 1. Data sets employed in the experiments

| Data set | # Instances | #Attributes | #Classes |
|---|---|---|---|
| Bupa | 345 | 6 | 2 |
| Dermatology | 297 | 33 | 6 |
| Glass | 366 | 9 | 7 |
| Haberman | 336 | 3 | 2 |
| Heart | 270 | 13 | 2 |
| Hepatitis | 155 | 19 | 2 |
| Iris | 150 | 4 | 3 |
| Monk-2 | 432 | 6 | 2 |
| Spectfheart | 267 | 44 | 2 |
| Tae | 151 | 5 | 3 |

## 6.3  Validation Setup

The validation set up is the one from [17]. All the classifiers are evaluated using 10-fold cross-validation. 40% of data in each fold is labeled and 60% is not. To ensure that the obtained results for HSVSTW and CVSVSTW are comparable to the 14 classifiers from [17], the same labeled, unlabeled and test folds were used (as indicated in [17])[3]. The metric for classifier validation is accuracy.

## 6.4  Results and Discussion

Table 2 presents the accuracy results of the 16 semi-supervised methods on the 10 data sets. The results show that: HSVSTW and CVSVSTW belong to the best four semi-supervised methods in terms of averaged accuracy: HSVSTW has the fourth place and CVSVSTW the second. The first and third places belong

---

[3] Due to the exactly same cross-validation we do not perform experiments with the 14 classifiers from [17]. We process the experimental data from [17] together with ours.

**Table 2.** Accuracy in percentages and standard deviations (in parentheses) of the 16 semi-supervised methods on all the data sets from Table 1. The methods are ordered according to their averaged accuracy. The best accuracy results are in bold. All classifier-agnostic methods employ nearest-neighbor classifier indicated with (NN).

| Algorithms | Bupa | Dermatology | Glass | Haberman | Heart |
|---|---|---|---|---|---|
| CoForest | 62,67 (5,59) | **95,21** (4,59) | 67,96 (9,82) | 60,75 (8,20) | 77,04 (8,41) |
| CVSVSTW | **67,24** (6,80) | 82,12 (6,82) | **69,84** (8,32) | 74,12 (4,83) | 68,22 (6,59) |
| Democratic-Co | 60,35 (5,93) | 94,39 (4,21) | 55,35 (12,85) | 74,49 (2,12) | 82,59 (10,22) |
| HSVSTW | 65,85 (6,83) | 77,97 (6,73) | 64,45 (9,60) | **74,65** (5,31) | 67,75 (7,84) |
| SETRED | 58,74 (9,47) | 94,66 (4,08) | 64,87 (6,75) | 69,24 (7,44) | 77,41 (10,79) |
| Co-Bagging (NN) | 57,24 (11,34) | 93,83 (3,72) | 64,89 (7,99) | 67,62 (8,97) | 78,15 (11,17) |
| Self-Trainer(NN) | 57,60 (9,50) | 94,93 (4,36) | 65,55 (5,69) | 67,91 (7,33) | 78,15 (10,27) |
| ADE-CoForest | 59,00 (9,27) | 94,67 (3,63) | 61,16 (11,21) | 64,42 (7,23) | 72,22 (6,88) |
| SNNRCE | 58,07 (8,21) | 94,94 (3,33) | 64,38 (9,57) | 69,59 (3,41) | 77,78 (10,86) |
| DE-TriTraining (NN) | 54,10 (6,79) | 92,97 (4,78) | 60,32 (8,35) | 69,63 (6,70) | 79,63 (9,26) |
| TriTraining (NN) | 53,25 (11,64) | 92,69 (5,24) | 65,81 (9,26) | 66,66 (5,78) | 76,30 (10,37) |
| CLCC | 57,32 (3,71) | 85,79 (5,54) | 56,25 (11,69) | 73,53 (0,94) | 78,52 (5,93) |
| APSSC | 56,04 (7,52) | 93,26 (3,81) | 49,22 (9,82) | 63,39 (6,71) | **83,70** (7,63) |
| Co-Training (NN) | 57,05 (8,53) | 92,15 (44,9) | 59,20 (6,52) | 58,10 (9,73) | 76,30 (8,80) |
| Rasco (NN) | 59,66 (9,19) | 57,53 (7,02) | 54,99 (7,20) | 65,95 (9,28) | 74,44 (10,53) |
| Rel-Rasco (NN) | 60,79 (7,32) | 60,57 (5,58) | 54,73 (7,71) | 66,61 (5,70) | 71,85 (10,76) |

(a) Results for the data sets Bupa, Dermatology, Glass, Haberman, and Heart.

| Algorithms | Hepatitis | Iris | Monk-2 | Spectfheart | Tae |
|---|---|---|---|---|---|
| CoForest | 84,03 (11,69) | 92,67 (6,29) | **99,77** (0,68) | 77,91 (6,93) | 43,08 (7,60) |
| CVSVSTW | 83,16 (13,38) | 96,03 (3,47) | 96,47 (3,44) | 79,49 (1,78) | 39,91 (10,07) |
| Democratic-Co | **85,43** (12,24) | **97,33** (3,27) | 95,43 (3,38) | 69,72 (13,34) | 41,04 (8,18) |
| HSVSTW | 82,83 (11,59) | 95,48 (5,34) | 95,27 (3,26) | **79,72** (3,73) | 39,66 (11,68) |
| SETRED | 83,92 (12,83) | 93,33 (4,22) | 77,61 (5,67) | 75,00 (7,21) | 41,75 (9,06) |
| Co-Bagging (NN) | 84,28 (10,37) | 94,00 (3,59) | 78,38 (5,76) | 70,46 (7,75) | 47,63 (8,91) |
| Self-Trainer(NN) | 79,34 (15,52) | 92,67 (4,67) | 77,61 (5,94) | 73,48 (8,03) | 41,75 (8,01) |
| ADE-CoForest | 85,09 (15,29) | 90,67 (6,11) | 85,49 (5,30) | 77,19 (8,21) | 39,00 (11,36) |
| SNNRCE | 79,50 (12,96) | 92,00 (8,33) | 74,19 (6,04) | 75,74 (6,50) | 39,71 (8,77) |
| DE-TriTraining (NN) | 82,61 (12,99) | 92,67 (9,17) | 75,02 (6,67) | 75,30 (5,22) | 38,42 (12,57) |
| TriTraining (NN) | 80,92 (14,84) | 93,33 (4,22) | 70,31 (6,00) | 74,91 (6,26) | 45,08 (9,96) |
| CLCC | 83,43 (11,29) | 92,00 (8,84) | 73,47 (8,09) | 79,42 (1,66) | 38,33 (10,67) |
| APSSC | 84,59 (15,17) | 92,00 (6,53) | 78,76 (7,56) | 45,71 (6,25) | 44,42 (8,63) |
| Co-Training (NN) | 84,03 (14,17) | 87,33 (7,57) | 72,78 (8,28) | 58,11 (9,81) | 41,71 (12,59) |
| Rasco (NN) | 78,53 (16,69) | 66,67 (11,16) | 76,19 (5,29) | 74,17 (5,80) | **46,42** (13,48) |
| Rel-Rasco (NN) | 83,51 (16,35) | 67,33 (10,93) | 76,02 (6,76) | 71,62 (8,34) | 37,12 (11,71) |

(b) Results for the data sets Hepatitis, Iris, Monk-2, Spectfheart, and Tae.

to CoForest and Democratic-Co that are both *not* based on nearest-neighbor classification. In addition, we see that HSVSTW and CVSVSTW have both two wins as CoForest and Democratic-Co. Thus, HSVSTW and CVSVSTW are the best semi-supervised classifier-agnostic methods that employ nearest-neighbor classification in terms of averaged accuracy and win numbers.

CVSVSTW outperforms HSVSTW on 8 data sets. Thus, the experiments confirm that cross-validation estimation of Shapley values improves the wrapper accuracy. However, this improvement can diminish for large data. Thus, in this case HSVSTW can be more preferable due to its computational efficiency.

CVSVSTW and HSVSTW outperform the standard self-training wrapper (with a nearest-neighbor base classifier) on 6 (out of 10) data sets with big accuracy margins. This shows that selecting pseudo-labeled data based on Shapley values results in rather different and eventually more accurate wrappers than selecting based on pseudo-label confidence values.

## 7   Conclusion

This paper introduced an alternative approach to selecting pseudo-labeled data for semi-supervised learning. The approach selects pseudo-labeled instances based on their individual contributions for the final-model performance. It was shown experimentally that it is capable of outperforming the standard selection approach that is based on the pseudo-label confidence values.

Estimating instance contributions was considered as a data-valuation problem, and, therefore, Shapley-value methods were proposed for this problem. The applicability of data valuation to semi-supervised learning implies two unresearched possibilities. The first one is that we can perform data valuation for different performance metrics such as ROC AUC, F1 score, $R^2$ etc. and, thus, we can perform semi-supervised learning for different aspects of the task in hand.

The second possibility is based on the fact that we can provide a value for any individual unlabeled instance for a model performance. Thus, it can be a first step for the eminent commoditization of unlabeled data.

## References

1. Berthelot, D., Carlini, N., Goodfellow, I.J., Papernot, N., Oliver, A., Raffel, C.: MixMatch: a holistic approach to semi-supervised learning. In: Advances in Neural Information Processing Systems 32, pp. 5050–5060 (2019)
2. Blum, A., Mitchell, T.: Combining labeled and unlabeled data with co-training. In: Proceedings of the Eleventh Annual Conference on Computational Learning Theory, COLT 1998, pp. 92–100. ACM (1998)
3. Deng, C., Guo, M.Z.: Tri-training and data editing based semi-supervised clustering algorithm. In: Gelbukh, A., Reyes-Garcia, C.A. (eds.) MICAI 2006. LNCS (LNAI), vol. 4293, pp. 641–651. Springer, Heidelberg (2006). https://doi.org/10.1007/11925231_61
4. Deng, C., Guo, M.: A new co-training-style random forest for computer aided diagnosis. J. Intell. Inf. Syst. **36**(3), 253–281 (2011)
5. Dheeru, D., Casey, G.: UCI machine learning repository (2017)
6. Ghorbani, A., Zou, J.: Data shapley: equitable valuation of data for machine learning. In: Proceedings of the 36th International Conference on Machine Learning, ICML 2019. Proceedings of Machine Learning Research, vol. 97, pp. 2242–2251. PMLR (2019)

7. Hady, M., Schwenker, F.: Co-training by committee: a new semi-supervised learning framework. In: Proceedings of the 8th IEEE International Conference on Data Mining (ICDM 2008), pp. 563–572. IEEE Computer Society (2008)

8. Halder, A., Ghosh, S., Ghosh, A.: Ant based semi-supervised classification. In: Dorigo, M., et al. (eds.) ANTS 2010. LNCS, vol. 6234, pp. 376–383. Springer, Heidelberg (2010). https://doi.org/10.1007/978-3-642-15461-4_34

9. He, J., Gu, J., Shen, J., Aurelio Ranzato, M.: Revisiting self-training for neural sequence generation. In: Proceedings of the 8th International Conference on Learning Representations, ICLR 2020 (2020). OpenReview.net

10. He, W., Jiang, Z.: Semi-supervised learning with the EM algorithm: a comparative study between unstructured and structured prediction. CoRR, abs/2008.12442 (2020)

11. Huang, T., Yu, Y., Guo, G., Li, K.: A classification algorithm based on local cluster centers with a few labeled training examples. Knowl. Based Syst. $23$(6), 563–571 (2010)

12. Ishii, M.: Semi-supervised learning by selective training with pseudo labels via confidence estimation. CoRR, abs/2103.08193 (2021)

13. Jia, R., et al.: Efficient task-specific data valuation for nearest neighbor algorithms. Proc. VLDB Endow. $12$(11), 1610–1623 (2019)

14. Li, M., Zhou, Z.-H.: SETRED: self-training with editing. In: Ho, T.B., Cheung, D., Liu, H. (eds.) PAKDD 2005. LNCS (LNAI), vol. 3518, pp. 611–621. Springer, Heidelberg (2005). https://doi.org/10.1007/11430919_71

15. Li, M., Zhou, Z.H.: Improve computer-aided diagnosis with machine learning techniques using undiagnosed samples. IEEE Trans. Syst. Man Cybern. Part A $37$(6), 1088–1098 (2007)

16. Shapley, L.S.: A value for n-person games. Ann. Math. Stud. $28$, 307–317 (1953)

17. Triguero, I., García, S., Herrera, F.: Self-labeled techniques for semi-supervised learning: taxonomy, software and empirical study. Knowl. Inf. Syst. $42$(2), 245–284 (2013). https://doi.org/10.1007/s10115-013-0706-y

18. van Engelen, J.E., Hoos, H.H.: A survey on semi-supervised learning. Mach. Learn. $109$(2), 373–440 (2019). https://doi.org/10.1007/s10994-019-05855-6

19. Wang, J., Luo, S., Zeng, X.: A random subspace method for co-training. In: Proceedings of the International Joint Conference on Neural Networks, IJCNN 2008, pp. 195–200. IEEE (2008)

20. Wang, Y., Xu, X., Zhao, H., Hua, Z.: Semi-supervised learning based on nearest neighbor rule and cut edges. Knowl. Based Syst. $23$(6), 547–554 (2010)

21. Yarowsky, D.: Unsupervised word sense disambiguation rivaling supervised methods. In: Proceedings of the 33rd Annual Meeting of the Association for Computational Linguistics, pp. 189–196. Morgan Kaufmann Publishers/ACL (1995)

22. Yaslan, Y., Cataltepe, Z.: Co-training with relevant random subspaces. Neurocomputing $73$(10–12), 1652–1661 (2010)

23. Zhou, Y., Goldman, S.: Democratic co-learning. In: Proceedings of the 16th IEEE International Conference on Tools with Artificial Intelligence (ICTAI 2004), pp. 594–602. IEEE Computer Society (2004)

24. Zhou, Z.H., Li, M.: Tri-training: exploiting unlabeled data using three classifiers. IEEE Trans. Knowl. Data Eng. $17$(11), 1529–1541 (2005)

25. Zhu, X., Schleif, F.M., Hammer, B.: Adaptive conformal semi-supervised vector quantization for dissimilarity data. Pattern Recogn. Lett. $49$, 138–145 (2014)

# Data Streams

# A Network Intrusion Detection System for Concept Drifting Network Traffic Data

Giuseppina Andresini[1](✉)🆔, Annalisa Appice[1,2]🆔, Corrado Loglisci[1,2]🆔,
Vincenzo Belvedere[1], Domenico Redavid[1], and Donato Malerba[1,2]🆔

[1] Dipartimento di Informatica, Università degli Studi di Bari Aldo Moro,
Via Orabona, 4-70125 Bari, Italy
{giuseppina.andresini,annalisa.appice,corrado.loglisci,domenico.redavid1,
donato.malerba}@uniba.it, v.belvedere@studenti.uniba.it
[2] CINI - Consorzio Interuniversitario Nazionale per l'Informatica, Bari, Italy

**Abstract.** Deep neural network architectures have recently achieved state-of-the-art results learning flexible and effective intrusion detection models. Since attackers constantly use new attack vectors to avoid being detected, concept drift commonly occurs in the network traffic by degrading the effect of the detection model over time also when deep neural networks are used for intrusion detection. To combat concept drift, we describe a methodology to update a deep neural network architecture over a network traffic data stream. It integrates a concept drift detection mechanism to discover incoming traffic that deviates from the past and triggers the fine-tuning of the deep neural network architecture to fit the drifted data. The methodology leads to high predictive accuracy in presence of network traffic data with zero-day attacks.

**Keywords:** Network intrusion detection · Deep learning · Data stream · Concept drift detection · Transfer learning

## 1 Introduction

A Network Intrusion Detection System (NIDS) aims to protect information security by detecting malicious activities (attacks) in computer networks. In recent studies, advances in Deep Learning (DL) have been widely exploited to design accurate neural network (NN) models that turn out robust and effective also in detecting zero-day attacks [2–6]. However a limit of current DL-based NIDSs is that they follow the assumption of stationary traffic data distribution, while this assumption is ineffective in modern network traffic environments, where the malicious activities are often polymorphic and evolve continuously. The desirable behaviour is that NIDS models built around to "normal connections" or "intrusions" change over time to deal with the "concept drift" of the network traffic characteristics [13]. This behaviour can be actually achieved by resorting to data stream learning, in order to incorporate the most recent behaviours of intruders into NIDS models.

© Springer Nature Switzerland AG 2021
C. Soares and L. Torgo (Eds.): DS 2021, LNAI 12986, pp. 111–121, 2021.
https://doi.org/10.1007/978-3-030-88942-5_9

In this paper, we define the stream version of the DL-based network intrusion detection method described in [3]. We select this method for the stream upgrade described in this study as experiments [3] have proved that it outperforms several, recent state-of-the-art intrusion detection methods evaluated in various static settings. We describe a stream learning methodology that resorts to the Page-Hinkley test (PHT) [15] to capture possible drifts in traffic data. When these changes get being as significant, we fine-tune the current DL model on the most recent normal and attack data that have been processed to alert the drifts. This corresponds to transfer the previously trained DL models to new data, updating only the layer weights, while avoiding the effect of catastrophic forgetting [11]. We investigate the effectiveness of the presented data stream methodology in a benchmark stream of network flows.

The rest of this paper is organised as follows. Related works are presented in Sect. 2. The basic DL architecture is described in Sect. 3. The proposed stream learning DL methodology is described in Sect. 4. The results of the evaluation are discussed in Sect. 5. Finally, Sect. 6 refocuses on the purpose of the research and illustrates possible future developments.

## 2   Related Works

The problem of learning intrusion detection models able to handle streaming traffic data has not still attracted large interest. A few studies that have investigated the task basically resort to sequential learning approaches. The authors of [17] explore the capability of a RNN to build unsupervised autoencoders. Once these models learn the distribution of the normal network traffic, they are used to recognise anomalous data that are restored with difficulty, since they come from a different distribution. However, to assess the reconstruction quality and the detection rate, these methods need ranking metrics that deem as true intrusions only samples with higher scores. Autoencoders for anomaly detection are also explored in [14], while the ranking is also investigated in [19] with an unsupervised approach based on feed-forward and recurrent deep networks.

On the other hand, the availability of labelled public network traffic datasets has prompted more studies of DL focused on supervised classification approaches, than studies that discard training activities. For instance, the authors of [12] propose a CNN-LSTM architecture working also on networking metadata besides connection features. The model is continuously re-trained, after the validation of the malicious events by the human analyst. However, in a streaming scenario, the human validation turns out to be unfeasible. The authors of [1] focus on the generalisation properties of the model rather than on a way to re-train it. They learn a RNN based on a new regularizer that decays the weights of the hidden layers according to their standard deviation in the weight matrices. In any case, the above-mentioned studies have been proved as adequate for real-time environments that do not necessarily exhibit data distribution drift. Instead, handling concept drift has been extensively studied coupled with conventional machine learning approaches [9] that do not offer the great detection performance of DL techniques like CNN-based architectures [2,3,5].

# 3   Background: MINDFUL

MINDFUL [3] is a DL-based NIDS that is trained processing a flow-level historical network traffic training set $\mathcal{D} = \{(\mathbf{x}_i, y_i)\}_{i=1}^N$ composed of $N$ training network flows. Each training sample $\mathbf{x}_i \in \mathbb{R}^D$ is a row vector corresponding to an input sample defined over $D$ flow-level attributes, while $y_i$ is the corresponding binary label denoting a *normal* or an *attack* sample. The intrusion detection model is trained combining an unsupervised DL architecture—based on two autoencoder—with a supervised DL architecture—based on a 1D CNN. Specifically, two independent autoencoders—$z_n$ and $z_a$—are separately learned from the subset of training samples, whose label is normal, resp. attack., respectively. These autoencoders are used to map single-channel training samples to a multi-channel representation that is used as input to a 1D CNN. In particular, each training sample $\mathbf{x}_i \in \mathcal{D}$ is replaced by 3-channel sample $\hat{\mathbf{x}}_i = [\mathbf{x}_i, z_n(\mathbf{x}_i), z_a(\mathbf{x}_i)]^\top \in \mathbb{R}^{D \times 3}$, where $z_n(\mathbf{x}_i)$ and $z_a(\mathbf{x}_i)$ correspond to the reconstructed representations of the single-channel sample $\mathbf{x}_i$ in both autoencoders $z_n$ and $z_a$. We note that when the samples belong to two different distributions, samples $\mathbf{x}_i$, labelled as normal should be more similar to the representation $z_n(\mathbf{x}_i)$ than that of $z_a(\mathbf{x}_i)$, or equivalently $||\mathbf{x}_i - z_n(\mathbf{x}_i)||^2 < ||\mathbf{x}_i - z_a(\mathbf{x}_i)||^2$, and viceversa. A 1D CNN—*cnn*—is trained from the multi-channel representation of the training samples.

# 4   Data Stream Methodology: Str-MINDFUL

In this Section we present Str-MINDFUL—a supervised stream learning algorithm—that trains the MINDFUL architecture over the network traffic data stream by incrementally updating the weights of the trained architecture to fit possible changes occurring in the network traffic. Str-MINDFUL initialises the MINDFUL architecture using the initial labelled samples recorded in the stream. At the completion of this initialisation step, as a new unlabelled sample is recorded in the stream, Str-MINDFUL uses the current MINDFUL architecture to yield the class prediction. After the true class label of the sample is available in the stream, Str-MINDFUL processes this information to identify possible drifts in network data that, if neglected, may worsen the intrusion detection ability of the current MINDFUL architecture. The drift detection triggers the update of the MINDFUL architecture for the subsequent predictions. The block diagram of Str-MINDFUL algorithm is reported in Fig. 1. The algorithm uses the Page-Hinkley test (PHT) to detect possible drifts in the network traffic data and trigger the operation of incremental updating of the MINDFUL architecture as requested. We briefly describe basics of the Page-Hinkley test in Sect. 4.1, while we specify the details of the initialisation step and the incremental learning step of Str-MINDFUL in Sects. 4.2 and 4.3, respectively.

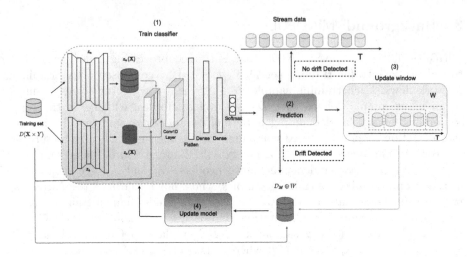

**Fig. 1.** Architecture of Str-MINDFUL.

## 4.1 Page-Hinkley Test

The Page-Hinkley test (PHT) [15] is commonly used for monitoring data drift detection in the average of a timestamped Gaussian variable $v_t$ [10]. As a new value $v_t$ is acquired at time $t$, the test updates a variable $m_t$, defined as the cumulative difference between the observed values and the mean up until the current time $t$. Formally, $m_0 = 0$ and $m_t = \alpha \times m_{t-1} + (v_t - \overline{v_t} - \delta)$, where $\overline{v_t} = \frac{1}{t} \sum_{i=1}^{t} v_i$, $\delta$ is the tolerable magnitude of the drifts and $\alpha$ is the fading factor introducing a forgetting mechanism for accumulating the oldest data. In addition, the test updates $M_t = \min_{i=1,...,t} m_i$, i.e. the minimum of $m_t$. The PHT detects a drift by monitoring the difference $PHT_t = m_t - M_t$. When this difference is greater than $\lambda$ a change is flagged. According to comments reported in [10], larger $\lambda$ will entail fewer false alarms, but may miss some changes. In Str-MINDFUL, we use the PHT to monitor the classification errors, as well as the reconstruction errors of the samples restored through autoencoders.

## 4.2 Initialisation Step

Let $\mathcal{S}$ be a stream of labelled network flows (samples), that is $\mathcal{S} = \{(\mathbf{x_t}, y_t)\}$ with $t = 1, 2, \ldots, N, \ldots$, where $\mathbf{x_t}$ is a vector of flow-level attribute values and $y_t$ is the corresponding binary label denoting a *normal* or an *attack* sample. The initialisation step of Str-MINDFUL starts as a historical training set is acquired from $\mathcal{S}$. This corresponds to acquire the $N$ initial labelled samples $(\mathbf{x_t}, y_t)$ of $\mathcal{S}$ (with $t = 1, \ldots, N$) and record these labelled samples in the $N$-sized data synopsis $\mathcal{D}$. The training samples recorded in $\mathcal{D}$ are processed to complete the initialization phase in three steps:

1. The MINDFUL architecture is initialized by learning both the weights of $z_n$ and $z_a$—the autoencoder NNs—and the weights of $cnn$—the multi-channel 1D CNN. This is done by running the DL algorithm described in [3].
2. The cumulative variables of the PHT, $PHT_y$, $PHT_{z_n}$ and $PHT_{z_a}$, are initialised, so that: (1) the cumulative variables of $PHT_y$ are initialised with the sequence of classification errors $||y_t - y_t'||$ with $y_t'$ is the label predicted with the multi-channel CNN $cnn$, that is, $y_t' = cnn([\mathbf{x_t}, z_n(\mathbf{x_t}), z_a(\mathbf{x_t})])$; (2) the cumulative variables of $PHT_{z_n}$ and $PHT_{z_a}$ are initialised with the reconstruction errors $||\mathbf{x_t} - z_n(\mathbf{x_t})||$ and $||\mathbf{x_t} - z_a(\mathbf{x_t})||$ of the samples restored through the autoencoders $z_n$ and $z_a$, respectively.
3. The stratified sampling algorithm is used to retain a $M$-sized representative sub-sample of original $\mathcal{D}$ for the next incremental learning stage. This sample is maintained in $\mathcal{D}$ and used as a historical background for the update of the weights of $z_n$, $z_a$ and $cnn$ during the incremental learning step. This step allows us to speed-up any subsequent update operations involving the processing of historical data.

### 4.3  Incremental Learning Step

For each new sample $\mathbf{x_t}$ recorded at the time step $t$ of $\mathcal{S}$, the sample reconstructions $z_n(\mathbf{x_t})$ and $z_a(\mathbf{x_t})$ of $\mathbf{x_t}$ are restored through the current autoencoders $z_n$ and $z_a$, respectively. Both reconstructions are used to derive the multi-channel representation of the sample and predict the label $y_t' = cnn([\mathbf{x_t}, z_n(\mathbf{x_t}), z_a(\mathbf{x_t})])$ by using the current $cnn$.

As soon as the class label $y_t$ is available, this information is used to understand if a drift has occurred in the network traffic. To perform the drift detection, Str-MINDFUL updates the cumulative variables of $PHT_y$ according to the classification error value $||y_t - y_t'||$. In addition, if the label $y_t$ is equal to normal then Str-MINDFUL updates the cumulative variables of $PHT_{z_n}$ according to the sample reconstruction error $||\mathbf{x_t} - z_n(\mathbf{x_t})||$ and records the sample $(\mathbf{x_i}, y_i)$ in the sliding window synopsis $W_n$. Otherwise, if the label $y_t$ is equal to attack then Str-MINDFUL updates the cumulative variables of $PHT_{z_a}$ according to sample reconstruction error $||\mathbf{x_t} - z_a(\mathbf{x_t})||$ and records the sample $(\mathbf{x_i}, y_i)$ in the sliding window $W_a$. Both $W_n$ and $W_a$ are windows with size $w$, which record the last $w$ normal samples and the last $w$ attack samples acquired in $\mathcal{S}$, respectively.

By analysing the updated cumulative variables of $PHT_y$, $PHT_{z_n}$ and $PHT_{z_a}$, Str-MINDFUL can raise alerts on drifts detected either testing the classification error (test on $PHT_y$) or testing the reconstruction error of the normal samples (test on $PHT_{z_n}$) and the attack samples (test on $PHT_{z_a}$). The detection of a drift event triggers the refresher of the historical training data actually recorded in $\mathcal{D}$. In particular, if a drift alert is raised by the test on $PHT_{z_n}$ then Str-MINDFUL updates $\mathcal{D}$ with the samples currently recorded in $W_n$. Let $|W_n|$ be the number of normal samples currently recorded in $W_n$. Str-MINDFUL first removes $|W_n|$ random normal samples from $\mathcal{D}$. Then it adds all the samples recorded in $W_n$ to $\mathcal{D}$ and empties $W_n$. Similarly, if a drift alert is raised by the test on $PHT_{z_a}$ then Str-MINDFUL updates $\mathcal{D}$ with the samples currently recorded in $W_a$. Finally, if neither the test on $PHT_{z_n}$ nor the test on $PHT_{z_n}$ raise a drift alert, while the test on $PHT_y$ raises a drift alert then Str-MINDFUL updates $\mathcal{D}$ with the samples that are currently recorded in both $W_n$ and $W_a$.

Finally, based on the alerts raised from the drift detection module, Str-MINDFUL starts the adaptation of $z_n$, $z_a$ and $cnn$ to fit these NNs to the change occurred in the network traffic data. The autoencoder NN $z_n$ is updated if an alert is raised from $PHT_{z_n}$; the autoencoder NN $z_a$ is updated if an alert is raised from $PHT_{z_a}$; the multi-channel 1D CNN $cnn$ is updated if an alert is raised by either $PHT_{z_n}$ or $PHT_{z_a}$ or $PHT_y$. For re-training $z_n$, $z_a$ and $cnn$, Str-MINDFUL runs the algorithm in [3] with current $\mathcal{D}$ as training set. In this incremental stage, $z_n$, $z_a$ and $cnn$ are re-trained by using the weights saved in the initial networks as starting point. This is a simple application of a transfer learning principle in DL [18]. In fact, the structure and the weights of $z_n$, $z_a$ and $cnn$ are transferred from the past data to the new data. The weights saved from the previous networks are fine-tuned on the refined input-output pair data available for the intrusion detection task. Final concerns regard the fact that the fine-tuning operation may be completed only after that a few new samples are acquired in $\mathcal{S}$. In this case, old weights of $z_n$, $z_a$ and $cnn$ are still used to classify a few incoming samples even after the drift detection. This happens until the fine-tuning of the new weights of these architectures have not been completed.

## 5   Empirical Evaluation

This empirical evaluation is conducted using a stream of timestamped network flows to verify the accuracy and efficiency of the proposed IDS methodology.

### 5.1   Implementation Details

St-MINDFUL is implemented in Python 3.6 using the Keras 2.4 library with TensorFlow as back-end. The source code is available online.[1] The autoencoder architectures and the multi-channel architecture are implemented as described in [3]. In the initialization step, the weights are initialized following the Xavier scheme. In the incremental learning step, the weights saved from the previous network are used as a starting point of the new fine-tuning operation. During the initialization step, a hyper-parameter optimization is conducted following the description reported in [3]. Finally, the Page Hinkley test is that implemented in the Scikit-multiflow library. It is used with the default parameter set-up, that is, $\lambda = 50$, $\delta = 0.005$ and $\alpha = 1 - 0.0001$. The minimum number of samples before detecting a data drift is set equal to 30.

### 5.2   Dataset Description

CICIDS2017 was collected by the Canadian Institute for Cybersecurity in 2017. The original dataset is a 5-day labelled log collected from Monday July 3, 2017

---

[1] https://github.com/gsndr/Str-MINDFUL

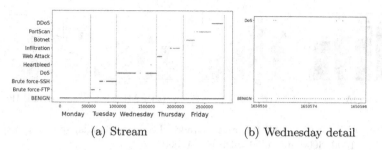

(a) Stream                           (b) Wednesday detail

**Fig. 2.** Label distribution of the entire network flow stream (Fig. 2a) and zooming in a portion of data streamed on Wednesday (Fig. 2b).

to Friday July 7, 2017 [16]. The first day (Monday) contains only benign traffic, while the other days contain various types of attacks, in addition to normal network flows. Every network flow sample is spanned over 79 attributes extracted with CICFlowMeter processing the real PCAPs. We consider the stream of data with labels distributed as reported in Fig. 2. We note that new types of attacks appear over time in the captured network traffic.

### 5.3 Experimental Setting

The labelled data recorded on both Monday and Tuesday are processed in the initialisation step, while the data streamed on Wednesday, Thursday and Friday are processed in the incremental learning step. The effectiveness of Str-MINDFUL is measured by analysing the performance of the algorithm in monitoring the data streamed on Wednesday, Thursday and Friday. When a drift is detected, we consider the past model to predict incoming samples until the update of the new model has been completed. The accuracy performance is measured by analysing the Overall Accuracy (OA), F1-score (F1) and False Alarm Rate (FAR) of the classifications. The efficiency performance is evaluated with the computation time spent processing each new sample monitored on Wednesday, Thursday and Friday. The computation time is measured in seconds on a Linux machine with an Intel(R) Core(TM) i7-9700F CPU @ 3.00 GHz and 32 GB RAM. All the experiments are executed on a single GeForce RTX 2080.

### 5.4 Results

Figures 3a, 3b, 3c and 3d report the OA, F1, FAR and average TIME spent (in seconds) by monitoring each sample recorded on Wednesday, Thursday and Friday by varying $w$ among 750, 1500 (default), 3000 and 4500 with $M = 12000$. Figures 4a, 4b, 4c and 4d report the OA, F1, FAR and average TIME spent (in seconds) by monitoring each sample recorded on Wednesday, Thursday and Friday by varying $M$ among 6000, 12000 (default) and 24000 with $w = 1500$. These results show that both the accuracy and efficiency performances vary

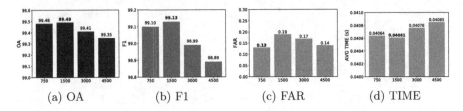

(a) OA            (b) F1            (c) FAR            (d) TIME

**Fig. 3.** OA (%), F1 (%), FAR (%) and average TIME (in seconds) spent by varying $w$ among 750, 1500, 3000 and 4500 with $M = 12000$

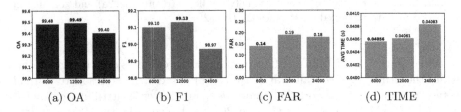

(a) OA            (b) F1            (c) FAR            (d) TIME

**Fig. 4.** OA (%), F1 (%), FAR (%) and average TIME (in seconds) by varying $M$ among 6000, 12000 and 24000 with $w = 1500$

slightly with $M$ and $w$. The default configuration ($M = 12000$ and $w = 1500$) achieves the best trade-off of accuracy and efficiency of the in-stream monitoring.

Figure 5 reports the TIME spent in seconds processing each sample recorded on Wednesday, Thursday and Friday with the default configuration of Str-MINDFUL ($M = 12000$ and $w = 1500$). This computation time ranges between 0.04 and 22.85 s. The computation peaks are achieved as the drifts alerts are raised. This is because drifts trigger the fine-tuning of the weights of the autoencoders and/or the multi-channel 1D CNN. In any case, the time spent completing this fine-tuning stage is low (less than 23 s). This is thanks to the application of a transfer learning approach that starts from weights saved from the previous NNs coupled with the use of a sample set of the historical data sample that is appropriately updated with a sufficient amount of the newest drifted samples.

Finally, Table 1 compares the accuracy performance of Str-MINDFUL to that of the baseline MINDFUL and the competitor described in [7]. In MINDFUL the intrusion detection model learned during the initialisation step with the samples recorded on Monday and Tuesday is consider to predict all the samples recorded on Wednesday, Thursday and Friday without triggering any NN update operation. The competitor described in [7] handles the intrusion detection task as an anomaly detection problem. These results confirm that Str-MINDFUL leverages the ability to deal with concept drift outperforming significantly MINDFUL. In addition, Str-MINDFUL takes advantage of a supervised learning process completed with normal and attack samples achieving better performance than [7] that completes the learning stage with the normal samples only.

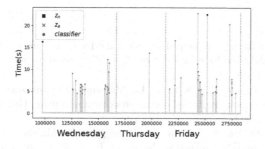

**Fig. 5.** The TIME spent in seconds. Drifts on reconstruction errors with $z_n$ and $z_a$ are denoted with blue squares and red crosses, respectively. Drifts on classification errors of predictions yielded with $cnn$ are denoted with green circles. (Color figure online)

**Table 1.** OA, F1 and FAR of Str-MINDFUL ($M = 12000$ and $w = 1500$), its baselines MINDFUL and the anomaly detector described in [7].

| Method | OA (%) | F1 (%) | FAR (%) |
|---|---|---|---|
| Str-MINDFUL | 99.49 | 99.13 | 0.19 |
| MINDFUL | 88.01 | 74.90 | 0.80 |
| [7] | – | 89.89 | 6.15 |

# 6  Conclusion

In this study, we have presented a DL-based network intrusion detection methodology that takes advantage of the PHT technique to detect events of concept drift in the monitored traffic network data stream and applies a transfer learning technique to fine-tune the DL architecture to the drifted data. The experimental analysis confirms the effectiveness of the proposed methodology.

One research direction is investigating a strategy to add the ability to classify the attack category (e.g., Dos, Port Scan). Another direction is that of exploring the use of count-based windowing [8] to update the intrusion detection model as a fixed number of connections has been streamed in. Then, we plan to explore different stream learning mechanisms, in alternative to the PHT test, in order to detect the concept drift in network traffic data. Finally, we intend to validate the effectiveness of the proposed methodology with datasets comprising different attack categories.

**Acknowledgment.** We acknowledge the support of the MIUR through the project "TALIsMan -Tecnologie di Assistenza personALizzata per il Miglioramento della quAlità della vitA" (Grant ID: ARS01_01116), funding scheme PON RI 2014-2020 and the project "Modelli e tecniche di data science per la analisi di dati strutturati" funded by the University of Bari "Aldo Moro".

# References

1. Albahar, M.A.: Recurrent neural network model based on a new regularization technique for real-time intrusion detection in SDN environments. Secur. Commun. Netw. **2019**, 1–9 (2019)
2. Andresini, G., Appice, A., De Rose, L., Malerba, D.: Gan augmentation to deal with imbalance in imaging-based intrusion detection. Future Gener. Comput. Syst. **123**, 108–127 (2021)
3. Andresini, G., Appice, A., Di Mauro, N., Loglisci, C., Malerba, D.: Multi-channel deep feature learning for intrusion detection. IEEE Access **8**, 53346–53359 (2020)
4. Andresini, G., Appice, A., Malerba, D.: Autoencoder-based deep metric learning for network intrusion detection. Inf. Sci. **569**, 706–727 (2021)
5. Andresini, G., Appice, A., Malerba, D.: Nearest cluster-based intrusion detection through convolutional neural networks. Knowl. Based Syst. **216**, 106798 (2021)
6. Andresini, G., Appice, A., Mauro, N.D., Loglisci, C., Malerba, D.: Exploiting the auto-encoder residual error for intrusion detection. In: 2019 IEEE European Symposium on Security and Privacy Workshops, EuroS&P Workshops 2019, Stockholm, Sweden, 17–19 June 2019, pp. 281–290. IEEE (2019)
7. Angelo, P., Costa Drummond, A.: Adaptive anomaly-based intrusion detection system using genetic algorithm and profiling. Secur. Priv. **1**(4), 1–13 (2018)
8. Appice, A., Ciampi, A., Malerba, D.: Summarizing numeric spatial data streams by trend cluster discovery. Data Min. Knowl. Disc. **29**(1), 84–136 (2013). https://doi.org/10.1007/s10618-013-0337-7
9. Buczak, A.L., Guven, E.: A survey of data mining and machine learning methods for cyber security intrusion detection. IEEE Commun. Surv. Tutor. **18**(2), 1153–1176 (2016)
10. Gama, J., Zliobaite, I., Bifet, A., Pechenizkiy, M., Bouchachia, A.: A survey on concept drift adaptation. ACM Comput. Surv. **46**(4), 44:1–44:37 (2014)
11. Goodfellow, I.J., Mirza, M., Xiao, D., Courville, A., Bengio, Y.: An empirical investigation of catastrophic forgeting in gradient based neural networks. In: International Conference on Learning Representations, ICLR 2014 (2014)
12. Kim, A.C., Park, M., Lee, D.H.: AI-IDS: application of deep learning to real-time web intrusion detection. IEEE Access **8**, 70245–70261 (2020)
13. Lu, J., Liu, A., Dong, F., Gu, F., Gama, J., Zhang, G.: Learning under concept drift: a review. IEEE Trans. Knowl. Data Eng. **31**(12), 2346–2363 (2019)
14. Madani, P., Vlajic, N.: Robustness of deep autoencoder in intrusion detection under adversarial contamination. In: Proceedings of the 5th Annual Symposium and Bootcamp on Hot Topics in the Science of Security, HoTSoS 2018. Association for Computing Machinery (2018)
15. Page, E.S.: Continuous inspection schemes. Biometrika **41**(1/2), 100–115 (1954)
16. Sharafaldin, I., Habibi Lashkari, A., Ghorbani, A.: Toward generating a new intrusion detection dataset and intrusion traffic characterization. In: 4th International Conference on Information Systems Security and Privacy, pp. 108–116 (2018)

17. Sovilj, D., Budnarain, P., Sanner, S., Salmon, G., Rao, M.: A comparative evaluation of unsupervised deep architectures for intrusion detection in sequential data streams. Expert Syst. Appl. **159**, 113577 (2020)
18. Tan, C., Sun, F., Kong, T., Zhang, W., Yang, C., Liu, C.: A survey on deep transfer learning. In: Kůrková, V., Manolopoulos, Y., Hammer, B., Iliadis, L., Maglogiannis, I. (eds.) ICANN 2018. LNCS, vol. 11141, pp. 270–279. Springer, Cham (2018). https://doi.org/10.1007/978-3-030-01424-7_27
19. Tuor, A., Kaplan, S., Hutchinson, B., Nichols, N., Robinson, S.: Deep learning for unsupervised insider threat detection in structured cybersecurity data streams. In: The Workshops of the 31st AAAI Conference on Artificial Intelligence (2017)

# Incremental $k$-Nearest Neighbors Using Reservoir Sampling for Data Streams

Maroua Bahri[1(✉)] and Albert Bifet[1,2(✉)]

[1] LTCI, Télécom Paris, IP-Paris, Paris, France
{maroua.bahri,albert.bifet}@telecom-paris.fr
[2] University of Waikato, Hamilton, New Zealand

**Abstract.** The online and potentially infinite nature of data streams leads to the inability to store the flow in its entirety and thus restricts the storage to a part of – and/or synopsis information from – the stream. To process these evolving data, we need efficient and accurate methodologies and systems, such as window models (e.g., sliding windows) and summarization techniques (e.g., sampling, sketching, dimensionality reduction). In this paper, we propose, RW-$k$NN, a $k$-Nearest Neighbors ($k$NN) algorithm that employs a practical way to store information about past instances using the *biased reservoir sampling* to sample the input instances along with a *sliding window* to maintain the most recent instances from the stream. We evaluate our proposal on a diverse set of synthetic and real datasets and compare against state-of-the-art algorithms in a traditional test-then-train evaluation. Results show how our proposed RW-$k$NN approach produces high-predictive performance for both real and synthetic datasets while using a feasible amount of resources.

**Keywords:** Data stream classification · K-nearest neighbors · Reservoir sampling · Sliding window

## 1 Introduction

Data have become ubiquitous in today's fast-paced world. The evolution of technology has invaded our lives in multiple domains and changed the way in which we generate and manage data. These data can be transformed into valuable information and insightful decisions through machine learning tools and techniques. Several emerging applications and devices generate an overwhelming volume of data that are continuously arriving in an online fashion as *"streams"*. The application Internet of Things (IoT) is a good example, where connected devices and sensors yield to a massive amount of data [11,14].

Research in data mining is mainly devoted to static and offline environments where patterns hidden in data are fixed and instances[1] can be accessed several times. Classification is one of the most popular tasks with widely used data

---

[1] In the sequel, we use the terms *instance* or *observation* interchangeably.

© Springer Nature Switzerland AG 2021
C. Soares and L. Torgo (Eds.): DS 2021, LNAI 12986, pp. 122–137, 2021.
https://doi.org/10.1007/978-3-030-88942-5_10

mining methods that consists in building a model based on the attribute values of existing instances and seeks to predict the class labels of a given test set of unlabelled observations [25]. Multiple static classifiers have been proposed which build and test the model in the traditional batch setting by accessing the data more than once. However, when applied on streams, static algorithms fail to process potentially infinite sequence of data because of its evolving nature that requires methods to adapt automatically. Moreover, with the data stream model, we have resource limitations, so we need to address the space and time restrictions. Under these constraints, an efficient classification algorithm should have the following properties: (i) obtain a high – relatively good – accuracy; and (ii) use a low computational cost, in terms of memory and time.

These properties are somewhat correlated because reducing the memory and time used by a classifier will lead to a loss in information that can impact the accuracy. An algorithm can be fast by processing less information and using less space, on the other hand, the accuracy can increase if more information are stored. To sum up, it is a resource-accuracy tradeoff that depends on the application purposes.

To cope with the main data stream challenges (e.g., the memory and time constraints) and address the stream framework requirements while processing evolving data, stream algorithms use well-established manners [2]. The latter techniques include, but not limited to, *one-pass* processing where instances should be processed only once, *sliding window* [27] of a fixed size, where only the most recent instances from the stream are stored, *reservoir sampling* [32] which is a probabilistic method for stream synopsis construction through sampling, and *dimensionality reduction* [5] to reduce the number of attributes of data.

Several classification algorithms have been proposed to deal with evolving data streams, mostly derived from the traditional algorithms for the offline setting. For instance, decision trees [16,19,20], naive Bayes [7,17], $k$-Nearest Neighbors ($k$NN) [6,10,26], and the ensemble-based methods [21–23,28].

The $k$NN is a well-known algorithm that has been adapted to the stream setting by maintaining a sliding window of a fixed size since it is impossible to store the entire stream in memory [10]. Despite the fact that this stream version of $k$NN is limited by the size of the moving window, previous empirical studies [10,29] and our analysis (in Sect. 4) show that the standard stream $k$NN is still costly in terms of memory usage and time.

To cope with this issue, we investigate the benefits of using reservoir sampling with the $k$NN algorithm. We also investigate the predictive performance gains of using this technique.

The main contributions of this work are the following:

- We propose a $k$NN algorithm for evolving data streams that uses a small sliding window to maintain the recent observations from the stream, along with a reservoir sampling to keep track of old observations that are removed constantly from the window.

– We provide an experimental study where we evaluate our proposed approach and compare it against state-of-the-art classification algorithms using a diverse set of real and synthetic datasets.

The remainder of this work is organized as follows. In Sect. 2, we detail related work for comparison. Section 3 contains the description of the proposed $k$NN approach for evolving data streams. Section 4 details the datasets used and the comparison results of our proposal with the state-of-the-art methods. We finally make concluding remarks and poses directions for future work in Sect. 5.

## 2   Related Work

The evolving data stream mining has gained in popularity during the past decade because of the overwhelming volume of data generated daily in different domains. Classification is one of the most widely used tasks for data stream mining, where several classifiers have been thoroughly studied and used with evolving data streams. A stream algorithm must be able to process instances as fast as possible by reading every instance only once (one-pass) or a small number of times using limited resource capabilities [18].

For instance, the Naive Bayes (NB) [17] is the simplest classifier that updates counters with each observation and uses the assumption "all the attributes are independent of each other given the class label". In order to make prediction, the NB classifier computes the Bayes theorem using the stored counters which makes it useful with massive data streams. This naive assumption between attributes does not always hold in practice, which can lead to (potentially) bad results.

Unlike naive Bayes, the $k$NN algorithm does not learn any model, since it maintains all instances in order to find the neighbors for every test instance. A basic implementation consists in keeping a moving window that stores the most recent instances from the stream. Self-Adjusting Memory $k$NN (Sam$k$NN) [26] algorithm is another $k$NN variation that builds an ensemble of models to deal with concept drifts. For this to happen, the Sam$k$NN algorithm uses two memories: short-term memory to target current concept, the long-term memory to keep track about the past concepts.

In [10], Bifet et al. proposed a Probabilistic Adaptive Window (PAW) which is based on the approximate counting of Morris. PAW includes older instances as well as the most recent ones from the stream, and therefore maintains somewhat information about past concept drifts and adapts to new ones. PAW has been used with the $k$NN as a window to maintain instances. In order to add an explicit change detection mechanism to the aforementioned $k$NN, authors in [10] used, on top of PAW, ADWIN [8], a change detector that keeps a variable-length window of recently seen instances to handle drifts in the distribution.

The aforementioned single algorithms serve the purpose of common baselines since they are used for comparison in the data stream classification.

# 3  Reservoir Stream $k$NN

We start by defining the notation used throughout this paper and the data stream classification problem. Let $S = x_1, x_2, x_3, \ldots$ be an open-ended sequence of observations arriving over time. Each observation is composed of a vector of $d$ attributes $x \in \mathbb{R}^d$. Unlike the offline classification, the stream classification task must build the model incrementally with each incoming instances from the stream. It starts by predicting to which class a new arrived instance belongs, then uses it to update (and build) the model. Filtering spam emails is a good example for classification where we predict if an email is a spam or not based on the text contents (attributes).

More formally, the classifier builds a model $f$ and given an instance $x$ we want to predict its discrete class $y$ from a set of class labels $C$, st. $y = f(x)$. Back to our previous example, $x$ could be the email text and $y$ determines whether the email is spam or not.

## 3.1  Background

The unbounded nature of data streams requires some innovative adaptations over the traditional setting in order to extend the offline algorithms to the streaming framework. Many synopsis techniques such as sketches, histograms and wavelets are designed for approximations and to be used with particular applications. We will give an overview of some techniques, considered in this paper, which have been used with some stream classification algorithms.

Window models are a very popular way to keep instances in the memory for data streams. For instance, the sliding window with a fixed size that moves forward as time progresses. It maintains the most recent observations from the stream consisting of a smaller number than the real size of the huge stream (which is potentially infinite). Several stream mining algorithms uses windows, e.g., the $k$NN and SAM$k$NN algorithms. Nevertheless, the use of sliding window uniquely may represent an unstable solution because we lose the entire history of the past data stream.

Reservoir Sampling (RS) [32] is an important class of synopsis construction techniques from evolving streams that processes data in one-pass. The basic idea of RS is to maintain an unbiased sample through the probabilistic insertions and deletions. In the offline setting, a dataset is composed of $N$ instances, so it is trivial to construct a sample from this dataset of size $r$ where all instances have an equal probability of $r/N$ to be added to the sample.

However, in the continuous stream process, the size of the stream "$N$" cannot be known in advance and is potentially infinite. In fact, the probability of insertion reduces with the stream progression since $N$ increases, i.e., recent observations will not be added to the reservoir. So, we may have another extreme unstable solution that will not be efficient with applications where recent data may be more relevant. In very general terms, the quality of learning will degrade when the stream progresses since few instances in the reservoir remain relevant with time.

## 3.2 Algorithm

The $k$NN is one of the fundamental algorithms used for classification. In the offline setting, the $k$NN algorithm stores all the instances and searches for the $k$ closet instances to an unlabeled instance by computing a distance metric (e.g., the Euclidean distance). The prediction is therefore made by taking the most frequent class label over the $k$-nearest neighbors in the dataset. In data stream learning, since it is infeasible to maintain of all the instances from the stream due to eventual resource limitations, a basic adaptation of the $k$NN algorithm to the stream setting consists in an implementation that maintains a sliding window of a fixed size moving over the stream. The class label of a test instance has to be predicted by taking the majority vote of its $k$ closest instances from the sliding window instead of the entire instances seen so far.

As mentioned before, the sliding window is indeed an efficient technique to make some static algorithms applicable on evolving data streams. However, it only maintains the most recent instances in the stream and forgets about the history of the past instances which may potentially impact the predictive performance of some stream algorithms that uses sliding windows, such as the $k$NN.

On the other hand, a different solution where we use the $k$NN with exclusively the RS technique will lead to another problem where only historical instances will be maintained in the reservoir because, as explained before (in Sect. 3.1), the probability of insertion reduces over the progression of stream [24].

To cope with these issue, we propose in this paper a $k$NN that uses a RS technique in conjunction with a sliding window. Besides, the standard reservoir sampling technique [32] assumes a reservoir of size $r$ stores the first $r$ instances from the stream ($r$ is a predefined parameter). After that, when an instance $n$ arrives, with probability $r/n$, we insert it to the reservoir, otherwise we discard it. If this instance is kept, it will replace another one, in the reservoir, picked randomly. Consequently, the instances are inserted into the reservoir with a decreasing probability, $r/n$, that reduces with time (when $n$ increases). By the end, most of the observations in the reservoir will represent the very old data from the stream [1], since recent instances will mostly be ignored because of their low insertion probabilities.

In order to regulate the choice of the stream sampling, in [31], a RS extension has been proposed, called Biased Random/Reservoir Sampling (BRS), that ensures the insertion of recently arrived instances from the stream over instances that arrived before. Similar to the RS, the BRS technique starts by filling the reservoir with the first instances from the data stream. Then, it does not need to compute the insertion probability of later instances because they will definitely enter the reservoir. The insertion is made by replacing the instance at the position of a random number (generated randomly in the range $[0, r]$) with the new instance from the stream.

Taking into consideration these techniques, we use particularly (i) a sliding window in order to keep the most recent instances from the stream; and (ii) a biased reservoir sampling to keep track of the old instances from the stream.

---

**Algorithm 1.** RW-$k$NN algorithm.

**Symbols:** $S$: data stream; $C$: set of class labels; $k$: number of neighbors; $W$: sliding window; $w$: maximum window size; $R$: biased reservoir; $r$: maximum reservoir size.

---

1:  **function** RW-$k$NN$(S, w, r, k)$
2:      $W \leftarrow \emptyset$
3:      $R \leftarrow \emptyset$
4:      **while** $HasNext(S)$ **do**
5:          $(x, y) \leftarrow Next(S)$
6:          **if** $size(W) > 0$ **then**
7:              $N \leftarrow D_{W,k}(x)$                                          ▷ $k$NN in $W$
8:              $N \leftarrow D_{R,k}(x)$                                          ▷ Add the $k$NN in $R$
9:          **end if**
10:         $\hat{y} \leftarrow Predict_{\hat{y} \in C} D_{N,k}(x)$                  ▷ Predict the class label for $x$
11:         **if** $size(W) < w$ **then**
12:             $W \leftarrow Add((x, y))$                                       ▷ Add the instance in $W$
13:         **else**                                                              ▷ If $size(W) \geq w$
14:             $(x_0, y_0) = Remove(W[0])$                              ▷ Delete the oldest instance in $W$
15:             $W \leftarrow Add((x, y))$                                       ▷ Add the recent instance to $W$
16:             **if** $size(R) < r$ **then**
17:                 $R \leftarrow Add((x_0, y_0))$
18:             **else**                                                          ▷ If the reservoir is entirely filled
19:                 $i = Random(0, r)$                                  ▷ Pick a random index in the reservoir
20:                 $R \leftarrow Add(i, (x_0, y_0))$                        ▷ Replace the instance of index $i$
21:             **end if**
22:         **end if**
23:     **end while**
24: **end function**

---

Hence, we aim to obtain a stable $k$NN solution that takes into account, not only the recent instances for prediction, as the standard stream $k$NN, but also historical instances received so far from the stream.

In the following, we propose a $k$NN algorithm that uses the Euclidean distance function to obtain the nearest neighbors. Let us consider a window $W$, the Euclidean distance between pairs of instances, $x_i$ and $x_j$, is computed as follows:

$$D_{x_j}(x_i) = \sqrt{\|x_i - x_j\|^2}. \tag{1}$$

Likewise, the $k$-nearest neighbors distance is defined as follows:

$$D_{W,k}(x_i) = \min_{\binom{W}{k}, x_j \in W} \sum_{j=1}^{k} D_{x_j}(x_i), \tag{2}$$

where $\binom{W}{k}$ stands for the subset of $k$-nearest neighbors to the instance $x_i$ in $W$.

The overall pseudocode for the Reservoir Window $k$NN (RW-$k$NN) is presented in Algorithm 1. The RW-$k$NN is focused on the well-known test-then-train

Table 1. The algorithms we consider

| Abbr. | Classifier | Parameters |
|---|---|---|
| NB | Naive Bayes | |
| $k$NN | $k$-Nearest Neighbors | $w = 1000, k = 10$ |
| Sam$k$NN | Self adjusting memory $k$NN | $w = 1000, k = 10$ |
| $k$NN$_W$ | $k$NN with PAW | $w = 1000, k = 10$ |
| $k$NN$_W^A$ | $k$NN with PAW + ADWIN | $w = 1000, k = 10$ |
| RW-$k$NN | Reservoir-Window $k$NN | $w = 500, r = 500, k = 10$ |

setting, where every instance is firstly used for testing (prediction) and then for training. Besides, the RW-$k$NN algorithm does not build a model, hence the training instances are used to evaluate its classification performance before being used for training. The latter consists of maintaining the instances inside the window and the reservoir (rather than building a model as learners do, such as naive Bayes and decision trees).

When a new instance arrives from the stream, the prediction is made by taking the most frequent label over the nearest neighbors (line 10, Algorithm 1) retrieved from the moving window $W$ and the biased reservoir $R$ (line 7–8, Algorithm 1) using the $k$NN distance (Eq. (2)). Assuming that the class label $y$ for an instance is available before the next instance arrives, we use this information to maintain in memory the instance with its the true label.

The strategy used in the RW-$k$NN approach to keep instances begins by adding the first $w$ instances from the stream into the sliding window (line 12, Algorithm 1) and whenever the window is filled, we remove the oldest instance from the window and add the new one to it (line 14–15, Algorithm 1) in order to maintain a window that slides over the stream with the same size. To keep track of the old instances, we add each instance removed from the window (line 14, Algorithm 1) to the biased reservoir sampling $R$ and when it becomes full, we generate a random number between 0 and the size of the reservoir $r$. The instance at the index of the generated number will be replaced by the new instance (line 19–20, Algorithm 1).

## 4   Experimental Study

In this section, we look at the performance offered by our proposed RW-$k$NN algorithm against its competitors (presented in Sect. 2), in terms of predictive performance and resources usage. All the algorithms evaluated in this paper were implemented in Java using the Massive Online Analysis (MOA) software [9]. These algorithms and their parameterization are displayed in Table 1.

Based on the empirical results in [6, 10, 26], we select $w = 1000$ and $k = 10$ for all the $k$NN-based algorithms except our proposed reservoir window $k$NN (RW-$k$NN), where we divide the 1000 instances between the reservoir and the window and fix $w = 500, r = 500$.

**Table 2.** Overview of the datasets

| Dataset | #Instances | #Attributes | #Classes | Type | MF label | LF label |
|---------|-----------|-------------|----------|------|----------|----------|
| RTG | 1,000,000 | 200 | 5 | Synthetic | 46.89 | 2.70 |
| Hyper | 1,000,000 | 10 | 2 | Synthetic | 50.00 | 50.00 |
| $LED_a$ | 1,000,000 | 24 | 10 | Synthetic | 10.08 | 9.94 |
| $LED_g$ | 1,000,000 | 24 | 10 | Synthetic | 10.08 | 9.94 |
| $SEA_a$ | 1,000,000 | 3 | 2 | Synthetic | 57.55 | 42.45 |
| $SEA_g$ | 1,000,000 | 3 | 2 | Synthetic | 57.55 | 42.45 |
| $AGR_a$ | 1,000,000 | 9 | 2 | Synthetic | 52.83 | 47.17 |
| $AGR_g$ | 1,000,000 | 9 | 2 | Synthetic | 52.83 | 47.17 |
| IMDB | 120,919 | 1,001 | 2 | Real | 72.96 | 27.04 |
| Nomao | 34,465 | 119 | 2 | Real | 71.44 | 28.56 |
| Poker | 829,201 | 10 | 10 | Real | 41.55 | 2.00 |
| CNAE | 1,080 | 856 | 9 | Real | 12.00 | 12.00 |
| Har | 10,299 | 561 | 6 | Real | 19.44 | 14.06 |

MF and LF labels (in %) stands for the Most and Less Frequent class label, respectively.

## 4.1 Datasets

In our experiments, we used a diverse set of both synthetic and real datasets in different scenarios. For this paper, we used data generators and real data, where most of them have been thoroughly used in the literature to assess the performance of data stream classification algorithms. Table 2 shows an overview of the datasets used while further details are provided in the rest of this section.

**RTG** The random tree generator builds a decision tree by randomly choosing attributes as split nodes. This dataset allows customizing the number of attributes as well as the number of classes. RTG generates instances with 200 attributes.

**Hyper** The hyperplane generator used to generate streams with incremental concept drift by changing the values of its weights. We parameterize Hyper with 10 attributes and a magnitude of change equals to 0.001.

**LED** The LED generator produces 24 attributes, where 17 are considered irrelevant. The goal is to predict the digit displayed on the LED display. $LED_a$ simulates 3 abrupt drifts, while $LED_g$ simulates 3 gradual drifts.

**SEA** The SEA Generator proposed by [30] is generated with 3 attributes, where only 2 are relevant, and 2 decision classes. $SEA_a$ simulates three abrupt drifts while $SEA_g$ simulates 3 gradual drifts.

**AGR** The AGRAWAL generator [3] creates data stream with 9 attributes and 2 classes. A perturbation factor is used to add noise to the data, both $AGR_a$ and $AGR_g$ includes 10% perturbation factor. $AGR_a$ simulates 3 abrupt drifts in the generated stream while $AGR_g$ simulates 3 gradual drifts.

**IMDB** It is movie reviews dataset[2] that was first proposed for sentiment analysis, where reviews have been pre-processed, and each review is encoded as a sequence of word indexes (integers).

**Nomao** It is a large dataset that has been provided by Nomao Labs [12]. This dataset contains data that arrive from multiple sources on the web about places (name, website, localization, address, fax, etc. $\cdots$).

**Poker** The poker-hand dataset[3] consists of 829.201 instances and 10 attributes describing each hand. The class indicates the value of a hand.

**CNAE** CNAE is the national classification of economic activities dataset, initially used in [13]. It contains 1,080 instances, each of 856 attributes, representing descriptions of Brazilian companies categorized into 9 classes. The original texts were preprocessed to obtain the current highly sparse dataset.

**Har** Human Activity Recognition dataset [4] built from several subjects performing daily living activities, such as walking, walking upstairs/downstairs, sitting, standing and laying, while wearing a waist-mounted smartphone equipped with sensors. The sensor signals were preprocessed using noise filters and attributes were normalized and bounded within $[-1, 1]$.

### 4.2    Results

In our experiments, we used the test-then-train evaluation methodology [15], where every instance is used for prediction and then used for training. For fair comparison, we used the configuration previously stated in Table 1.

Table 3 reports the predictive performance results of all algorithms stated in Table 1 with the different datasets in Table 2. The accuracy is measured as the final percentage of correctly classified instances over the test-then-train evaluation. Table 4 shows the memory (measured in MB) used by the neighborhood-based classification methods, induced on synthetic and real datasets, in order to maintain the instances and/or statistical information from the stream. Table 5 reports the running time (in seconds) of the classification algorithms, which consists in the time used in order to make prediction.

Figure 1 shows the standard deviation based on the accuracies obtained over several runs with the different datasets. Figure 2 depicts the sensitivity of our proposed RW-$k$NN approach to the sizes of the biased reservoir $R$ and the sliding window $W$, $r$ and $w$ respectively.

### 4.3    Discussions

Table 3 shows that, on the overall average of all the methods, our RW-$k$NN performs good by obtaining accurate results in comparison with its competitors. We notice that the $k$NN-based methods perform much better than the baseline naive Bayes because of the naive assumption between attributes that does not hold always. One exception of naive Bayes, where it performs better, is obtained

---

[2] http://waikato.github.io/meka/datasets/.

[3] https://archive.ics.uci.edu/ml/datasets/Poker+Hand.

**Table 3.** Accuracy comparison (%)

| Dataset | RW-$k$NN | $k$NN | Sam$k$NN | $k$NN$_W$ | $k$NN$_W^A$ | NB |
|---|---|---|---|---|---|---|
| RTG | 51.03 | 48.62 | 47.00 | 49.03 | 49.03 | 74.75 |
| Hyper | 85.12 | 83.32 | 84.02 | 83.87 | 83.87 | 70.90 |
| LED$_a$ | 64.72 | 64.24 | 59.61 | 65.94 | 66.01 | 53.96 |
| LED$_g$ | 63.62 | 63.19 | 58.49 | 64.99 | 64.86 | 54.02 |
| SEA$_a$ | 87.10 | 86.79 | 85.46 | 87.17 | 87.17 | 85.37 |
| SEA$_g$ | 86.85 | 86.54 | 85.21 | 86.93 | 86.93 | 85.37 |
| AGR$_a$ | 69.29 | 62.44 | 67.43 | 64.87 | 64.88 | 65.73 |
| AGR$_g$ | 67.92 | 61.26 | 65.73 | 63.60 | 63.54 | 65.75 |
| IMBD | 71.49 | 70.23 | 72.64 | 70.07 | 70.99 | 67.61 |
| Nomao | 95.83 | 96.09 | 96.68 | 96.03 | 96.23 | 86.86 |
| Poker | 75.97 | 69.34 | 78.32 | 66.81 | 68.78 | 59.55 |
| CNAE | 78.80 | 71.75 | 81.75 | 69.07 | 69.07 | 55.92 |
| Har | 90.87 | 92.33 | 87.43 | 93.07 | 93.03 | 73.36 |
| *Overall avg* | 76.05 | 73.55 | 74.60 | 73.96 | 74.18 | 69.17 |

with the RTG dataset which is associated to the fact that the underlying data generator is based on a tree structure constructed with a random assignment of values to instances. This structure leads sometimes to uncorrelated features (200) with the class label making it a good fit to NB, which has the assumption that attributes are independent of each other knowing the class label, but not to neighborhood-based algorithms.

Surprisingly, the strategy employed in our RW-$k$NN approach is very competitive to the baselines which are coupled with an explicit concept drift mechanism, notably the Sam$k$NN and the $k$NN$_W^A$. The specific strategy, used to keep track of old instances (in the biased reservoir) while maintaining a sliding window of the most recent instances, performs well with datasets that contain drifts (e.g., with AGR$_a$, AGR$_g$). It is worth it to point out that the Sam$k$NN showed limited capabilities, in comparison to our proposal, on tracking the drifts on these datasets despite its explicit strategy employed to handle drifts.

Our approach achieves better accuracy than all the competitors on the overall average, and is slightly defeated by $k$NN$_W$ and $k$NN$_W^A$ on some datasets (e.g., SEA$_a$, SEA$_g$) where the difference is mostly after the decimal point. We also achieved better accuracy than vanilla $k$NN that only uses a sliding window (on 84.62% of the datasets), we however are defeated by only two datasets, such as the Har dataset, where we are predicting the human current activity based on information transmitted from his smartphone. So the most recent instances are more relevant for such task making this dataset a good fit for baselines that exclusively use sliding window.

For fair comparison, we assess the performance of our proposal against the neighborhood-based competitors using the same values for $k$ and $w$. In Table 4,

**Table 4.** Memory comparison (MB)

| Dataset | RW-$k$NN | $k$NN | Sam$k$NN | $k$NN$_W$ | $k$NN$_W^A$ |
|---|---|---|---|---|---|
| RTG | 20.01 | 19.99 | 11.70 | 28.25 | 28.30 |
| Hyper | 1.32 | 1.32 | 8.88 | 1.91 | 1.94 |
| LED$_a$ | 2.67 | 2.67 | 10.75 | 3.89 | 3.90 |
| LED$_g$ | 2.67 | 2.67 | 10.13 | 3.81 | 3.77 |
| SEA$_a$ | 0.84 | 0.83 | 8.81 | 1.23 | 1.18 |
| SEA$_g$ | 0.84 | 0.83 | 8.87 | 1.23 | 1.18 |
| AGR$_a$ | 1.25 | 1.25 | 9.34 | 1.86 | 1.76 |
| AGR$_g$ | 1.25 | 1.25 | 9.30 | 1.82 | 1.76 |
| IMBD | 106.19 | 106.09 | 88.00 | 151.55 | 152.27 |
| Nomao | 10.63 | 10.62 | 13.76 | 11.47 | 14.99 |
| Poker | 1.32 | 1.32 | 8.80 | 1.94 | 1.94 |
| CNAE | 92.82 | 92.73 | 86.19 | 71.25 | 71.28 |
| Har | 57.40 | 57.35 | 58.27 | 82.71 | 82.36 |
| *Overall avg* | 23.27 | 22.99 | 25.60 | 27.92 | 28.20 |

**Table 5.** Runtime comparison (s)

| Dataset | RW-$k$NN | $k$NN | Sam$k$NN | $k$NN$_W$ | $k$NN$_W^A$ |
|---|---|---|---|---|---|
| RTG | 560.63 | 6911.97 | 769.20 | 16333.34 | 29205.99 |
| Hyper | 81.64 | 265.06 | 148.81 | 635.00 | 1207.41 |
| LED$_a$ | 113.98 | 545.69 | 306.42 | 886.38 | 1709.50 |
| LED$_g$ | 112.90 | 537.36 | 305.10 | 866.66 | 2391.58 |
| SEA$_a$ | 55.71 | 105.44 | 110.15 | 257.83 | 412.63 |
| SEA$_g$ | 55.84 | 104.28 | 109.59 | 208.56 | 436.85 |
| AGR$_a$ | 72.36 | 237.68 | 84.76 | 406.85 | 747.15 |
| AGR$_g$ | 71.36 | 242.62 | 81.04 | 410.18 | 851.20 |
| IMBD | 242.16 | 5672.10 | 558.07 | 10657.48 | 18116.53 |
| Nomao | 9.76 | 71.83 | 33.86 | 99.05 | 189.47 |
| Poker | 59.40 | 223.10 | 99.01 | 374.72 | 541.89 |
| CNAE | 1.60 | 28.97 | 2.09 | 24.35 | 28.80 |
| Har | 12.99 | 195.56 | 23.84 | 227.58 | 249.70 |
| *Overall avg* | 111.56 | 1164.74 | 202.46 | 2414.46 | 4314.52 |

we observe that the RW-$k$NN uses less memory than its competitors on the overall average except for the standard $k$NN where the difference is very small.

Actually, the Sam$k$NN uses a dual memory (short-term and long-term memories) to maintain models for current and past concepts which makes it memory inefficient on several datasets. Nevertheless, on other datasets, such as RTG and

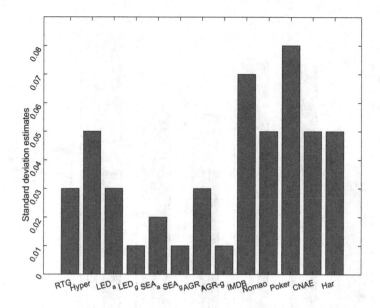

**Fig. 1.** Standard deviation of the RW-$k$NN with different datasets.

IMDB mainly the high-dimensional ones, it is more efficient due to its cleaning process that could decide to remove instances from the short-term and the long-term memories as the distribution may change which leads to an effective smaller window size than the one kept by the RW-$k$NN and the other $k$NN-based methods [26]. Besides, the $k$NN$_W$ maintains, other than the probabilistic approximate window, statistics related to the probability of insertion of a new instance to the window, and the $k$NN$_W^A$, on top of that, uses the drift detector, ADWIN, to handle concept drifts. Therefore, with its simple strategy of keeping a small reservoir and window, our proposed RW-$k$NN approach is memory-efficient.

The results in Table 5 show that the RW-$k$NN outperforms the standard $k$NN, Sam$k$NN, $k$NN$_W$, and $k$NN$_W^A$ in terms of execution time thanks to the strategy of using a window and a reservoir instead of one big window. In fact, at a prediction time, the pairwise distance calculations in a big window (of 1000 instances in the case of the baselines) are more costly, i.e., slower, than calculations in a small window and a reservoir (each of 500 instances). Thus, computing the $k$NN distance in two small data structures, such as the reservoir and window, provides faster result than extracting the $k$ nearest neighbors from one big window. We observe that with high-dimensional datasets (e.g., RTG, IMDB, CNAE), our RW-$k$NN offers a very significant gain.

Similar behavior, in terms of accuracy and computational demand, has been obtained with different configurations (different values of $k$, $w$, and $r$) in comparison with the competitors. Our RW-$k$NN uses feasible computational resources and produces good accurate results while implicitly dealing with drifts.

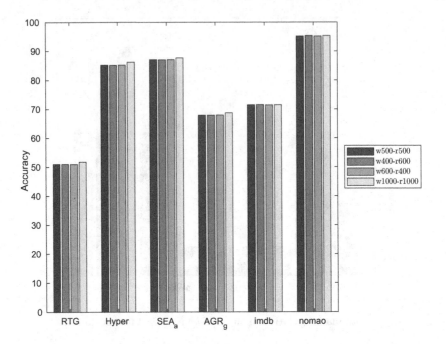

**Fig. 2.** Sensitivity of the RW-$k$NN approach to parameters $r$ and $w$.

Due to the stochastic nature of the biased reservoir and therefore our RW-$k$NN approach, we study the standard deviation in Fig. 1. For all the datasets, the proposed approach has a small standard deviation (very close to zero), i.e., for all the runs, the achieved accuracies are close to the mean reported in this paper.

To evaluate the sensitivity of our proposed RW-$k$NN approach to parameters, we performed evaluation with some datasets using different size values of the window and the reservoir as shown in Fig. 2. We remark that the best accuracy is obtained using the biggest reservoir and window sizes (1000 each: the yellow bar), which slightly more accurate than the performance with smaller reservoir and window. This is an expected result because the more instances we have, the better is the predictive performance with the $k$NN algorithms. Still, the difference with the other configurations ($w = 500, r = 500$; $w = 600, r = 400$; and $w = 400, r = 600$) is not huge which reveals the insensitivity of RW-$k$NN to the sizes of $R$ and $W$. Based on the results of different sizes with all the datasets, we concluded that the size of the reservoir should be at least equal to the window size in order to keep track of more instances from the past.

# 5   Conclusions

In this paper, we introduced the reservoir window $k$NN approach, RW-$k$NN, for data stream classification that is built upon the use of a sliding window, to maintain recent instances from the stream, and biased random sampling, to keep track of old instances from the stream in a reservoir. The prediction for incoming instances is done by taking the majority vote over the nearest neighbors extracted from the window and the biased reservoir sampling.

We used a range of different datasets to show empirical evidence that our proposed approach is very competitive and tends to improve the predictive performance while consuming less computational resources in comparison with state-of-the-art baselines. On top of that, we showed its ability in handling the presence of concept drifts without being coupled with a particular drift detector, thanks to its reservoir-window strategy that keeps track of old and recent instances from the stream.

We want to pursue our promising approach further by investigating the development of an ensemble method that uses $k$NN with multiple reservoirs which might lead to better predictive performance.

**Acknowledgements.** This work has been carried out in the frame of a cooperation between Huawei Technologies France SASU and Télécom Paris (Grant no. YBN2018125164).

# References

1. Aggarwal, C.C.: On biased reservoir sampling in the presence of stream evolution. In: Very Large Data Bases (VLDB), pp. 607–618 (2006)
2. Aggarwal, C.C., Philip, S.Y.: A survey of synopsis construction in data streams. In: Aggarwal, C.C. (eds.) Data Streams. ADBS, vol. 31, pp. 169–207. Springer, Boston (2007). https://doi.org/10.1007/978-0-387-47534-9_9
3. Agrawal, R., Imielinski, T., Swami, A.: Database mining: a performance perspective. Trans. Knowl. Data Eng. (TKDE) 5(6), 914–925 (1993)
4. Anguita, D., Ghio, A., Oneto, L., Parra, X., Reyes-Ortiz, J.L.: Human activity recognition on smartphones using a multiclass hardware-friendly support vector machine. In: Bravo, J., Hervás, R., Rodríguez, M. (eds.) IWAAL 2012. LNCS, vol. 7657, pp. 216–223. Springer, Heidelberg (2012). https://doi.org/10.1007/978-3-642-35395-6_30
5. Bahri, M., Bifet, A., Maniu, S., Gomes, H.M.: Survey on feature transformation techniques for data streams. In: International Joint Conference on Artificial Intelligence (2020)
6. Bahri, M., Bifet, A., Maniu, S., de Mello, R., Tziortziotis, N.: Compressed k-nearest neighbors ensembles for evolving data streams. In: European Conference on Artificial Intelligence (ECAI). IEEE (2020)
7. Bahri, M., Maniu, S., Bifet, A.: Sketch-based Naive Bayes algorithms for evolving data streams. In: International Conference on Big Data, pp. 604–613. IEEE (2018)
8. Bifet, A., Gavalda, R.: Learning from time-changing data with adaptive windowing. In: International Conference on Data Mining (ICDM), pp. 443–448. SIAM (2007)

9.  Bifet, A., Holmes, G., Kirkby, R., Pfahringer, B.: MOA: massive online analysis. J. Mach. Learn. Res. (JMLR) **11**(May), 1601–1604 (2010)
10. Bifet, A., Pfahringer, B., Read, J., Holmes, G.: Efficient data stream classification via probabilistic adaptive windows. In: Symposium On Applied Computing (SIGAPP), pp. 801–806. ACM (2013)
11. Caiming, Z., Yong, C.: A review of research relevant to the emerging industry trends: industry 4.0, IoT, blockchain, and business analytics. J. Ind. Integr. Manag. **5**, 165–180 (2020)
12. Candillier, L., Lemaire, V.: Design and analysis of the Nomao challenge active learning in the real-world. In: ALRA, Workshop ECML-PKDD. sn (2012)
13. Ciarelli, P.M., Oliveira, E.: Agglomeration and elimination of terms for dimensionality reduction. In: International Conference on Intelligent Systems Design and Applications, pp. 547–552. IEEE (2009)
14. Da Xu, L., He, W., Li, S.: Internet of things in industries: a survey. IEEE Trans. Industr. Inf. **10**(4), 2233–2243 (2014)
15. Dawid, A.P.: Present position and potential developments: some personal views statistical theory the prequential approach. J. R. Stat. Soc. Ser. A (General) **147**(2), 278–290 (1984)
16. Domingos, P., Hulten, G.: Mining high-speed data streams. In: SIGKDD International Conference on Knowledge Discovery & Data Mining (2000)
17. Friedman, N., Geiger, D., Goldszmidt, M.: Bayesian network classifiers. Mach. Learn. **29**(2–3), 131–163 (1997)
18. Gama, J.: Knowledge Discovery from Data Streams. CRC Press, Boca Raton (2010)
19. Gama, J., Fernandes, R., Rocha, R.: Decision trees for mining data streams. Intell. Data Anal. (IDA) **10**(1), 23–45 (2006)
20. Gama, J., Rocha, R., Medas, P.: Accurate decision trees for mining high-speed data streams. In: SIGKDD International Conference on Knowledge Discovery & Data Mining, pp. 523–528. ACM (2003)
21. Gomes, H.M., et al.: Adaptive random forests for evolving data stream classification. Mach. Learn. **106**, 1469–1495 (2017). https://doi.org/10.1007/s10994-017-5642-8
22. Gomes, H.M., Barddal, J.P., Enembreck, F., Bifet, A.: A survey on ensemble learning for data stream classification. Comput. Surv. (CSUR) **50**(2), 23 (2017)
23. Gomes, H.M., Read, J., Bifet, A.: Streaming random patches for evolving data stream classification. In: International Conference on Data Mining (ICDM). IEEE (2019)
24. Haas, P.J.: Data-stream sampling: basic techniques and results. In: Data Stream Management. DSA, pp. 13–44. Springer, Heidelberg (2016). https://doi.org/10.1007/978-3-540-28608-0_2
25. Hand, D.J., Mannila, H., Smyth, P.: Principles of Data Mining. MIT Press, Cambridge (2001)
26. Losing, V., Hammer, B., Wersing, H.: KNN classifier with self adjusting memory for heterogeneous concept drift. In: International Conference on Data Mining (ICDM), pp. 291–300. IEEE (2016)
27. Ng, W., Dash, M.: Discovery of frequent patterns in transactional data streams. In: Hameurlain, A., Küng, J., Wagner, R., Bach Pedersen, T., Tjoa, A.M. (eds.) Transactions on Large-Scale Data- and Knowledge-Centered Systems II. LNCS, vol. 6380, pp. 1–30. Springer, Heidelberg (2010). https://doi.org/10.1007/978-3-642-16175-9_1

28. Oza, N.C., Russell, S.: Experimental comparisons of online and batch versions of bagging and boosting. In: SIGKDD International Conference on Knowledge Discovery & Data Mining, pp. 359–364 (2001)
29. Read, J., Bifet, A., Pfahringer, B., Holmes, G.: Batch-incremental versus instance-incremental learning in dynamic and evolving data. In: Hollmén, J., Klawonn, F., Tucker, A. (eds.) IDA 2012. LNCS, vol. 7619, pp. 313–323. Springer, Heidelberg (2012). https://doi.org/10.1007/978-3-642-34156-4_29
30. Street, W.N., Kim, Y.: A streaming ensemble algorithm (SEA) for large-scale classification. In: SIGKDD International Conference on Knowledge Discovery & Data Mining, pp. 377–382. ACM (2001)
31. Tabassum, S., Gama, J.: Sampling massive streaming call graphs. In: ACM Symposium on Applied Computing, pp. 923–928 (2016)
32. Vitter, J.S.: Random sampling with a reservoir. Trans. Math. Softw. (TOMS) 11(1), 37–57 (1985)

# Statistical Analysis of Pairwise Connectivity

Georg Krempl[1]($\boxtimes$)(iD), Daniel Kottke[2](iD), and Tuan Pham[2](iD)

[1] Utrecht University, 3584 CC Utrecht, The Netherlands
g.m.krempl@uu.nl
[2] Kassel University, 34121 Kassel, Germany
{daniel.kottke,tuan.pham}@uni-kassel.de
https://www.uu.nl/staff/GMKrempl

**Abstract.** Analysing correlations between streams of events is an important problem. It arises for example in Neurosciences, when the connectivity of neurons should be inferred from spike trains that record neurons' individual spiking activity. While recently some approaches for inferring delayed synaptic connections have been proposed, they are limited in the types of connectivities and delays they are able to handle, or require computation-intensive procedures. This paper proposes a faster and more flexible approach for analysing such delayed correlated activity: a statistical **A**nalysis of the **C**onnectivity of spiking **E**vents (ACE), based on the idea of hypothesis testing. It first computes for any pair of a source and a target neuron the inter-spike delays between subsequent source- and target-spikes. Then, it derives a null model for the distribution of inter-spike delays for *uncorrelated* neurons. Finally, it compares the observed distribution of inter-spike delays to this null model and infers pairwise connectivity based on the Pearson's $\chi^2$ test statistic. Thus, ACE is capable to detect connections with a priori unknown, non-discrete (and potentially large) inter-spike delays, which might vary between pairs of neurons. Since ACE works incrementally, it has potential for being used in online processing. In an experimental evaluation, ACE is faster and performs comparable or better than four baseline approaches, in terms of AUPRC (reported here), F1, and AUROC (reported on our website), for the majority of the 11 evaluated scenarios.

**Keywords:** Machine learning from complex data · Event streams · Neurosciences · Neural connectomics · Connectivity inference

## 1 Introduction

An important problem in various applications is detecting correlations between streams of events. This is of particular importance in Neurosciences, where it arises for example when inferring the functional connectivity of neurons [PGM67]. Given spike trains with recordings of the neuron's individual spike activity, the objective is to detect correlations between the spike activities of pairs or networks of neurons. Most of the existing approaches are designed for

ⓒ Springer Nature Switzerland AG 2021
C. Soares and L. Torgo (Eds.): DS 2021, LNAI 12986, pp. 138–148, 2021.
https://doi.org/10.1007/978-3-030-88942-5_11

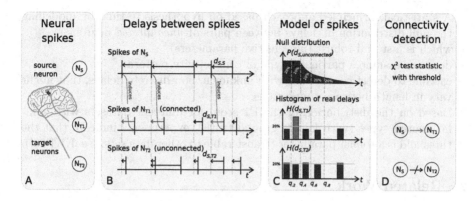

**Fig. 1.** Schematic visualisation of ACE (**A**nalysis of **C**onnectivity of spiking **E**vents).

detection of correlated synchronous activity, considering solely events within the same discretised time interval. More recently, the detection of delayed synaptic connections has gained attention. However, existing methods have limitations in the types of connectivities and delays they are able to handle, for example due to requiring range-parameters for expected delays, or they require computation-intensive procedures, such as computing cross-correlation histograms or performing cross-evaluations of parameter values.

We propose a faster and more robust approach for the detection of delayed correlated activity. The principle of this statistical **A**nalysis of **C**onnectivity of spiking **E**vents (ACE) of neural spikes is illustrated in Fig. 1. This statistical approach follows the idea of hypothesis testing: Starting (A) with data in the form of spike trains that are recorded for several neurons, the aim is to infer for any pair of source ($N_S$) and target neuron ($N_T$) the pairwise connectivity between them. For this purpose, we compute in step (B) for any pair of neurons the inter-spike intervals, i.e. the delays between their subsequent spikes ($d_{S,S}$, $d_{S,T1}$ and $d_{S,T2}$). In the third step (C), we use the inter-spike intervals of a potential source neuron to determine the null distribution of delays for *unconnected* target neurons (shown on top as $P(d_{S,unconnected})$). If a target neuron is *not* connected to the source neuron, the observed distribution of the inter-spike intervals should follow this distribution (shown on the bottom as $H(d_{S,T2})$). In contrast, if this observed distribution (shown at the center as $H(d_{S,T1})$) differs sufficiently, we assume that these two neurons are connected. Thus, in step (D) we use the Pearson's $\chi^2$ test statistic to determine the connectivity for each pair of neurons.

As a consequence, and in contrast to existing methods, our approach works incrementally. It requires neither a cross-correlation histogram (faster), nor pre-specified range-parameters for expected delays (more flexible), but assumes that the signal reaches the target neuron faster than the source neuron fires again. This makes it particularly interesting for online processing. Summarising, the contributions of our statistical **ACE** approach are:

- A statistical, principled approach based on hypothesis testing, by modelling the null distribution of delays between pairs of *unconnected* neurons,
- which is fast and robust (no sensitive parameters),
- does not assume a particular type of connectivity pattern,
- capable to detect connections with unknown, non-discrete delays, that might vary in length between connections.
- Based on the distribution of the F1 score for different datasets with similar neuron types (with known ground truth), experiments indicate that the threshold is a robust parameter (transferable to other data sets, see [KKP20]).

## 2   Related Work

There is a rich literature on analysing neuronal spike train data for inferring connectivity. This comprises recent reviews, e.g., [MYD18], and a recent machine learning challenge on neural connectomics [BGL+17]. Following [MYD18], we distinguish model-based approaches from model-free ones. An important limitation of model-based approaches is that they rely on assumptions on the data generating process. Their structure and function depend on a large number of factors, resulting in a variety of models and approaches [MYD18]. These include autoregressive models, which are fast but assume a directed linear interaction and generalised linear models, which despite recent extensions to handle transmission delays remain limited to small and uniform delays within the network.

In contrast, model-free approaches rely on principles from descriptive statistics, information theory, and supervised learning. This comprises approaches based on the correlation between the activities of neurons, which was the key component in the winning approach of the neural connectomics challenge [BGL+17]. While in the simplest form only simultaneous spikes are considered, the extension to cross correlation allows for a delay $\tau$ between spike times [IHH+11]. Extending the idea presented at the neural connectomics challenge, [Moh14] suggested to use inverse covariance estimates together with an initial convolution filter to preprocess the data. The convolution kernel and other parameter are learned by optimising the binomial log-likelihood function [Moh14] on a training data set, where ground truth is known. In the above mentioned challenge they report comparable AUC scores and accuracy compared to the winner, while being noticeably faster.

A further group of model-free approaches are based on information theoretic approaches. Their applicability for neural connectomics was investigated in [GNMM09], comparing methods based on Mutual Information, Joint-Entropy, Transfer Entropy (TE) and Cross-Correlation. This study revealed Transfer Entropy and Joint-Entropy being the best of the aforementioned methods. Transfer Entropy (TE) is equivalent to Granger causality for Gaussian variables [BBS09], which describes a statistical hypothesis test that measures the ability of predicting future events of a time series based on past events of related time series. However, pairwise Granger causality only detects direct correlations. This leads to problems when two neurons are driven by a common third neuron with different delays [CDHD08]. In [IHH+11], two different TE-based approaches are proposed, one using Delayed TE (DTE) and one using Higher Order TE (HOTE). DTE calculates the TE for multiple delays (e.g. 1–30 ms), which is extended in HOTE by considering multiple bins for each delay.

Recently, statistical methods for analysing synchrony across neurons were reviewed in [HAK13]. Therein, it is emphasised that the statistical identification of synchronous spiking presumes a null model that describes spiking without synchony. Then, if the observed synchrony is not consistent with the distribution under the null model, i.e. there is more synchrony than expected by chance, the null hypothesis of no synchrony is rejected. For hypothesis testing, [HAK13] focus their discussion on deriving the test statistics from a cross-correlation histogram (CCH). This shows the observed frequency of different time delays between two spike trains, but is scaling dependent and computationally costly.

Convolutional Neural Networks have also been used [Rom17] to learn connectivity directly from calcium imaging data. This scored worse than the correlation-based approaches, and required extremely high computation time [BGL+17].

## 3    A Method for Analysing Potential Connectivity in Events

The new method ACE (**A**nalyser for **C**orrelated spiking **E**vents) is a statistical approach following the idea of hypothesis testing. Observing the inter-spike intervals of a potential source neuron, we determine the null distribution of delays for *unconnected* target neurons. If the real observed distribution differ sufficiently (using the Pearson's $\chi^2$ test statistic), we infer that these two neurons are connected. In contrast to existing methods, our approach works incrementally without using a cross-correlation histogram (faster) and without providing range-parameters for expected delays (more flexible).

The complete analysis pipeline of our algorithm is provided in Fig. 1: In the first step, we compute the delays of two consecutive spikes of all neurons to determine the neuron's parameters $\lambda$ and $RP$ (Sect. 3.1, Fig. 1B (top)). Thereby, we are able to reconstruct the neuron's delay distribution and to determine the null model for the delay distribution between two neurons (Sect. 3.2, Fig. 1C (top)). As we estimate the null model in advance, we have information on the expected delay distribution and can use the histogram with intervals according to the quantiles to estimate the distribution of observed delays (Sect. 3.3). Then, the real delays from the source neurons to the target neurons are determined and the histogram is completed (Fig. 1C (bottom)). The $\chi^2$ statistic provides a score for distinguishing connected and unconnected neurons (Sect. 3.4, Fig. 1D). A threshold is used to determine if the connection score was sufficiently large. A detailed description follows in the next subsections, while the threshold specification is discussed in the evaluation section.

### 3.1    Modeling Spiking Behaviour of a Single Neuron

As we will show in Sect. 3.2, the null-distribution for the delays between unconnected source and target neurons depends on the distribution of time intervals (delays) $d$ between consecutive spikes of the source neuron $(N_S)$. To model the random variable $(X_{N_S \to N_S})$ corresponding to these delays, we use an exponential distribution as a simplification of the gamma distribution, in accordance to [Pil09]. This yields two parameters: the refractory period $RP$, describing the time a neuron is inhibited to spike again, and the firing rate $\lambda$, defining the shape of the exponential distribution. The probability density function (pdf) of $X_{N_S \to N_S} = EXP(\lambda) + RP$ and its first two moments are given as:

$$f_{N_S \to N_S}(d) = \begin{cases} \lambda \exp(-\lambda(d - RP)) & d \geq RP \\ 0 & d < RP \end{cases} \tag{1}$$

$$E(X_{N_S \to N_S}) = E(EXP(\lambda) + RP) = E(EXP(\lambda)) + RP = 1/\lambda + RP \tag{2}$$

$$V(X_{N_S \to N_S}) = V(EXP(\lambda) + RP) = V(EXP(\lambda)) = 1/\lambda^2 \tag{3}$$

The expectation value $E(X_{N_S \to N_S})$ and the variance $V(X_{N_S \to N_S})$ can be incrementally calculated [BDMO03] to find the values for both parameters $RP$ and $\lambda$.

$$\lambda = 1/\sqrt{V(X_{N_S \to N_S})} \qquad\qquad RP = E(X_{N_S \to N_S}) - 1/\lambda \tag{4}$$

## 3.2   Determining the Null-Distribution for Uncorrelated Neurons

Our approach follows the idea of a statistical test: Instead of deriving models for cases when a source neuron $(N_S)$ is connected to a target neuron $(N_T)$, we develop a model to describe the delays $d$ if $N_S$ and $N_T$ are *unconnected*. Observing a spike at the target neuron at time $t$, and knowing the time $t_S$ of the source neuron's last spike, this delay is $d = t - t_S$.

If the source neuron $(N_S)$ is not connected to the target neuron $(N_T)$, spikes of $N_T$ seem to appear randomly from the perspective of $N_S$ as they are *independent*. Instead of using the real spike time points, we could also use an equal number of randomly chosen time points. Thus, the null-distribution solely depends on the firing frequency of the source neuron, which is defined by its parameters $RP$ and $\lambda$. Determining the distribution of delays $d = t - t_S$ corresponds to estimating the probability $P(X_{N_S \to N_S} > d)$ that $N_S$ has not spiked again within $[t_s, t]$:

$$P(X_{N_S \to N_S} > d) = 1 - \int_0^d f_{N_S \to N_S}(d')\, \mathrm{d}d' = \begin{cases} \exp(-\lambda(d - RP)) & d \geq RP \\ 1 & 0 \leq d < RP \end{cases} \tag{5}$$

$$\int_0^\infty P(X_{N_S \to N_S} > d)\, \mathrm{d}d = RP + \frac{1}{\lambda} \int_0^\infty \lambda \exp(-\lambda d)\, \mathrm{d}d = RP + \frac{1}{\lambda} \tag{6}$$

Using the normalised probability from above, we obtain the distribution $X_{N_S \to N_?}$ of delays between $N_S$ and an unconnected neuron $N_?$:

$$f_{N_S \to N_?}(d) = \frac{P(X_{N_S \to N_S} > d)}{\int_0^\infty P(X_{N_S \to N_S} > d')\, \mathrm{d}d'} = \begin{cases} \frac{\exp(-\lambda(d - RP))}{RP + 1/\lambda} & d \geq RP \\ \frac{1}{RP + 1/\lambda} & 0 \leq d < RP \end{cases} \tag{7}$$

Summarising, this null model describes the distribution of delays between two unconnected neurons. Hence, our model is based on (but not similar) to the interspike intervals of neuron $N_S$ which depends on the refractory period ($RP$) and the firing rate ($\lambda$).

### 3.3 Estimating the Distribution of Observed Delays

To compare the true distribution of observed delays to the distribution under the null model, we build a histogram with $B$ bins such that every bin should contain the same amount of delays following the null distribution. The delay interval of bin $b \in \{1, \ldots, B\}$ is given in Eq. 8 with $F^{-1}$ being the quantile function (inverse cumulative distribution function) of the null distribution:

$$\mathcal{I}_b = \left[ F^{-1}\left(\frac{b-1}{B}\right), F^{-1}\left(\frac{b}{B}\right) \right[ \tag{8}$$

Given the source neuron's $RP$ and $\lambda$, this quantile function is:

$$F^{-1}(q) = \begin{cases} RP - \frac{\ln\left(1-(q-\frac{RP}{RP+1/\lambda})\cdot(\lambda RP+1)\right)}{\lambda} & q > \frac{RP}{RP+1/\lambda} \\ q \cdot (RP + 1/\lambda) & q \leq \frac{RP}{RP+1/\lambda} \end{cases} \tag{9}$$

### 3.4 Infering Connectivity Using the Pearson's $\chi^2$-test Statistic

Following the null hypothesis (neurons are not connected), the previously mentioned histogram should be uniformly distributed. Hence, the frequencies $H_b$ of bin $b$ should be similar to $H_b \approx N/B$ ($N = \sum H_b$, which is the total number of delays). Our method uses the Pearson's $\chi^2$-Test statistic to find a threshold for distinguishing connected and unconnected neurons.

$$\chi^2 = \sum_{b=1}^{B} \frac{(H_b - \frac{N}{B})^2}{\frac{N}{B}} \tag{10}$$

Instead of calculating the $p$ value, we directly use the $\chi^2$ statistic to determine a threshold as the degrees of freedom ($B-1$) are similar for every pair of neurons.

## 4 Experimental Evaluation

We evaluate our algorithm to the most used baseline techniques [MYD18] regarding its detection quality and its robustness to parameters like the detection threshold. All code and data are available at our repository[1].

---

[1] https://bitbucket.org/geos/ace-public.

### 4.1  Baseline Algorithms and Performance Scores

We compare our algorithm with the method proposed by [Moh14] (denoted as
$IC$), which is as the winner of the Neural Connectomics Challenge based on
inverse covariance but faster, and the higher order transfer entropy ($HOTE$)
approach [IHH+11]. Besides the threshold that all connectivity detection algorithms have in common, both methods require an additional binning parameter
that influences the performance. Following the suggestions, we use realistic standard parameters and two variants with 20 and 50 bins. The experiments were
implemented in MATLAB (HOTE was provided by the author of [IHH+11]).

As in the challenge and common in literature, we compare the algorithms' connection scores using the Area Under the Precision-Recall Curve
(AUPRC) (see [Pow11] and [MYD18]). This describes the relationship between
precision $\frac{TP}{TP+FP}$ and recall $\frac{TP}{TP+FN}$ at different thresholds [MYD18].

### 4.2  Sensitivity of the Algorithms

To evaluate the sensitivity of algorithms to different neural patterns, in our first
series of experiments we used artificially generated data with varying neural
parameters, which are summarised in Table 1. Each parameter range has been
chosen according to animal studies [Izh06] and related comparisons [MYD18].

To evaluate the detection capabilities for each algorithm, we show the
**AUPRC scores** in Table 1. Our algorithm outperforms all competitors except
for the data set NU_H with higher number of neurons (equal performances with
the HOTE approach) and the data set with high delays (DE_H) which is more
difficult for all algorithms. To explain our poor performance on the latter, we
need to recall that the self-initiated firing rate is between 17 ms and 36 ms,
calculated as the interval given in expected latency (default $[10, 25)$) plus the
RP ($[7, 11)$) for the DE_H data set. If we observe delays longer than the source
neuron's inter-spike intervals (here: $9\,\text{ms} \leq d < 120\,\text{ms}$) it is likely that the
source neuron spiked again before its signal reaches the target neuron. Hence,
we are not able to link the spike of the source with the spike of the target neuron which makes it impossible to find the respective connection. Fortunately,
this behaviour is rare in real neural systems [Izh06].

Unfortunately, it is not possible to set the detection threshold of the algorithms to a fixed value. In this section, we aim to evaluate the sensitivity of the
respective parameter mentioned in Table 1. Therefore, we use each of the three
different configurations (low, mid, high) as one fold of an experiment. For each
fold, we tune the detection threshold on the remaining folds and calculate the
*F1 score* accordingly. Except for the data set with varying delays, our approach
shows superior performance than the baseline algorithms although the AUPRC
score differences have not been that large. This indicates that our detection score
is more robust to changes in the number of neurons (NU), the latency (LA), the
number of connections (CO) and noise (NO). The F1 scores for the delay (DE)
data sets for our approach ACE are: DE_L 0.6164, DE_M = ST 0.7539 and

DE_H $= 0.1100$. We see that the previously discussed low performance on DE_H is the reason for the low mean score.

In the experiments, we presented two variants of IC and HOTE as they need an additional parameter which highly influences the results. Our **parameter** $B$ has not such an influence on the performance and the runtime (see also Sect. 4.3) of ACE. This is visualised in Fig. 3 which plots the AUPRC w.r.t. the number of bins $B$ on the ST data set. The plot shows that the performance generally increases with higher resolution, but only marginally beyond $B = 100$, our default for further experiments.

**Table 1.** Area Under the Precision-Recall Curve values for all algorithms on data sets with varying characteristics. The default values for unvaried characteristics are 100 neurons (in the dataset), [10, 25] ms + RP expected latency (between two consecutive spikes), 1% as relative number of connections, [5, 9] ms delay (between two connected neurons), and +0 ms noise (when determining spike times). The length of the spike stream is $30s$ and the refractory period (RP) is uniformly between 7 and 11 ms. The data set name is composed by the characteristic abbreviation and a suffix for either low, mid or high. Thus, the setting with 200 Neurons is called NU_H.

|       | Characteristic | ACE | IC20 | IC50 | HOTE20 | HOTE50 |
|-------|----------------|-----|------|------|--------|--------|
| ST    | Defaults | **0.8626** | 0.2028 | 0.1409 | 0.7832 | 0.7761 |
| NU_L  | 50 neurons | **0.8519** | 0.0834 | 0.0616 | 0.6310 | 0.6186 |
| NU_H  | 200 neurons | 0.9504 | 0.2586 | 0.1038 | 0.9552 | **0.9553** |
| LA_L  | [1, 10) ms + RP latency | **0.8652** | 0.0219 | 0.0138 | 0.5579 | 0.5546 |
| LA_H  | [25, 50) ms + RP latency | **0.8135** | 0.2500 | 0.1430 | 0.8126 | 0.8119 |
| CO_L  | 5‰ connections | **0.8850** | 0.2656 | 0.1620 | 0.8738 | 0.8739 |
| CO_H  | 2% connections | **0.8086** | 0.0406 | 0.0317 | 0.5952 | 0.5894 |
| DE_L  | [2, 5) ms delay | 0.7598 | 0.2770 | 0.1528 | **0.7671** | 0.7670 |
| DE_H  | [9, 120) ms delay | 0.1334 | 0.1687 | 0.1053 | 0.1505 | **0.3669** |
| NO_M  | +[0, 3) ms noise | **0.7838** | 0.2248 | 0.1225 | 0.7246 | 0.7224 |
| NO_H  | +[0, 5) ms noise | **0.8518** | 0.2246 | 0.1021 | 0.8322 | 0.8309 |

## 4.3 Computational Complexity

To provide a computational run time complexity bound of ACE, let $N$ denote the number of neurons, $M$ the number of spikes over all neurons, and $B$ the number of bins used in the histogram. ACE's first step is estimating the refractory period $RP$ and firing rate $\lambda$ for each neuron by iterating once over all its spikes, suming up to $O(M)$ constant time operations for all neurons. Second, for each neuron's bin the frequencies according to its null model are computed, resulting overall in $O(N \cdot B)$. Third, the histograms of all neurons are updated after each spike, requiring to insert the observed delay into the corresponding bin. Using a k-d-tree, this requires overall $O(M \cdot N \cdot \log(B))$. Fourth, the $\chi^2$-test statistic is computed and tested for each pair of neurons, requiring $O(N^2 \cdot B)$ operations. The last two steps dominate, giving an overall complexity of $O(M \cdot N \cdot \log(B))$ or $O(N^2 \cdot B)$, respectively. In practice, the third step might be the bottleneck, as the number of spikes depends on the number of neurons, i.e., $M > N$, and the number of bins is small (e.g., $B = 100$). In contrast, HOTE's time complexity is $O(N^2 \cdot (F \cdot D \cdot R + 2^R))$, with firing rate $F$, recording duration after discretisation $D$, and $R$ being the total order $(k + l + 1)$ used in the calculations. IC's time complexity is $O(N^2 \cdot T + N^2 \cdot \log(N))$, with $T$ as number of considered time lags.

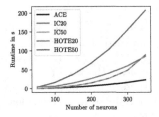
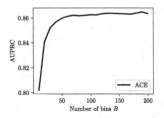

**Fig. 2.** Runtime over data sets with varying number of neurons.

**Fig. 3.** AUPRC scores of ACE with varying bin sizes on the ST dataset.

Figure 2 shows the empirical runtime[2] of all strategies according to the number of neurons in the data set. Those runtimes are obtained by creating datasets that duplicate the spike trains from NE_L and evaluating them 10 times. One can see that ACE is the fastest. IC20 and IC50 only differ slightly in runtime. HOTE20's runtimes are higher than the ones from IC20 and IC50 for $N \leq 300$. The number of bins affects HOTE's runtime, making HOTE50 the slowest.

## 5    Conclusion

This paper proposed ACE for detecting correlated, but delayed event patterns in streams, e.g., for detecting delayed connectivity of neurons. Using a null model for the distribution of inter-spike delays of *uncorrelated* neurons, ACE employs principles from hypothesis testing. Against this null-distribution, the distribution of observed inter-spike delays is compared using a Pearson's $\chi^2$ test statistic.

In an experimental evaluation, this algorithm was compared against recently proposed approaches based on inverse covariance and higher order transfer entropy, on data sets with varying characteristics based on our own data generator. On all data sets, ACE is faster and performs comparable or better in terms of AUPRC, F1 and AUROC (see [KKP20]) score, except for networks with very long inter-spike delays that interfere with uncorrelated spike activity. In particular, ACE performs also better on the publicly available, state-of-the-art benchmark data generator with realistic spike characteristics. ACE has only

---

[2] Experiments were performed using a Intel(R) Core(TM) i7-6820HK CPU @ 2.70 GHz, 16 GB RAM.

two parameters, both being very robust and transferable between data sets of similar characteristic. ACE is fast and flexible, allowing to detect connections with a priori unknown, non-discrete delays, that might vary in length between connections. Furthermore, due to its incremental nature, ACE has potential for being used in online processing.

# References

BBS09. Barnett, L., Barrett, A.B., Seth, A.K.: Granger causality and transfer entropy are equivalent for Gaussian variables. Phys. Rev. Lett. **103**(23), 238701 (2009)

BDMO03. Babcock, B., Datar, M., Motwani, R., O'Callaghan, L.: Maintaining variance and k-medians over data stream windows. In: Proceedings of the 22nd ACM SIGMOD-SIGACT-SIGART Symposium on Principles of Database Systems, pp. 234–243. ACM (2003)

BGL+17. Battaglia, D., Guyon, I., Lemaire, V., Orlandi, J., Ray, B., Soriano, J. (eds.): Neural Connectomics Challenge. TSSCML, Springer, Cham (2017). https://doi.org/10.1007/978-3-319-53070-3

CDHD08. Cadotte, A.J., DeMarse, T.B., He, P., Ding, M.: Causal measures of structure and plasticity in simulated and living neural networks. PloS One **3**(10), e3355 (2008)

GNMM09. Garofalo, M., Nieus, T., Massobrio, P., Martinoia, S.: Evaluation of the performance of information theory-based methods and cross-correlation to estimate the functional connectivity in cortical networks. PloS One **4**(8), e6482 (2009)

HAK13. Harrison, M.T., Amarasingham, A., Kass, R.E.: Statistical identification of synchronous spiking. In: Di Lorenzo, P.M., Victor, J.D. (eds.) Spike Timing: Mechanisms and Function, pp. 77–120 (2013)

IHH+11. Ito, S., Hansen, M.E., Heiland, R., Lumsdaine, A., Litke, A.M., Beggs, J.M.: Extending transfer entropy improves identification of effective connectivity in a spiking cortical network model. PLOS One **6**(11), e27431 (2011)

Izh06. Izhikevich, E.M.: Polychronization: computation with spikes. Neural Comput. **18**(2), 245–282 (2006)

KKP20. Krempl, G., Kottke, D., Minh, T.P.: ACE: a novel approach for the statistical analysis of pairwise connectivity. arXiV (2020)

Moh14. Mohler, G.: Learning convolution filters for inverse covariance estimation of neural network connectivity. In: Ghahramani, Z., Welling, M., Cortes, C., Lawrence, N.D., Weinberger, K.Q. (eds.) Advances in Neural Information Processing Systems 27, pp. 891–899 (2014)

MYD18. de Abril, I.M., Yoshimoto, J., Doya, K.: Connectivity inference from neural recording data: challenges, mathematical bases and research directions. Neural Netw. **102**, 120–137 (2018)

PGM67. Perkel, D.H., Gerstein, G.L., Moore, G.P.: Neuronal spike trains and stochastic point processes: II. Simultaneous spike trains. Biophys. J. **7**(4), 419–440 (1967)

Pil09. Pillow, J.W.: Time-rescaling methods for the estimation and assessment of non-Poisson neural encoding models. In: Advances in Neural Information Processing Systems, pp. 1473–1481 (2009)

Pow11. Powers, D.M.: Evaluation: from precision, recall and f-measure to ROC, informedness, markedness and correlation. J. Mach. Learn. Technol. $2(1)$, 37–63 (2011)

Rom17. Romaszko, L.: Signal correlation prediction using convolutional neural networks. In: Battaglia, D., Guyon, I., Lemaire, V., Orlandi, J., Ray, B., Soriano, J. (eds.) Neural Connectomics Challenge. TSSCML, pp. 47–60. Springer, Cham (2017). https://doi.org/10.1007/978-3-319-53070-3_4

# Graph and Network Mining

# FHA: Fast Heuristic Attack Against Graph Convolutional Networks

Haoxi Zhan[✉] and Xiaobing Pei

Huazhong University of Science and Technology, Wuhan, China
hz12@hampshire.edu, xiaobingp@hust.edu.cn

**Abstract.** Recent years have witnessed a significant growth in Graph Convolutional Networks (GCNs). Being widely applied in a number of tasks, the safety issues of GCNs have draw the attention of many researchers. Recent studies have demonstrated that GCNs are vulnerable to adversarial attacks such that they are easily fooled by deliberate perturbations and a number of attacking methods have been proposed. However, state-of-the-art methods, which incorporate meta learning techniques, suffer from high computational costs. On the other hand, heuristic methods, which excel in efficiency, are in lack of satisfactory attacking performance. In order to solve this problem, it is supposed to find the patterns of gradient-based attacks to improve the performance of heuristic algorithms.

In this paper, we take advantage of several patterns discovered in untargeted attacks to propose a novel heuristic strategy to attack GCNs via creating viscous edges. We introduce the Fast Heuristic Attack (FHA) algorithm, which deliberately links training nodes to nodes of different classes. Instead of linking nodes fully randomly, the algorithm picks a batch of training nodes, which are of the same type, and links them to another class each time. Experimental studies show that our proposed method is able to achieve competitive attacking performance when attacking against various GCN models while significantly outperforming Mettack, which is the state-of-the-art untargeted structure attack, in terms of runtime.

**Keywords:** Graph Convolutional Network · Graph deep learning · Adversarial attack

## 1 Introduction

Graphs are widely used in a variety of domains such as academic citation analysis [1], scientific knowledge graphs [6], social network analysis [21], pandemic forecasting [10] and on-line banking [16], which is highly safety-critical. As a result, graph-based deep learning methods have received significant research attention. Graph Convolutional Networks (GCNs), in particular, have achieved state-of-the-art performance in a variety of graph learning tasks and have become one of the most popular research topics in the field of artificial intelligence.

© Springer Nature Switzerland AG 2021
C. Soares and L. Torgo (Eds.): DS 2021, LNAI 12986, pp. 151–165, 2021.
https://doi.org/10.1007/978-3-030-88942-5_12

In order to apply GCNs in real world scenarios, especially those safety-critical ones, it's important to investigate the robustness of GCN models. On one hand, attackers aim to degrade the performance of GCNs. On the other hand, users of GCNs want to ensure the performance against such adversarial attacks. Studies on deep learning models have demonstrated the vulnerability of Deep Neural Networks (DNNs) under adversarial attacks [7]. Although GCNs have a discrete nature and take advantage of the structure information on graphs, recent studies show that they are also easily fooled by adversarial perturbations [29].

Graph adversarial attacks, which deliberately modify the graph structures or contaminate the node features, are able to successfully inject adversarial information around the node neighborhoods and thus reduce the classification accuracies of GCNs. Various kinds of techniques such as heuristic algorithm [28], gradient decent [23], meta learning [29] and exploratory strategy [15] have been applied in graph adversarial attacks. Heuristic Approaches such as Random Attack and DICE [28] Attack are computationally efficient but their attacking performance are not as satisfactory as gradient-based attacks. On the other hand, meta-learning based algorithms, which use a surrogate model to simulate the GCN and compute meta-gradients iteratively when choosing perturbations, achieve state-of-the-art performance in grey-box scenarios while costing numerous computational resources. Especially for untargeted attacks, which aim to reduce the classification accuracies on the whole graph instead of a certain victim vertex, the algorithm need to compute the gradients for all possible edges on the entire graph, resulting in a long running time.

To remedy the above problems and to find a balance between performance and efficiency, this paper proposes to attack the graph topology with a heuristic algorithm which is based upon the patterns of gradient attacks. Several patterns have been discussed by previous studies. At first, attacking algorithms prefer to add edges instead of removing ones [22]. Secondly, graph structure attacks tend to connect nodes from different classes [30]. Finally, untargeted attacks tend to perturb the graph unevenly such that the algorithms modify a higher ratio of edges near the training set [25].

Carefully taking advantage the above patterns, we construct the Fast Heuristic Attack (FHA) algorithm. Instead of randomly linking dissimilar nodes, FHA divides the node labels into pairs. For each pair of labels, FHA randomly connects the nodes from the selected labels and ensures that at least one of the node is in the training set. Experiments show that when comparing with the state-of-the-art Mettack algorithm, FHA has competitive attacking performance while being much faster. When comparing with other attacking algorithms, FHA has better performance.

The contributions of this paper can be summarized as follows:

- We propose a novel heuristic algorithm to perform untargeted GCN attacks. To our best knowledge, it's the first untargeted structure attack to take advantage of discovered patterns to improve the efficiency.

- We conduct experiments on three widely-used graph datasets to verify the effectiveness of FHA. Experiments show that our proposed method excels in both attacking performance and runtime measurements.
- Besides the representative GCN model proposed by Kipf et al. [11], we show that FHA is also effective against various graph defense methods.

For the rest of the paper, we cover related works in Sect. 2. Then we introduce mathematical preliminaries in Sect. 3 before introducing our proposed method in Sect. 4. Experimental results are reported in Sect. 5 while we conclude our paper in Sect. 6.

## 2 Related Works

### 2.1 Graph Convolutional Networks (GCNs)

Firstly introduced by Bruna et al. [2], Graph Convolutional Networks have became a research hot spot since the introduction of the representative GCN model [11]. Generalizing the convolution operation from Convolutional Neural Networks to the graph domain, GCNs have a "message-passing" mechanism which aggregates information according to the graph topology. According to the methods used to aggregate messages, GCNs are divided into two families: spectral GCNs and spatial GCNs. Spectral GCNs are based on spectral representations of graphs. Typical spectral GCNs include ChebNet [4] and CayleyNet [13]. Spatial GCNs directly define the convolution operations on the graph topology. The representative GCN model proposed by Kipf et al. [11] is considered to be both spectral and spatial since it simplifies the spectral convolution defined in ChebNet into a spatial operation. Most recent GCN methods such as GAT [18], JK-Net [24] and FastGCN [3] are of the spatial family.

### 2.2 Graph Adversarial Attacks and Defenses

Various taxonomies of graph adversarial attacks have been discussed [17]. In order to review the related works, we briefly introduce three fundamental taxonomies. For more information on graph adversarial attacks, we refer the readers to recent surveys such as [17] and [9].

- Goal of the attacker: *targeted attacks* aim to fool the GCN to misclassify a set of victim vertices, *untargeted attacks* aim to reduce the classification accuracies of all testing nodes.
- Accessible information of the attacker: *white-box attacks* allow the attacker to access all possible information. In the settings of *grey-box attacks*, the training data is hacked but no model parameters are available. For *black-box attacks*, training data is also not accessible.
- Type of perturbations: *feature attacks* modify the node features to inject adversarial information. *Structure attacks* add or delete edges in the graph. Other kinds of attacks include node-injection attack [19], which adds viscous nodes to the graph, and label-flipping attack [26], which modifies the ground-truth labels of the training nodes.

In this work, we focus on untargeted grey-box structure attacks. White-box untargeted structure attacks include PGD Attack, which utilize projected gradient descent, and Min-Max Attack, which considers the attacking problem as a bi-level optimization problem [23]. In grey-box scenarios, model parameters are not accessible so a surrogate model is used to approximate parameters and gradients. Mettack [28] chooses perturbations by computing the meta-gradients of the surrogate iteratively and reaches the state-of-the-art performance.

Following the efforts to develop graph attacking algorithms, researches on the methods to defend against the attacks and to increase the robustness of GCNs are also conducted. Pre-processing methods detect and delete adversarial edges [22] or utilize low-rank approximation [5] before the training of GCN. Robust GCN (R-GCN) [27], which is a robust training method based on the attention mechanism, learns the hidden representations of nodes as Gaussian distributions and reduces the weights of high-variance nodes. Graph structure learning methods such as Pro-GNN [8] is also proposed to improve the robustness of GCNs.

## 3  Notations and Preliminaries

### 3.1  Mathematical Notations

We denote an undirected graph as $G = (V, E)$ where $V$ is the set of $N$ vertices while $E \subseteq V \times V$ is the set of edges. The adjacency matrix of $G$ is denoted as $A \in \{0,1\}^{N \times N}$ and the features of nodes are encoded in one-hot vectors, which form a feature matrix $X \in \{0,1\}^{N \times F}$. Each node $V_i$ has a ground-truth label and we denote it as a one-hot vector $\overrightarrow{y}_i \in \{0,1\}^C$ where $C$ is the number of classes. The $N \times N$ identity matrix is denoted as $I_N$. The indicator function is denoted as $\mathbb{I}(\cdot)$ such that

$$\mathbb{I}(p) = \begin{cases} 1 & \text{if } p \text{ is true} \\ 0 & \text{if } p \text{ is false} \end{cases}. \tag{1}$$

### 3.2  Graph Convolutional Networks

In this work, we focus on the GCN model introduced by Kipf et al. [11]. This model is used as the surrogate model in our proposed method. Consisting of 2 layers, each layer is defined as:

$$H^{(l+1)} = \sigma(\hat{A} H^{(l)} \theta^{(l)}), \tag{2}$$

where $\sigma$ is an activation function such as $ReLU$, $\hat{A}$ is the normalized adjacency matrix and $H^{(l)}$ is the hidden representations in the $l^{th}$ layer. The trainable weights are denoted as $\theta^{(l)}$. For the normalization of $A$, the renormalization trick [11] is used such that $\hat{A} = \tilde{D}^{-\frac{1}{2}} \tilde{A} \tilde{D}^{-\frac{1}{2}}$, where $\tilde{A} = A + I_N$ and $\tilde{D}_{ii} = \sum_j \tilde{A}_{ij}$.

The full network is defined as:

$$f(X, A) = \text{softmax}(\hat{A} \sigma(\hat{A} X \theta^{(0)}) \theta^{(1)}) \in \mathbb{R}^{N \times C}, \tag{3}$$

where $f_\theta(X, A)_i \in \{0, 1\}^C$, the $i^{th}$ row of the output, is the prediction of the $i^{th}$ node. The whole network is trained with a cross-entropy loss:

$$\mathcal{L}_{train} = - \sum_{v \in V_{train}} \sum_{i=1}^{C} \sum_{i=1}^{C} \overrightarrow{y}_v \log f(X, A)_v. \tag{4}$$

The performance of GCN is usually evaluated by the classification accuracies on the testing set [11]. The accuracy is defined as $\mathrm{acc}_{test}(f_\theta(A, X)) = 1 - E_{test}(f_\theta(A, X))$ where

$$E_{test}(f_\theta(A, X)) = \frac{1}{|V_{test}|} \sum_{i \in V_{test}} \mathbb{I}(\max f_\theta(X, A)_i \neq \max \overrightarrow{y}_i). \tag{5}$$

### 3.3 Problem Definition

We now define the attacking problem and its conditions. Given a graph $G = (V, E)$. The attacker is allowed to access the adjacency matrix $A$, feature matrix $X$. In addition, both the training set $V_{train}$ and validation set $V_{val}$ are hacked such that their labels are also accessible. The algorithm is supposed to return a new adjacency matrix $A'$ such that for a GCN network $f(X, A)$:

$$A' = \mathrm{argmax}_{A'} E_{test}(f_{\theta^*}(A', X))$$
$$\text{s.t. } \theta^* = \mathrm{argmin}_\theta \mathcal{L}_{train}(f_\theta(A', X)). \tag{6}$$

The number of perturbations is also restricted by a budget $\Delta$ such that at most $\Delta$ edges are allowed to be created. Mathematically, the restriction is stated as:

$$\|A' - A\|_F^2 \leq 2\Delta. \tag{7}$$

## 4   Proposed Method

Equation (6) is a bi-level optimization problem. In order to solve the bi-level optimization problem, Mettack [28] incorporates a meta learning framework. While achieving promising performance, the meta learning process requires a numerous amount of computational resources. In this section, we describe our proposed method to relieve this problem.

### 4.1   Exploring the Patterns of Attacks

We take advantage of three significant patterns discovered in graph structure attacks. In order to demonstrate the patterns, we conduct a case study on a Cora graph which is attacked by Mettack with a perturbation rate of 25%. We use the Meta-Self variant of Mettack without the log-likelihood restraint.

**Tendency to Add Edges.** While both adding or deleting edges are possible in graph structure attacks, Wu et al. [22] report that attacking methods tend to add fake edges instead of removing existed edges. Intuitively, creating fake edges will inject new information from newly linked neighbors while removing edges only influences the weights of existed neighbors during the message passing. In our case study, the statistics show that among the 1267 perturbations, 1230 of them are additions while only 37 deletions are crafted.

**Connect Different Classes.** Wu et al. [22] and Jin et al. [8] find that dissimilar nodes are more likely to be connected by attacking algorithms. Zügner et al. [30] report that nodes from different classes are more likely to be connected. In our case study on the attacked graph, we find that 92.93% newly connected pairs of nodes are from different classes.

**Perturb Around the Training Set.** In targeted attacks, it has been revealed that directed attack, which directed modifies edges that are incident to the target node, is the more effective than indirect attacks [29]. Similar patterns are also discussed for untargeted attacks. Wang et al. [20] suggest that in graph adversarial learning perturbations are mostly crafted in the neighborhoods of training nodes. Zhan et al. [25] find that perturbations crafted by Mettack are uneven such that a higher perturbation rate is observed near the training set. In our case study, we find that 1130 out of 1267 perturbations are focused on the connections between the training set and the rest of the graph.

## 4.2   The FHA Algorithm

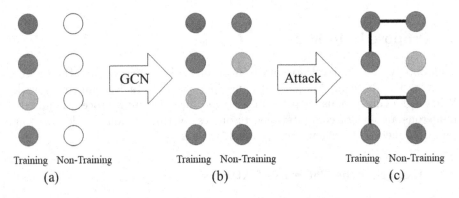

**Fig. 1.** The illustration of the FHA algorithm. (a) Before the training, (b) After the training of GCN, the pseudo-labels are generated, (c) the FHA connects nodes in a paired manner.

Although it's not hard to incorporate the discovered patterns, a major challenge is to take advantage of them to achieve competitive attacking performance. Simply utilizing them as restrictions will only lead to an improved version of random attack. In order to guarantee promising performance, we propose a novel Fast Heuristic Attack (FHA) algorithm.

The general framework of FHA is illustrated in Fig. 1. Before the training of FHA, the labels of the testing nodes are not known. In order to take advantage of the second pattern to connect nodes from different classes, we use GCN to pre-train the graph in order to get pseudo-labels for testing nodes. The pseudo-labels are used as the estimations of the real labels for testing nodes and the GCN is trained for only once.

After the pseudo-labels are generated, the algorithms pairs the labels to connect instead of adding edges randomly. Intuitively, connecting nodes from fixed pairs of labels will be more misleading for the algorithm since the decision boundaries between the paired classes will be blurred. Each pair contains one **training-specific label** and one **universal label**. For the training-specific label, we only connect training nodes of this label. For the universal label, we connect both the training nodes of this label and non-training nodes with the same pseudo-label. For instance, as shown in Fig. 1(c), a total number of 4 classes are in the dataset and the algorithm divides them into two pairs. In the first pair, blue is the training-specific label while red is the universal label. When FHA is crafting adversarial connections, blue nodes in the training set are connected to arbitrary red nodes, i.e., the red nodes being connected could be in either the training set, the validation set or the testing set.

How many blue training nodes should be linked to red nodes? Instead of randomly picking the pair to link each time a perturbation is being created, we propose a cyclic perturbation strategy. Each time, the algorithm creates a batch of $k$ edges between a pair of classes. For example, in the case described in Fig. 1, $k$ fake edges, which link blue training nodes with arbitrary red nodes, are created. If the attacking budget is not used up, then the algorithm traverses for the next pair of classes. If only $\Delta' = \Delta - nk < k$ perturbations are allowed after traversing through $n$ pairs of labels, the algorithm will only create a batch of $\Delta'$ perturbations for the final pair of labels. In the FHA algorithm, the label pairs are selected with two pointers in the list of labels.

To determine the batch sizes, we introduce two parameters. At first, the sizes of all batches are bounded by a parameter $\alpha$. Secondly, to prevent from saturation when the algorithm is dealing with rarely-appeared labels, a parameter $\eta$ is introduced such that if the training-specific label appears $x$ times in $V_{train}$ and $V_{val}$, then the batch size of this label pair is bounded by $\eta \cdot x$.

The pseudocode of FHA is reported in Algorithm 1. In line 1, FHA utilizes GCN to generate pseudo-labels for testing nodes. In lines 2–5, FHA initializes the modified adjacency matrix $A'$ and label pointers. $\Delta$ perturbations are selected via lines 7–16. Notice that in line 14 we require that $A'_{tu} = 0$ so the algorithm will only choose unconnected pairs of node hence no deletion of edges will be crafted.

**Algorithm 1.** Fast Heuristic Attack

---

**Input:** Adjacency matrix $A$, node feature $X$, number of classes $C$, attack budget $\Delta$,
    training set $V_{train}$, validation set $V_{val}$, ground-truth labels $\overrightarrow{y}_i$ for $i$ in $V_{train}$ or
    $V_{val}$, parameters $\alpha$ and $\eta$.
**Output:** Modified adjacency matrix $A'$.
1: Train GCN to get pseudo-labels for testing nodes.
2: $F \in \mathbb{N}^C \leftarrow$ Frequencies of the labels in $V_{train}$ and $V_{val}$.
3: $A' \leftarrow A$;
4: Training-specific pointer $pt \leftarrow 0$;
5: Universal pointer $pu \leftarrow 1$;
6: Batch counter $c \leftarrow 0$;
7: **for** $i \in [1, \Delta]$ **do**
8:    **if** $c > \eta \cdot F[pt \bmod C] \vee c > \alpha$ **then**
9:        $pt \leftarrow (pt + 2) \bmod C$;
10:       $pu \leftarrow (pu + 2) \bmod C$;
11:       $c \leftarrow 0$;
12:   **end if**
13:   Choose node $t$ in the training set such that $t$ is of the class $pt$;
14:   Choose node $u$ in the graph such that $u$ is of class $pu$ and $A'_{tu} = 0$.
15:   $A'_{tu} = A'_{ut} = 1$.
16: **end for**
    **return:** $A'$.

---

## 5    Experiments

In this section, we evaluate our proposed FHA algorithm against several graph defense algorithms. We want to answer the following research questions.

- **RQ1**: Does FHA has competitive attacking performance when comparing with state-of-the-art methods?
- **RQ2**: Is FHA as efficient as expected?
- **RQ3**: How does FHA differ from the naive combination of the attacking patterns? How much does the cyclic perturbation strategy contributes to FHA?

### 5.1    Experimental Settings

Before reporting the experimental results, we introduce the experimental conditions and settings at first. The experiments are conducted on a Windows 10 workstation with an Intel i9-10980XE CPU and two Nvidia RTX6000 GPUs. We use CUDA 10.2, Python 3.7.6, Numpy 1.18.5, Scipy 1.3.1, Pytorch 1.5.0, Tensorflow 1.15.0 and DeepRobust [14] (commit id 456f8b5).

**Datasets.** In our experiments, we used three publicly available datasets that are commonly used in previous studies such as [12,27,28]. All the three datasets are citation networks in which each node represents an academic paper and if

**Table 1.** Statistics of datasets. We only consider the largest connected components.

| Dataset | $|V|$ | $|E|$ | Classes | Features |
|---------|-------|-------|---------|----------|
| Cora | 2485 | 5069 | 7 | 1433 |
| Citeseer | 2110 | 3668 | 6 | 3703 |
| Cora-ML | 2810 | 7981 | 7 | 2879 |

two nodes are connected they have a citation relationship. Following [28], we only consider the largest connected components of the graphs. The statistics of the datasets are reported in Table 1. For each dataset, we follow [8] to randomly split the nodes into three subsets. For each graph, the training set contains 10% of the nodes, the validation set contains another 10%. The remaining nodes, which occupy 80% of the vertex set, form the testing set.

**Baselines.** In order to demonstrative the performance of FHA, we compare it with 4 baselines, which include both grey-box methods, white-box methods and random attack, which is black-box.

- **Mettack:** Mettack [28] is a state-of-the-art untargeted structure attacking method. Being a grey-box algorithm it requires the same information as FHA. Various variants of Mettack have been proposed and we use the Meta-Self variant, which is suggested by [28]. The log-likelihood constraint is released for fair comparison.
- **PGD Attack:** PGD Attack [23] is a white-box method which utilizes projected gradient descent and it assumes that the GCN will not be retrained after the initial training.
- **DICE:** DICE is a baseline method introduced by [28]. It's a white-box method such that it randomly connects nodes from different classes and disconnects vertices of the same label. In our study, we only allows DICE to add edges.
- **Random:** Random Attack simply flips the edges randomly on the graph. In our comparison, Random Attack is only allowed to add edges.

**Defense Methods.** The vanilla GCN proposed by Kipf et al. [11] is the major target of our study. However, we also employ several representative defense methods in our study.

- **GCN:** The representative model introduced in Sect. 3.2.
- **R-GCN:** An attention-based defense method introduced by Zhu et al. [27].
- **GCN-Jaccard:** A pre-processing method introduced by [22]. It deletes edges which connect dissimilar nodes before the training of GCN.
- **GCN-SVD:** Another pre-processing method which performs low-rank approximation on the adjacency matrix [5].
- **Pro-GNN:** A robust training method which incorporates graph structure learning with graph representation learning [8].

**Implementation and Parameters.** We use the official PyTorch implementation of GCN and the official implementation of R-GCN, which is Tensorflow-based. For Random Attack, DICE, GCN-Jaccard, GCN-SVD and Pro-GNN we use the implementations provided by DeepRobust [14]. As for parameters, we adopt the default parameters in the implementations for all baselines and defense methods. For our proposed FHA algorithm, we set $\alpha = 200$ and $\eta = 2.0$ in our experiments.

## 5.2 Attack Performance

**Table 2.** Attack performance against the original GCN (Classification accuracy $\pm$ Std).

| Dataset | Ptb | FHA | Mettack | PGD | DICE | Random |
|---|---|---|---|---|---|---|
| Cora | 0% | $83.59 \pm 0.17$ | $83.59 \pm 0.17$ | $83.59 \pm 0.17$ | $83.59 \pm 0.17$ | $83.59 \pm 0.17$ |
| | 5% | $\mathbf{75.48 \pm 1.54}$ | $75.97 \pm 1.54$ | $80.45 \pm 1.69$ | $81.27 \pm 1.06$ | $82.25 \pm 1.13$ |
| | 10% | $\mathbf{64.72 \pm 1.04}$ | $71.33 \pm 1.81$ | $76.88 \pm 1.87$ | $79.55 \pm 1.06$ | $80.56 \pm 1.18$ |
| | 15% | $\mathbf{58.77 \pm 1.49}$ | $63.73 \pm 2.44$ | $73.53 \pm 1.55$ | $77.97 \pm 0.99$ | $79.28 \pm 0.88$ |
| | 20% | $\mathbf{54.19 \pm 1.27}$ | $57.15 \pm 4.42$ | $72.76 \pm 2.29$ | $74.99 \pm 2.04$ | $78.48 \pm 0.76$ |
| | 25% | $\mathbf{50.9 \pm 1.49}$ | $51.99 \pm 4.97$ | $70.01 \pm 2.32$ | $72.99 \pm 1.26$ | $76.55 \pm 0.83$ |
| Citeseer | 0% | $73.90 \pm 0.41$ | $73.90 \pm 0.41$ | $73.90 \pm 0.41$ | $73.90 \pm 0.41$ | $73.90 \pm 0.41$ |
| | 5% | $\mathbf{72.19 \pm 0.76}$ | $74.07 \pm 0.80$ | $75.88 \pm 0.58$ | $75.17 \pm 0.53$ | $75.29 \pm 0.52$ |
| | 10% | $\mathbf{68.89 \pm 0.84}$ | $70.47 \pm 1.61$ | $75.83 \pm 0.53$ | $73.95 \pm 0.46$ | $74.90 \pm 0.46$ |
| | 15% | $\mathbf{65.08 \pm 1.14}$ | $65.28 \pm 2.22$ | $75.54 \pm 0.56$ | $72.78 \pm 0.74$ | $74.18 \pm 0.40$ |
| | 20% | $\mathbf{59.72 \pm 1.57}$ | $62.03 \pm 2.82$ | $74.77 \pm 1.06$ | $71.50 \pm 0.34$ | $73.78 \pm 0.69$ |
| | 25% | $57.1 \pm 1.83$ | $\mathbf{55.57 \pm 2.99}$ | $74.28 \pm 0.58$ | $70.24 \pm 0.65$ | $72.34 \pm 0.55$ |
| Cora-ML | 0% | $83.59 \pm 0.17$ | $83.59 \pm 0.17$ | $83.59 \pm 0.17$ | $83.59 \pm 0.17$ | $83.59 \pm 0.17$ |
| | 5% | $\mathbf{77.68 \pm 0.76}$ | $78.55 \pm 1.02$ | $81.36 \pm 0.90$ | $80.91 \pm 0.84$ | $82.11 \pm 0.88$ |
| | 10% | $72.82 \pm 1.20$ | $\mathbf{67.28 \pm 2.15}$ | $76.00 \pm 2.04$ | $79.03 \pm 1.06$ | $79.99 \pm 1.51$ |
| | 15% | $65.12 \pm 1.58$ | $\mathbf{58.47 \pm 2.87}$ | $71.21 \pm 1.99$ | $76.38 \pm 0.72$ | $77.53 \pm 2.97$ |
| | 20% | $59.05 \pm 1.89$ | $\mathbf{47.00 \pm 3.49}$ | $66.50 \pm 2.10$ | $73.90 \pm 1.20$ | $77.39 \pm 1.30$ |
| | 25% | $52.19 \pm 2.02$ | $\mathbf{42.63 \pm 3.67}$ | $58.76 \pm 2.34$ | $72.65 \pm 1.35$ | $75.38 \pm 1.06$ |

In this subsection, we answer the research question RQ1. The performance of FHA and the baselines against GCN, R-GCN, GCN-Jaccard, GCN-SVD and Pro-GNN are reported in Table 2, 3, 4, 5 and 6 respectively. Following both GCN [11] and Mettack [28], we evaluate the attacking performance via comparing the classification accuracies after attacks. For each combination of attacking method, defense method and perturbation rate, we run the experiments for 10 times and we report the average accuracies along with the standard deviations. 75 different settings are tested for each attacking method. Hence, we conduct experiments on 375 different experimental combinations and 3750 runs are conducted. For each setting, the best performance is highlighted in bold.

As revealed in the tables, among the 75 experimental settings, our proposed FHA algorithm achieves the best performance in 55 different settings. Mettack

**Table 3.** Attack performance against R-GCN (classification accuracy ± Std).

| Dataset | Ptb | FHA | Mettack | PGD | DICE | Random |
|---|---|---|---|---|---|---|
| Cora | 0% | 85.48 ± 0.27 | 85.48 ± 0.27 | 85.48 ± 0.27 | 85.48 ± 0.27 | 85.48 ± 0.27 |
| | 5% | 78.46 ± 0.87 | **78.27 ± 0.82** | 83.81 ± 0.90 | 83.68 ± 0.39 | 84.55 ± 0.34 |
| | 10% | **67.17 ± 1.22** | 73.88 ± 1.03 | 82.28 ± 1.08 | 82.01 ± 0.74 | 83.11 ± 0.52 |
| | 15% | **60.47 ± 1.07** | 67.60 ± 1.86 | 81.19 ± 0.92 | 80.46 ± 0.42 | 82.23 ± 0.52 |
| | 20% | **53.79 ± 0.98** | 61.18 ± 2.99 | 79.40 ± 1.30 | 78.77 ± 0.73 | 81.23 ± 0.61 |
| | 25% | **49.49 ± 1.78** | 57.10 ± 3.25 | 77.92 ± 1.07 | 76.87 ± 0.86 | 80.38 ± 0.53 |
| Citeseer | 0% | 76.39 ± 0.13 | 76.39 ± 0.13 | 76.39 ± 0.13 | 76.39 ± 0.13 | 76.39 ± 0.13 |
| | 5% | **70.94 ± 1.39** | 73.70v0.80 | 76.52 ± 0.48 | 75.24 ± 0.48 | 75.62 ± 0.39 |
| | 10% | **66.68 ± 0.65** | 70.30 ± 1.46 | 76.08 ± 0.56 | 73.87 ± 0.43 | 74.86 ± 0.63 |
| | 15% | **62.64 ± 0.78** | 65.70 ± 1.48 | 75.51 ± 0.66 | 72.97 ± 0.65 | 74.25 ± 0.42 |
| | 20% | **60.13 ± 1.10** | 61.22 ± 2.07 | 74.92 ± 0.83 | 71.64 ± 0.34 | 73.93 ± 0.78 |
| | 25% | 58.14 ± 1.30 | **56.74 ± 2.26** | 74.16 ± 1.14 | 70.44 ± 0.42 | 72.87 ± 0.83 |
| Cora-ML | 0% | 95.66 ± 0.15 | 95.66 ± 0.15 | 95.66 ± 0.15 | 95.66 ± 0.15 | 95.66 ± 0.15 |
| | 5% | 81.54 ± 0.60 | **76.51 ± 0.68** | 85.51 ± 0.49 | 84.31 ± 0.27 | 85.03 ± 0.36 |
| | 10% | 76.43 ± 1.03 | **70.64 ± 0.70** | 83.81 ± 0.63 | 82.75 ± 0.57 | 83.62 ± 0.51 |
| | 15% | **69.20 ± 1.31** | 69.58 ± 0.80 | 82.03 ± 0.98 | 80.68 ± 0.36 | 82.32 ± 0.62 |
| | 20% | **62.70 ± 1.60** | 68.24 ± 1.03 | 80.25 ± 1.48 | 78.75 ± 0.51 | 81.76 ± 0.35 |
| | 25% | **50.78 ± 1.86** | 66.27 ± 0.57 | 77.58 ± 1.94 | 77.24 ± 0.59 | 79.96 ± 0.55 |

**Table 4.** Attack performance against GCN-Jaccard (classification accuracy ± Std).

| Dataset | Ptb | FHA | Mettack | PGD | DICE | Random |
|---|---|---|---|---|---|---|
| Cora | 0% | 82.32 ± 0.50 | 82.32 ± 0.50 | 82.32 ± 0.50 | 82.32 ± 0.50 | 82.32 ± 0.50 |
| | 5% | 79.50 ± 0.85 | **77.62 ± 1.13** | 79.77 ± 1.05 | 80.97 ± 0.23 | 81.40 ± 0.64 |
| | 10% | 74.73 ± 0.76 | **74.05 ± 1.06** | 77.83 ± 1.36 | 79.76 ± 0.41 | 80.59 ± 0.66 |
| | 15% | **70.43 ± 1.09** | 71.51 ± 1.05 | 76.57 ± 1.37 | 78.79 ± 0.53 | 80.15 ± 0.31 |
| | 20% | **65.86 ± 1.32** | 68.31 ± 2.28 | 73.62 ± 2.49 | 77.77 ± 0.55 | 79.27 ± 0.46 |
| | 25% | **61.00 ± 1.12** | 66.02 ± 1.99 | 73.49 ± 1.68 | 77.05 ± 1.07 | 78.94 ± 0.58 |
| Citeseer | 0% | 73.73 ± 0.93 | 73.73 ± 0.93 | 73.73 ± 0.93 | 73.73 ± 0.93 | 73.73 ± 0.93 |
| | 5% | 70.34 ± 0.43 | **70.03 ± 1.87** | 72.70 ± 0.97 | 72.87 ± 0.44 | 72.61 ± 0.70 |
| | 10% | **66.09 ± 1.19** | 66.79 ± 1.08 | 71.76 ± 0.91 | 71.04 ± 0.57 | 72.17 ± 0.80 |
| | 15% | **63.98 ± 1.37** | 65.69 ± 1.50 | 70.50 ± 0.83 | 69.90 ± 0.77 | 71.30 ± 0.54 |
| | 20% | 63.15 ± 1.16 | **62.66 ± 1.29** | 70.13 ± 1.21 | 69.13 ± 0.82 | 70.93 ± 0.84 |
| | 25% | 62.04 ± 1.74 | **61.05 ± 1.90** | 69.36 ± 1.44 | 67.69 ± 0.91 | 69.72 ± 0.97 |
| Cora-ML | 0% | 85.38 ± 0.30 | 85.38 ± 0.30 | 85.38 ± 0.30 | 85.38 ± 0.30 | 85.38 ± 0.30 |
| | 5% | 81.93 ± 0.71 | **80.56 ± 0.39** | 80.58 ± 0.88 | 83.68 ± 0.36 | 84.07 ± 0.53 |
| | 10% | 75.61 ± 0.86 | 75.82 ± 0.45 | **75.55 ± 1.33** | 82.02 ± 0.43 | 83.12 ± 0.54 |
| | 15% | **68.39 ± 1.12** | 73.14 ± 0.45 | 71.80 ± 1.70 | 80.63 ± 0.38 | 82.64 ± 0.69 |
| | 20% | **62.22 ± 1.72** | 70.32 ± 0.75 | 69.03 ± 1.77 | 79.50 ± 0.34 | 81.71 ± 0.38 |
| | 25% | **51.44 ± 2.50** | 65.16 ± 1.05 | 67.48 ± 2.00 | 77.83 ± 0.54 | 80.96 ± 0.58 |

**Table 5.** Attack performance against GCN-SVD (classification accuracy $\pm$ Std).

| Dataset | Ptb | FHA | Mettack | PGD | DICE | Random |
|---|---|---|---|---|---|---|
| Cora | 0% | $73.03 \pm 0.64$ | $73.03 \pm 0.64$ | $73.03 \pm 0.64$ | $73.03 \pm 0.64$ | $73.03 \pm 0.64$ |
| | 5% | $\mathbf{71.50 \pm 1.02}$ | $72.51 \pm 0.41$ | $72.17 \pm 1.04$ | $71.68 \pm 0.74$ | $72.15 \pm 0.69$ |
| | 10% | $\mathbf{68.93 \pm 1.09}$ | $69.54 \pm 1.15$ | $72.65 \pm 1.71$ | $70.15 \pm 0.52$ | $70.84 \pm 0.54$ |
| | 15% | $\mathbf{64.37 \pm 1.71}$ | $67.18 \pm 1.14$ | $72.54 \pm 0.81$ | $68.77 \pm 0.98$ | $70.57 \pm 0.75$ |
| | 20% | $\mathbf{57.41 \pm 2.19}$ | $63.13 \pm 3.16$ | $72.42 \pm 1.14$ | $67.56 \pm 1.19$ | $69.94 \pm 0.43$ |
| | 25% | $\mathbf{51.67 \pm 2.36}$ | $60.89 \pm 2.64$ | $72.13 \pm 1.62$ | $66.72 \pm 0.86$ | $68.96 \pm 1.15$ |
| Citeseer | 0% | $70.97 \pm 1.52$ | $70.97 \pm 1.52$ | $70.97 \pm 1.52$ | $70.97 \pm 1.52$ | $70.97 \pm 1.52$ |
| | 5% | $\mathbf{69.55 \pm 1.25}$ | $71.27 \pm 0.97$ | $70.01 \pm 1.16$ | $70.27 \pm 0.94$ | $70.30 \pm 1.38$ |
| | 10% | $69.92 \pm 0.95$ | $\mathbf{68.79 \pm 1.68}$ | $69.54 \pm 1.09$ | $68.89 \pm 0.93$ | $69.63 \pm 1.03$ |
| | 15% | $\mathbf{59.32 \pm 19.16}$ | $68.56 \pm 1.72$ | $69.67 \pm 1.50$ | $68.37 \pm 0.80$ | $68.32 \pm 1.05$ |
| | 20% | $\mathbf{57.52 \pm 18.42}$ | $67.90 \pm 0.87$ | $69.18 \pm 1.76$ | $67.23 \pm 1.12$ | $67.52 \pm 1.32$ |
| | 25% | $\mathbf{59.36 \pm 2.87}$ | $65.95 \pm 1.47$ | $69.24 \pm 1.78$ | $66.38 \pm 1.23$ | $66.66 \pm 1.27$ |
| Cora-ML | 0% | $79.84 \pm 0.16$ | $79.84 \pm 0.16$ | $79.84 \pm 0.16$ | $79.84 \pm 0.16$ | $79.84 \pm 0.16$ |
| | 5% | $\mathbf{77.76 \pm 0.72}$ | $79.69 \pm 0.32$ | $79.60 \pm 0.35$ | $78.55 \pm 0.57$ | $79.17 \pm 0.36$ |
| | 10% | $\mathbf{73.04 \pm 1.15}$ | $78.96 \pm 0.45$ | $79.97 \pm 0.41$ | $77.43 \pm 0.41$ | $78.17 \pm 0.66$ |
| | 15% | $\mathbf{67.88 \pm 1.46}$ | $78.35 \pm 0.11$ | $79.27 \pm 0.84$ | $75.84 \pm 0.86$ | $77.47 \pm 0.49$ |
| | 20% | $\mathbf{64.02 \pm 2.22}$ | $77.01 \pm 0.69$ | $70.24 \pm 20.35$ | $74.73 \pm 0.70$ | $76.14 \pm 0.68$ |
| | 25% | $\mathbf{57.72 \pm 2.43}$ | $75.71 \pm 0.39$ | $73.23 \pm 3.15$ | $73.70 \pm 0.79$ | $75.29 \pm 0.50$ |

**Table 6.** Attack performance against Pro-GNN (classification accuracy $\pm$ Std).

| Dataset | Ptb | FHA | Mettack | PGD | DICE | Random |
|---|---|---|---|---|---|---|
| Cora | 0% | $84.95 \pm 0.82$ | $84.95 \pm 0.82$ | $84.95 \pm 0.82$ | $84.95 \pm 0.82$ | $84.95 \pm 0.82$ |
| | 5% | $80.81 \pm 1.19$ | $\mathbf{77.81 \pm 0.90}$ | $81.49 \pm 1.16$ | $83.04 \pm 0.75$ | $83.88 \pm 0.39$ |
| | 10% | $\mathbf{72.46 \pm 1.39}$ | $73.03 \pm 1.10$ | $79.57 \pm 1.54$ | $81.20 \pm 0.46$ | $82.50 \pm 0.56$ |
| | 15% | $\mathbf{63.12 \pm 2.58}$ | $68.33 \pm 1.24$ | $77.34 \pm 2.21$ | $79.65 \pm 0.72$ | $81.75 \pm 0.58$ |
| | 20% | $\mathbf{51.17 \pm 1.59}$ | $62.00 \pm 3.04$ | $75.57 \pm 1.83$ | $78.00 \pm 0.54$ | $80.76 \pm 0.58$ |
| | 25% | $\mathbf{46.55 \pm 1.88}$ | $58.17 \pm 4.09$ | $73.36 \pm 2.60$ | $76.64 \pm 0.78$ | $80.40 \pm 0.60$ |
| Citeseer | 0% | $73.79 \pm 0.52$ | $73.79 \pm 0.52$ | $73.79 \pm 0.52$ | $73.79 \pm 0.52$ | $73.79 \pm 0.52$ |
| | 5% | $\mathbf{70.10 \pm 0.99}$ | $72.18 \pm 1.36$ | $71.69 \pm 1.27$ | $73.03 \pm 0.72$ | $73.08 \pm 0.95$ |
| | 10% | $\mathbf{68.26 \pm 1.67}$ | $68.89 \pm 0.82$ | $69.85 \pm 1.78$ | $71.90 \pm 0.66$ | $72.88 \pm 0.84$ |
| | 15% | $\mathbf{65.47 \pm 2.30}$ | $65.53 \pm 1.24$ | $70.24 \pm 2.36$ | $71.02 \pm 0.89$ | $72.01 \pm 1.33$ |
| | 20% | $63.71 \pm 1.56$ | $\mathbf{60.63 \pm 1.89}$ | $70.08 \pm 2.06$ | $70.37 \pm 0.60$ | $71.71 \pm 0.83$ |
| | 25% | $63.16 \pm 2.63$ | $\mathbf{57.53 \pm 2.77}$ | $69.31 \pm 1.95$ | $69.46 \pm 0.45$ | $71.08 \pm 0.73$ |
| Cora-ML | 0% | $84.96 \pm 0.46$ | $84.96 \pm 0.46$ | $84.96 \pm 0.46$ | $84.96 \pm 0.46$ | $84.96 \pm 0.46$ |
| | 5% | $\mathbf{82.24 \pm 0.65}$ | $82.80 \pm 0.63$ | $84.10 \pm 0.75$ | $82.41 \pm 0.49$ | $83.12 \pm 0.46$ |
| | 10% | $\mathbf{76.74 \pm 1.34}$ | $81.47 \pm 0.42$ | $81.91 \pm 1.74$ | $81.05 \pm 0.39$ | $82.11 \pm 0.52$ |
| | 15% | $\mathbf{71.18 \pm 1.75}$ | $80.59 \pm 0.59$ | $79.53 \pm 1.50$ | $79.98 \pm 0.42$ | $81.30 \pm 0.49$ |
| | 20% | $\mathbf{64.47 \pm 2.80}$ | $78.09 \pm 0.92$ | $77.03 \pm 1.37$ | $77.98 \pm 0.59$ | $80.74 \pm 0.76$ |
| | 25% | $\mathbf{48.10 \pm 4.33}$ | $72.11 \pm 2.84$ | $72.67 \pm 3.16$ | $77.21 \pm 0.60$ | $79.51 \pm 0.44$ |

wins 19 settings while PGD Attack wins exactly one of them. This demonstrates the efficacy of the FHA algorithm.

## 5.3   Runtime Analysis

In this subsection, we answer the research question RQ2. As revealed in Sect. 5.1, our proposed method has competitive attacking performance when comparing with Mettack. *As a heuristic method which doesn't need to compute gradients iteratively, is FHA faster than Mettack?*

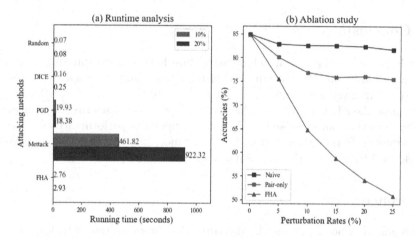

**Fig. 2.** (a) Runtime analysis of five attacking methods; (b) Ablation study on FHA.

We report the running times of the attacking algorithms on the Cora dataset in Fig. 2(a). Two perturbation rates are selected. As shown in Fig. 2(a), FHA is faster than Mettack and PGD Attack, which are gradient-based. Especially when the perturbation rate is 20%, FHA is 314 times faster than Mettack while achieving better performance. Random Attack and DICE are faster than FHA but they have the worst performance. Hence, FHA reaches a balance between performance and efficiency.

## 5.4   Ablation Study

In this subsection, we aim to answer the research question RQ3. A naive combination of the three patterns discussed in Sect. 4 is possible such that in each step, the naive algorithm links a training node to a non-training node from a different class. Our FHA algorithms extend the naive approach in two aspects. At first, we introduce a paired linking strategy which fixes the pairs of labels to be linked. Secondly, we propose a cyclic attacking strategy such that the label pairs are traversed in a specific order and adversarial edges are added by batches.

Hence, we compare the FHA algorithm with two model variants. The first variant is the naive algorithm. The second variant only has the paired linking strategy but it doesn't traverse among different label pairs, i.e. it only focuses on a first pair of labels. The Cora dataset is used in the ablation study. The defending model is the GCN described in Sect. 3.2.

As shown in Fig. 2(b), the performance of the naive algorithm is even worse than Random Attack. The Pair-only algorithm performs better than Random Attack when the perturbation rate is low but it becomes saturated when perturbation rate reaches 10%. Hence, both the paired linking strategy and the cyclic attacking strategy play vital roles in the success of the FHA algorithm.

# 6  Conclusions

Meta-learning based graph attacking algorithms have achieved promising attacking performance but they are in need of numerous computational resources. In this work, we investigate three different patterns of graph structure attacks and we propose the FHA algorithm, which is based upon the patterns. Experimental results show that our proposed method has competitive performance while being more efficient. In the future we aim to explore more patterns in graph adversarial attacks and to improve the performance of defense methods.

# References

1. Bioglio, L., Rho, V., Pensa, R.: Measuring the inspiration rate of topics in bibliographic networks. In: DS (2017)
2. Bruna, J., Zaremba, W., Szlam, A.D., LeCun, Y.: Spectral networks and locally connected networks on graphs. CoRR abs/1312.6203 (2014)
3. Chen, J.J., Ma, T., Xiao, C.: FastGCN: fast learning with graph convolutional networks via importance sampling. arXiv:1801.10247 (2018)
4. Defferrard, M., Bresson, X., Vandergheynst, P.: Convolutional neural networks on graphs with fast localized spectral filtering. In: NIPS (2016)
5. Entezari, N., Al-Sayouri, S.A., Darvishzadeh, A., Papalexakis, E.E.: All you need is low (rank): defending against adversarial attacks on graphs. In: Proceedings of the 13th International Conference on Web Search and Data Mining (2020)
6. Giarelis, N., Kanakaris, N., Karacapilidis, N.: On the utilization of structural and textual information of a scientific knowledge graph to discover future research collaborations: a link prediction perspective. In: DS (2020)
7. Goodfellow, I., Shlens, J., Szegedy, C.: Explaining and harnessing adversarial examples. CoRR abs/1412.6572 (2015)
8. Jin, W., Ma, Y., Liu, X., Tang, X.F., Wang, S., Tang, J.: Graph structure learning for robust graph neural networks. In: Proceedings of the 26th ACM SIGKDD International Conference on Knowledge Discovery & Data Mining (2020)
9. Jin, W., Li, Y., Xu, H., Wang, Y., Tang, J.: Adversarial attacks and defenses on graphs: a review and empirical study. arXiv:2003.00653 (2020)
10. Kapoor, A., et al.: Examining COVID-19 forecasting using spatio-temporal graph neural networks. arXiv:2007.03113 (2020)

11. Kipf, T., Welling, M.: Semi-supervised classification with graph convolutional networks. arXiv:abs/1609.02907 (2017)
12. Klicpera, J., Bojchevski, A., Günnemann, S.: Predict then propagate: graph neural networks meet personalized pagerank. In: ICLR (2019)
13. Levie, R., Monti, F., Bresson, X., Bronstein, M.: CayleyNets: graph convolutional neural networks with complex rational spectral filters. IEEE Trans. Sig. Process. **67**, 97–109 (2019)
14. Li, Y., Jin, W., Xu, H., Tang, J.: DeepRobust: a Pytorch library for adversarial attacks and defenses. arXiv:abs/2005.06149 (2020)
15. Lin, X., et al.: Exploratory adversarial attacks on graph neural networks. In: 2020 IEEE International Conference on Data Mining (ICDM), pp. 1136–1141 (2020)
16. Shumovskaia, V., Fedyanin, K., Sukharev, I., Berestnev, D., Panov, M.: Linking bank clients using graph neural networks powered by rich transactional data: Extended abstract. In: 2020 IEEE 7th International Conference on Data Science and Advanced Analytics (DSAA), pp. 787–788 (2020)
17. Sun, L., et al.: Adversarial attack and defense on graph data: a survey. arXiv preprint arXiv:1812.10528 (2018)
18. Velickovic, P., Cucurull, G., Casanova, A., Romero, A., Lio', P., Bengio, Y.: Graph attention networks. arXiv:1710.10903 (2018)
19. Wang, J., Luo, M., Suya, F., Li, J., Yang, Z., Zheng, Q.: Scalable attack on graph data by injecting vicious nodes. Data Min. Knowl. Disc. **34**(5), 1363–1389 (2020). https://doi.org/10.1007/s10618-020-00696-7
20. Wang, X., Liu, X., Hsieh, C.J.: GraphDefense: towards robust graph convolutional networks. arXiv:1911.04429 (2019)
21. Wasserman, S., Faust, K.: Social Network Analysis - Methods and Applications. Structural Analysis in the Social Sciences (2007)
22. Wu, H., Wang, C., Tyshetskiy, Y., Docherty, A., Lu, K., Zhu, L.: Adversarial examples for graph data: deep insights into attack and defense. In: IJCAI (2019)
23. Xu, K., et al.: Topology attack and defense for graph neural networks: an optimization perspective. arXiv:1906.04214 (2019)
24. Xu, K., Li, C., Tian, Y., Sonobe, T., Kawarabayashi, K., Jegelka, S.: Representation learning on graphs with jumping knowledge networks. In: ICML (2018)
25. Zhan, H., Pei, X.: Black-box gradient attack on graph neural networks: deeper insights in graph-based attack and defense. arXiv:2104.15061 (2021)
26. Zhang, M., Hu, L., Shi, C., Wang, X.: Adversarial label-flipping attack and defense for graph neural networks. In: 2020 IEEE International Conference on Data Mining (ICDM), pp. 791–800 (2020)
27. Zhu, D., Zhang, Z., Cui, P., Zhu, W.: Robust graph convolutional networks against adversarial attacks. In: Proceedings of the 25th ACM SIGKDD International Conference on Knowledge Discovery & Data Mining (2019)
28. Zugner, D., Gunnemann, S.: Adversarial attacks on graph neural networks via meta learning (2019)
29. Zügner, D., Akbarnejad, A., Günnemann, S.: Adversarial attacks on neural networks for graph data. In: Proceedings of the 24th ACM SIGKDD International Conference on Knowledge Discovery & Data Mining (2018)
30. Zügner, D., Borchert, O., Akbarnejad, A., Günnemann, S.: Adversarial attacks on graph neural networks: perturbations and their patterns. ACM Trans. Knowl. Disc. Data (TKDD) **14**, 1–31 (2020)

# Ranking Structured Objects with Graph Neural Networks

Clemens Damke[1(✉)] and Eyke Hüllermeier[2]

[1] Heinz Nixdorf Institute, Paderborn University, Paderborn, Germany
cdamke@mail.upb.de
[2] University of Munich, Munich, Germany
eyke@ifi.lmu.de

**Abstract.** *Graph neural networks* (GNNs) have been successfully applied in many structured data domains, with applications ranging from molecular property prediction to the analysis of social networks. Motivated by the broad applicability of GNNs, we propose the family of so-called *RankGNNs*, a combination of *neural Learning to Rank* (LtR) methods and GNNs. RankGNNs are trained with a set of pair-wise preferences between graphs, suggesting that one of them is preferred over the other. One practical application of this problem is drug screening, where an expert wants to find the most promising molecules in a large collection of drug candidates. We empirically demonstrate that our proposed pair-wise RankGNN approach either significantly outperforms or at least matches the ranking performance of the naïve point-wise baseline approach, in which the LtR problem is solved via GNN-based graph regression.

**Keywords:** Graph-structured data · Graph neural networks · Preference learning · Learning to rank

## 1 Introduction

Bringing a set of objects $o_1, \ldots, o_N$ into a particular order is an important problem with many applications, ranging from task planning to recommender systems. In such domains, the criterion defining the underlying order relation $\succeq$ typically depends on properties (features) of the objects (for example the price and quality of a product). If the sorting criterion (and hence the relation $\succeq$) is not explicitly given, one may think of inferring it from exemplary data, often provided in the form of a set of pair-wise orderings $o_i \succeq o_j$ (e.g., representing that the user prefers product $o_i$ over product $o_j$). This gives rise to a machine learning task often referred to as *Learning to Rank* (LtR). Thus, the goal is to learn a general ordering strategy (preference model) from sample data of the above kind, which can then be used to sort any new (previously unseen) set of objects.

While existing state-of-the-art LtR approaches assume that objects $o_i$ are represented by feature vectors $x_i \in \mathbb{R}^n$, in this paper, we will consider the LtR problem for another quite natural and practically important representation, namely the domain of finite graphs. Methods for learning to rank objects

© Springer Nature Switzerland AG 2021
C. Soares and L. Torgo (Eds.): DS 2021, LNAI 12986, pp. 166–180, 2021.
https://doi.org/10.1007/978-3-030-88942-5_13

represented in the form of graphs can, for example, be used in applications such as drug screening, where the ranked objects are the molecular structures of drug candidates.

To support the ranking of structured objects such as graphs, existing LtR methods need to be adapted. Previously, Agarwal [1] has considered the problem of ranking the vertices within a given graph. However, to the best of our knowledge, the graph-wise LtR problem has so far only been described in the context of specific domains, such as drug discovery, where manually chosen graph feature representations were used [29]. Motivated by the success of *graph neural networks* (GNNs) in graph representation learning, we propose a simple architecture that combines GNNs with neural LtR approaches. The proposed approach allows for training ranking functions in an end-to-end fashion and can be applied to arbitrary graphs without the need to manually choose a domain-specific graph feature representation.

Our neural graph ranking architecture will be introduced in Sect. 4. Before, the LtR and GNN models that are used in this architecture are described in Sect. 2 and Sect. 3, respectively. In Sect. 5, we evaluate our approach on a selection of graph benchmark datasets.

## 2 Object Ranking

LtR approaches are often categorized as point-wise, pair-wise, and list-wise methods. We begin with a short overview of these families. Afterwards, a more in-depth introduction is given to a selection of neural pair-wise approaches that we shall built upon in Sect. 4.

### 2.1 Overview of LtR Approaches

*Point-wise methods* assume the existence of a (latent) utility function representing the sought preference relation $\succeq$, i.e., that an ordinal or numeric utility score $u_i \in \mathbb{R}$ can be assigned to each object $o_i \in O$ such that

$$\forall o_i, o_j \in O : u_i \geq u_j \Leftrightarrow o_i \succeq o_j .$$

Based on training data in the form of exemplary (and possibly noisy) ratings, i.e., object/utility pairs $\{(x_i, u_i)\}_{i=1}^N \subset X \times \mathbb{R}$, where $x_i \in X$ is the feature representation of $o_i$, the LtR problem can be solved by fitting a model $f_u : X \to \mathbb{R}$ using standard ordinal or numeric regression methods. Given a new set of objects $\{o'_j\}_{j=1}^M$ to be ranked, these objects are then sorted in decreasing order of their estimated utilities $f_u(o'_j)$. Note that point-wise methods are restricted to linear orders but cannot represent more general relations, such as partial orders.

*Pair-wise methods* proceed from training data in the form of a set of ordered object pairs $S = \{o_{a_i} \succeq o_{b_i}\}_{i=1}^N$, i.e., *relative* training information in the form of pair-wise comparisons rather than absolute assessments. Based on such training samples $S$, the goal is to learn the underlying preference relation $\succeq$. The resulting model $f_{\succeq} : O \times O \to \{0, 1\}$ is a binary classifier, which is supposed to return $f_{\succeq}(o_i, o_j) = 1$ iff $o_i \succeq o_j$.

One of the first pair-wise preference methods was the Ranking SVM [13]— essentially a standard *support vector machine* (SVM) trained on the differences between vector representations of object preference pairs. Later, Burges et al. [5] proposed the RankNet architecture, which is also trained using feature vector differences but uses a *multilayer perceptron* (MLP) instead of an SVM. Since then, multiple extensions of those approaches have been developed [4]. One commonality between all of them is their training optimization target, namely to minimize the number of predicted inversions, i.e., the number of pairs $o_i \succeq o_j$ with $f_{\succeq}(o_i, o_j) = 0$. An important difference between existing pair-wise approaches concerns the properties they guarantee for the learned preference relation; three properties commonly considered are

- reflexivity ($\forall x : x \succeq x$),
- antisymmetry ($\forall x, y : x \not\succeq y \Rightarrow y \succeq x$), and
- transitivity ($\forall x, y, z : (x \succeq y \land y \succeq z) \Rightarrow x \succeq z$).

The set of desirable properties depends on the domain. While some approaches guarantee that the learned relation fulfills all three properties [16], others, for example, explicitly allow for non-transitivity [22].

Assuming a suitable pair-wise ranking model $f_{\succeq}$ was selected and trained, one then typically wants to produce a ranking for some set of objects $\{o_i'\}_{i=1}^{M}$. To this end, a ranking (rank aggregation) procedure is applied to the preferences predicted for all pairs $(o_i', o_j')$. A simple example of such a procedure is to sort objects $o_i$ by their Borda count $c_i = \sum_{j \neq i} f_{\succeq}(o_i, o_j)$, i.e., by counting how often each object $o_i$ is preferred over another object. Alternatively, the classifier $f_{\succeq}$ can also be used directly as the comparator function in a sorting algorithm; this reduces the number of comparisons from $\mathcal{O}(M^2)$ to $\mathcal{O}(M \log M)$. While the latter approach is much more efficient, it implicitly assumes that $f_{\succeq}$ is transitive. The rankings produced by an intransitive sorting comparator are generally unstable, because they depend on the order in which the sorting algorithm compares the objects [19]. This might not be desirable in some domains.

*List-wise methods* generalize the pair-wise setting. Instead of determining the ordering of object pairs, they directly operate on complete rankings (lists) of objects, training a model based on a list-wise ranking loss function. One of the first list-wise losses was proposed by Cao et al. [7]. Given a set $S$ of objects, their ListNet approach uses a probability distribution over all possible rankings of $S$ and is trained by minimizing the cross-entropy between the model's current ranking distribution and some target distribution. Compared to pair-wise approaches, list-wise methods exhibit a higher expressivity, which can be useful to capture effects such as context-dependence of preferences [21]. In general, however, if this level of expressiveness is not required, recent results by Köppel et al. [16] suggest that the list-wise approaches have no general advantage over the (typically simpler) pair-wise methods. To tackle the graph LtR problem in Sect. 4, we will therefore focus on the pair-wise approach.

## 2.2  Neural Pair-Wise Ranking Models

As already stated, we propose a combination of existing LtR methods and GNNs to solve graph ranking problems. Due to the large number of existing LtR approaches, we will however not evaluate all possible combinations with GNNs, but instead focus on the following two representatives:

1. **DirectRanker** [16]: A recently proposed generalization of the already mentioned pair-wise RankNet architecture [5]. It guarantees the reflexivity, antisymmetry, and transitivity of the learned preference relation and achieves state-of-the-art performance on multiple common LtR benchmarks.
2. **CmpNN** [22]: Unlike DirectRanker, this pair-wise architecture does not enforce transitivity. The authors suggest that this can, for example, be useful to model certain non-transitive voting criteria.

Formally, the DirectRanker architecture is defined as

$$f_{\succeq}^{\mathrm{DR}}(o_i, o_j) := \sigma\left(w^\top (h(x_i) - h(x_j))\right), \tag{1}$$

where $x_i, x_j \in \mathbb{R}^n$ are feature vectors representing the compared objects $o_i, o_j$, the function $h : \mathbb{R}^n \to \mathbb{R}^d$ being a standard MLP, $w \in \mathbb{R}^d$ a learned weight vector and an activation function $\sigma : \mathbb{R} \to \mathbb{R}$ such that $\sigma(-x) = -\sigma(x)$ and $\mathrm{sign}(x) = \mathrm{sign}(\sigma(x))$ for all $x \in \mathbb{R}$. One could, for example, use $\sigma = \tanh$ and interpret negative outputs of $f_{\succeq}^{\mathrm{DR}}(o_i, o_j)$ as $o_j \succeq o_i$ and positive outputs as $o_i \succeq o_j$. This model can be trained in an end-to-end fashion using gradient descent with the standard binary cross-entropy loss. Note that $f_{\succeq}^{\mathrm{DR}}$ can be rewritten as $\sigma(f_u^{\mathrm{DR}}(x_i) - f_u^{\mathrm{DR}}(x_j))$, with $f_u^{\mathrm{DR}}(x) := w^\top h(x)$. DirectRanker therefore effectively learns an object utility function $f_u^{\mathrm{DR}}$ and predicts $o_i \succeq o_j$ iff $f_u^{\mathrm{DR}}(x_i) \geq f_u^{\mathrm{DR}}(x_j)$. Thus, the learned preference relation $f_{\succeq}^{\mathrm{DR}}$ directly inherits the reflexivity, antisymmetry and transitivity of the $\geq$ relation. The main difference between DirectRanker and a point-wise regression model is that DirectRanker learns $f_u^{\mathrm{DR}}$ indirectly from a set of object preference pairs. Consequently, DirectRanker is not penalized if it learns some order-preserving transformation of $f_u^{\mathrm{DR}}$. We will come back to this point in Sect. 5.3.

Let us now look at the so-called *Comparative Neural Network* (CmpNN) architecture, which generalizes the DirectRanker approach. The main difference between both is that CmpNN does not implicitly assign a score $f_u(x_i)$ to each object $o_i$. This allows it to learn non-transitive preferences. CmpNNs are defined as follows:

$$f_{\succeq}^{\mathrm{Cmp}}(o_i, o_j) := \sigma(z_{\succeq} - z_{\preceq}), \text{ with} \tag{2}$$

$$z_{\succeq} := \tau(w_1^\top z_1 + w_2^\top z_2 + b'), \quad z_1 := \tau(W_1 x_i + W_2 x_j + b),$$

$$z_{\preceq} := \tau(w_2^\top z_1 + w_1^\top z_2 + b'), \quad z_2 := \tau(W_2 x_i + W_1 x_j + b).$$

Here, $w_1, w_2 \in \mathbb{R}^d$ and $W_1, W_2 \in \mathbb{R}^{d \times n}$ are shared weight matrices, $b, b'$ bias terms, and $\sigma, \tau$ activation functions. Intuitively, $z_\succeq \in \mathbb{R}$ and $z_\preceq \in \mathbb{R}$ can be interpreted as weighted votes towards the predictions $o_i \succeq o_j$ and $o_j \succeq o_i$, respectively. A CmpNN will simply choose the alternative with the largest weight. The key idea behind the definitions in (2) is that the pairs $z_\succeq, z_\preceq$ and $z_1, z_2$ will swap values when swapping the compared objects $o_i, o_j$. Consequently, $f_\succeq^{\mathrm{Cmp}}$ must be reflexive and antisymmetric [see 22]. If we set $W_1 = w_1 = 0$, the voting weights $z_\succeq, z_\preceq \in \mathbb{R}$ reduce to the predictions of a standard MLP $h$ with the input $o_i$ and $o_j$, respectively, i.e., $z_\succeq = h(x_i)$ and $z_\preceq = h(x_j)$. In this case, the CmpNN effectively becomes a DirectRanker model. By choosing non-zero weights for $W_1$ and $w_1$, the model can however also learn non-transitive dependencies between objects. In fact, Rigutini et al. have shown that CmpNNs are able to approximate almost all useful pair-wise preference relations [22, Theorem 1].

## 3    Graph Neural Networks

Over the recent years, GNNs have been successfully employed for a variety of graph ML tasks, with applications ranging from graph classification and regression to edge prediction and graph synthesis. Early GNN architectures were motivated by spectral graph theory and the idea of learning eigenvalue filters of graph Laplacians [3,10]. Those spectral GNNs take a graph $G = (V, E)$ with vertex feature vectors $x_i \in \mathbb{R}^n$ as input and iteratively transform those vertex features by applying a filtered version of the Laplacian $L$ of $G$. Formally, the filtered Laplacian is defined as $\hat{L} = U^\top g(\Lambda) U$, where $L = U^\top \Lambda U$ is an eigendecomposition of $L$ and $g$ is a learned eigenvalue filter function that can amplify or attenuate the eigenvectors $U$. Intuitively, spectral GNNs learn which structural features of a graph are important and iteratively aggregate the feature vectors of the vertices that are part of a common important structural graph feature. Each of those aggregations is mathematically equivalent to a convolution operation. This is why they are referred to as *(graph) convolution layers*.

One important disadvantage of spectral convolutions is their computational complexity, making them especially unsuitable for large graphs. To overcome this limitation, Kipf and Welling proposed the so-called *graph convolutional network* (GCN) architecture [15], which restricts the eigenvalue filters $g$ to be linear. As a consequence of this simplification, only adjacent vertices need to be aggregated in each convolution. Formally, the simplified GCN convolution can be expressed as follows:

$$x_i' = \sigma \left( W \left( \eta_{ii} x_i + \sum_{v_j \in \Gamma(v_i)} \eta_{ij} x_j \right) \right) \qquad (3)$$

Here, $x_i, x_i' \in \mathbb{R}^d$ are the feature vectors of $v_i \in V$ before and after applying the convolution, $\Gamma(v_i)$ is the set of neighbors of $v_i$, $W \in \mathbb{R}^{d \times n}$ is a learned linear operator representing the filter $g$, $\sigma$ some activation function, and $\eta_{ii}, \eta_{ij} \in [0, 1]$ normalization terms that will not be discussed here. After applying a series of such convolutions to the vertices of a graph, the resulting convolved vertex features can be used directly to solve vertex-level prediction tasks, e.g. vertex classification. To solve graph-level problems, such as graph classification or graph ranking, the vertex features must be combined into a single graph vector representation. This is typically achieved via a pooling layer, which could, for example, simply compute the component-wise mean or sum of all vertex features. More advanced graph pooling approaches use sorting or attention mechanisms in order to focus on the most informative vertices [17,28].

Xu et al. [27] show that restricting the spectral filter $g$ to be linear not only reduces the computational complexity but also the discriminative power of the GCN architecture. More precisely, they prove that any GNN using a vertex neighborhood aggregation scheme such as (3) can *at most* distinguish those graphs that are distinguishable via the so-called 1-dimensional *Weisfeiler-Lehman* (WL) graph isomorphism test [6]. GCNs do, in fact, have a strictly lower discriminative power than 1-WL, i.e., there are 1-WL distinguishable graphs, which will always be mapped to the same graph feature vector by a GCN model. In addition to this bound, Xu et al. [27] also propose the *graph isomorphism network* (GIN) architecture, which is able to distinguish *all* 1-WL distinguishable graphs. Recently, multiple approaches going beyond the 1-WL bound have been proposed. The so-called 2-WL-GNN architecture, for example, is directly based on the 2-dimensional (Folklore) WL test [8]. Other current approaches use higher-order substructure counts [2] or so-called $k$-order invariant networks [18].

# 4    Neural Graph Ranking

To tackle the graph LtR problem, we propose the family of *RankGNN* models. A RankGNN is a combination of a GNN and one of the existing neural LtR methods. The GNN component is used to embed graphs into a feature space. The embedded graphs can then be used directly as the input for a comparator network, such as DirectRanker [16] or CmpNN [22]. Formally, a RankGNN is obtained by simply using a GNN to produce the feature vectors $x_i, x_j$ in (1) and (2) for a given pair of graphs $G_i, G_j$. Since all components of such a combined model are differentiable, the proposed RankGNN architecture can be trained in an end-to-end fashion. Despite the simplicity of this approach, there are a few details to consider when implementing it; these will be discussed in the following sections.

## 4.1    Efficient Batching for RankGNNs

In the existing neural LtR approaches for objects $o_i$ that are represented by features $x_i \in \mathbb{R}^n$, efficient batch training is possible by encoding a batch of $k$ relations $\{o_{a_i} \succeq o_{b_i}\}_{i=1}^k$ with two matrices

$$A := \begin{pmatrix} x_{a_1} \\ \vdots \\ x_{a_k} \end{pmatrix} \in \mathbb{R}^{k \times n}, \quad B := \begin{pmatrix} x_{b_1} \\ \vdots \\ x_{b_k} \end{pmatrix} \in \mathbb{R}^{k \times n}.$$

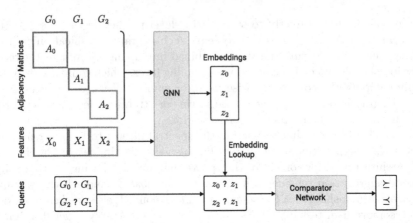

**Fig. 1.** General architecture of the proposed family of RankGNNs. Here the common sparse adjacency representation for message-passing GNNs is shown; different types of graph batch encodings can of course also be used.

and using

$$Y := \begin{pmatrix} 1 \\ \vdots \\ 1 \end{pmatrix} \in \mathbb{R}^k$$

as the target prediction of the model. However, this approach is suboptimal in the graph LtR setting. Given the relations $\{G_0 \succeq G_1, G_1 \succeq G_2\}$, the graph $G_1$ would for example have to be encoded twice. When dealing with datasets that consist of possibly large graphs, such redundant encodings quickly become infeasible due to the additional memory and runtime requirements incurred by the GNN having to embed the same graph multiple times. To prevent this redundancy, each graph occurring on the left or the right side of a relation should instead only be encoded once as part of a single graph batch. This graph batch can be fed directly into a GNN to produce a matrix $Z$ of graph feature embeddings. The individual graph relation pairs $G_i \succeq G_j$ can then be simply represented as pairs of indices $(i, j)$ pointing to the corresponding feature vectors in the embedding matrix $Z$. Using those pointers, the graph vector representations for each pair can be looked up in $Z$. Figure 1 illustrates this idea.

### 4.2   Sorting Graphs with RankGNNs

After training a RankGNN model using a set of graph relation pairs, the model can be used to compare arbitrary graph pairs. Following the approach of Köppel et al. [16] and Rigutini et al. [22], a set of graphs can then be ordered by using the RankGNN as the comparator function in a standard sorting algorithm. We propose a simple parallelized quicksort-based scheme to sort graphs. When implementing a RankGNN model on a parallel compute device, such as a GPU, there is a constant time overhead for each call to the model. To reduce the total cost of this call overhead, we suggest that all pivot comparison queries in one layer of the recursive quicksort call tree should be evaluated by the RankGNN in parallel. Using this parallel comparison approach, only one model invocation is required for each layer of the call tree, i.e., the asymptotic model call overhead for sorting $n$ graphs is in $O(\log n)$. Additionally, a more efficient approach is available for DirectRanker-based models. There, the implicitly learned utility function $f_u^{DR}$ can be computed directly for a set of graphs. A standard sorting algorithm can then be applied without any further calls to the model, which reduces the call overhead to $O(1)$.

## 5   Evaluation

To evaluate the family of RankGNNs described in Sect. 4, we choose six different combinations of GNNs and comparator networks. The evaluated graph embedding modules are GCN [15], GIN [27], and 2-WL-GNN [8]. Those three GNN methods are combined with the previously described DirectRanker [16] and the CmpNN [22] comparator. Because there are currently no common graph ranking benchmark datasets, we instead convert a selection of graph regression benchmarks into ranking problems by interpreting the numeric regression targets as utility values, which are used to determine the target orderings. The following five graph regression datasets are used:

1. **TRIANGLES:** This is a synthetic dataset that we created. It consists of 778 randomly sampled graphs, each of which contains 3 to 85 unlabeled vertices. The regression target is to learn how many triangles, i.e. 3-cliques, a given graph contains. The triangle counts in the sampled graphs vary between 0 and 9. The sampled graphs are partitioned into 80%/10%/10% training/validation/test splits.
2. **OGB-molesol, -mollipo and -molfreesolv:** These three datasets are provided as part of the *Open Graph Benchmark* (OGB) project [11]. They contain 1128, 4200, and 642 molecular structure graphs, respectively. The regression task is to predict the solubility of a molecule in different substances. We use the dataset splits that are provided by OGB.
3. **ZINC:** This dataset contains the molecular structures of 250k commercially available chemicals from the ZINC database [24]. The regression task is to predict the so-called *octanol-water partition coefficients*. We use the preprocessed and presplit graphs from the TUDataset collection [20].

To train the proposed pair-wise graph ranking network architecture, a subset of graph pairs from the training split is sampled uniformly at random. The size of a training sample is $M = \alpha N$, where $N$ is the number of graphs in the training split of a dataset and $\alpha \in \mathbb{R}^+$ is a constant factor. We use a sampling factor of $\alpha = 20$ for all datasets except ZINC, where we use $\alpha = 3$ due to the large number of graphs in the training split ($N_{\text{ZINC}} = 220011$, whereas e.g. $N_{\text{OGB-mollipo}} = 3360$). This sampling strategy guarantees that each training graph occurs in at least one sampled pair with a probability of at least $1 - e^{-2\alpha}$; thus, for both $\alpha = 20$ and even $\alpha = 3$, all graphs are considered with high probability (>99.75%).

In addition to the six pair-wise RankGNN model variants, we also evaluate the ranking performance of standard point-wise GNN graph regression models, which are trained directly on graph utility values. We use two different target graph utilities: The original regression target $y_i \in \mathbb{R}$ for each training graph $G_i$, and the *normalized graph rank* $\bar{r}_i \in [0, 1]$, i.e. the normalized ordinal index of each training graph $G_i$ when sorted by $y_i$.

## 5.1   Experimental Setup

We evaluate the performance of the different RankGNN variants via Kendall's $\tau_B$ rank correlation coefficient. Given two graph rankings $r_1 : \mathcal{G} \to \mathbb{N}$, $r_2 : \mathcal{G} \to \mathbb{N}$, this coefficient is defined as

$$\tau_B := \frac{C - D}{\sqrt{(C + D + T_1)(C + D + T_2)}},$$

where $C$ is the number of concordant pairs

$$\{\{G_i, G_j\} \mid i \neq j \wedge r_1(G_i) < r_1(G_j) \wedge r_2(G_i) < r_2(G_j)\},$$

$D$ is the number of discordant pairs

$$\{\{G_i, G_j\} \mid i \neq j \wedge r_1(G_i) < r_1(G_j) \wedge r_2(G_i) > r_2(G_j)\},$$

and $T_{1,2}$ are the numbers of tied graph pairs, which have the same rank in $r_1$ and $r_2$, respectively. Kendall's $\tau_B$ rank coefficient ranges between $-1$ and $+1$, where $\tau_B = +1$ indicates that the two compared rankings are perfectly aligned, whereas $\tau_B = -1$ means that one rankings is the reversal of the other.

Another commonly used metric in the LtR literature is the *normalized discounted cummulative gain* (NDCG), which penalizes rank differences at the beginning of a ranking more than differences at the end. This is motivated by the idea that typically only the top-$k$ items in a ranking are of interest. We do not employ the NDCG metric because this motivation does not hold for the used target graph rankings. Since the target rankings are derived from regression targets, such as the water solubility of a molecule, both, the beginning and the end of a ranking are of interest and should therefore be weighted equally.

To train the evaluated point- and pair-wise models, we use the standard Adam optimizer [14]. The *mean squared error* (MSE) loss is used for the point-wise regression models, while the pair-wise variants of those GNNs are optimized via binary cross-entropy. All models were tuned via a simple hyperparameter grid search over the following configurations:

1. **Layer widths:** $\{32, 64\}$. The width of both, the convolutional layers, as well as the fully-connected MLP layers that are applied after graph pooling.
2. **Number of graph convolutions:** $\{3, 5\}$. A fixed number of two hidden layers was used for the MLP that is applied after the pooling layer.

3. **Pooling layers:** {mean, sum, softmax}. Here, "mean" and "sum" refer to the standard arithmetic mean and sum operators, as described by Xu et al. [27], while "softmax" refers to the weighted mean operator described by Damke et al. [8].

4. **Learning rates:** $\{10^{-2}, 10^{-3}, 10^{-4}\}$.

We used standard sigmoid activations for all models and trained each hyperparameter configuration for up to 2000 epochs with early stopping if the validation loss did not improve by at least $10^{-4}$ for 100 epochs. The configuration with the highest $\tau_B$ coefficient on the validation split was chosen for each model/dataset pair. To account for differences caused by random weight initialization, the training was repeated three times; 10 repeats were used for the TRIANGLES dataset due to its small size and fast training times. Note that, depending on the type of GNN, the pair-wise models can have between 3% and 10% more trainable weights than their point-wise counterparts, due to the added comparator network. All models were implemented in Tensorflow and trained using a single Nvidia GTX 1080Ti GPU. The code is available on GitHub[1].

## 5.2 Discussion of Results

Table 1 shows the ranking performance of the evaluated point- and pair-wise approaches on the test splits of the previously described benchmark datasets. Each group of rows corresponds to one of the three evaluated GNN variants. The first two rows in each group show the results for the point-wise models that are trained directly on the original regression targets and on the normalized ranks, respectively. The last two rows in each group hold the results for the pair-wise DirectRanker- and CmpNN-based models. Generally speaking, the pair-wise approaches either significantly outperform or at least match the performance of the point-wise regression models. The most significant performance delta between the point- and pair-wise approaches can be observed on the ZINC and OGB-mollipo datasets. Only on the OGB-molesol dataset, the point-wise models achieve a slightly higher average $\tau_B$ value than the pair-wise models, which is however not significant when considering the standard deviations. Overall, we find that the pair-wise rank loss that directly penalizes inversions is much better suited for the evaluated graph ranking problems than the point-wise MSE loss.

Comparing the two evaluated variants of point-wise regression models, we find that the ones trained on normalized graph ranks generally either have a similar or significantly better ranking performance than the regression models with the original targets. We will come back to this difference in Sect. 5.3.

Looking at the results for the synthetic TRIANGLES dataset, we find that only the higher-order 2-WL-GNN is able to reliably rank graphs by their triangle counts. This is plausible, because architectures bounded by the 1-WL test, such as GCN and GIN, are unable to detect cycles in graphs [9]. While both the point-and the pair-wise 2-WL-GNN models achieve perfect or near-perfect $\tau_B$ scores

---

[1] https://github.com/Cortys/rankgnn.

**Table 1.** Mean Kendall's $\tau_B$ coefficients with standard deviations for the rankings produced by point- and pair-wise models on unseen test graphs.

|  |  | TRIANGLES | OGB-molesol | -mollipo | -molfreesolv | ZINC |
|---|---|---|---|---|---|---|
| GCN | Utility regr. | **0.273** ± 0.004 | 0.706 ± 0.001 | 0.232 ± 0.002 | 0.015 ± 0.242 | 0.547 ± 0.414 |
|  | Rank regr. | 0.172 ± 0.040 | 0.702 ± 0.012 | 0.224 ± 0.006 | 0.446 ± 0.038 | 0.823± 0.005 |
|  | DirectRanker | 0.234 ± 0.002 | **0.714** ± 0.003 | 0.327 ± 0.006 | **0.483** ± 0.011 | **0.879** ± 0.002 |
|  | CmpNN | 0.195 ± 0.007 | 0.632 ± 0.076 | **0.381** ± 0.008 | 0.351 ± 0.055 | 0.819 ± 0.002 |
| GIN | Utility regr. | 0.469 ± 0.056 | **0.729**± 0.006 | 0.353 ± 0.052 | 0.243 ± 0.125 | 0.790± 0.003 |
|  | Rank regr. | 0.481 ± 0.014 | 0.717 ± 0.011 | 0.310 ± 0.017 | 0.495 ± 0.021 | 0.827 ± 0.011 |
|  | DirectRanker | 0.502 ± 0.028 | 0.712 ± 0.007 | 0.429 ± 0.026 | 0.439 ± 0.065 | **0.894** ± 0.012 |
|  | CmpNN | **0.520** ± 0.070 | 0.710 ± 0.007 | **0.506** ± 0.013 | **0.518** ± 0.018 | 0.891 ± 0.006 |
| 2-WL | Utility regr. | 0.997 ± 0.006 | **0.747** ± 0.007 | 0.318 ± 0.017 | 0.379 ± 0.207 | 0.803 ± 0.006 |
|  | Rank regr. | 0.972 ± 0.017 | 0.720 ± 0.019 | 0.332 ± 0.083 | 0.524 ± 0.020 | 0.810 ± 0.003 |
|  | DirectRanker | **1.000** ± 0.000 | 0.745± 0.009 | **0.505** ± 0.012 | 0.525 ± 0.010 | **0.894** ± 0.008 |
|  | CmpNN | **1.000** ± 0.000 | 0.718± 0.020 | 0.503 ± 0.010 | **0.527** ± 0.064 | 0.873 ± 0.002 |

on this task, the pair-wise approaches did perform more consistently, without a single inversion on the test graphs over 10 iterations of retraining.

Since the target graph rankings for all evaluated datasets are derived from regression values, all models have to learn a transitive preference relation. Consequently, the ability of CmpNN-based RankGNNs to learn non-transitive preferences is, in theory, not required to achieve optimal ranking performance. If the sample size of training graph pairs is too small, such that it contains few transitivity-indicating subsets, e.g. $\{G_1 \succeq G_2, G_2 \succeq G_3, G_1 \succeq G_3\}$, the higher expressiveness of CmpNNs could even lead to overfitting and therefore worse generalization performance compared to DirectRanker. Nonetheless, with the used sampling factor of $\alpha = 20$ (and $\alpha = 3$ for ZINC), each graph is, in expectation, sampled 40 times (6 for ZINC). This appears to be sufficient to prevent overfitting. In fact, the CmpNN-based RankGNNs perform very similarly to their DirectRanker-based counterparts. However, since DirectRanker-based models allow for a more efficient sorting implementation than CmpNN-based ones (cf. Sect. 4.2), we suggest the use of DirectRanker for problems where transitivity can be assumed.

## 5.3   Analysis of the Implicit Utilities of DirectRanker GNNs

As described in Sect. 2.2, a DirectRanker model $f_{\succeq}^{\mathrm{DR}} : O \times O \to \{0, 1\}$ implicitly learns a utility function $f_u^{\mathrm{DR}} : O \to \mathbb{R}$ from the set of pairs it sees during training. We will now take a closer look at this implicitly learned utility function $f_u^{\mathrm{DR}}$ and compare it to the explicitly learned utilities $f_u^{\mathrm{util.}}$ and $f_u^{\mathrm{rank}}$ of the point-wise GNN regression models. Figure 2 shows the values of all three, $f_u^{\mathrm{DR}}$ (in red), $f_u^{\mathrm{util.}}$ (in blue) and $f_u^{\mathrm{rank}}$ (in gray), normalized to the unit interval. Any monotonically increasing curve corresponds to a perfect ranking ($\tau_B = +1$), while a monotonically decreasing curve would signify an inverse ranking ($\tau_B = -1$).

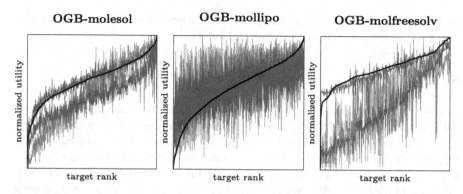

**Fig. 2.** Normalized learned utility values of the point-wise GNN regression model trained on the original utilities (in blue), the point-wise model trained on normalized ranks (in gray) and the pair-wise DirectRanker model (in red). For each dataset, we plot the predicted utilities of the GNN architecture that achieved the best point-wise ranking performance in Table 1, i.e. 2-WL-GNN for OGB-molesol and -molfreesolv and GIN for OGB-mollipo. Each point along the horizontal axes corresponds to a graph in the training split of a dataset. The graphs are sorted in ascending order by the ground truth utility values (shown in black) from which the target rankings are derived. (Color figure online)

As expected, the blue utility curves of the point-wise approaches align with the black target utility curves, while the gray curve more closely follows the 45° diagonal line on which the normalized graph ranks would lie. However, this alignment does not necessarily imply good ranking performance. For example, on the OGB-molfreesolv dataset, the blue utility curve of the point-wise 2-WLGNN model fits the black target curve fairly well for the graphs in the middle of the ranking. However, near the low and the high graph ranks, the target curve abruptly falls/rises to its minimum and maximum values; the point-wise regression model that is trained on the original utilities ignores those outliers. By instead training a point-wise model on the normalized ranks, outliers in the original utility values are effectively smoothed out, as can be seen in the gray OGB-molfreesolv utility curve. Looking at Table 1, we find that this corresponds to a significantly higher mean $\tau_B$ coefficient and a lower variance on the OGB-molfreesolv dataset. The pair-wise DirectRanker-based approach solves the problem of outliers in a more general fashion. It uses a loss function that does not penalize for learning a monotonous, rank-preserving transformation of the target utility curve. This allows it to effectively "stretch" the target utilities into a linearly growing curve with fewer abrupt changes, which results in a similar performance to that of the regression model trained on normalized ranks.

The target utilities of the OGB-molesol dataset are distributed more smoothly, without any outliers. There the advantage of approaches that work well with outliers (e.g. pair-wise models) over the ones that do not is less pronounced. Lastly, looking at the OGB-mollipo dataset, we also do not find outliers in the target utility curve. However, there the pair-wise RankGNN models perform significantly better than the point-wise approaches. The reason for this performance difference is not yet fully understood.

## 6    Conclusion

In this paper, we addressed the problem of learning to rank graph-structured data and proposed RankGNNs, a combination of neural pair-wise ranking models and GNNs. When compared with the naïve approach of using a point-wise GNN regression model for ranking, we found that RankGNNs achieve a significantly higher or at least similar ranking performance on a variety of synthetic and real-world graph datasets. We therefore conclude that RankGNNs are a promising approach for solving graph ranking problems.

There are various directions for future research. First, due to the lack of graph ranking benchmark datasets, we had to use graph regression datasets in our evaluation instead. For a more thorough analysis of the practical applicability of graph ranking models, a collection of real-world graph ranking benchmarks should be created. One potential benchmark domain could, for example, be the drug screening problem we described in the introduction, where the training data consists of drug candidate pairs ranked by a human expert.

Second, list-wise graph ranking approaches could be evaluated in addition to the point- and pair-wise models considered in this paper. Such list-wise models can be useful to learn a human's individual preferences for structured objects, such as task schedules or organizational hierarchies, represented as directed acyclic graphs or trees, respectively. A list-wise ranking approach [e.g. 21] would be able to consider context-dependent preferences in such scenarios [12]. Yet another interesting idea, motivated by the behavior we observed for the point- and pair-wise 2-WL-GNN-based models on the OGB-molfreesolv dataset (cf. Fig. 2), is a hybrid approach that combines regression and ranking, that is, point-wise and pair-wise learning [23].

Third, although graph neural networks are quite popular these days, the problem of graph ranking could also be tackled by well-established kernel-based methods. In the past, there has been a lot of work on graph kernels [25], making graph-structured data amenable to kernel-based learning methods. In principle, one may hence think of combining graph kernels with learning-to-rank methods such as RankSVM. However, our first experiences with an approach of that kind suggest that kernel-based approaches are computationally complex and do not scale sufficiently well, even for point-wise implementations—for larger data sets, the running time as well as the memory requirements are extremely high (which is also the reason why we excluded them from the experiments). Although they can be reduced using suitable approximation techniques, complexity clearly remains an issue. Besides, the ranking performance turned out to be rather poor. For pair-wise approaches, not only the complexity further increases, but the problem also becomes conceptually non-trivial. This is because the simple reduction of ranking to classification, on which RankSVM is based, no longer works (this reduction takes differences between feature vectors, an operation that cannot be applied to graphs). Instead, a (preference) kernel function on pairs of pairs of objects, i.e. on quadruples, has to be used [26]. Nevertheless, this does of course not exclude the existence of more efficient (approximate) algorithms operating on kernel-representation for graphs.

# References

1. Agarwal, S.: Learning to rank on graphs. Mach. Learn. **81**(3), 333–357 (2010)
2. Bouritsas, G., Frasca, F., Zafeiriou, S., Bronstein, M.M.: Improving graph neural network expressivity via subgraph isomorphism counting (2020)
3. Bruna, J., Zaremba, W., Szlam, A., LeCun, Y.: Spectral networks and locally connected networks on graphs (2013)
4. Burges, C.: From RankNet to LambdaRank to LambdaMART: an overview. Technical report, MSR-TR-2010-82, Microsoft Research (2010)
5. Burges, C., et al.: Learning to rank using gradient descent. In: ICML (2005)
6. Cai, J., Fürer, M., Immerman, N.: An optimal lower bound on the number of variables for graph identification. Combinatorica **12**(4), 389–410 (1992)
7. Cao, Z., Qin, T., Liu, T.Y., Tsai, M.F., Li, H.: Learning to rank: from pairwise approach to listwise approach. In: Proceedings of the 24th International Conference on Machine Learning. ACM Press (2007)
8. Damke, C., Melnikov, V., Hüllermeier, E.: A novel higher-order Weisfeiler-Lehman graph convolution. In: Proceedings of the 12th Asian Conference on Machine Learning (ACML 2020). Proceedings of Machine Learning Research, vol. 129. PMLR (2020)
9. Fürer, M.: On the combinatorial power of the Weisfeiler-Lehman algorithm. In: Fotakis, D., Pagourtzis, A., Paschos, V.T. (eds.) CIAC 2017. LNCS, vol. 10236, pp. 260–271. Springer, Cham (2017). https://doi.org/10.1007/978-3-319-57586-5_22
10. Henaff, M., Bruna, J., LeCun, Y.: Deep convolutional networks on graph-structured data (2015)
11. Hu, W., Fey, M., Zitnik, M., Dong, Y., Ren, H., et al.: Open graph benchmark: datasets for machine learning on graphs (2020)
12. Huber, J., Payne, J.W., Puto, C.: Adding asymmetrically dominated alternatives: violations of regularity and the similarity hypothesis. J. Consum. Res. **9**(1), 90 (1982)
13. Joachims, T.: Optimizing search engines using clickthrough data. In: Proceedings of the 8th ACM SIGKDD International Conference on Knowledge Discovery and Data Mining, KDD 2002. ACM Press (2002)
14. Kingma, D.P., Ba, J.: Adam: a method for stochastic optimization. In: ICLR (2015)
15. Kipf, T.N., Welling, M.: Semi-supervised classification with graph convolutional networks. In: ICLR (2017)
16. Köppel, M., Segner, A., Wagener, M., Pensel, L., Karwath, A., Kramer, S.: Pairwise learning to rank by neural networks revisited: reconstruction, theoretical analysis and practical performance. In: Brefeld, U., Fromont, E., Hotho, A., Knobbe, A., Maathuis, M., Robardet, C. (eds.) ECML PKDD 2019. LNCS (LNAI), vol. 11908, pp. 237–252. Springer, Cham (2020). https://doi.org/10.1007/978-3-030-46133-1_15
17. Lee, J., Lee, I., Kang, J.: Self-attention graph pooling. In: ICML, pp. 6661–6670 (2019)

18. Maron, H., Ben-Hamu, H., Serviansky, H., Lipman, Y.: Provably powerful graph networks. In: NeurIPS 2019, pp. 2153–2164 (2019)
19. Mesaoudi-Paul, A.E., Hüllermeier, E., Busa-Fekete, R.: Ranking distributions based on noisy sorting. In: Proceedings of the 35th International Conference on Machine Learning, ICML 2018, Stockholm, Sweden, pp. 3469–3477 (2018)
20. Morris, C., Kriege, N.M., Bause, F., Kersting, K., Mutzel, P., Neumann, M.: TUDataset: a collection of benchmark datasets for learning with graphs (2020)
21. Pfannschmidt, K., Gupta, P., Hüllermeier, E.: Deep architectures for learning context-dependent ranking functions (March 2018)
22. Rigutini, L., Papini, T., Maggini, M., Scarselli, F.: SortNet: learning to rank by a neural preference function. IEEE Trans. Neural Netw. $22(9)$, 1368–1380 (2011)
23. Sculley, D.: Combined regression and ranking. In: Proceedings of the 16th ACM SIGKDD International Conference on Knowledge Discovery and Data Mining, Washington, DC, USA, 25–28 July 2010, pp. 979–988 (2010)
24. Sterling, T., Irwin, J.J.: ZINC 15 – Ligand discovery for everyone. J. Chem. Inf. Model. $55(11)$, 2324–2337 (2015)
25. Vishwanathan, S., Schraudolph, N., Kondor, R., Borgwardt, K.: Graph kernels. J. Mach. Learn. Res. $11$, 1201–1242 (2010)
26. Waegeman, W., Baets, B.D., Boullart, L.: Kernel-based learning methods for preference aggregation. 4OR $7$, 169–189 (2009)
27. Xu, K., Hu, W., Leskovec, J., Jegelka, S.: How powerful are graph neural networks? In: ICLR (2019)
28. Zhang, M., Cui, Z., Neumann, M., Chen, Y.: An end-to-end deep learning architecture for graph classification. In: 32nd AAAI Conference on Artificial Intelligence (2018)
29. Zhang, W., et al.: When drug discovery meets web search: learning to rank for Ligand-based virtual screening. J. Cheminf. $7(1)$, 1–13 (2015)

# Machine Learning for COVID-19

# Knowledge Discovery of the Delays Experienced in Reporting COVID-19 Confirmed Positive Cases Using Time to Event Models

Aleksandar Novakovic[1,2(✉)], Adele H. Marshall[1,2(✉)], and Carolyn McGregor[2,3]

[1] School of Mathematics and Physics, Queen's University Belfast, Belfast, UK
{a.novakovic,a.h.marshall}@qub.ac.uk
[2] Faculty of Business and IT, Ontario Tech University, Oshawa, Canada
c.mcgregor@ieee.org
[3] Faculty of Engineering and IT, University of Technology, Sydney, Australia

**Abstract.** Survival analysis techniques model the time to an event where the event of interest traditionally is recovery or death from a disease. The distribution of survival data is generally highly skewed in nature and characteristically can include patients in the study who never experience the event of interest. Such censored patients can be accommodated in survival analysis approaches. During the COVID-19 pandemic, the rapid reporting of positive cases is critical in providing insight to understand the level of infection while also informing policy. In this research, we introduce the very novel application of survival models to the time that suspected COVID-19 patients wait to receive their positive diagnosis. In fact, this paper not only considers the application of survival techniques for the time period from symptom onset to notification of the positive result but also demonstrates the application of survival analysis for multiple time points in the diagnosis pathway. The approach is illustrated using publicly available data for Ontario, Canada for one year of the pandemic beginning in March 2020.

**Keywords:** COVID-19 · Survival analysis · Process mining · Knowledge discovery · Process discovery and analysis

## 1 Introduction

On the 11th March 2020, the World Health Organisation declared, after reporting cases in 114 different countries, that the Coronavirus 2019 disease outbreak had become a global pandemic. Like many other countries, COVID-19 has had a significant impact on Canada with the first reported case in January 2020 and the first community acquired case in early March 2020. The total number of COVID-19 positive cases in Canada now is almost 1.4 million (as of 10th June 2021), with most found to be in Ontario and Quebec. In Ontario there have been 538,651 cases (as of 10th June, 2021) representing a rate of 3623.8 per 100,000 population with 8,935 deaths representing 60.1 per 100,000

A. Novakovic and A.H. Marshall—Sharing First Authorship

population. The spread of the virus in Ontario has resulted in significant deaths, the significant disruption of learning in schools and an impact on business and the economy through government mandated closures.

Like many other countries, Canada introduced non-pharmaceutical interventions (NPIs) to fight the pandemic and control the spread of the disease infection in the population. As one of its NPIs, Ontario applied school closures during the first wave of the COVID-19 pandemic. While children were seen to be less likely to manifest symptoms, there was concern that they could still transmit the virus to older more vulnerable members of their household or high risk members of the community [1]. As a result, school children attended school online for the remainder of the school year to June 2020. Prior to schools returning, Phillips et al. [2] proposed an Agent Based Model of transmission within a childcare centre and households. This considered family clustering of students along with student-to-educator ratios to propose approaches for childcare and primary school reopening. In September, schools reopened in Ontario for face to face learning with significantly altered organizational structures to reduce student interactions.

Naimark et al. [1] created an agent-based transmission model that clustered a synthetic population of 1 million individuals into households, and neighbourhoods, within either rural districts, cities or rural regions. They further allocated the synesthetic population based on a life stage construct to either attending day care facilities, classrooms (elementary/high school), colleges or universities, and workplaces. Based on their modelling, they concluded that school closures were a lower priority NPI than more broader community encompassing NPIs such as reducing contacts outside the household and closing non-essential workplaces. However, their research proposed projections through to 31$^{st}$ October. While this model considered the potential rate of transmission within schools and the follow on effect of the risk of spread, they did not consider another important aspect of the increased burden on testing due to schools returning.

When children returned to school in Ontario on the 9$^{th}$ September 2020, new public health guidelines were in place requiring any child exhibiting any symptoms, from a defined list, to not be allowed to return to school unless they were symptom free or had a negative COVID-19 test. In addition, children wishing to play hockey were also required to be tested. This placed a significant load on testing both within laboratories and within primary care locations who were required to assess and request the test. As a result, significant delays of several days were experienced before a specimen for testing was collected. Alongside NPIs, considerations for diagnostic COVID-19 tests must also be made with rapid accurate diagnostic tests for rapid detection of patients actively pursued to improve on diagnostic preparedness and deal with future global waves or regional outbreaks [3]. These are commonly referred to as Test and Trace or Test-Trace-Isolate or Test-Trace-Quarantine strategies where success is dependent on "high rates of routine testing, rapid return of test results, high rates of contact tracing, and social support for people who have been diagnosed or quarantined" [4]. The implementation of each of these components is critical in delivering a successful strategy, one which has already presented challenges and criticism for the United States and United Kingdom. Torres et al. [5] conducted analysis on delays in processing COVID-19 tests for Ecuador where they report an average national processing time of 3 days until results are communicated to the patient and local authorities and approximately 12% of patients not receiving results within 10 days. They conclude that such delays may mask a case burden higher than what is reported, that it could impede timely awareness, adequate clinical care provision and vaccination strategies and subsequent monitoring. Meanwhile in Australia, a public inquiry into the Victorian Government's COVID-19 contact tracing system and testing regime concluded that 90% of tests and reporting positive cases should be completed within 24 hours' and positive case's close contacts should be notified within 48 hours in order to reduce onward transmission by up to 40 per cent [6]. This suggests a benchmark for countries when assessing their strategies. The timely detection and isolation of positive cases is of utmost importance in preventing the spread of disease in the community [6]. This is key for informing policy for the introduction of NPIs and preparedness for future infection waves. In this paper we explore the time to reporting of COVID-19 positive cases in Ontario, Canada. To date, such data reporting has used descriptive statistics of average times for reporting visualized through graphical methods. However, the time to reporting is highly skewed in nature with large variability in the data [5]. Hence, in this research, we provide novel knowledge discovery of delays in COVID-19 reporting using a robust statistical technique. Survival analysis is designed to model time to event skewed data and hence ideally placed to make a robust assessment of case reporting data.

We propose creating a statistical distribution to represent the delay times and explore how this can be developed and use the limited number of other information about the patients who are delayed to inform the model. We do this for Ontario, Canada which has openly available data for the times between the key events in the testing process. Due to the skewness of the data and the time to event nature, we use the survival analysis approaches of Kaplan-Meier to estimate survival curves to evaluate the percentage completed within certain time limits and the Cox proportional hazards method to explore the other information on the tests/patients to evaluate statistical significance.

The remainder of the paper has the following structure. In Sect. 2 we describe the publicly available dataset that has been used in the study and provide the theoretical foundations for using survival analysis for modelling the delay in information flow between different stages of the COVID-19 reporting process. In Sect. 3 we provide a detailed presentation and discussion of the results obtained. Finally, in Sect. 4 we provide our concluding remarks in addition to the directions for our future research.

## 2 Methods

### 2.1 An Overview of COVID-19 Dataset

In this study we use publicly available data on positive COVID-19 cases in Ontario, Canada from the Ontario Ministry of Health and Long-Term Care's consolidation of Public Health Unit (PHU) data [7]. For each positive case the information provided includes, the estimated date of the symptom onset of disease, the exact dates for when their specimen was collected and processed in the lab, and when the lab notified the PHU about the positive case. We assume that each of these four dates represent one stage in the COVID-19 reporting process, with the entire timeline illustrated in Fig. 1.

To investigate the effectiveness of the COVID-19 reporting process during the time that encompasses all three infection waves, we focus on analysing the incidence records that were collected in the period between 1st April 2020 and 31st March 2021. In addition to the four previously mentioned dates, we aggregated the dataset to also include the information on patients' age and outcome. The patient ages are represented in 10 year age bands ("20s", "30s", ..., "90+"), except for those younger than 20 and those whose age was not recorded or unknown, who are labelled as " <20" and "Unknown", respectively. The patient outcomes are classified as fatal, resolved or not resolved.

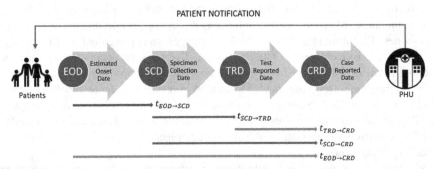

**Fig. 1.** End to end COVID-19 reporting process, with $t_{EOD \to SCD}$, $t_{SCD \to TRD}$, $t_{TRD \to CRD}$, $t_{SCD \to CRD}$, and $t_{EOD \to CRD}$, denoting the elapsed time in days between: symptoms onset and specimen collection, specimen collection and lab processing, lab processing and PHU notification, specimen collection and PHU notification and symptoms onset and PHU notification, respectively.

## 2.2    Survival Analysis Approach for Modelling the Delays in the COVID-19 Information Reporting Process

Survival analysis is a collection of methods used to model the time until an event of interest occurs (survival time). Typical to survival analysis is the highly skewed nature of the survival times when plotted on a graph which generally will peak at the beginning and slowly tail off to the right as fewer individuals will experience the event as time progresses. A common feature in survival analysis is that of censoring where the patient may not have experienced the event over the course of the study period [8, 9].

Traditionally, survival analysis models the time to an event occurring where the event of interest is the patient experiencing an event such as death or recovery [9]. In the context of this paper, the event of interest is the patient experiencing a delay of at least one day in the overall end-to-end COVID-19 reporting process. Our aim is to investigate which stages of this process takes excessive time which may suggest a back log or bottle neck in the process. Hence the survival time is the time elapsed in days between the different stages in the reporting process with longer survival times indicating slower information flow from one stage to another. Of note, a record is considered to be censored when the reporting from one stage to another occurred on the same day, as this would suggest no delay so the event of interest was not experienced.

In this manuscript we propose an approach in which the delay in information flow between different stages of reporting process is modeled with a survival function that is computed according to the Kaplan-Meier (KM) estimator [10]. This survival function, denoted as $S_{x \to y}(t)$, can be interpreted as the function that gives the probability that the delay in information flow from stage $x$ to stage $y$ (i.e. $x \to y$) in the record processing pipeline will be experienced after at least the specified amount of $t$ days, where $x \to y \in \{EOD \to SCD, SCD \to TRD, TRD \to CRD, SCD \to CRD, EOD \to CRD\}$. Using the KM estimator the survival function is approximated as follows:

$$\hat{S}_{x \to y}(t) = \prod_{k:t_k \le t} \left( 1 - \frac{d(t_k)}{n(t_k)} \right) \tag{1}$$

where $t_k$ indicates the time in days when at least one record experienced a delay, $d(t_k)$ indicates the number of records that experienced a reporting delay of $t_k$ days, and $n(t_k)$ indicates the number of records that are still at risk of experiencing a reporting delay at time $t_k$ (i.e. the records that are neither censored nor delayed prior to time $t_k$).

To investigate how the delays in the information flow between different stages affected the entire COVID-19 reporting process, and how these delays evolved over the analysed time period, we fitted survival curves for each calendar month separately. We then performed pairwise comparisons using log-rank tests to investigate statistically significant differences in the distributions of their survival times [11]. Given that the multiple pairwise comparisons increase the chance of committing Type I errors, we used the Bonferroni correction to adjust the test significance level. This information allows us to better understand which months the effect of the reporting delay on the survival times was the strongest, and whether it has the tendency to improve over time.

One of the limitations of the KM estimator is that it cannot account for the simultaneous effects of multiple covariates on the survival function. To overcome this problem we used the Cox proportional hazards model which is linear and semi-parametric technique that allows the assessment of the effect that each covariate has on the rate of occurrence of the event of interest (i.e. experiencing delay in reporting process) at a specific point of time [12]. This rate is also known as the hazard rate [13], calculated as follows:

$$h_{x \to y}(t) = h_{x \to y}^0(t) \cdot e^{\sum_{i \in \{month, age, outcome\}} \beta_i X_i} \tag{2}$$

where $h_{x \to y}^0(t)$ indicates the baseline hazard (the probability that the delay will be experienced at time $t$ if all covariates $X_i$ are equal to 0), and $\beta_i$ the coefficients that measure the effect of covariates $X_i$. If any of the coefficients $e^{\beta_i}$, the hazard ratios equal one, this will indicate that the corresponding covariate $X_i$ has no effect on the length of survival time, the reporting process delay length. Values greater than one are associated with shorter delay times, and vice versa. The entire analysis presented in this paper has been performed using the R v4.0.3 programming language and its *survival* library [14].

## 3   Results and Discussions

The publicly available dataset described in Sect. 2.1 consists of 351,419 instances of reported positive COVID-19 cases. Of those, 25,923 cases were omitted from the analysis because at least one of the dates did not appear ranked in the order required by the COVID-19 reporting process shown in Fig. 1 (eg TRD occurred before SCD).

The mean overall reporting delays between different stages of the COVID-19 reporting process are depicted in Fig. 2 a). We can see that as time progresses the mean overall time from symptoms onset to reporting to PHU (i.e. $EOD \rightarrow CRD$) significantly decreased, from 6.26 (95% CI = [6.16, 6.37]) days in April 2020 to 3.56 (95% CI = [3.52, 3.60]) days in March 2021. As this overall delay time can be expressed as the sum of the delay times between each individual stage, we can see that the largest contributing factors to such long reporting times are the delays experienced between the stage symptom onset to specimen collection (i.e. $EOD \rightarrow SCD$) and from specimen collection and processing in the lab (i.e. $SCD \rightarrow TRD$). This suggests that those infected could continue to spread the disease possibly also reflecting a lack of knowledge of the general public for the importance of reporting early, which has improved as time has progressed (as more public health awareness campaigns are rolled out).

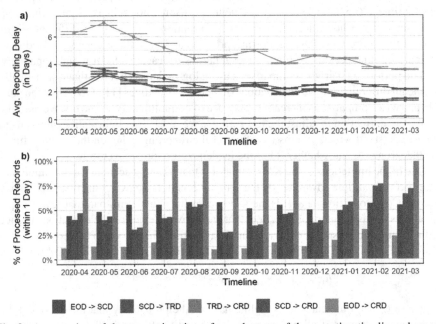

**Fig. 2.** An overview of the processing times for each stage of the reporting timeline where **a)** represents the average reporting time delay and **b)** represents the percentage of processed records within one day for each stage for each month of the period under investigation.

The longest delays were in both cases in April and May 2020, which have been significantly reduced by the end of the study period. The experienced delays in information processing in the $EOD \rightarrow SCD$ stage were 4.01 (95% CI = [3.91, 4.11]) and 3.6 (95% CI = [3.48, 3.73]) in April and May 2020 respectively, and 2.12 (95% CI = [2.09, 2.15]) days, a year later in March 2021. In the same time period, delays experienced in the $SCD \rightarrow TRD$ stage were 2 (95% CI = [1.96, 2.04]), 3.22 (95% CI = [3.1, 3.34]) and 1.31 (95% CI = [1.3, 1.32]) days, for April 2020, May 2020 and March

2021 respectively. Figure 2 b) shows that the main reason for significant improvement in delay times is due to the significant increase in records being processed within one day. Fluctuations in processing times drive the mean reporting times. Interestingly, PHU notification is almost instant. The average delay times in information flow between lab reporting and PHU notification ($TRD \rightarrow CRD$) are uniformly distributed with records processed within one day increasing from 95% to 99%, when comparing April 2020 with March 2021 and the average processing times for $TRD \rightarrow CRD$, 0.255 (95% CI = [0.237, 0.273]) days and 0.13 (95% CI = [0.11, 0.15]) days, respectively.

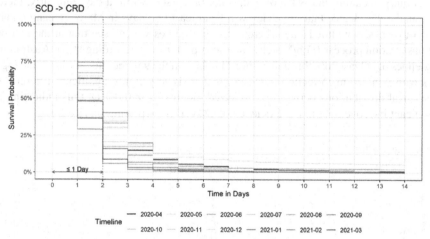

**Fig. 3.** The Kaplan-Meier estimated survival curve for delays in the information flow in the COVID-19 reporting process for the overall time from specimen collection to PHU notification.

With the $TRD \rightarrow CRD$ reporting delays being very short and almost constant during the observed time period, we can observe that the delays in information flow between $SCD \rightarrow TRD$ stages are the main drivers behind increased processing times between specimen collection and reporting to PHU ($SCD \rightarrow CRD$). The average delay between $SCD \rightarrow CRD$ stages followed the same trend as the delay between $SCD \rightarrow TRD$ stages and was 2.25 (95% CI = [2.22, 2.29]) days in April 2020 and 1.44 (95% CI = [1.42, 1.47]) days in March 2021.

In order to analyse how the quality of information flow between the stages of the COVID-19 reporting process changes over time, we apply survival analysis as introduced in Sect. 2.2. Consider Fig. 3. which represents the Kaplan-Meier estimate of the survival distribution for the probability that the delay in information flow from stage SCD to stage CRD will be experienced after at least $t$ days. The vertical dashed line indicates the time window by which [7] proposes 90% of the cases should have reported within 24 hours. From this data set, this is only the case for February and March 2021.

It is apparent that as time progresses, the delays become smaller and hence the system appears to becoming more efficient. It is also clear that there is a lot of variation between months so to further explore these differences we consider the most extreme months in Fig. 4 and examine the estimated survivals for the different component parts within

*SCD* → *CRD*. Figure 4 a) shows the largest delay occurring in the pipeline between *SCD* → *TRD* for September 2020, whereas the delay for *TRD* → *CRD* (Fig. 4 b), shows September 2020 performing as good as the best performing months of February and March 2021. Although this may appear odd especially given there are fewer numbers of COVID-19 confirmed cases in September 2020, it is likely that the change in policy outlined in Sect. 1 could have impacted on the processing times for *SCD* → *TRD* where there was quite likely many more school children presenting for COVID-19 in order to enable them to return to School in September. There is around a 6% chance that the delay in the information flow between *SCD* → *CRD* stages will be at least two days in February, meaning that 94% of delays get processed within a day. Similarly, there is around a 9% chance that this delay will be at least two days in March, meaning that 91% of delays is within 1 day. Therefore in both cases we can see that at the end of the observation process (Feb/March) Ontario is performing according to the Australian guidelines of at least 90% through within 24 hours. At the 95% confidence level, the log-rank with Bonferroni correction suggests that there is very highly significant differences in survival distributions between the selected months ($p < 0.0001$), thus indicating a significant improvement in the efficiency of COVID-19 reporting services.

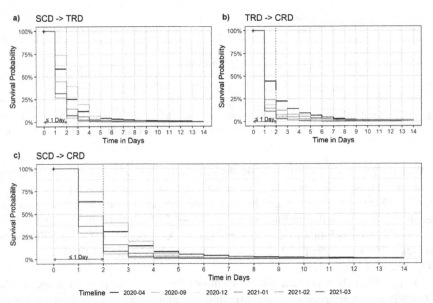

**Fig. 4.** The Kaplan-Meier estimated survival curve for delays in the COVID-19 reporting process from **a)** specimen collection to lab test processing, **b)** lab test processing to PHU notification and **c)** the overall time from specimen collection to PHU notification, for selected months.

The results of the Cox proportional hazards test also highlight some interesting insights where the baseline reference group represent those cases reported positive in April 2020, who are less than 20 years and whose outcome is fatal. With all else remaining constant, if we consider different age groups, most hazard ratios are close if not exactly one. The last column in Fig. 5 provides the p values resulting from testing the null hypothesis that the hazard ratio value is not significantly different to one. However, the 20–40 years age group are showing a significantly different delay than the under 20 year olds and likewise, there is a very highly significant difference for patients who are over 80 years of age with a much higher delay for the older age groups, when all other factors remain constant. This may be explained by the mechanism in which the specimens are collected and delivered to the testing lab for instance differences in mobility may have an influence where those aged over 80 years may have to wait to get their specimens collected and taken to the labs for testing or be in nursing homes where specimens may only be collected in batches at specific times for all residents.

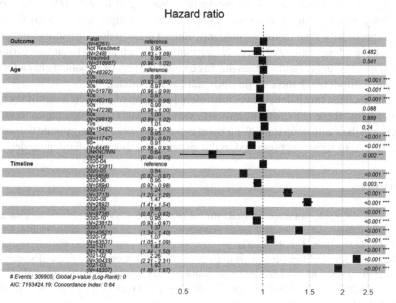

**Fig. 5.** Hazard ratio values calculated using the Cox Proportional Hazards approach for patient characteristics age group, timeline and outcome.

It is also clear from Fig. 5, that there is significant variation in the delay times compared to the baseline of April 2020. This again confirms earlier analysis that there is a large amount of variation in delay times depending on the time (month) when the test is conducted. It seems reasonable that the delays in April 2020 (not long after the pandemic had spread and grown in the community) would be significantly longer than later in the pandemic when it would be expected that testing and healthcare facilities would have had time to allocate resources to the testing process and to organize and co-ordinate testing. As would be expected, there is significant improvement in processing times in July and August 2020 when the case numbers had reduced placing less demand on the facilities while again in December 2020, we see significantly slower processing times when Ontario was experiencing another wave (peak in cases), with more demand placed on resources. Interestingly, for September 2020, Cox proportional hazards shows a significant increase in processing similar to that observed in earlier analyses.

# 4  Conclusions and Further Research

Rapid reporting of COVID-19 positive cases is key in providing governments with insight on the level of infection in the country and inform policy through the introduction of NPIs and preparedness for future infection waves. Failure to report quickly means a delay in critical decision making at the policy level and may lead to an escalation in cases. The time to reporting data is typically highly skewed in nature with large variability thus providing motivation for the research in this paper where we provide novel knowledge discovery of delays in COVID-19 reporting using the very novel application of robust survival models. The approach has been illustrated using publicly available data for Ontario, Canada for one year of the pandemic beginning in March 2020. The survival methodology used in this paper garners additional insights regarding the delay times and identify specifically what factors significantly impact these. It does so using visual graphs of the Kaplan-Meier estimated survival distributions which are useful at comparing the different months' delays and extends this to consider the impact of patient characteristics on the delay times using the Cox proportional hazards model.

There has been much debate over the timely follow up of contacts hence if such data were available, there is the opportunity to extend the analysis in this paper to include contacts and their follow up time. We wish to expand our work to incorporate the survival model of time to reporting presented in this paper into a disease infection model.

# References

1. Naimark, D., et al.: Simulation-based estimation of SARS-CoV-2 infections associated with school closures and community-based nonpharmaceutical interventions in Ontario, Canada. JAMA Netw. Open. **4**, 213793 (2021)
2. Phillips, B., et al.: Model-based projections for COVID-19 outbreak size and student-days lost to closure in Ontario childcare centres and primary schools. Sci. Rep. **11**, 6402 (2021)
3. Vandenberg, O., et al.: Considerations for diagnostic COVID-19 tests. Nat. Rev. Microbiol. **19**, 171–183 (2021)
4. Kerr, C.C., et al.: Controlling COVID-19 via test-trace-quarantine. Nat. Commun. **12**, 2993 (2021)
5. Torres, I., Sippy, R., Sacoto, F.: Assessing critical gaps in COVID-19 testing capacity: the case of delayed results in Ecuador. BMC Pub. Health **21**, 1–8 (2021)
6. Office of the Chief Scientist: National Contact Tracing Review A report for Australia's National Cabinet (2020)
7. Ontario Ministry of Health: Confirmed Positive Cases of COVID-19 in Ontario - Datasets - Ontario Data Catalogue. https://bit.ly/3yCiFai
8. Austin, P.C.: A tutorial on multilevel survival analysis: methods, models and applications. Int. Stat. Rev. **85**, 185–203 (2017)
9. Marshall, A.H., Novakovic, A.: Analysing the performance of a real-time healthcare 4.0 system using shared frailty time to event models. In: 2019 IEEE 32nd International Symposium on Computer-Based Medical Systems (CBMS), pp. 622–627 (2019)
10. Kaplan, E.L., Meier, P.: Nonparametric estimation from incomplete observations. Source J. Am. Stat. Assoc. **53**, 457–481 (1958)
11. Kartsonaki, C.: Survival analysis. Diagn. Histopathol. **22**, 263–270 (2016)
12. Nemati, M., Ansary, J., Nemati, N.: Machine-Learning Approaches in COVID-19 Survival Analysis and Discharge-Time Likelihood Prediction Using Clinical Data. Patterns 1 (2020)

13. Marshall, A.H., Zenga, M., Kalamatianou, A.: Academic students' progress indicators and gender gaps based on survival analysis and data mining frameworks. Soc. Indic. Res. **151**(3), 1097–1128 (2020). https://doi.org/10.1007/s11205-020-02416-6
14. Therneau, T.M., Grambsch, P.M.: Modeling Survival Data: Extending the Cox Model. Springer, New York (2000)

# Multi-scale Sentiment Analysis of Location-Enriched COVID-19 Arabic Social Data

Tarek Elsaka[1,2]([⊠]), Imad Afyouni[1], Ibrahim Abaker Targio Hashem[1], and Zaher AL-Aghbari[1]

[1] Computer Science Department, University of Sharjah, Sharjah, UAE
telsaka@shaarjah.ac.ae, {iafyouni,ihashem,zaher}@sharjah.ac.ae
[2] Agricultural Research Center, Cairo, Egypt

**Abstract.** After the recent outbreak of COVID-19, researchers have risen working on several challenges related to the mining of social data to learn about people's reactions to the epidemic. Recent studies have largely focused on extracting current themes and inferring broad attitudes, with a particular emphasis on the English language. This study presents various perspective for Arabic social data mining to provide in-depth insights related to the COVID-19 pandemic. We initially devised a method for inferring geographical whereabouts from Arabic tweets not initially geotagged. Secondly, a sentiment analysis mechanism based on Arabic word embeddings is introduced, with several levels of geographical granularity (regions/countries) considered. Sentiment-based classifications of topics and subtopics related to COVID-19 will also be presented. According to our findings, the overall percentage of location-enabled tweets has increased from 2% to 46% (about 2.5M tweets). During the pandemic, Arab Twitter users' negative emotions about lockdown, restriction, and law enforcement were also widely expressed.

**Keywords:** Arabic tweets · COVID-19 pandemic · Sentiment Analysis · Social data mining · Arabic COVID-19

## 1 Introduction

Social media has turned into a home for a variety of real-life events that may occur in our daily lives (such as today's top trending issues, the COVID-19 epidemic) [15]. Specifically, millions of Arab users utilize social media to communicate and contribute a significant amount of daily Arabic material, particularly on Twitter. As a result, researchers used the Arabic material on Twitter to learn more about people's opinions, thoughts, and feelings related to the COVID-19 epidemic. Sentiment Analysis (SA) is one of the areas related to social data mining [21]. Recent research on SA has centered on evaluating social data by identifying current themes and inferring general attitudes from related subjects, with a particular focus on the English language. However, sentiment research connected to COVID-19 on Arabic social media has not been properly addressed.

© Springer Nature Switzerland AG 2021
C. Soares and L. Torgo (Eds.): DS 2021, LNAI 12986, pp. 194–203, 2021.
https://doi.org/10.1007/978-3-030-88942-5_15

Several researchers have been interested in natural language processing (NLP) approaches such as word embedding models and n-gram feature weighted by TF-IDF for topic analysis and classification, even for non-COVID-19 related information. Other researchers, on the other hand, used feature extraction in feature-based sentiment analysis to determine sentiment polarity and predict sentiments in social data, with the use of Machine Learning (ML) classifiers to evaluate their findings. However, there has been some investigation into the correlation between official health statistics and social media content.

In this study, we analyze Arabic social data from Twitter related to the COVID-19 pandemic to identify people's sentiments. We want to examine the effects of the worldwide pandemic on a variety of factors at many spatial and temporal levels. This study presents a comprehensive social data mining approach for the Arabic language, by employing Arabic-specific word embedding techniques. Our approach presents several unique contributions over existing works as follows:

1. From January 2020 to November 2020, we gathered Arabic tweets regarding COVID-19 from publicly available datasets (about 5.5M tweets). Then, based on user profiles and textual content, we created a location inference approach for non-geotagged tweets, which increased the total percentage of location-enabled tweets from 2% to 46%. (about 2.5M tweets). We created a Geo-Database that includes bi-lingual (English and Arabic) names of international countries and capitals, as well as well-known Arab cities.
2. Developed an innovative sentiment analysis technique that uses unique insights from users' responses on important topics (like lockdown and vaccination) at multiple degrees of geographic granularity (regions, nations, and cities).
3. Based on the produced geo-social dataset, sentiment analysis, official health records, and lockdown data, a set of studies was conducted.

The following is a description of the paper's structure: A review of some related work is included in Sect. 2. Section 3 presents the suggested methodology's description and execution. The suggested methodology's results and findings are described in Sect. 4. Section 5 concludes with some last comments and future research objectives.

## 2    Related Work

The SA process reflects the users' opinions in a variety of ways and a variety of linguistic styles. Researchers began examining COVID-19-related social media content in the early days of 2020. Previous research on social media content has mostly focused on English tweets regarding COVID-19 or other Latin languages while few have looked at Arabic. Many research works prompted topic analysis to demonstrate the popular subjects talked on social media. Other researchers used feature extraction in feature-based sentiment analysis to detect sentiment polarity and predict sentiment in social data, while others used feature extraction in feature-based sentiment analysis to assess sentiment polarity and forecast sentiment in social data [6]. Many of them utilized machine learning classifiers to improve the results of semantic analysis. Our evaluation of the most research papers on social streams, notably in Arabic, is organized in the sections below.

### 2.1 Collection and Classification of Social Data

Recent research has mostly focused on analyzing social data by extracting trending topics and inferring general opinions from related subjects, with a particular emphasis on English material and less on Arabic content. Many research works have been driven by the desire to acquire social datasets to be shared. Furthermore, these datasets were utilized in statistical analysis studies such as Alanazi et al. [2] and Haouari et al. [13]. Some researchers, like Alharbi [3], discovered a coronavirus collection of Arabic tweets, mostly from three Saudi social media streams. Furthermore, several research works focused on the study of Twitter datasets for classification, such as Hamdy et al. [11]. Some researchers, such as Qazi et al. [23], researched location-enabled elements of social data and released the GeoCoV19, a large-scale Twitter dataset connected to the COVID-19 epidemic. They derived their geolocation information using a gazetteer-based technique to extract toponyms from user location and tweet content using Nominatim (Open Street Maps) data at geolocation granularity levels. Lamsal [17] also released the COV19Tweets Dataset, a large-scale English language tweets dataset including sentiment ratings. They created the GeoCOV19Tweets Dataset by filtering the COV19Tweets Dataset's geotagged tweets, which comprises just 141k tweets (0.045%).

### 2.2 Topic and Semantic Analysis

Alshalan et al. [5] analyzed hate speech related to the COVID-19 epidemic in the Arab world using the ArCov-19 dataset [13]. Likewise, Alsafari et al. [4] developed an Arabic hate and offensive speech detection system in response to the growing prevalence of hate speech on social media. As Well, Hamoui et al. [12] examined the Arabic content on Twitter to see what the most popular topics were among Arabic users. Similarly, Al-Laith et al. [1] used tweets and a rule-based approach to categorize 300,000 tweets to assess people's emotional reactions during the COVID-19 epidemic. While, Bahja et al. [7] showed the preliminary findings of determining the relevance of tweets and what Arab people wrote regarding COVID-19 feelings/emotions (Safety, Worry, and Irony). However Essam and Abdo [10] looked into how Arabs on Twitter are dealing with the COVID-19 epidemic. Manguri et al. [18] collected Twitter data from Twitter social media and then did sentiment analysis. Furthermore, Chakraborty et al. [8] showed that tweets containing all important handles for COVID-19 and WHO failed to properly lead people through the pandemic outbreak. In addition, Kabir et al. [14] developed a neural network model and trained it using manually labeled data to recognize various emotions in Covid-19 tweets at fine-grained labeling.

## 2.3   Discussion:

To summarize, the authors attempted to analyze COVID-19-related social data many times. They used numerous Arabic-language social media datasets collected over a while, mostly in the year 2020. Researchers looked at social media postings to see how people are reacting to the COVID-19 epidemic. The current research is based on English data, with Arabic data receiving fewer contributions. Furthermore, previous research on Arabic social data does not take into account the spatial-temporal component of COVID-19 material. In addition, the relationship between official health statistics and social media content has not been thoroughly investigated.

# 3   Sentiment Analysis of Arabic COVID-19 Tweets

Using machine learning models and topic detection and tracking techniques, we propose a method for automatically detecting and processing social datasets including Arabic tweets related to the COVID-19 epidemic. The workflow of our technique is depicted in Fig. 1, and the subsections following discuss each phase in greater depth.

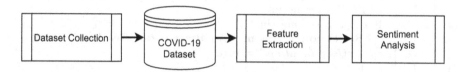

**Fig. 1.** The Workflow of our methodology to analyze the social data.

## 3.1   Dataset Collection

We began by collecting Arabic tweet data from the two openly available Arabic datasets for COVID-19 tweets, [2] (3,314,859 tweets) and [13] (2,111,650 tweets). Both datasets were collected by matching Arabic tweets with a list of COVID-19-related keywords (such as Coronavirus, Corona, and Pandemic) often used by ordinary people, news organizations, and government agencies. We filtered tweets in both datasets using the time frame of January 1, 2020, to November 30, 2020. We gathered tweets from both datasets and created a new geo-tagged dataset from them to feed our studies. Unfortunately, the provided datasets only contain tweet IDs due to Twitter's privacy policy. As a result, we hydrated (recollected) the dataset to retrieve the complete tweet objects from Twitter. The new dataset comprises about 5.5 million tweets (5,054,141 unique tweets and around 115,561 geo-tweets - about 2% only), with an average of 21 words per tweet.

## 3.2   Features Extraction

In our approach, we developed in the "Feature Extraction" module some processes such as "Prepare Dataset", "Infer Location-Enabled Tweets" and "Extract Location Features". The process "Prepare Dataset" contains some sub-processes like "Clean Dataset", "Filter Fields" and "Prepare Arabic Text".

---

**Algorithm 1:** Location Extraction

---

1  **begin**
2    *"Tweets Dataset"* = loadTweetsCorpus()
3  **for** *each tweet in Tweets Dataset* **do**
4     **if** *"Tweets Place" is not **None*** **then**
5       country = *"Tweet Country"*;
6       **if** *country is not **None*** **then**
7         place name = *"Tweets Place"*;
8         country code = *"Country Code"*;
9         coordinates = *"Tweet Coordinates"*;
10      **end**
11    **else**
12      **if** *"Tweets Place" is **None** OR country is **None*** **then**
13        **Select country from** *"Countries-Cities Database"* **where** *"TU"* = country;
14        **if** *country is **None*** **then**
15          **Select country from** *"Countries-Cities Database"* **where** *"User Location"* = city;
16        **end**
17        country code, coordinates = retrieveData("Geo-Location Database");
18      **end**
19    **end**
20 **end**

---

Each tweet object includes numerous metadata elements that provide a wealth of information. The geo-location information, such as "Place" and "User" Information, are essential pieces of information that identify the tweet's origin. Unfortunately, they always need refining because they are dependent on optional entry by users. Furthermore, it might be written in a variety of languages or contains inaccurate information. As a result, the two used datasets has a tiny percentage of geotagged tweets as 2% and 2.4% for the COVID-19 Arabic dataset [2] and the ArCov-19 [13] respectively.

Therefore, we designed a mechanism to generate "Location-Enabled Tweets" from non-geotagged tweets as outlined in Algorithm 1. It requires the use of GeoDB, our manually created geo-location database that comprises bilingual names (English and Arabic) of international country names, capital cities, and well-known towns in the Arab world. To extract the tweet source, this method examines all data fields found in the tweet object, such as "Place name", "Country", and "User location". We extracted the Chrononyms and Astionyms [16] from user location. Therefore, the merged dataset contained location-enabled tweets around 46% (approximately 2.5 million tweets) more than the original combined dataset, which enriches the experiments applied on it.

### 3.3 Sentiment Analysis

In our model, we developed Unsupervised SA processes that applied lexicon-based method. We created our Arabic sentiment lexicon by combining nine Arabic lexicons previously tested by the research community (Bing Liu Lexicon; NRC Emotion Lexicon; MPQA Subjectivity Lexicon; SemEval-2016 Arabic Lexicon; AEWNA Lexicon; NileULex Lexicon [9,19,20,22,24]. Then we utilized polarized Bag of Words (BoW) technique, a word-frequency approach, that counts positive, negative, and neutral words in each tweet to assign the tweet polarity as "positive", "negative", or "neutral". Algorithm 2 shows the SA processes such as: 1) loads the Arabic corpus of text gathered from Arabic geo-tweets dataset, 2) prepares the Arabic corpus by employing some sub-processes, 3) determines polarity of each tweet's words, 4) uses four machine learning (ML) classifiers to evaluate the sentiment analysis' classification performance: Linear Support Vector Machine (SVM), K-Nearest Neighbor (KNN), Multinomial Nave Bayes (NB), and Random Forest (RF).

The KNN is a scalable approach capable of handling training data that are too large to fit in memory. The most important parameters of KNN classifier are neighbors: {5}, weights: {uniform}, leaf-size: {30}, Power: {2}, metric: {minkowski}, and n-jobs: {None}. The SVM is used for Classification as well as Regression to create the best line or decision boundary that can segregate n-dimensional space into classes. It uses the parameters penalization: {standard}, Tolerance: {1e-4}, multi-class-trains: {n-classes}, intercept-scaling: {1}, class-weight: {1}, Random-State: {None}, and max-iterations: {1000}. Multinomial NB is a powerful algorithm that is used for text data analysis and with multiple classes problems. The default parameters used with this classifier are Additive: {smoothing}, fit-prior: {True}, and class-prior: {None}. Random forests is the most flexible and easy to use classification and regression algorithm. It creates decision trees on randomly selected data samples, gets prediction from each tree and selects the best solution by means of voting. We applied with RF the following parameters estimators: {10}, max-depth: {5}, and max-features: {1}. Our algorithm uses the Arabic polarity corpus to extract characteristics and labels as the source for the classifiers' data. The corpus is then divided into train and test datasets, with 80% and 20% split ratios, respectively. Then it uses the TF-IDF Vectorizer to vectorize the training dataset. Following that, the algorithm uses the classifier model applied to the test dataset using the classifier's scores, including precision, recall, f1-score, and accuracy.

---

**Algorithm 2:** Sentiment Extraction

---

1 **begin**
2    *"Tweets Corpus"* = loadTweetsCorpus();
3    RemoveDuplicatedTweets(TC);
4 **for** *each tweet in "Tweets Corpus"* **do**
5        FilterTweetsFields(tweet);
6        CleanTweets(tweet);
7        NormalizeTweets(tweet);
8        RemoveArabicStopWords(tweet);
9        RemoveduplicatedWords(tweet);
10       TokenizeTweetsText(tweet);
11       StemTweetsWords(tweet);
12       GetTweetPolarity(tweet, *"Arabic Lexicon"*);
13       SetTweetPolarity(tweet);
14 **end**
15 "Classifiers List" = {SVM, KNN, NB, RF};
16 **for** *each classifier in "Classifiers List"* **do**
17       ExtractFeaturesLabels(TC);
18       Train, Test = SplitDataset(TC);
19   **for** *n in uni-gram* **do**
20           Vectorising(Train, TF-IDF);
21           Prediction = Predict("Classifiers List", Test);
22           ConfusionMatrix = (precision, recall, f1-score);
23       **end**
24 **end**

---

## 4 Results and Discussion

Researchers may examine the influence of health prevention measures, environmental variables, and conversation topics on the spread of the COVID-19 pandemic using spatial-temporal analysis of social media posts. The geo-tagging of social media postings is required for this spatial-temporal analysis. Only a small portion of the recorded stream of postings (4.5% at most) has been geotagged on previous research. As a result, we used our mechanism to increase the size of the experiment's dataset from about 115K (2%) to 2.5M (46%) geotagged tweets. We used the occurrence-based approach to quantitatively assess the location-enabled Arabic tweets in the new geo-tweets dataset (2.5M tweets). The vast majority of Arabic tweets originate from Arab users all over the world, with the vast majority of them based in the Arab world. As a result, we compared the two locations (Arab and Non-Arab) in terms of the origins of the Arabic tweets that fuel our experiments. From January to November 2020, Saudi Arabia and Kuwait are the two Arab countries with the most tweets and hashtags. The United Kingdom and France, on the other hand, were the top two nations in the non-Arab area.

The monthly results of the sentiment analysis of COVID-19 Arabic tweets are shown in Fig. 2a. Figure 2b compares the results obtained after using the four

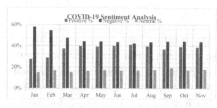

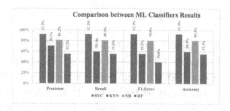

(a) SA of COVID-19 Arabic Tweets          (b) Classifiers' performance

**Fig. 2.** Sentiment analysis

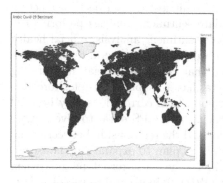

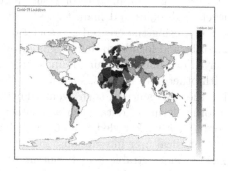

(a) Sentiment Analysis                    (b) Lockdown Days

**Fig. 3.** Visualization of experiments' results

machine learning classifiers. The SVM classifier scored about 92% in all four categories: Precision, Recall, F1, and Accuracy. This is a higher score than prior research projects have attained. Other classifiers obtain similar results, but the TF-IDF unigram achieves a better performance. Finally, to explain the insight gained from the analysis of progressively huge datasets, we ran numerous experiments using our Arabic tweets COVID-19 dataset and additional data acquired from earlier studies. The findings might be crucial in representing our large-scale dataset with official data that came in early. We offered figures such as Fig. 3a and 3b to help us extract information, better comprehend the data, and make more effective decisions by displaying data and results in an understandable and engaging way using the world maps. Therefore, we demonstrated that there is a link between the number of COVID-19 confirmed cases and the issues addressed in Arabic social media, such as government-imposed lockdown and travel restrictions. They clarify that the majority of them are from Saudi Arabia and Egypt, which have the largest populations and Internet users, respectively [15]. Some correlation research suggests that during the epidemic, unfavorable emotions among Arab Twitter users increased. Furthermore, the research revealed that there is a link between Arab users' unfavorable feelings and the number of daily confirmed cases of COVID-19.

# 5   Conclusion

This article presented a complete social data mining technique in Arabic for obtaining COVID-19-related insights, with an emphasis on the connection between spatio-temporal social data and health data. It also included a sentiment analysis technique that could be used to a variety of geographical granularities and subject scales. Furthermore, a strategy for inferring geo-information from non-geotagged tweets was created, which boosted the overall percentage of location-enabled tweets from 2% to 46%, much above the results of most prior studies. We used sentiment-based classifications to generate people's opinions at various geographical resolutions (regions/countries) and subject abstraction levels (subtopics and main topics) to validate sentiment analysis performance. Finally, using the created geo-social dataset, sentiment analysis, official health records, and lockdown data from across the world, we ran several tests and displayed our findings. Our findings suggested that combining social data mining with other data sources, such as health data, has a lot of promise for forecasting the emergence of such events. Furthermore, such correlations may be used in other forms of data, such as contact tracing and GPS data, to give a more comprehensive knowledge of human behavior and the relationship between social and physical user interactions. In the future, we plan to expand the dataset to include more Arabic social content to analyze the most recent periods when social media users' attention shifted from COVID-19 in general to vaccines. Furthermore, we want to undertake additional research to enhance the suggested location inferencing approach by using a natural language processing (NLP) procedure to process the tweet's text to boost the percentage of geotagged tweets created.

# References

1. Al-Laith, A., Alenezi, M.: Monitoring people's emotions and symptoms from Arabic tweets during the COVID-19 pandemic. Information **12**(2), 86 (2021)
2. Alanazi, E., Alashaikh, A., Alqurashi, S., Alanazi, A.: Identifying and ranking common COVID-19 symptoms from tweets in Arabic: content analysis. J. Med. Internet Res. **22**(11), e21329 (2020)
3. Alharbi, A., Lee, M.: Kawarith: an Arabic Twitter corpus for crisis events. In: Proceedings of the Sixth Arabic Natural Language Processing Workshop, pp. 42–52 (2021)
4. Alsafari, S., Sadaoui, S., Mouhoub, M.: Hate and offensive speech detection on Arabic social media. Online Soc. Netw. Media **19**, 100096 (2020)
5. Alshalan, R., Al-Khalifa, H., Alsaeed, D., Al-Baity, H., Alshalan, S.: Detection of hate speech in COVID-19-related tweets in the Arab region: deep learning and topic modeling approach. J. Med. Internet Res. **22**(12), e22609 (2020)
6. Asghar, M.Z., Khan, A., Ahmad, S., Kundi, F.M.: A review of feature extraction in sentiment analysis. J. Basic Appl. Sci. Res. **4**(3), 181–186 (2014)
7. Bahja, M., Hammad, R., Kuhail, M.A.: Capturing public concerns about coronavirus using Arabic tweets: an NLP-driven approach. In: 2020 IEEE/ACM 13th International Conference on Utility and Cloud Computing (UCC), pp. 310–315. IEEE (2020)

8. Chakraborty, K., Bhatia, S., Bhattacharyya, S., Platos, J., Bag, R., Hassanien, A.E.: Sentiment analysis of COVID-19 tweets by deep learning classifiers–a study to show how popularity is affecting accuracy in social media. Appl. Soft Comput. **97**, 106754 (2020)

9. El-Beltagy, S.R., Ali, A.: Open issues in the sentiment analysis of Arabic social media: a case study. In: 2013 9th International Conference on Innovations in Information Technology (IIT), pp. 215–220. IEEE (2013)

10. Essam, B.A., Abdo, M.S.: How do Arab tweeters perceive the COVID-19 pandemic? J. Psycholinguist. Res. **50**, 507–521 (2021)

11. Hamdy, A., Youssef, A., Ryan, C.: Arabic hands-on analysis, clustering and classification of large Arabic Twitter data set on COVID-19. Int. J. Simul.-Syst. Sci. Technol. **22**(1), 6.1–6.6 (2021)

12. Hamoui, B., Alashaikh, A., Alanazi, E.: COVID-19: what are Arabic tweeters talking about? In: Chellappan, S., Choo, K.-K.R., Phan, N.H. (eds.) CSoNet 2020. LNCS, vol. 12575, pp. 425–436. Springer, Cham (2020). https://doi.org/10.1007/978-3-030-66046-8_35

13. Haouari, F., Hasanain, M., Suwaileh, R., Elsayed, T.: ArCOV-19: the first Arabic COVID-19 Twitter dataset with propagation networks. In: Proceedings of the Sixth Arabic Natural Language Processing Workshop, pp. 82–91 (2021)

14. Kabir, M.Y., Madria, S.: Emocov: Machine learning for emotion detection, analysis and visualization using COVID-19 tweets. Online Soc. Netw. Media **23**, 100135 (2021)

15. Kemp, S.: Digital 2021: Global Overview Report, April 2021. https://datareportal.com/reports/digital-2021-global-overview-report

16. Khomutnikova, E., Gunbina, E., Zhurkova, M., Fetyukov, F.: Semantics and etymology of English astionyms in the aspect of linguistic geography. In: European Proceedings of Social and Behavioural Sciences EpSBS, pp. 505–513 (2020)

17. Lamsal, R.: Design and analysis of a large-scale COVID-19 tweets dataset. Appl. Intell. **51**(5), 2790–2804 (2021)

18. Manguri, K.H., Ramadhan, R.N., Amin, P.R.M.: Twitter sentiment analysis on worldwide COVID-19 outbreaks. Kurdistan J. Appl. Res. **5**(3), 54–65 (2020)

19. Mohammad, S., Salameh, M., Kiritchenko, S.: Sentiment lexicons for Arabic social media. In: Proceedings of the Tenth International Conference on Language Resources and Evaluation (LREC 2016), pp. 33–37 (2016)

20. Mohammad, S.M., Salameh, M., Kiritchenko, S.: How translation alters sentiment. J. Artif. Intell. Res. **55**, 95–130 (2016)

21. Mohsen, A.M., Hassan, H.A., Idrees, A.M.: A proposed approach for emotion lexicon enrichment. Int. J. Comput. Electr. Autom. Control Inf. Eng. **10**(1), 242–251 (2016)

22. Ptáček, T., Habernal, I., Hong, J.: Sarcasm detection on Czech and English Twitter. In: Proceedings of COLING 2014, the 25th International Conference on Computational Linguistics: Technical Papers, pp. 213–223 (2014)

23. Qazi, U., Imran, M., Ofli, F.: Geocov19: a dataset of hundreds of millions of multilingual COVID-19 tweets with location information. SIGSPATIAL Special **12**(1), 6–15 (2020)

24. Salameh, M., Mohammad, S., Kiritchenko, S.: Sentiment after translation: a case-study on Arabic social media posts. In: Proceedings of the 2015 Conference of the North American Chapter of the Association for Computational Linguistics: Human Language Technologies, pp. 767–777 (2015)

# Prioritization of COVID-19-Related Literature via Unsupervised Keyphrase Extraction and Document Representation Learning

Blaž Škrlj[1,2(✉)] ⓘ, Marko Jukič[3], Nika Eržen[1,2,3], Senja Pollak[1] ⓘ,
and Nada Lavrač[1,2]

[1] Jožef Stefan Institute, Ljubljana, Slovenia
blaz.skrlj@ijs.si
[2] Jožef Stefan International Postgraduate School, Ljubljana, Slovenia
[3] Faculty of Chemistry and Chemical Technology,
University of Maribor, Maribor, Slovenia

**Abstract.** The COVID-19 pandemic triggered a wave of novel scientific literature that is impossible to inspect and study in a reasonable time frame manually. Current machine learning methods offer to project such body of literature into the vector space, where similar documents are located close to each other, offering an insightful exploration of scientific papers and other knowledge sources associated with COVID-19. However, to start searching, such texts need to be appropriately annotated, which is seldom the case due to the lack of human resources. In our system, the current body of COVID-19-related literature is annotated using unsupervised keyphrase extraction, facilitating the initial queries to the latent space containing the learned document embeddings (low-dimensional representations). The solution is accessible through a web server capable of interactive search, term ranking, and exploration of potentially interesting literature. We demonstrate the usefulness of the approach via case studies from the medicinal chemistry domain.

**Keywords:** COVID-19 · Literature-based discovery · Representation learning

## 1 Introduction

Severe acute respiratory syndrome coronavirus 2 or SARS-CoV-2 is a coronavirus, member of the *Coronaviridae* family, a positive-sense single-stranded (+ssRNA) RNA virus [1]. The novel virus (initially 2019-$n$CoV now named SARS-CoV-2; '$n$' - novel) was reported in December of 2019 to be originating from Wuhan, Hubei China [40]. In the closing of 2019-early 2020, the virus caused a global pandemic of the COVID-19 disease [35]. The latter is of grave concern, as the majority of cases display mild symptoms, but up to 15% of patients progress to pneumonia and multi-organ failure leading to potential death, especially without medical assistance [36].

ⓒ Springer Nature Switzerland AG 2021
C. Soares and L. Torgo (Eds.): DS 2021, LNAI 12986, pp. 204–217, 2021.
https://doi.org/10.1007/978-3-030-88942-5_16

While there are no registered drugs, but several drug and vaccine discovery programs are being actively developed and scaled up, the scientific community coherently responded to the COVID-19 pandemic resulting in an increasing amount of literature that is beyond the search capabilities of individual medical professionals [38]. Exploration of scientific literature can be facilitated by computationally feasible approaches to summarizing a large amount of text [17]. This work explores how unsupervised document representation learning and keyphrase extraction methodologies [5,11] can be used to build a fast, scalable web server suitable for **literature prioritization**. To this end, we implemented a web server tool and showcase the solution's scalability on one of the largest currently known collections of COVID-19-related full medical document databases – CORD19 [38].

We next present the related work, followed by the developed web server and its use cases. We conclude with a discussion of the developed tool and further work.

## 2  Related Work

With the introduction of freely available literature, multiple tools have been recently developed [14].

Bras et al. [21] propose bubble-like visualization of the COVID-19 literature using keyword groups, resulting in hundreds of documents retrieved. The tool offers search based on pre-defined sets of keywords, which can be ambiguous and potentially result in papers not directly related to a given query, offering a fast overview of key topics.

Another interesting project is the Watson Annotator of Clinical data[1], capable of highlighting key terms within a given document. This tool aims not to provide the global search across the literature but to annotate an e.g., copy-pasted document with named entities. Such annotation can be very useful for medical professionals, as it offers, similarly to this work, quick insights into the key concepts appearing in a given document. A substantially different approach was undertaken by Google[2], where a question answering regime was adopted. Their search engine can identify publications based on a natural language-based query, e.g., "What is the medical care for patients during COVID-19 epidemic?". The engine recommends (according to its internal ranking) the documents that are of potential interest.

An interesting approach is also CADTH COVID-19 pandemic online tool[3], which offers string search to topics related to COVID-19. Another recently released tool is the COVIDScholar[4] whose core functionality is the most similar to the tool presented in this paper. It is based on word and document embedding techniques used for semantic search. It leverages open data from

---

[1] https://www.ibm.com/cloud/watson-annotator-for-clinical-data.
[2] https://covid19-research-explorer.appspot.com/.
[3] https://covid.cadth.ca/literature-searching-tools/cadth-covid-19-search-strings/.
[4] https://covidscholar.org/.

various data sources. The main results are links to full papers with abstracts and the most similar documents via document embeddings. The tool, however, does not explore the possibility of full-text annotation via keyphrase extraction, which is among the key functionalities of our tool. Finally, the SPIKE tool by the Allen Institute also offers an exploration of documents at scale, offering insight into named entities and their relations within documents, which can be very useful when attempting to answer specific queries based on literature[5]. The data set that gave rise to the tools was initially offered at Kaggle[6]. The majority of the available data on the heavily studied SARS-CoV-2 topic and related COVID-19 pandemic resides on traditional literature bodies that employ heavy user involvement and literature study. A few examples of the literature bodies besides the aforementioned CORD19 database are offered by NIH as SARS-CoV-2 Resources[7], NCBI as LitCovid[8], Rutgers university as COVID-19 Information Resources[9], ECDC[10], WHO[11,12], USCF[13], Wiley[14], ACS[15] and others. A more comprehensive overview of related literature databases and tools [37] is provided by CDC[16].

## 3   COVID-19 Explorer Design, Implementation and Functionality

The proposed COVID-19 Explorer webserver architecture, shown in Fig. 1, is comprised of two main parts.

First, the raw body of COVID-19-related literature is preprocessed and stored in the form, suitable for the two subsequent machine learning tasks. The first task, keyphrase extraction, is conducted with the recently introduced RaKUn algorithm [32], additionally equipped with scientific stop-word lists to prevent noisy keyphrases from being detected. The second task, document representation learning, is conducted by using the widely adopted doc2vec document embedding algorithm [20], used to learn representations of abstracts of individual documents. Once keyphrases and document embeddings are obtained, they are stored in a form suitable for fast access. The document embeddings are also projected to 2D with UMAP [25], a non-linear dimensionality reduction tool, as the front end part of the webserver offers **interactive exploration** also by querying the semantic (2D) space directly.

---

[5]  https://spike.covid-19.apps.allenai.org/datasets/covid19/search.
[6]  https://www.kaggle.com/allen-institute-for-ai/CORD-19-research-challenge.
[7]  https://ncbi.nlm.nih.gov/sars-cov-2/.
[8]  https://nncbi.nlm.nih.gov/research/coronavirus/.
[9]  libguides.rutgers.edu/covid19_resources/.
[10]  ecdc.europa.eu/en/coronavirus.
[11]  search.bvsalud.org/global-literature-on-novel-coronavirus-2019-ncov/.
[12]  who.int/emergencies/diseases/novel-coronavirus-2019.
[13]  guides.ucsf.edu/COVID19/literature.
[14]  novel-coronavirus.onlinelibrary.wiley.com.
[15]  acs.org/content/acs/en/covid-19.html.
[16]  cdc.gov/library/researchguides/2019novelcoronavirus.

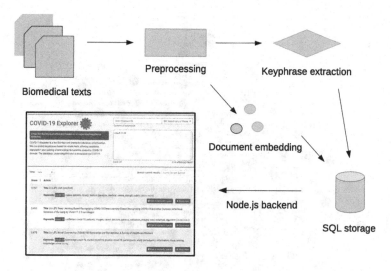

**Fig. 1.** Visualization of the main processing steps considered by the proposed solution.

All the information is presented in the form of a responsive and fast front end, requiring minimal computational resources on the client-side.

### 3.1 Keyphrase Extraction

One of the key functionalities of the COVID-19 Explorer is that it enables a direct search via keyphrases, computed from *whole* scientific documents (papers, reports, etc.). The extraction method is the in-house developed RaKUn algorithm [32]. The algorithm first transforms a given collection of sentences (a document) into a *document graph* – a graph comprised of key tokens, linked via the co-occurrence relation. An example graph is shown in Fig. 2.

Once the graph is constructed for a given collection of text, *ranking* of nodes is performed to identify single, two, and three-term *keyphrases.* The webserver also implements an auto-suggestion option, which offers interactive exploration of possible search queries in real-time. The current implementation of RaKUn employs *load centrality*, a centrality measure based on the amount of shortest paths that pass through a given node. The keyphrase computation step is conducted in parallel for each of the considered documents. The resulting keyphrases and the underlying token graphs are stored and browsed interactively as a part of the front-end functionality. Further, the keyphrase extraction offers another functionality that is crucial in the considered document prioritization task – each keyphrase has a dedicated score for a given document, meaning that the documents themselves can be prioritized for the global keyphrase score. An example of how this space can be directly inspected is shown in Fig. 3. In addition to scoring a given keyphrase within a given document, the search results also show other keyphrases and the document title linked to the corresponding DOI.

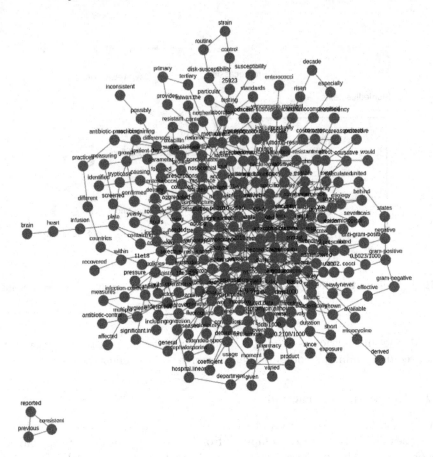

**Fig. 2.** An example RaKUn token graph. Each node represents a token in the document. Documents, when linked together, form the whole document graph suitable for identifying the keyphrases (as paths in this graph).

## 3.2    Global Document Representation Learning

A prominent capability of the natural language processing methods developed in recent years is that of *learning* the representation of a collection of texts, instead of merely considering the set of hand-crafted features. The current implementation of the COVID-19 Explorer exploits the widely used doc2vec algorithm [20] to learn the representations of *every* document abstract. The purpose of this step is to map the considered collection of documents into the same semantic space, offering the capability to explore the e.g., semantic neighborhoods of a given document, *interactively*. The current implementation of the COVID-19 Explorer first computes 256-dimensional representations of individual abstracts and next projects them to two dimensions via the UMAP [25] tool that approximates a low dimensional manifold representative of the learned high dimensional space. The implemented *semantic viewer* is shown in Fig. 4.

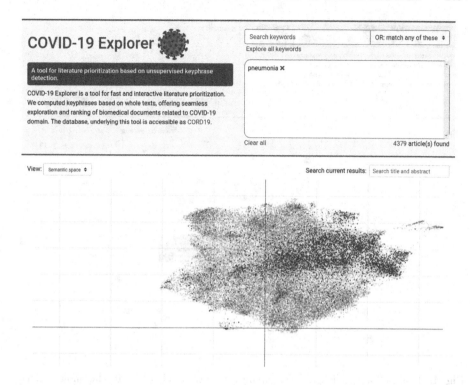

**Fig. 3.** Global interactive space of document embeddings.

## 4    Case Studies

This section presents the application of the reported webserver tool and its main functionalities. We also showcase its performance on multiple use-cases, aimed at the fast and efficient hypothesis elaboration for research on COVID-19 drug design. We present the general research-field examination together with key scientific questions regarding the development of novel drugs against the SARS-CoV-2 pathogen and SARS-CoV-2 therapeutic target examination. In the presented cases, we demonstrate how the COVID-19 Explorer effectively identifies the relevant semantically associated literature. Upon navigating to https:// covid19explorer.ijs.si/, the user is presented with a welcome screen where keyword(s) can be chosen (Fig. 5, subfigure 1) and their relationship using Boolean operators (Fig. 5, subfigure 2). The user is then presented with a list of examined keywords (Fig. 5, subfigure 3) and the list of semantically connected articles is dynamically updated in the output field below (Fig. 5, subfigure 4). Individual pinpointed articles can be examined in detail and its semantic space visualized (Fig. 5, subfigure 5).

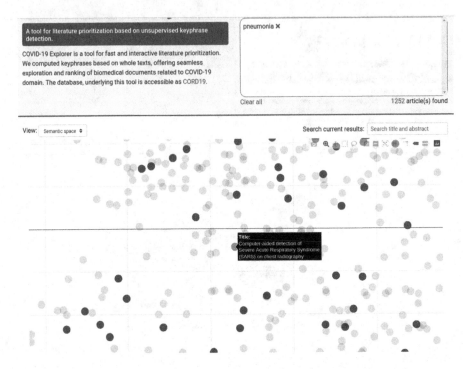

**Fig. 4.** The COVID-19 Explorer's semantic viewer. Each point in the shown space represents a document. The positions in the *global* space of documents are determined based on the distances between the representations of the documents' abstracts. Intuitively, the documents close to one another can offer insight into the semantically similar document. The implemented viewer offers a direct exploration of documents – each point is clickable and triggers an element with a detailed description of a given document.

## 4.1   Case Study 1: General COVID-19 Domain Inquiry

Single keyword examination using the term **pandemic** yields results where the top-scoring article (Score: 0.542) entitled *"The pandemic present"* immediately affords social anthropology discourse on current pandemic threats including COVID-19 [39], Amongst the 10 top scoring peer-review articles offered by the COVID-19 Explorer, 7 investigate the COVID-19 emergency and 3 articles offer information on the influenza pandemics. The former and latter are two major subjects found in modern medicinal literature regarding the general topic of pandemics [7,10,18]. The journals found by the COVID-19 Explorer are all of high-impact in the respective fields and encompass *Social Anthropology, Journal of Medical Humanities, Emerging Infectious Diseases, Public Health, The Canadian Journal of Addiction, The Lancet* and *British Journal of Surgery*. The keyword **pandemic** thus offered a balanced perspective from social studies (20%) and modern medical science perspective (80%) on the field of pandemic studies, especially focusing on COVID-19 on the first, and influenza in the second place.

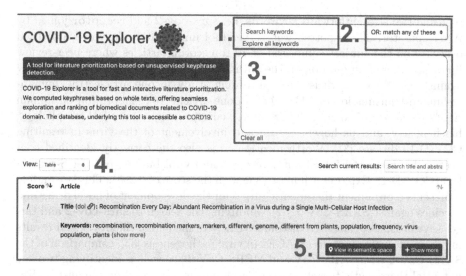

**Fig. 5.** The COVID-19 Explorer's welcome screen and user input fields. Relevant key sections of the online tool are emphasized and numbered in red color: 1. user keyword inspection and input field, 2. Boolean operators imposed on keywords, 3. selected keyword inspection window, 4. Dynamically updating result field displaying semantically related peer-review articles, 5. specific article detail button and 2D visualization of the corresponding semantic space. (Color figure online)

A similar broad, yet subject-focused outlook in correlation to the present-time problem would be difficult to impossible for identification using one search operation in other peer-review literature search engines. Supplementing the search with the **covid** keyword and using AND Boolean operator offered by COVID-19 Explorer shifts the result focus entirely to COVID-19 related peer-review articles. Top scoring hits offer the general outlook on COVID-19 health effects [30], possible treatments [13], promoting the mental healthcare during the COVID-19 pandemic [2,27], elaborating on children study problems [3] and clinical problems encountered during the COVID-19 pandemic [6,12]. The results in effect mirror the key media-reported problems and challenges imposed by the current global crisis and could be retrieved by COVID-19 Explorer in a single search operation.

### 4.2 Case Study 2: SARS-CoV-2 Potential Therapeutic Drug/target Identification

There are only a few therapeutic options for SARS-CoV-2, a pathogen causing worldwide havoc [16,34]. Therefore, novel drug design is paramount, as well as an inquiry into viral biochemistry along with the identification and assessment of potential novel therapeutic targets that could be of use for the development of novel drugs [40]. Using the COVID-19 Explorer, we examined the subject by using two straightforward keywords, i.e. **sars-cov-2** and **receptor** joined by AND operator. The reported web server tool immediately delivered a focused overview on the subject comprised of 10 top scoring articles where peer-review literature offered an outlook on the antimicrobial chemical matter with activity against SARS-CoV-2 virus - a repurposing study in one article [15], human-to-human transmission of COVID-19 in one article [26], elaboration on human ACE2 receptor in 7 articles [23,28,29,41], current insight into viral morphology, biochemistry, and pathogenesis [19] and involvement of the virus in resulting COVID-19 disease [31]. Worth mentioning is also the correctly identified connection between ACE2 host entry receptor and viral binding partner S-protein [41] (COVID-19 Explorer 4th hit article with the score of 0.295). This key finding represents a prominent therapeutic target for the development of novel drugs and vaccines against SARS-CoV-2 [42]. Modifying the search to **sars-cov-2** and target keywords, associates the discourse tightly to medicinal chemistry and results in peer-review article focus on ACE2 in viral pathogenesis [33], comparison of the SARS-CoV-2 with SARS-CoV and MERS-CoV [9] as well as identifying two key potential therapeutic targets for the development of novel drugs against SARS-CoV-2 - 3CLpro and RdRp [4,22]. Furthermore, the top 10 suggested articles of the COVID-19 Explorer tool also include elaboration on the viral entry mechanism involving TMPRSS2 extracellular protease (5th hit with a score of 0.285) [31]. TMPRSS2 protease is essential in understanding spike protein processing and the mechanism of viral cell entry [24].

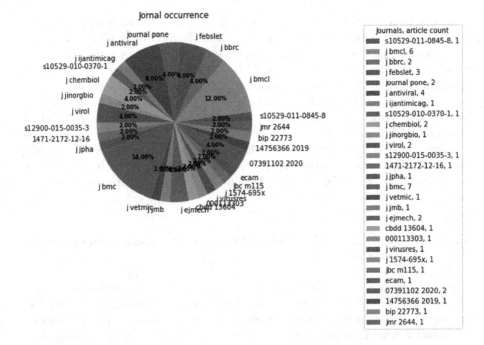

**Fig. 6.** Left: API hitlist for the keyword 3clpro; Right: article counts from specific journals

## 4.3   Using the COVID-19 Explorer API

For the user benefit, the COVID-19 Explorer web server tool also exposes a RESTful API with a simple syntax:

http://cord19explorer.ijs.si/gp/api?keyword=**query**

where **query** is a user-defined search term. Additional keywords can be added with *&query2&query3...* and the search output hitlist can be limited with a *limit* = *N* and **N** is the requested number of articles. To restraint the server load, the current hitlist is limited to 50 but this can be adjusted as needed. For example, using a scripting language (e.g., Python) and perhaps a notebook software (e.g., Jupyter) elaborate search patterns can be performed and results analyzed. For example upon searching for a simple **keyword** = **3clpro** a hitlist is obtained and can be readily analysed article/publisher-wise Fig. 6.

Similarly, the user can easily obtain articles with a specific term in the abstract. For example, a list of articles with a term inhibitor in the abstract as a subset of API hitlist, discern the publication year, field of study and so on. The exposed RESTful API is thus an easy approach towards automatization and incorporation of COVID-19 Explorer's searches into other tools and workflows.

## 5   Discussion

Let us discuss the usefulness of the proposed tool alongside its drawbacks. The developed COVID-19 Explorer offers scalable and highly efficient *summarization* of scientific documents via keyphrases, based on *whole* texts. Compared to e.g., conventional approaches are undertaken by large databases such as e.g., PubMed, where the keywords are determined by the authors themselves (and manually tagged by professionals), the purpose of this work was to demonstrate that at least to some extent, this process can be automated in and *unsupervised manner*, without any human interventions, and offers a scalable approach to the exploration of vast amounts of scientific literature. One of the key goals of the COVID-19 Explorer is to *filter* existing information and thus simplify arduous exploration (often random) of scientific literature to domain experts.

The proposed implementation offers two fundamentally different approaches to the exploration of the document, which we were able to link – namely, the tool offers exploration directly in the space of latent embeddings of documents via an interactive 2D visualization, but also exploration directly via *ranked* keywords, present throughout the documents. Exploration via keywords was specifically optimized by taking into account the existing lists of scientific stopwords. Thus the RaKUn keyphrase extraction phase was adopted for the scientific domain, which is also a contribution of this work. Furthermore, RaKUn was updated with the capability of detecting connectives – words that link multiple terms; for example "COVID-19 **in** the USA" represents a keyphrase where **in** was identified to fit between the two key phrases. The RaKUn achieves such behavior by backtracking back to the raw text and statistically identifying how suitable a given connective is. Therefore a suitable context can be tailored upon the search request. Finally, we believe that the existing semantic space could be improved by incorporating both the document embedding information, additionally equipped with the whole-document keywords. Finally, we plan to explore more sophisticated summarisation techniques, summarised for the interested reader in [8]; many of these techniques are based on computationally more expensive neural language models, which, however, could offer superior performance.

Using the reported tool for general COVID-19 subject outlook revealed a hit-list of semantically connected articles where a broad overview on the global COVID-19 crisis from the social studies and a modern medicinal perspective was obtained in a single search query. Furthermore, simple keyword exploration of the general subjects from the Medicinal chemistry domain quickly afforded relevant high-impacting peer-review articles from journals respective on the field. In essence, a review article deconstruction was achieved with key articles, ideas, and themes offered as top hits. Inspection of specific terms from SARS-CoV-2 antiviral drug design returned lists of relevant primary literature as well as associations to other complementary research approaches, e.g., exploration of 3CLpro target exposed associations to PLpro, RdRp, and 2'-O-MTase therapeutic targets. Furthermore, the reported tool helps the user with an implemented API in the background and precomputed lists of related keywords in the foreground. Upon inputting a specific keyword, a list of semantically related suggestions is offered, unaffected by the user preference but rather derived from the underlying body of data.

## 6    Conclusions

In this work, we presented an approach for summarization of large collections of scientific documents based on automatic keyphrase extraction. The approach was extended with a simple-to-use web interface, where users can explore the semantic space of COVID-19-related medical literature. As keyphrases were computed based on whole texts automatically, the proposed tool offers exploration capabilities beyond a few author-assigned keywords present in dominant search engines. Furthermore, the keyphrase extraction algorithm was specifically adapted for the biomedical domain via scientific stop word lists, which substantially improved the search performance and the quality of the results.

We demonstrated the usefulness of the proposed approach in different case studies, studying different aspects of the current COVID-19 pandemic, from molecular (receptor) level to more general, disease co-occurrence level. We demonstrated that the tool indeed offers a fast and intuitive exploration of the scientific literature as well as an alternative view on the underlying body of work. Furthermore, the proposed article ranking system, which assigns a score to each paper, was shown to prioritize the literature in a manner suitable for literature-based discovery and exploration. The article ranking idea is also a novelty of this paper. Finally, even though the existing web service offers an intuitive and fast exploration of existing literature, we believe the approach could be extended to incorporate contextual embeddings, which could further distill the relevant literature. Even though the focus of this work was the CORD-19 corpus, the authors are aware, the proposed approach can be generalized for *any* collection of relevant literature.

# 7  Availability and Requirements

COVID-19 Explorer is available at https://covid19explorer.ijs.si/ as a freely accessible webserver. The web server's landing page includes the links to the repository and the data used to completely reproduce the webserver locally.

**Acknowledgements.** This work was supported by the Slovenian Research Agency (ARRS) core research program P2-0103 and the CRP project V3-2033. The work of the first author was financed by the ARRS young researchers grant. The work was also supported by European Union's Horizon 2020 research and innovation programme under grant agreement No 825153, project EMBEDDIA (Cross-Lingual Embeddings for Less-Represented Languages in European News Media).

# References

1. The species severe acute respiratory syndrome-related coronavirus: classifying 2019-nCoV and naming it SARS-CoV-2. Nat. Microbiol. **5**(4), 536–544 (2020). https://doi.org/10.1038/s41564-020-0695-z
2. Advani, I., et al.: Is increased sleep responsible for reductions in myocardial infarction during the COVID-19 pandemic? Am. J. Cardiol. **131**, 128–130 (2020)
3. Agarwal, S., Kaushik, J.S.: Student's perception of online learning during COVID pandemic. Indian J. Pediatr. **87**(7), 554 (2020). https://doi.org/10.1007/s12098-020-03327-7
4. Buonaguro, L., Buonaguro, F.M.: Knowledge-based repositioning of the anti-HCV direct antiviral agent sofosbuvir as SARS-CoV-2 treatment. Infect. Agents Cancer **15**(1) (2020). https://doi.org/10.1186/s13027-020-00302-x
5. Campos, R., Mangaravite, V., Pasquali, A., Jorge, A., Nunes, C., Jatowt, A.: YAKE! Keyword extraction from single documents using multiple local features. Inf. Sci. **509**, 257–289 (2020)
6. Cattaneo, C.: Forensic medicine in the time of COVID 19: an editorial from Milano, Italy. Forensic Sci. Int. **312**, 110308 (2020)
7. Chew, C., Eysenbach, G.: Pandemics in the age of Twitter: content analysis of tweets during the 2009 H1N1 outbreak. PLoS ONE **5**(11), e14118 (2010)
8. El-Kassas, W.S., Salama, C.R., Rafea, A.A., Mohamed, H.K.: Automatic text summarization: a comprehensive survey. Expert Syst. Appl. **165**, 113679 (2021)
9. Fani, M., Teimoori, A., Ghafari, S.: Comparison of the COVID-2019 (SARS-CoV-2) pathogenesis with SARS-CoV and MERS-CoV infections. Future Virol. **15**(5), 317–323 (2020)
10. Gates, B.: Responding to COVID-19 – a once-in-a-century pandemic? N. Engl. J. Med. **382**(18), 1677–1679 (2020)
11. Hasan, K.S., Ng, V.: Automatic keyphrase extraction: a survey of the state of the art. In: Proceedings of the 52nd Annual Meeting of the Association for Computational Linguistics (Volume 1: Long Papers), pp. 1262–1273 (2014)

12. Hing, C., Al-Dadah, O.: Returning to elective surgery, the 'new normal'. Knee **27**(3), A1 (2020)
13. Honore, P.M., et al.: Therapeutic plasma exchange as a routine therapy in septic shock and as an experimental treatment for COVID-19: we are not sure. Critical Care **24**(1) (2020). https://doi.org/10.1186/s13054-020-02943-1
14. Hutson, M.: Artificial-intelligence tools aim to tame the coronavirus literature. Nature (2020). https://www.nature.com/articles/d41586-020-01733-7
15. Ijaz, M.K., et al.: Microbicidal actives with virucidal efficacy against SARS-CoV-2. Am. J. Infect. Control **48**(8), 972–973 (2020)
16. Jin, Z., et al.: Structure of Mpro from SARS-CoV-2 and discovery of its inhibitors. Nature **582**(7811), 289–293 (2020)
17. Jones, S., Lundy, S., Paynter, G.W.: Interactive document summarisation using automatically extracted keyphrases. In: Proceedings of the 35th Annual Hawaii International Conference on System Sciences, pp. 1160–1169. IEEE (2002)
18. Kilbourne, E.D.: Influenza pandemics of the 20th century. Emerg. Infect. Dis. **12**(1), 9–14 (2006)
19. Kumar, S., Nyodu, R., Maurya, V.K., Saxena, S.K.: Morphology, genome organization, replication, and pathogenesis of severe acute respiratory syndrome coronavirus 2 (SARS-CoV-2). In: Saxena, S.K. (ed.) Coronavirus Disease 2019 (COVID-19). MVFPDC, pp. 23–31. Springer, Singapore (2020). https://doi.org/10.1007/978-981-15-4814-7_3
20. Le, Q., Mikolov, T.: Distributed representations of sentences and documents. In: Proceedings of the 31st International Conference on International Conference on Machine Learning - Volume 32, ICML 2014, pp. II-1188–II-1196. JMLR.org (2014)
21. Le Bras, P., Gharavi, A., Robb, D., Vidal, A., Padilla, S., Chantler, M.: Visualising COVID-19 research. Working paper, arXiv, May 2020
22. Li, H., Zhou, Y., Zhang, M., Wang, H., Zhao, Q., Liu, J.: Updated approaches against SARS-CoV-2. Antimicrob. Agents Chemother. **64**(6) (2020). https://doi.org/10.1128/aac.00483-20
23. Lutchman, D.: Could the smoking gun in the fight against COVID-19 be the (rh)ACE-2? Eur. Respir. J. **56**(1), 2001560 (2020)
24. Matsuyama, S., et al.: Enhanced isolation of SARS-CoV-2 by TMPRSS2-expressing cells. Proc. Natl. Acad. Sci. **117**(13), 7001–7003 (2020)
25. McInnes, L., Healy, J., Saul, N., Großberger, L.: UMAP: uniform manifold approximation and projection. J. Open Source Softw. **3**(29), 861 (2018). https://doi.org/10.21105/joss.00861
26. Mohseni, A.H., Taghinezhad-S, S., Xu, Z., Fu, X.: Body fluids may contribute to human-to-human transmission of severe acute respiratory syndrome coronavirus 2: evidence and practical experience. Chin. Med. **15**(1) (2020). https://doi.org/10.1186/s13020-020-00337-7
27. Novins, D.K., et al.: JAACAP's role in advancing the science of pediatric mental health and promoting the care of youth and families during the COVID-19 pandemic. J. Am. Acad. Child Adolesc. Psychiatry **59**(6), 686–688 (2020)
28. Ortega, J.T., Serrano, M.L., Pujol, F.H., Rangel, H.R.: Role of changes in SARS-COV-2 spike protein in the interaction with the human ACE2 receptor: an in silico analysis. EXCLI J. **19**, Doc410 (2020). https://doi.org/10.17179/EXCLI2020-1167. ISSN 1611–2156, https://www.excli.de/vol19/Rangel_18032020_proof.pdf
29. Panciani, P.P., et al.: SARS-CoV-2: "three-steps" infection model and CSF diagnostic implication. Brain Behav. Immunity **87**, 128–129 (2020)

30. Randolph, G.W.: One virus, undivided ... equity, and the corona virus. Laryn-goscope Investigative Otolaryngol. **5**(3), 586–589 (2020). https://doi.org/10.1002/lio2.398

31. Saxena, S.K., Kumar, S., Maurya, V.K., Sharma, R., Dandu, H.R., Bhatt, M.L.B.: Current insight into the novel coronavirus disease 2019 (COVID-19). In: Saxena, S.K. (ed.) Coronavirus Disease 2019 (COVID-19). MVFPDC, pp. 1–8. Springer, Singapore (2020). https://doi.org/10.1007/978-981-15-4814-7_1

32. Škrlj, B., Repar, A., Pollak, S.: *RaKUn: Ra*nk-based *K*eyword extraction via *Un*supervised learning and meta vertex aggregation. In: Martín-Vide, C., Purver, M., Pollak, S. (eds.) SLSP 2019. LNCS (LNAI), vol. 11816, pp. 311–323. Springer, Cham (2019). https://doi.org/10.1007/978-3-030-31372-2_26

33. Su, S., Jiang, S.: A suspicious role of interferon in the pathogenesis of SARS-CoV-2 by enhancing expression of ACE2. Signal Transduction Targeted Therapy **5**(1) (2020). https://doi.org/10.1038/s41392-020-0185-z

34. Tiwari, V., Beer, J.C., Sankaranarayanan, N.V., Swanson-Mungerson, M., Desai, U.R.: Discovering small-molecule therapeutics against SARS-CoV-2. Drug Discov. Today **25**(8), 1535–1544 (2020)

35. Wang, C., Horby, P.W., Hayden, F.G., Gao, G.F.: A novel coronavirus outbreak of global health concern. Lancet **395**(10223), 470–473 (2020)

36. Wang, D., et al.: Clinical characteristics of 138 hospitalized patients with 2019 novel coronavirus-infected pneumonia in Wuhan, China. JAMA **323**(11), 1061 (2020). https://doi.org/10.1001/jama.2020.1585

37. Wang, L.L., Lo, K.: Text mining approaches for dealing with the rapidly expanding literature on COVID-19. Brief. Bioinform. **22**(2), 781–799 (2020). https://doi.org/10.1093/bib/bbaa296

38. Wang, L.L., et al.: CORD-19: the COVID-19 open research dataset. arXiv (2020)

39. Whitacre, R.P., Buchbinder, L.S., Holmes, S.M.: The pandemic present. Soc. Anthropol. **28**(2), 380–382 (2020)

40. Wu, C., et al.: Analysis of therapeutic targets for SARS-CoV-2 and discovery of potential drugs by computational methods. Acta Pharmaceutica Sinica B **10**(5), 766–788 (2020)

41. Zhang, H., Penninger, J.M., Li, Y., Zhong, N., Slutsky, A.S.: Angiotensin-converting enzyme 2 (ACE2) as a SARS-CoV-2 receptor: molecular mechanisms and potential therapeutic target. Intensive Care Med. **46**(4), 586–590 (2020)

42. Zhou, H., Fang, Y., Xu, T., Ni, W.J., Shen, A.Z., Meng, X.M.: Potential thera-peutic targets and promising drugs for combating SARS-CoV-2. Br. J. Pharmacol. **177**(14), 3147–3161 (2020)

# Sentiment Nowcasting During the COVID-19 Pandemic

Ioanna Miliou[(✉)], John Pavlopoulos, and Panagiotis Papapetrou

Stockholm University, Stockholm, Sweden
{ioanna.miliou,ioannis,panagiotis}@dsv.su.se

**Abstract.** In response to the COVID-19 pandemic, governments around the world are taking a wide range of measures. Previous research on COVID-19 has focused on disease spreading, epidemic curves, measures to contain it, confirmed cases, and deaths. In this work, we sought to explore another essential aspect of this pandemic, how do people feel and react to this reality and the impact on their emotional well-being. For that reason, we propose using epidemic indicators and government policy responses to estimate the sentiment, as this is expressed on Twitter. We develop a nowcasting approach that exploits the time series of epidemic indicators and the measures taken in response to the COVID-19 outbreak in the United States of America to predict the public sentiment at a daily frequency. Using machine learning models, we improve the short-term forecasting accuracy of autoregressive models, revealing the value of incorporating the additional data in the predictive models. We then provide explanations to the indicators and measures that drive the predictions for specific dates. Our work provides evidence that data about the way COVID-19 evolves along with the measures taken in response to the COVID-19 outbreak can be used effectively to improve sentiment nowcasting and gain insights into people's *current* emotional state.

**Keywords:** Sentiment · Nowcasting · COVID-19 · Twitter · Measures

## 1 Introduction

Epidemics of infectious diseases are triggered by factors such as changes in the ecology of a population or a novel pathogen. One such example is the outbreak of COVID-19, which resulted in a substantial burden to the world in terms of health risks and unnecessary deaths as well as financial risks and global economic turmoil. Identifying the optimal sequence of mitigation measures is always a challenge [17], with countries all over the world adopting different policies to control and limit the impact of the pandemic. Such decisions vastly rely on epidemic models (e.g., compartmental models [19]) that attempt to capture and reflect epidemic indicators, such as infection rate, recovery rate, deaths, and population mobility [2]. Recent work on reinforcement learning has also attempted to identify optimal mitigation policies [12,13] by defining reward functions considering the impact of the pandemic on public health and the economy.

© Springer Nature Switzerland AG 2021
C. Soares and L. Torgo (Eds.): DS 2021, LNAI 12986, pp. 218–228, 2021.
https://doi.org/10.1007/978-3-030-88942-5_17

Nonetheless, the attention to public sentiment has been limited as a result of the pandemic and the mitigation measures taken [11]. Sentiment analysis concerns the classification of the intentions of a text's author (e.g., of a tweet) as positive, negative, or neutral. This is a well-known field in Natural Language Processing [23], and it is often applied on social media texts [22]. When the outcome is emotions instead of a sentiment class, the task is called textual emotion recognition [6] (also known as emotion prediction, detection, or classification). Recent works have started to explore the automated analysis of sentiments of social media posts related to the recent COVID-19 pandemic as a means to understand people's behaviors and responses during the pandemic [11,21]. Previous research on nowcasting sentiment has employed different datasets. Either to measure consumer sentiment with the use of Google search data [5,7] or to predict people's mood using Twitter data [14] and several other heterogeneous data (Twitter, Facebook, mood forms, mobile phone use data, and sensor data) [20].

The effects of COVID-19 contact minimization, isolation measures, lockdowns, as well as the potential fear of infection and death can have an undoubtedly long-term negative psychological and emotional impact on the population. This can, in turn, lead to severe indirect socio-economical consequences [8]. In this paper, our goal is to explore the emotional well-being of the population and identify potential factors that contribute to negative emotions related to the COVID-19 pandemic. More concretely, we propose a workflow for nowcasting negative sentiment, as expressed by Twitter, using governmental mitigation policies and epidemic indicators as exogenous variables.

Our main contributions can be summarized as follows: (1) We propose a sentiment nowcasting workflow for predicting the daily sentiment in response to the mitigation measures and epidemic indicators related to the COVID-19 pandemic; (2) We employ a sentiment extraction approach from tweets using a transformer-based, multi-lingual, masked language model called XLM-R; (3) Our workflow supports both statistical as well as machine learning models. For the latter, it also provides explanations for the predictions in the form of local model agnostic explainable features, using LIME; (4) Our empirical evaluation on data including tweets, mitigation measures, and epidemic indicators, obtained over two periods during the development of the pandemic suggests that mitigation measures and epidemic indicators can potentially function as factors for predicting negative public sentiment.

## 2   Sentiment Nowcasting

We propose a workflow for estimating the negative sentiment value related to COVID-19 one day ahead of the latest ground-truth value by taking advantage of exogenous variables, such as epidemic indicators and mitigation measures. Let $T$ define a set of tweets written in natural language. The first step of our workflow is to convert $T$ to a time series of sentiment values by employing a function $g(\cdot)$. Let $Y = y_1, \ldots, y_t$ denote the time series of $t$ real-valued sentiment observations, with each $y_i \in \mathbb{R}$. We additionally consider a set of exogenous variables $\boldsymbol{X} = \{\mathcal{X}^j\}$

that can occur concurrently with $Y$. Each $\mathcal{X}^j$ comprises a set of variables that together correspond to some common exogenous factor that can contribute to the estimation of $Y$. The main goal of this paper is to define a function $f(\cdot)$ that predicts the next observation $y_{t+1} = f(Y, X)$ by taking into account both the historical observations of $Y$, as well as the sets of exogenous variables up to time $t$. In our setup, $Y$ models the degree of negative sentiment per day, which is extracted as an aggregate value from a set of tweets that are filtered based on language and location. Moreover, we employ two sets of exogenous variables. The first ($\mathcal{X}^1$) contains 20 indicators and four indices that correspond to government mitigation policies for COVID-19, while the second ($\mathcal{X}^2$) contains 55 epidemic metrics related to the development of COVID-19.

**Sentiment Extraction.** Each tweet was annotated regarding sentiment by XLM-R, a multilingual, Transformer-based model [3]. We fine-tuned XLM-R to extract sentiment for a tweet as a valence score from zero (very negative) to one (very positive), and we binarized that score by using a threshold (see Sect. 3.1). Then, $y_t$ is the fraction of the negative tweets out of the filtered tweets of day $t$.

**Data Smoothing.** We decided to smooth the sentiment data to eliminate noise and random fluctuations. This allows important patterns to more clearly stand out and is intended to ignore one-time outliers. We choose to apply a Trailing Moving Average. The value at time $t$ is calculated as the average of the raw observations over a time window of length $w = 3$ ending at time $t$.

**Statistical Models.** Models handling time series are used in order to predict future values of indices by extracting relevant information from historical data. Traditional time series models are based on various mathematical approaches, such as autoregression. For this study, we apply the models of **Autoregression**, **Exponential Smoothing**, **ARIMA**, and **ARIMAX**.

**Machine Learning Models.** We used regression models to assess whether the inclusion of mitigation measures and epidemic indicators can improve the accuracy of the classical methods. Regression analysis is a form of predictive technique that models the relationship between a dependent (target) and one or more independent variables (predictor). In our case, the target is the negative sentiment expressed on Twitter and the predictors are the epidemic indicators and the mitigation measures. For this study, we apply **Linear Regression**, **Ridge Regression**, **Lasso Regression**, **Random Forest**, and **eXtreme Gradient Boosting (XGBoost)**. For each model, the best hyperparameters are selected in each training phase by Grid Search and 10-fold Cross-validation.

# 3 Empirical Evaluation

## 3.1 Data Description

**Sentiment Evaluation Data.** To evaluate the performance of sentiment extraction through XML-R, we used the SemEval-2018 Affect in Tweets sentiment dataset (V-reg), which considers sentiment as a score from zero (very negative) to one (very positive) [15]. More specifically, scores below 42.9% indicate the negative sentiment class, scores above 61% indicate the positive sentiment class, and scores in between indicate the neutral sentiment class. The dataset consists of 2,567 tweets that were annotated by 175 annotators (49,856 annotations reported in total), and it is already split into 1,181 tweets for training, 449 for validation, and 937 for testing.

**Twitter COVID-19.** The tweets that were used in our study were obtained through the Twitter Streaming API. Considering we are interested in capturing the sentiment during the COVID-19 pandemic, we filtered the tweets that comprise COVID-19 related keywords. Our data spans two chronological periods. The first period is from 3/11/2020 to 17/12/2020, but unfortunately, we have a few missing dates for a total of 32 days of available data. The second period is from 20/4/2021 to 14/5/2021, for a total of 25 days. We have more than 13 million tweets and an average of 179,000 tweets per day. We filter the tweets based on their location to include only tweets from the USA, and we obtain almost 654,000 tweets. The fraction of negative tweets (threshold of 42.9%; see the above paragraph) is 317,000.

**Mitigation Measures.** The Oxford COVID-19 Government Response Tracker (OxCGRT) was designed to systematically collect information on different common policy responses taken by governments in response to the pandemic [9]. It contains data from 186 countries on various policies, including school closures, stay-at-home orders, economic support for households, and vaccination. The data is publicly available [1], and more concretely comprises 20 indicators of government responses that can be grouped into three categories: (1) Containment and closure policies (indicators C1-C8), such as school closures and restrictions in movement, (2) Economic policies (indicators E1-E4), such as income support to citizens or provision of foreign aid, and (3) Health system policies (indicators H1-H8), such as the COVID-19 testing regime, emergency investments into healthcare, and most recently, vaccination policies. The data from these 20 indicators is aggregated into a set of four indices: (1) Overall government response index; (2) Containment and health index; (3) Economic support index; (4) Stringency index.

**Epidemic Indicators.** We additionally use the COVID-19 dataset maintained by Our World in Data [18]. The data is updated daily throughout the COVID-19 pandemic covering 226 countries and territories on 55 metrics, including (1) confirmed cases and deaths, (2) hospitalizations and intensive care unit (ICU) admissions, (3) tests and positivity data, (4) vaccination data, (5) other variables of interest. We should note that due to the long reporting chain of new cases and deaths, the daily reported number does not necessarily represent the actual number on each day. For that reason, negative values in cases and deaths may appear if a country corrects previously overestimated historical data.

## 3.2  Setup

We choose to study the United States of America (USA), an English-speaking country, as it is better represented in the Twitter dataset. We construct the statistical and machine learning models to produce the sentiment predictions. Initially, we split our datasets into training and test sets (85%–15%). The training data is used to estimate and generate the models' parameters, and the test data is used to calculate the accuracy of the models. However, at every step of the training, we update the training set with the latest historical value, and the models are retrained (i.e., we employ dynamic training). Thus, the models are updated with the latest information available to include any fluctuation in the sentiment, indicating an increase in COVID-19 cases, a new measure taken, etc.

At each step, we obtain a new predicted value for the sentiment. Once the training is completed, we have our predictions according to the initial test set's length. Then, we evaluate the accuracy of the predictions with respect to the initial test set that contains the actual sentiment values. We consider standard performance indicators to evaluate the performance of the predictive models: the Pearson Correlation, the Mean Absolute Percentage Error (MAPE), and the Root Mean Square Error (RMSE) [4,10].

## 3.3  Results

**Sentiment Extraction.** We used XLM-R, which achieves a root mean square error (RMSE) of 0.015 and a mean absolute percentage error (MAPE) of 0.261 on the test set of the SemEval-2018 V-reg sentiment dataset. The high predictive power of the model, reflected by the low error, makes it a suitable candidate for our sentiment annotation task. We note that our data comprises more than a million tweets, making human annotation impossible. On the same evaluation dataset, NLTK's Sentiment Intensity Estimation baseline model achieves a much worse RMSE (0.053) and MAPE (0.529) score.[1]

**Statistical Models.** Table 1 presents the performance indicators for the statistical models that use only the endogenous variable, i.e., the sentiment. We notice that the three models, Autoregression, Exponential Smoothing, and ARIMA perform very similarly with respect to the three metrics, and they are able to capture somehow the way the sentiment evolves in time, but not very accurately.

We conclude that the sentiment time series itself is not sufficient to predict the future sentiment. For that reason, we attempt to estimate the future sentiment by including different sets of exogenous variables into the ARIMAX model. Table 2

---

[1] http://www.nltk.org/howto/sentiment.html.

**Table 1.** Performance indicators of the statistical models.

| Model | Pearson | MAPE | RMSE |
|---|---|---|---|
| Autoregression | 0.620 | 1.502 | 0.009 |
| Exp Smoothing | 0.626 | 1.560 | 0.009 |
| ARIMA | 0.610 | 1.580 | 0.009 |

**Table 2.** Performance indicators of the ARIMAX model. We use the sentiment variable along with the independent variables referring to the COVID-19 indicators (COV), mitigation measures (ME), and a combination of all (COV&ME).

| Model | Pearson | | | MAPE | | | RMSE | | |
|---|---|---|---|---|---|---|---|---|---|
| | COV | ME | COV&ME | COV | ME | COV&ME | COV | ME | COV&ME |
| ARIMAX | 0.258 | 0.563 | 0.051 | 3.750 | 2.021 | 4.484 | 0.021 | 0.011 | 0.025 |

**Table 3.** Performance indicators of the machine learning models. We make use of the independent variables referring to the COVID-19 indicators (COV), mitigation measures (ME) and a combination of all (COV&ME).

| Model | Pearson | | | MAPE | | | RMSE | | |
|---|---|---|---|---|---|---|---|---|---|
| | COV | ME | COV&ME | COV | ME | COV&ME | COV | ME | COV&ME |
| Linear | 0.232 | **0.598** | 0.520 | 4.581 | 2.017 | **1.763** | 0.027 | 0.014 | **0.011** |
| Ridge | **0.616** | 0.610 | 0.612 | 2.318 | **1.917** | 2.323 | **0.013** | **0.013** | 0.014 |
| Lasso | **0.716** | 0.355 | 0.714 | **1.623** | 2.185 | 1.721 | **0.009** | 0.013 | **0.009** |
| RF | 0.868 | 0.666 | **0.870** | 1.718 | **1.455** | 1.570 | **0.009** | **0.009** | **0.009** |
| **XGBoost** | 0.666 | 0.618 | **0.892** | 1.427 | 1.451 | **1.082** | 0.008 | 0.009 | **0.006** |

presents the performance indicators for the ARIMAX model that makes use of the sentiment variable along with the independent variables referring to the COVID-19 indicators (COV), mitigation measures (ME), and a combination of all (COV&ME). We see that not only the exogenous variables do not improve the performance, but in the case of COV&ME, where the number of features is very high, the model fails to predict the future sentiment. We explain such a result due to the nature of the model that assumes a linear relationship between the target variable and the various features. In this case, the model is incapable of selecting only the relevant features, resulting in the inclusion of noisy signals.

**Machine Learning Models.** At this point, we choose to explore the possibility of estimating the sentiment more accurately with the machine learning models that use the different sets of exogenous variables without incorporating any autoregressive behavior. Table 3 presents the performance indicators for the machine learning models. We test the performance of Linear, Ridge, and Lasso Regressions, as well as Random Forest (RF) and XGBoost. We test the models with the use of a) the COVID-19 indicators (COV), b) the mitigation measures (ME), c) a combination of the COVID-19 indicators and mitigation measures (COV&ME). Overall, the XGBoost outperforms the other prediction models in

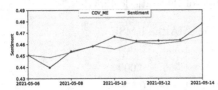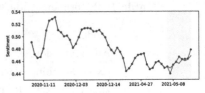

**Fig. 1.** Ground-truth sentiment time series along with the predictions from the XGBoost model with the COVID-19 indicators and mitigation measures (COV&ME). The left plot zooms in the predictions time frame of the right plot.

terms of Pearson correlation, MAPE, and RMSE. It outperforms all the statistical models, as well as the other machine learning models. More specifically, we observe that the best results are obtained when we make use of both the epidemic indicators and mitigation measures (COV&ME), with a Pearson correlation of 0.892, MAPE of 1.082, and RMSE of 0.006.

Figure 1 presents the ground-truth sentiment time series along with the predictions from the XGBoost model with the COVID-19 indicators and mitigation measures (COV&ME). We observe that the predictions are able to monitor the evolution of the sentiment accurately over time. Moreover, Table 3 and Fig. 1 reflect the added value of using the epidemic indicators and measure data over the historical autoregressive and Exponential Smoothing approaches. Forecasts obtained with XGBoost are significantly more accurate for COV&ME compared to the statistical models.

**Feature Importance.** In general, AI models make it difficult, even for the experts, to explain the rationale of their conclusions. For that reason, we consider it crucial to provide understandable results, not only to verify their correctness and quality but, above all, to explain what drives the sentiment of the people during the COVID-19 pandemic period. A benefit of using XGBoost is that it is easy to retrieve importance scores for each feature. Generally, importance provides a score indicating how useful or valuable each feature was in constructing the boosted decision trees within the model. The more an attribute is used to make key decisions with decision trees, the higher its relative importance. We notice in Fig. 1 that the biggest error is for the prediction of 2021-05-10. For that reason, we choose to analyze this particular record to explain what drives the prediction on this day.

Figure 2 (left) shows the 20 most important features (importance on the x-axis) that drive the prediction of 2021-05-10, as calculated from XGBoost with the epidemic indicators and mitigation measures (COV&ME). The most important feature is "stringency_index" with an importance of 0.461, which records the strictness of "lockdown style" policies that primarily restrict people's behavior. We have several COVID-19 indicators that score between the most important features related to tests, cases, deaths, and ICU patients. Additionally, we have the "GovernmentResponseIndex" which records the government's response throughout the outbreak, and the "H7_Vaccination policy" indicator that records the vaccination policy of the health system.

**Local Explanations.** Feature importance measures rarely provide insight into the average direction that a feature affects the response function. They state the magnitude of a feature's relationship with the response compared to other features used in the model. We cannot know specifically the influence of each factor for a single observation. We hence decided to use LIME, which stands for Local Interpretable Model-agnostic Explanations [16] to help us understand individually what features and how they influence the sentiment of each day. LIME is a novel explanation technique that explains the prediction of any classifier or regressor in an interpretable and faithful manner by approximating it locally with an interpretable model. LIME supports explanations for tabular models, text classifiers, and image classifiers.

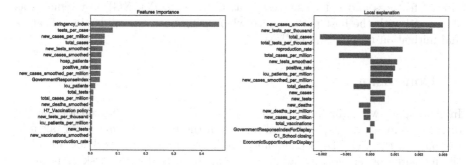

**Fig. 2.** Local explanation of the 20 most important features that drive the prediction of the 2021-05-10, as calculated by XGBoost (left) and with LIME for XGBoost (right) with the COVID-19 indicators and mitigation measures (COV&ME). (Color figure online)

Figure 2 (right) provides the local explanation for the most important features that drive the prediction of 2021-05-10 and their relative strength. Each feature is then color-coded to indicate the relative increase or decrease in the prediction probability, i.e., whether the feature supports or increases the prediction value (Green) or it has a negative effect or decreases the prediction value (Red), respectively. For example, "new_cases_smoothed" is the most important feature with a weight of 0.003, and "new_tests_per_thousand" the second most important with a weight of 0.0025, both of green color, which indicates that they increase the value of the prediction. On the contrary, "total_cases" and "total_tests_per_thousand" are red, indicating a decrease in the prediction value.

Comparing the most important features obtained from the XGBoost model and LIME, we see that the two approaches have 15 out of 20 features in common. That is a very good indicator of the goodness of this result. To further verify if those 15 features are indeed the most relevant for our predictions, we test the accuracy of the predictions with the `ARIMAX` model that suffered greatly from the increased number of features. In Table 4 we report the results from an `ARIMAX` model that uses all the features and an `ARIMAX` model that uses

**Table 4.** Performance indicators of the `ARIMAX` model with all the features and of the `ARIMAX` model with only the 15 (`ARIMAX-15`).

| Model | Pearson | MAPE | RMSE |
|---|---|---|---|
| ARIMAX-COV&ME | 0.051 | 4.484 | 0.025 |
| ARIMAX-15 | 0.604 | 1.774 | 0.009 |

only these 15 (`ARIMAX-15`). Comparing the two models, we observe that feature filtering improves the performance of `ARIMAX` in terms of the three metrics. We should note that `ARIMAX` does not outperform the simpler statistical models. Our objective, however, is to emphasize the improvement in its performance with the filtered features and not to suggest it as a better model. XGBoost remains as the model with the best performance when using both the epidemic indicators and mitigation measures (COV&ME).

## 4    Conclusions

In this paper, we highlighted the limitations of earlier work in considering the direct emotional and psychological impact of the mitigation measures taken in response to the COVID-19 pandemic. We hence proposed a workflow for identifying potential exogenous factors that can be used for the task of negative sentiment nowcasting, employing both statistical and machine learning models. Our results suggest that machine learning models, such as XGBoost, can substantially improve sentiment nowcasting compared to standard autoregressive models. Directions for future work include the exploration of multivariate statistical models, such as VARMAX, as well as RNN- and Transformer-based architectures. We will also explore more extensive feature selection technique as an assistive preprocessing step to statistical models, and the use of data collected over a more extended period, e.g., over the whole pandemic.

**Acknowledgements.** The work of IM and PP has been supported in part by the Digital Futures EXTREMUM project titled "Explainable and Ethical Machine Learning for Knowledge Discovery from Medical Data Sources". The work of PP has also been supported in part by the Vinnova project titled "Artificial Intelligence for Mitigation of Pandemics".

## References

1. Blavatnik School of Government, University of Oxford: Covid-19 government response tracker (2021). https://www.bsg.ox.ac.uk/research/research-projects/covid-19-government-response-tracker#data. Accessed June 2021
2. Cobey, S.: Modeling infectious disease dynamics. Science **368**(6492), 713–714 (2020)

3. Conneau, A., et al.: Unsupervised cross-lingual representation learning at scale. In: Proceedings of the 58th Annual Meeting of the Association for Computational Linguistics, pp. 8440–8451, July 2020
4. De Myttenaere, A., Golden, B., Le Grand, B., Rossi, F.: Mean absolute percentage error for regression models. Neurocomputing **192**, 38–48 (2016)
5. Della Penna, N., Huang, H., et al.: Constructing consumer sentiment index for us using google searches. Technical report (2010)
6. Deng, J., Ren, F.: A survey of textual emotion recognition and its challenges. IEEE Trans. Affect. Comput. (2021)
7. Duncan, B., Elkan, C.: Nowcasting with numerous candidate predictors. In: Calders, T., Esposito, F., Hüllermeier, E., Meo, R. (eds.) ECML PKDD 2014. LNCS (LNAI), vol. 8724, pp. 370–385. Springer, Heidelberg (2014). https://doi.org/10.1007/978-3-662-44848-9_24
8. Gismero-Gonzalez, E., Bermejo-Toro, L., Cagigal, V., Roldan, A., Martinez-Beltran, M.J., Halty, L.: Emotional impact of COVID-19 lockdown among the Spanish population. Front. Psychol. **11**, 3634 (2020)
9. Hale, T., et al.: A global panel database of pandemic policies (oxford COVID-19 government response tracker). Nat. Hum. Behav. **5**(4), 529–538 (2021)
10. James, G., Witten, D., Hastie, T., Tibshirani, R.: An Introduction to Statistical Learning. STS, vol. 103. Springer, New York (2013). https://doi.org/10.1007/978-1-4614-7138-7
11. Kabir, Y., Madria, S.: EMOCOV: machine learning for emotion detection, analysis and visualization using COVID-19 tweets. Online Soc. Netw. Media **23**, 100135 (2021)
12. Kompella, V., et al.: Reinforcement learning for optimization of COVID-19 mitigation policies. arXiv preprint arXiv:2010.10560 (2020)
13. Kwak, G.H., Ling, L., Hui, P.: Deep reinforcement learning approaches for global public health strategies for COVID-19 pandemic. Plos One **16**(5), e0251550 (2021)
14. Lansdall-Welfare, T., Lampos, V., Cristianini, N.: Nowcasting the mood of the nation. Significance **9**(4), 26–28 (2012)
15. Mohammad, S., Bravo-Marquez, F., Salameh, M., Kiritchenko, S.: SemEval-2018 task 1: affect in tweets. In: Proceedings of The 12th International Workshop on Semantic Evaluation, New Orleans, Louisiana, pp. 1–17. Association for Computational Linguistics, June 2018
16. Ribeiro, M.T., Singh, S., Guestrin, C.: "Why should I trust you?" explaining the predictions of any classifier. In: Proceedings of the 22nd ACM SIGKDD International Conference on Knowledge Discovery and Data Mining, pp. 1135–1144 (2016)
17. Richard, Q., Alizon, S., Choisy, M., Sofonea, M.T., Djidjou-Demasse, R.: Age-structured non-pharmaceutical interventions for optimal control of COVID-19 epidemic. PLOS Comput. Biol. **17**(3), 1–25 (2021). https://doi.org/10.1371/journal.pcbi.1008776
18. Ritchie, H., et al.: Coronavirus pandemic (COVID-19). Our World in Data (2020)
19. Tolles, J., Luong, T.: Modeling epidemics with compartmental models. Jama **323**(24), 2515–2516 (2020)
20. Tsakalidis, A., Liakata, M., Damoulas, T., Jellinek, B., Guo, W., Cristea, A.: Combining heterogeneous user generated data to sense well-being. In: Proceedings of COLING 2016, the 26th International Conference on Computational Linguistics: Technical Papers, pp. 3007–3018 (2016)
21. Yang, Q., et al.: SenWave: monitoring the global sentiments under the COVID-19 pandemic. JMIR Public Health Surveill. **6**, e19447 (2020). https://doi.org/10.2196/19447

22. Yue, L., Chen, W., Li, X., Zuo, W., Yin, M.: A survey of sentiment analysis in social media. Knowl. Inf. Syst. **60**(2), 617–663 (2019). https://doi.org/10.1007/s10115-018-1236-4
23. Zhang, L., Wang, S., Liu, B.: Deep learning for sentiment analysis: a survey. Wiley Interdiscip. Rev. Data Min. Knowl. Discov. **8**(4), e1253 (2018)

# Neural Networks and Deep Learning

# A Sentence-Level Hierarchical BERT Model for Document Classification with Limited Labelled Data

Jinghui Lu[1,4]($\boxtimes$) ⓘ, Maeve Henchion[2] ⓘ, Ivan Bacher[3],
and Brian Mac Namee[1,4] ⓘ

[1] School of Computer Science, University College Dublin, Dublin, Ireland
Jinghui.Lu@ucdconnect.ie, Brian.MacNamee@ucd.ie
[2] Teagasc Agriculture and Food Development Authority, Dublin, Ireland
Maeve.Henchion@teagasc.ie
[3] ADAPT Research Centre, Dublin, Ireland
[4] Insight Centre for Data Analytics, Dublin, Ireland

**Abstract.** The emergence of transformer models like BERT means that deep learning language models can achieve reasonably good performance in document classification with few labelled instances. However, there is a lack of evidence for the utility of applying BERT-like models on long document classification in few-shot scenarios. This paper introduces a long-text-specific model—the *Hierarchical BERT Model* (HBM)—that learns sentence-level features of a document and works well in few-shot scenarios. Evaluation experiments demonstrate that HBM can, with only 50 to 200 labelled instances, achieve higher document classification performance than existing state-of-the-art methods, especially when documents are long. Also, as an extra benefit of HBM, the salient sentences identified by a HBM are useful as explanations for document classifications. A user study demonstrates that highlighting these salient sentences is an effective way to speed up the document annotation required in interactive machine learning approaches like active learning.

**Keywords:** Document classification · Few-shot learning · BERT · Interactive machine learning

## 1 Introduction

In many real-world scenarios, it is difficult to access sufficiently large collections of labelled data to train deep learning models. However, recent advances in transformer model architectures, for example the Bidirectional Encoder Representations from Transformers (BERT) approach and its variants [4,7], provide language models that can extract extensive general language features from large corpora and transfer this learned knowledge to a target domain where labelled data is limited. Previous work has demonstrated that this *pre-training and fine-tuning* approach based on BERT outperforms existing approaches to document

© Springer Nature Switzerland AG 2021
C. Soares and L. Torgo (Eds.): DS 2021, LNAI 12986, pp. 231–241, 2021.
https://doi.org/10.1007/978-3-030-88942-5_18

classification when only 100–1,000 labelled examples per class are available [13]. However, in its basic form BERT does not handle long texts well [5] and there is a lack of evidence for the utility of BERT-based models for long text classification when labelled examples are scarce—for instance the datasets used in [13] are short Amazon reviews and Twitter posts.

To address this our work introduces the *Hierarchical BERT Model* (HBM), a sentence-level model based on BERT that handles long texts and is effective in few-shot learning scenarios. In the same way that BERT captures the connections between words, the HBM is designed to capture the connections between sentences, overcoming the loss of document structure information that occurs when simple word vector averaging (typical in BERT-based approaches) is used to form document representations.

The sentence attention mechanism in HBM means that information from sentences that receive higher attention scores are more likely to be aggregated into the representation of the document. We propose that these are important sentences in the document and so can be used as explanations for document classification. This is especially useful for tasks like automated literature screening [8] where a human annotator works in close collaboration with a document classification model, and highlighting salient sentences can make the annotator's job much more efficient.

Yang et al. [15] were the first to propose a sentence-level model for document classification, *Hierarchical Attention Network* (HAN), and they adopted the Bi-GRU with attention as their architecture. Other work investigating the effectiveness of sentence-level models based on BERT includes Yang et al. [14] who proposed a sentence-level BERT model integrated with audio features to forecast the price of a financial asset over a certain period, and Zhang et al. [16] who designed HIBERT for document summarisation. Though all use sentence-level information, our approach differs from these in its use of the pre-training/fine-tuning paradigm and our focus on scenarios when labelled data is scarce. The contributions of this paper are:

- the *Hierarchical BERT Model* (HBM), a sentence-level BERT-based model that is shown to be effective in few-shot, long text classification scenarios
- a demonstration that the salient sentences identified by the HBM are useful as explanations for document classification, and can help annotators label documents faster

The remainder of the paper is organised as follows: Sect. 2 describes the HBM method in detail; Sect. 3 describes an evaluation experiment comparing the HBM method with state-of-the-art methods under the few-shot learning scenario; Sect. 4 introduces the approach to identifying important sentences used in HBM; Sect. 5 describes a user study demonstrating that highlighting important sentences selected using the HBM method leads to faster annotation; and, finally, Sect. 6 draws conclusions.

## 2    The Hierarchical BERT Model

Figure 1 illustrates the Hierarchical BERT Model (HBM) architecture which consists of 3 components: (1) the token-level RoBERTa encoder [7]; (2) the sentence-level BERT encoder; and (3) the prediction layer. To make predictions using this model, raw text is first fed into the token-level RoBERTa encoder to extract text features and form the vector representation for each sentence. These sentence vectors are used as input for the sentence-level BERT encoder. The intermediate representation generated by the sentence-level BERT encoder is then input into the prediction layer to make a classification. The details of each component are described in the following sections.

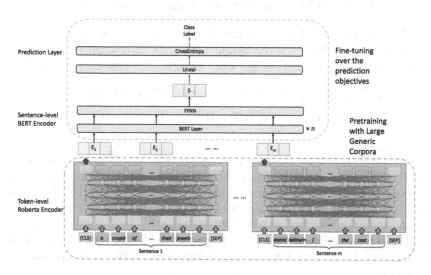

**Fig. 1.** The Hierarchical BERT Model for document classification. (Color figure online)

### 2.1    Token-Level RoBERTa Encoder

RoBERTa [7] uses the same architecture of BERT but optimises key hyper-parameters and pre-trains with larger datasets. It has achieved state-of-the-art results across many NLP tasks and is adopted as the word-level foundation of HBM. Briefly, the multi-head self-attention mechanism in RoBERTa can effectively capture the semantic and syntactic information between words in a document. We leverage a pre-trained RoBERTa model to extract raw text features to represent a document as $\mathcal{D} = (E_1, E_2, \ldots, E_m)$, where $E_i$ denotes the $i^{th}$ sentence vector of document $\mathcal{D} \in \mathbb{R}^{m \times d_e}$ generated by the token-level RoBERTa encoder, and $m$ is a specified maximum number of sentences (zero padding is used for shorter documents). Sentence vectors are generated by averaging the representations of each token in a sentence and are input to the sentence-level BERT encoder.

## 2.2  Sentence-Level BERT Encoder

The sentence-level BERT encoder generates an intermediate document representation, $\mathcal{S}$, based on the sentence vectors, $\mathcal{D}$, output by the token-level RoBERTa encoder. The sentence-level BERT encoder consists of several identical BERT layers and one feedforward neural network with a *tanh* activation. Initially, sentence vectors $\mathcal{D}$ from the token-level RoBERTa encoder are input together into a BERT layer, *BertAtt*, where the multi-head self-attention mechanism is applied. The output of *BertAtt* is calculated as:

$$\text{BertAtt}(\mathcal{D}) = \text{LayerNorm}\ (\mathcal{D} + \text{MultiHead}(\mathcal{D})) \qquad (1a)$$

$$\text{MultiHead}(\mathcal{D}) = \text{Concat}(\text{head}_1, \text{head}_2, \ldots, \text{head}_h) \times W^O \qquad (1b)$$

where $h$ is the number of heads in the *BertAtt* layer; LayerNorm is layer normalization; $W^O \in \mathbb{R}^{hd_e \times d_e}$ is the weight matrix for dimension transformation; and head$_i$ is the attention of the $i^{th}$ head. Each head$_i$ value is calculated as:

$$\text{head}_i = \text{Attention}(Q_i, K_i, V_i) = \text{Softmax}(Q_i \times K_i^\top / \sqrt{d_e}) \times V_i \qquad (2a)$$

$$Q_i = \mathcal{D} \times W_i^Q \qquad K_i = \mathcal{D} \times W_i^K \qquad V_i = \mathcal{D} \times W_i^V \qquad (2b)$$

where $W_i^Q, W_i^K, W_i^V \in \mathbb{R}^{d_e \times d_e}$ are weight matrices for the $i^{th}$ head; and, Softmax $(Q_i \times K_i^\top / \sqrt{d_e})$ is an $m \times m$ matrix in which the entry at the $a^{th}$ row and $j^{th}$ column denotes the attention weight that the $a^{th}$ sentence pays to the $j^{th}$ sentence. Here, $V_i$ contains information from sentences to pass through the following layers while the attention weights matrix, $\text{Softmax}(Q_i \times K_i^\top / \sqrt{d_e})$, acts as a gate to control how much information can be passed through (i.e. after the multiplication, it is hard for sentences with low attention score to further propagate their information).

The output of *BertAtt* is passed through a standard feedforward neural network with a residual mechanism and layer normalisation:

$$\mathcal{D}' = \text{LayerNorm}(\text{BertAtt}(\mathcal{D}) + \text{Relu}(\text{BertAtt}(\mathcal{D}) \times W^r) \times W^S) \qquad (3)$$

where $W^r \in \mathbb{R}^{d_e \times nd_e}$ is a weight matrix that transforms the dimension of BertAtt$(\mathcal{D})$ from $d_e$ to $nd_e$ ($n = 4$ in this work); and $W^S \in \mathbb{R}^{nd_e \times d_e}$ is the weight matrix with dropout to transform the dimension back to $d_e$. $\mathcal{D}'$ then goes into multiple identical BERT layers to form the matrix $\mathcal{Z}$ which is used for computing the intermediate document representation $\mathcal{S}$.

The intermediate document representation $\mathcal{S} \in \mathbb{R}^{1 \times d_e}$ output by the sentence-level BERT encoder can be computed as:

$$\mathcal{S} = \text{Tanh}(\text{Avg}(\mathcal{Z}) \times W^t) \qquad (4)$$

where $W^t \in \mathbb{R}^{d_e \times d_e}$ is a weight matrix; $\mathcal{Z} \in \mathbb{R}^{m \times d_e}$ is the output of the multiple BERT layers; and Avg denotes the mean pooling layer (i.e. $\text{Avg}(\mathcal{Z}) \in \mathbb{R}^{1 \times d_e}$). $\mathcal{S}$ is then passed to the prediction layer.

## 2.3   Prediction Layer

The prediction layer is appended to the sentence-level BERT encoder for final prediction as shown in Fig. 1. The output, $\mathcal{S}$, from the sentence-level BERT encoder is multiplied by a weight matrix $W \in \mathbb{R}^{d_e \times y}$ ($y$ is the number of classes), then fed into a CrossEntropy layer for computing loss.

## 2.4   Training a Hierarchical BERT Model (HBM)

BERT-like models usually exploit the pre-training + fine-tuning procedure to perform downstream tasks [4,7]. In pre-training the model is trained using self-supervised methods with various, large, unlabelled datasets. Then in fine-tuning, a task-specific head is added to the model to further update all parameters of the model by supervised learning over small local datasets. However, pre-training and fine-tuning are a little different in the HBM approach. Since the RoBERTa encoder used in HBM is pre-trained with large generic corpora, we assume that the RoBERTa encoder can encode the token-level information well. Hence, we do not further adjust the pre-trained RoBERTa model during fine-tuning so its weights are frozen. The weights in the sentence-level BERT encoder and the prediction layer are randomly initialised and updated based on our prediction objectives. This is illustrated in Fig. 1 with the dashed green and blue rectangles.

# 3   Experiment 1: Evaluating the Effectiveness of HBM

This section describes an experiment performed to compare the performance of HBM for document classification to state-of-the-art methods, especially when training datasets are small. The datasets used in the experiments, the set up of several baselines as well as HBM, and the experimental results are described.

## 3.1   Datasets

Six fully-labelled binary document classification datasets are used in this experiment (the average number of sentences per document in each dataset is shown in brackets): *Movie Reviews* (33.12) [12]; *Multi-domain Customer Review* (3.78) [2]; *Blog Author Gender* (22.83) [11]; *Guardian 2013 (politics vs business)* (27.88) [1]; *Reuters (acq vs earn)* (7.40); *20 Newsgroups (comp.sys.mac.hardware vs comp.sys.pc.hardware)* (10.61). Most of the datasets contain long documents (from news feeds, blogs, etc.), but datasets that consist of short documents are also included to investigate the performance of HBM on short texts.

## 3.2   Baselines and Setup

Partially following [13], we compare HBM with several baseline methods. We selected an SVM model using a document representation based on FastText (FastText+SVM), an SVM model using text representations based on a pre-trained RoBERTa model (RoBERTa+SVM) and a fine-tuned RoBERTa model as baselines due to their strong performance. We also used another sentence-level model, Hierarchical Attention Network (HAN) [15], as a baseline.[1]

**Table 1.** Performance measured using AUC score of different methods that are trained with limited labelled data across six datasets.

| Dataset | Movie review | | | | |
|---|---|---|---|---|---|
| #Instances | n = 50 | n = 100 | n = 150 | n = 200 | Avg rank |
| FastText+SVM | 0.6653 ± 0.171 (5) | 0.7942 ± 0.020 (4) | 0.8040 ± 0.018 (4) | 0.8260 ± 0.010 (4) | 4.25 |
| RoBERTa+SVM | 0.8743 ± 0.032 (3) | 0.9132 ± 0.025 (3) | 0.9397 ± 0.010 (3) | 0.9449 ± 0.008 (3) | 3.00 |
| Fine-tuned RoBERTa | 0.8878 ± 0.018 (2) | 0.9298 ± 0.013 (2) | 0.9451 ± 0.007 (2) | 0.9536 ± 0.005 (2) | 2.00 |
| HAN | 0.7013 ± 0.096 (4) | 0.7789 ± 0.014 (5) | 0.7504 ± 0.082 (5) | 0.8128 ± 0.011 (5) | 4.75 |
| HBM | **0.9139 ± 0.020 (1)** | **0.9420 ± 0.011 (1)** | **0.9572 ± 0.007 (1)** | **0.9638 ± 0.006 (1)** | **1.00** |
| | Blog author gender | | | | |
| FastText+SVM | 0.6400 ± 0.043 (4) | 0.6669 ± 0.021 (5) | 0.6899 ± 0.009 (4) | 0.6861 ± 0.023 (5) | 4.50 |
| RoBERTa+SVM | 0.5538 ± 0.153 (5) | 0.6783 ± 0.025 (3) | 0.7058 ± 0.017 (3) | 0.7213 ± 0.012 (3) | 3.50 |
| Fine-tuned RoBERTa | 0.6462 ± 0.036 (2) | 0.6892 ± 0.021 (2) | 0.7177 ± 0.019 (2) | 0.7295 ± 0.024 (2) | 2.00 |
| HAN | 0.6402 ± 0.017 (3) | 0.6670 ± 0.010 (4) | 0.6845 ± 0.008 (5) | 0.6876 ± 0.014 (4) | 4.00 |
| HBM | **0.6820 ± 0.025 (1)** | **0.7150 ± 0.031 (1)** | **0.7371 ± 0.007 (1)** | **0.7488 ± 0.013 (1)** | **1.00** |
| | Reuters | | | | |
| FastText+SVM | 0.9757 ± 0.010 (4) | 0.9795 ± 0.005 (5) | 0.9851 ± 0.004 (5) | 0.9862 ± 0.003 (5) | 4.75 |
| RoBERTa+SVM | 0.9838 ± 0.007 (2) | 0.9890 ± 0.003 (3) | 0.9931 ± 0.001 (3) | 0.9930 ± 0.001 (3) | 2.75 |
| Fine-tuned RoBERTa | **0.9885 ± 0.005 (1)** | **0.9933 ± 0.002 (1)** | 0.9953 ± 0.001 (2) | 0.9955 ± 0.001 (2) | **1.50** |
| HAN | 0.9270 ± 0.038 (5) | 0.9804 ± 0.005 (4) | 0.9865 ± 0.003 (4) | 0.9897 ± 0.002 (4) | 4.25 |
| HBM | 0.9825 ± 0.008 (3) | 0.9917 ± 0.004 (2) | **0.9980 ± 0.003 (1)** | **0.9990 ± 0.001 (1)** | 1.75 |
| | Multi-domain customer review | | | | |
| FastText+SVM | 0.6694 ± 0.047 (4) | 0.6927 ± 0.030 (4) | 0.7226 ± 0.030 (5) | 0.7471 ± 0.018 (5) | 4.50 |
| RoBERTa+SVM | 0.8317 ± 0.020 (2) | 0.8558 ± 0.036 (2) | 0.8976 ± 0.019 (2) | 0.9190 ± 0.006 (2) | 2.00 |
| Fine-tuned RoBERTa | **0.9110 ± 0.036 (1)** | **0.9437 ± 0.007 (1)** | **0.9534 ± 0.003 (1)** | **0.9565 ± 0.004 (1)** | **1.00** |
| HAN | 0.6497 ± 0.021 (5) | 0.6907 ± 0.011 (5) | 0.7312 ± 0.014 (4) | 0.7739 ± 0.034 (4) | 4.50 |
| HBM | 0.7669 ± 0.024 (3) | 0.8342 ± 0.014 (3) | 0.8615 ± 0.010 (3) | 0.8913 ± 0.004 (3) | 3.00 |
| | Gurdian 2013 | | | | |
| FastText+SVM | 0.9720 ± 0.003 (3) | 0.9789 ± 0.005 (5) | 0.9794 ± 0.004 (5) | 0.9789 ± 0.005 (5) | 4.50 |
| RoBERTa+SVM | 0.9694 ± 0.011 (4) | 0.9814 ± 0.003 (3) | 0.9860 ± 0.003 (2) | 0.9852 ± 0.002 (3) | 3.00 |
| Fine-tuned RoBERTa | 0.9727 ± 0.010 (2) | 0.9848 ± 0.002 (2) | 0.9854 ± 0.003 (3) | 0.9864 ± 0.002 (2) | 2.25 |
| HAN | 0.9684 ± 0.005 (5) | 0.9794 ± 0.001 (4) | 0.9850 ± 0.002 (4) | 0.9849 ± 0.002 (4) | 4.25 |
| HBM | **0.9740 ± 0.013 (1)** | **0.9862 ± 0.007 (1)** | **0.9904 ± 0.003 (1)** | **0.9925 ± 0.001 (1)** | **1.00** |
| | 20Newsgroups | | | | |
| FastText+SVM | **0.7052 ± 0.030 (1)** | 0.7516 ± 0.033 (3) | 0.7827 ± 0.012 (3) | 0.8094 ± 0.013 (3) | 2.50 |
| RoBERTa+SVM | 0.5969 ± 0.100 (5) | 0.6988 ± 0.024 (5) | 0.7117 ± 0.030 (5) | 0.7436 ± 0.011 (4) | 4.75 |
| Fine-tuned RoBERTa | 0.6098 ± 0.051 (4) | 0.7555 ± 0.041 (2) | 0.8576 ± 0.060 (2) | 0.8838 ± 0.045 (2) | 2.50 |
| HAN | 0.6296 ± 0.030 (3) | 0.7142 ± 0.005 (4) | 0.7295 ± 0.012 (4) | 0.7417 ± 0.029 (5) | 4.00 |
| HBM | 0.6883 ± 0.094 (2) | **0.8168 ± 0.024 (1)** | **0.8579 ± 0.021 (1)** | **0.9158 ± 0.018 (1)** | 1.25 |

To simulate the few-shot training scenario, we randomly subsample a small set of data from the fully labelled dataset as the training data. For each dataset, we subsample 50, 100, 150 and 200 instances respectively and we always use the

---

[1] Setup of all parameters, experiment code, the user study platform, and all user study samples are available at https://github.com/GeorgeLuImmortal/Hierarchical-BERT-Model-with-Limited-Labelled-Data.

full dataset minus 200 instances as the test set. For each experiment, the training is repeated 10 times with training datasets subsampled by 10 different random seeds and we report the averaged result of these 10 repetitions. We use mean area under the ROC curve (AUC) to measure the classification performance.

## 3.3   Results and Analysis

Table 1 summarises the results of the evaluation experiments performed. The leftmost column denotes the modelling approach, the topmost row denotes the relevant dataset, and $n$ is the number of instances in the training set. The numbers in brackets stand for the ranking of each method when compared to the performance of other approaches for a dataset with a specific number of labelled instances. Avg Rank denotes the average ranking of each model for a specific dataset across training set sizes. The best performance in each column is highlighted in bold. It is clear from Table 1 that the performance of each method improves as the number of instances in the training set increases. The HBM method is shown to outperform the state-of-the-art methods in the majority of experiments. Following [3], Friedman tests were also performed on ranks for mean AUC scores across the models and the $p$-values were 0.0382 ($n = 50$), 0.0010 ($n = 100$), 0.0014 ($n = 150$), and 0.0006 ($n = 200$) respectively, indicating that differences in ranks for mean AUC scores were significant. This clearly demonstrates the utility of the HBM approach for few-shot learning.

The performance of HBM is not good for the *Multi-domain Customer Review* and *Reuters* datasets (even lower than that of the approach based on the pre-trained RoBERTa model for the former). We believe that this is because number of sentences per document in these datasets is so low—3.78 and 7.40 on average for the *Multi-domain Customer Review* and *Reuters* datasets respectively—and that the HBM approach cannot effectively extract any sentence structure information from such short documents. This highlights that the HBM approach works best in domains with longer documents.

## 4   Extracting Salient Sentences Using a HBM Model

As mentioned previously, one advantage of the HBM approach is that the attention scores assigned to sentences in a document can be used as an indication of the saliency of those sentences in determining document class. The most salient sentences in a document can be used as an explanation of the classifications that a HBM model produces. This section describes how this can be done.

As described in Eq. 2, $\text{Softmax}(Q_i \times K_i^\top / \sqrt{d_e})$ is a matrix of attention weights used to compute the weighted average of the self-attention heads in the *BertAtt* layer of HBM. In this matrix, entry $e^{aj}$ (at the $a^{th}$ row and $j^{th}$ column) is the attention weight that the $a^{th}$ sentence pays to the $j^{th}$ sentence. In other words, the $j^{th}$ column represents all the attention weights the $j^{th}$ sentence received from other sentences. We use the sum of all entries in the $j^{th}$ column as the saliency score of the $j^{th}$ sentence. Therefore, we compute the saliency score of each sentence in a document by summing up all entries in the corresponding column of the attention weights matrix of the last sentence-level BERT layer.

After obtaining the saliency score of all sentences in a document, we select the salient sentences based on the ratio of the sentence saliency score to the highest saliency score. We regard all sentences with a ratio bigger than 0.9 as salient sentences in a document and disregard other sentences. We believe that, besides the sentence with the highest saliency score, those with a saliency score close to the highest saliency score also contribute to the document representation and should be highlighted. In the next subsection, we will demonstrate the usefulness of highlighting salient sentences in a user study.

## 5    User Study

It is interesting to understand to what extent we can take advantage of important sentences identified by the HBM. In many few-shot learning scenarios human annotators are used to generate labelled data using approaches such as active learning [6,9]. The objective of this user study is to investigate whether or not highlighting the important sentences in a text makes it easier for annotators to provide class labels, thus facilitating labelling of larger numbers of texts in a short time.

### 5.1    Experiment Setup

The user study was run on an online platform and data collected is stored anonymously. We recruited participants by circulating an advertisement among social media such as Twitter and Facebook, and internal email lists of two Irish research centres.[2] In this study, all participants were required to be fluent English speakers. Participants were shown texts and asked to indicate which category, from a list of options provided, best describes it. We use ten documents from the *Movie Review* and *Multi-domain Customer Review* datasets. We assume these two sentiment classification datasets would be easy for participants to annotate for both datasets only two label categories could be chosen: *positive* or *negative*. Ground truth categories for the texts used in the study exist, and these were used to measure the accuracy of the category labels provided by participants. The time taken for participants to select a category for each document were also collected to measure the annotation efficiency.

A between-subject design was used for the experiment. Half of the participants were shown texts in which important sentences have been highlighted based on HBM (highlighting condition group) to make it easier for them to come to a judgement on the category that best describes the text. Half of the participants were shown texts without highlighting (no-highlighting condition group). All documents were presented to the participants in a random sequence. The accuracy of labels provided and the time taken to provide them were compared across the two groups to measure the impact of providing highlighting.

---

[2] The Insight Centre for Data Analytics (https://www.insight-centre.org/) and ADAPT Centre (https://www.adaptcentre.ie/).

## 5.2 Results and Analysis

We had 75 participants in total: 38 participants in the highlighting condition group and 37 participants in the no-highlighting condition group. To mitigate against participants who didn't pay careful attention to the tasks in the study, we removed the records of participants who achieved accuracy of less than 60% or who spent less than 90 seconds in total on all ten questions. Similarly, to mitigate against participants who left the study for a long period of time, we also removed data from participants who spent more than 420 s providing the answer for a single document. After this cleaning, we retained 57 valid participants, among these participants 32 were in the highlighting condition group and 25 were in the no-highlighting condition group.

Figure 2 shows a box plot of the time spent per user in providing category labels for each document, where red boxes denote the performance of the highlighting condition group, and blue boxes denote the performance of the no-highlighting condition group. It is obvious that except for documents CR-2 and MR-3, the time spent by the highlighting condition group is much less than that spent by the no-highlighting condition group.

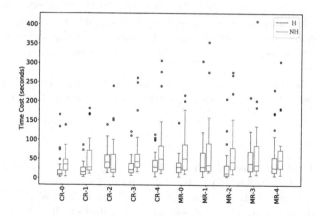

**Fig. 2.** A box plot of the time cost per user on each document. The vertical axis is the time cost per document (unit second); the horizontal axis represents the documents.

We also analyse the results from the user perspective, the average total time spent labelling all ten documents per user in the highlighting condition group is 375.50 s, and in the no-highlighting condition group is 645.54 s. Following [10] we analyse these results using a Mann Whitney U test in which the $p$-value is 0.01939, indicating that we can reject the null hypothesis at the 0.05 confidence level and consider a significant difference between the time spent by members of the two groups. The average accuracy achieved by participants in the highlighting condition group is 0.9575, while the average accuracy achieved by participants in the no-highlighting condition group is 0.9448. A Mann Whitney U test on average accuracy gives a $p$-value of 0.3187, which indicates no significant difference between the average accuracy of the two groups. Based on the previous analysis, we conclude that highlighting important sentences identified by the HBM can significantly improve the human's annotation efficiency while still maintaining a high annotation quality.

# 6 Conclusions

In this work we have proposed a novel text classification approach, the Hierarchical BERT Model (HBM), which has been demonstrated to perform well with limited labelled data especially when documents are long. Experiment results show the superior performance of the HBM approach compared to other state-of-the-art methods under a few-shot training scenario. We also demonstrated that using a HBM facilitates extracting salient sentences from a document for classification explanation and showed through a user study that highlighting these sentences leads to more efficient labelling.

However, as demonstrated previously, HBM is limited when documents are short, necessitating a further exploration of using HBM on short-texts. An ablation study would also be useful to understand the contribution each component of HBM makes to its performance. We also plan to integrate the HBM method into an active-learning based document screening platform taking advantage of its strong performance with small labelled datasets, and the ability to use HBM to highlight salient sentences to make the annotation task easier.

**Acknowledgements.** This research was kindly supported by a Teagasc Walshe Scholarship award (2016053) and Science Foundation Ireland (12/RC/2289_P2).

# References

1. Belford, M., Mac Namee, B., Greene, D.: Stability of topic modeling via matrix factorization. Expert Syst. Appl. **91**, 159–169 (2018)
2. Blitzer, J., Dredze, M., Pereira, F.: Biographies, bollywood, boom-boxes and blenders: domain adaptation for sentiment classification. In: 45th Annual Meeting of the Association of Computational Linguistics (2007)
3. Demšar, J.: Statistical comparisons of classifiers over multiple data sets. J. Mach. Learn. Res. **7**, 1–30 (2006)
4. Devlin, J., Chang, M.W., Lee, K., Toutanova, K.: BERT: pre-training of deep bidirectional transformers for language understanding. arXiv:1810.04805 (2018)
5. Ding, M., Zhou, C., Yang, H., Tang, J.: CogLTX: applying BERT to long texts. In: Advances in Neural Information Processing Systems, vol. 33 (2020)
6. Hu, R., Mac Namee, B., Delany, S.J.: Active learning for text classification with reusability. Expert Syst. Appl. **45**, 438–449 (2016)
7. Liu, Y., et al.: RoBERTa: a robustly optimized BERT pretraining approach. arXiv preprint arXiv:1907.11692 (2019)
8. Lu, J., Henchion, M., Mac Namee, B.: Effect of combination of HBM and certainty sampling on workload of semi-automated grey literature screening. In: 3rd Workshop on Human in the Loop Learning at 38th International Conference on Machine Learning (2021)

9. Lu, J., MacNamee, B.: Investigating the effectiveness of representations based on pretrained transformer-based language models in active learning for labelling text datasets. arXiv preprint arXiv:2004.13138 (2020)

10. Marusteri, M., Bacarea, V.: Comparing groups for statistical differences: how to choose the right statistical test? Biochemia medica: Biochemia medica 20(1), 15–32 (2010)

11. Mukherjee, A., Liu, B.: Improving gender classification of blog authors. In: 2010 Conference on Empirical Methods in Natural Language Processing (2010)

12. Pang, B., Lee, L.: A sentimental education: sentiment analysis using subjectivity summarization based on minimum cuts. In: 42nd Annual Meeting on Association for Computational Linguistics (2004)

13. Usherwood, P., Smit, S.: Low-shot classification: a comparison of classical and deep transfer machine learning approaches. arXiv preprint arXiv:1907.07543 (2019)

14. Yang, L., Ng, T.L.J., Smyth, B., Dong, R.: HTML: hierarchical transformer-based multi-task learning for volatility prediction. In: The Web Conference 2020 (2020)

15. Yang, Z., Yang, D., Dyer, C., He, X., Smola, A., Hovy, E.: Hierarchical attention networks for document classification. In: Conference of the North American Chapter of the Association for Computational Linguistics: Human Language Technologies (2016)

16. Zhang, X., Wei, F., Zhou, M.: HIBERT: document level pre-training of hierarchical bidirectional transformers for document summarization. arXiv preprint arXiv:1905.06566 (2019)

# Calibrated Resampling for Imbalanced and Long-Tails in Deep Learning

Colin Bellinger[1]([✉]), Roberto Corizzo[2], and Nathalie Japkowicz[2]

[1] National Research Council of Canada, Ottawa, Canada
colin.bellinger@nrc-cnrc.gc.ca
[2] American University, Washington, D.C., USA
{rcorizzo,japkowicz}@american.edu

**Abstract.** Long-tailed distributions and class imbalance are problems of significant importance in applied deep learning where trained models are exploited for decision support and decision automation in critical areas such as health and medicine, transportation and finance. The challenge of learning deep models from such data remains high, and the state-of-the-art solutions are typically data dependent and primarily focused on images. Important real-world problems, however, are much more diverse thus necessitating a general solution that can be applied to diverse data types. In this paper, we propose ReMix, a training technique that seamlessly leverages batch resampling, instance mixing and soft-labels to efficiently enable the induction of robust deep models from imbalanced and long-tailed datasets. Our results show that fully connected neural networks and Convolutional Neural Networks (CNNs) trained with ReMix generally outperform the alternatives according to the g-mean and are better calibrated according to the balanced Brier score.

**Keywords:** Deep learning · Calibration · Class imbalance · Long-tail distribution

## 1 Introduction

There is a growing amount of interest in applying deep learning to complex and critical domains, such has medicine, health and safety and finance [22,23], that exhibit long tails, imbalanced class priors and asynchronous misclassification costs. To be safely applied, the learned predictive models must achieve high recall on the minority classes, and be well-calibrated. Deep learning algorithms, however, have been shown to exhibit unsatisfactory predictive performance on poorly represented classes [5,6], suffer from poor calibration [13] and drastically shift their predictions with small changes in the input space. These facts lead to safety concerns related to the use of deep learning in real-world applications

**Electronic supplementary material** The online version of this chapter (https://doi.org/10.1007/978-3-030-88942-5_19) contains supplementary material, which is available to authorized users.

© National Research Council Canada 2021
C. Soares and L. Torgo (Eds.): DS 2021, LNAI 12986, pp. 242–252, 2021.
https://doi.org/10.1007/978-3-030-88942-5_19

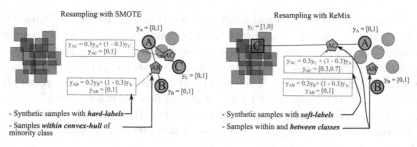

**Fig. 1.** A comparison on the generation process and the distribution of samples generate by ReMix and SMOTE. The figure demonstrates that ReMix expands the minority class space and adds structure between the classes, which reduces predictive bias and improves generalization.

[1,11]. Although there has been a great deal of research into imbalanced learning, existing techniques do not fully ameliorate the problem of deep learning from imbalanced and long tailed data.

Traditional methods to deal with class imbalance involve resampling (random undersampling the majority class, random oversampling the minority class and generating additional synthetic minority samples) or cost-adjustment [4,17]. Synthetic resampling methods based on SMOTE are generally preferred as they are simple-to-apply preprocessing steps that are classifier independent [8]. Whilst these methods have been shown to improve the predictive performance of shallow models, they do not improve calibration with respect to underrepresented classes [25]. In addition, they are not well-suited for batch training in deep learning [20].

In the context of deep learning, recently proposed strategies have focused on the generation of additional synthetic samples via GANs and VAEs to balance the training set [10,20,27]. The existing work, however, is primarily intended for image classification problems, and its efficacy on other data types is not well established. Moreover, these methods necessitate the learning of additional models and/or significantly more parameters. Training with GAMO [20], a recently proposed GAN-based synthetic oversampling technique, for example, requires one generator and discriminator per class, plus the classifier, which can quickly become prohibitive in domains with a large number of classes. Thus, the size and complexity of existing methods renders inappropriate in many cases. Finally, these methods are limited in their ability to expand the minority space and do not improve model calibration because, *a*) they focus the generation of synthetic training samples within the manifold of the minority class(es), and *b*) they assign hard class labels for training.

We propose ReMix, a data-independent algorithm, for integrated training of deep models. ReMix is an advancement of the Mixup algorithm [29] for domains with imbalanced class priors and long tails. ReMix improves the predictive performance and model calibration on imbalanced and long-tailed datasets via a prioritized mixing strategy that efficiently balances the classes in each batch of training data and expands the minority class space to reduce predictive bias.

A comparison of SMOTE-based sample generation and ReMix is provided in Fig. 1. ReMix is unique from previous resampling strategies as it generates soft labels in addition to the synthetic feature vectors. The soft labels are approximate class probabilities that account for label uncertainty due to the synthetic resampling process. The use of soft labels as part of the resampling regime increases the learnt model's robustness to noisy synthetic samples. Moreover, training deep learning models with soft labels has been shown to improve calibration, which we postulate should be a key focus of imbalanced learning solutions.

Our empirical results on imbalanced binary and multi-class tabular and long-tailed image datasets show that deep learning models training using ReMix, achieve equivalent or better predictive performance, and better calibration than models trained with deep generation, traditional imbalanced learning techniques and cost-adjustment methods.

## 2    Related Work

The authors in [2], studied the impact of training NNs on imbalanced classification data and found that the majority class errors dominate the gradient-based weight updates during training. This results in a predictive bias towards the majority classes. Cost-adjustment and re-sampling are the standard techniques to deal with the predictive bias during NN training [5,9,15,16]. In highly parameterized deep models, however, these approaches can cause the models to over-fit the limited information in the minority classes [6]. ReMix addresses these issues by balances the training batches, increasing the diversity in the minority samples and incorporating soft labels.

Within long-tailed and imbalanced image classification, there is a strong emphasis on generation based methods that exploit autoencoders, variational autoencoders and generative adverserial networks [3,10,20,27]. These techniques have been shown to improve performance on image domains. However, they are not designed for general data formats that are common in most real-world applications. Moreover, the rely on large, highly parameterize generator models that are likely to suffer from a lack of training data in highly imbalanced domains.

In [25], the authors found that for models trained on imbalanced data the minority class predictions were poorly calibrated, and that standard resampling methods do not improve model calibration. Moreover, recent studies of deep learning models have shown that they have worse calibration than shallow models [21]. Model calibration in deep learning applied to balanced classification has been addressed by training with soft-labels [13], however, it has not been explored for class imbalance. ReMix offers a solution that uses soft-labels with resampling.

## 3    ReMix: Resampling MixUp

### 3.1    Sample Generation

As proposed in [29], ReMix utilizes the principle of Vicinal Risk Minimization (VRM) [7]. Under this principle, the model is trained on batches

$\mathcal{B}_\nu := \{(\tilde{x}_i, \tilde{y}_i)\}_{i=1}^m$ of synthetic feature-target pairs $(\tilde{x}, \tilde{y})$ drawn vicinity of the true data $X$ and labels $Y$. This is known as the vicinal distribution. This mechanism enables us to sample an infinite supply of training samples $(\tilde{x}, \tilde{y}) \sim \nu_{X,Y}(\cdot)$. The vicinal distribution is approximated by random convex combinations between the seeds in $\mathcal{S} := \{(x_i, y_i)\}_{i=1}^m$, with balanced class representation, and a random subset of the training data $X' := \{(x_j, y_j)\}_{j=1}^m$. Specifically, give a mixing parameter $\lambda$, the seed data and labels $S_X, S_Y$ and the alternate samples $X'_X, X'_Y$, a batch is generated as:

$$B_X = \lambda S_X + (1 - \lambda)X'_X$$
$$B_Y = \lambda S_Y + (1 - \lambda)X'_Y,$$

where the true labels $S_Y$ and $X'_Y$ are label matrices of one-hot vectors and the resulting $B_Y$ is a soft label matrix (*i.e.*, the probability mass spread over multiple classes rather than placed on a single class). The $\lambda \in [0, 1]$ is independently sampled from a Beta distribution $Beta(\alpha, \alpha)$ with $\alpha \in [0, \infty]$ for each batch of training. Sampling a new $\lambda$ value for each batch reduces the risk of memorization and improves generalization by driving down the chance of training on the same synthetic sample more than once. The $\alpha$-value is a user-specified as a hyperparameter that controls the amount of mixing between training samples. The details of the ReMix algorithm are presented in Algorithm 1.

---

**Algorithm 1. ReMix Algorithm.**

---

**Input:** Beta parameter $\alpha \geq 0$, mini-batch size $B$
**Output:** Balanced random mini-batch $X''$ sampled from $\nu(X, y)$
**Algorithm:**

1: Sample the next seed set $S_{X,Y}$ with probability inversely proportional to the class priors
2: Sample the next mixing set $X'_{X,Y}$ with uniform probability
3: Sample a mixing parameter $\lambda = \text{Beta}(\alpha, \alpha)$
4: MixUp features $X'' = \lambda \times S_X + (1.0 - \lambda) \times X'_X]$
5: MixUp labels $Y'' = \lambda \times S_Y + (1.0 - \lambda) \times X'_Y$
6: **return** ReMixed mini-batch $X'', Y''$

---

## 4 Experimental Setup

### 4.1 Datasets

We conducted classification experiments on 2 long-tailed image datasets and 7 tabular imbalanced binary and multi-class datasets. The datasets were created by randomly removing samples from the original datasets in order to create imbalanced and long-tailed problems. For the tabular data, class imbalance ratios ($IR = \frac{N^-}{N^+}$) in the range of 0.1 to 0.01 were created. For each dataset, the lower limit was determined by the original class sizes.

**Table 1.** Details of the binary and multi-class datasets used in the following experiments. All datasets are available at [12].

| Dataset | Classes | Dim | Inst | IRs |
|---------|---------|-----|------|-----|
| Musk | 0/1 | 168 | 6, 597 | 0.01, 0.025, 0.05 |
| Segment | 1/2..7 | 19 | 2, 310 | 0.01, 0.025, 0.05 |
| Statlog | 1/2..7 | 36 | 6, 435 | 0.01, 0.025, 0.05 |
| Seizure | 1/2..5 | 179 | 11, 500 | 0.05, 0.025, 0.01 |
| coil2000 | 0/1 | 86 | 9, 000 | 0.05, 0.025, 0.01 |
| Ozone | 0/1 | 73 | 2, 536 | 0.025, 0.01 |
| APS | −1/1 | 171 | 60, 000 | 0.01 |
| Seizure | 1, 2/3...5 | 179 | 11, 500 | 0.25, 0.1, 0.05 |
| Digits | 1...3/0, 4...9 | 64 | 5, 620 | 0.25, 0.1, 0.05 |
| Landsat | 2...4/1, 5...7 | 36 | 6, 435 | 0.25, 0.1, 0.05 |

**Table 2.** Samples per-class in MNIST Fashion [28] and CIFAR10 datasts [18].

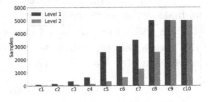

The long-tailed distributions were motivated by the work of [20]. The smallest class in Level 1 has 20 samples, and subsequent classes double until 5,000. Level 2 starts at 80 and increases in non-uniform steps. The specifics of the tabular datasets are outlined in Table 1 and the long-tailed image data are in Table 2. We examine standard imbalance in tabular data and long-tailed image datasets in order to understand on the effectiveness of ReMix on a wide variety of problems of interest to the imbalanced learning community.

## 4.2  Deep Learning Models

In the tabular data experiments, we utilize a three layer deep NN with 0.1 dropout and relu activation. For the image data, we employ a convolutional NN with relu activation, two $3 \times 3$ convolutional layers and two $2 \times 2$ max pooling layers interleaved, and a dense layer. Both network architectures utilize categorical cross-entropy loss with ADAM optimization. These represent common baseline architectures for the target datasets.

## 4.3  Resampling and Cost-Adjustment Methods

Standard domain-independent resampling methods are applied to the tabular data. This includes adjusting the misclassification costs to account for the class imbalance (cost adjustment), generating additional minority samples by random interpolation (SMOTE), and ReMix. For comparison on the image datasets, DEnoising Autoncoder Oversampling (DEAGO) [3], Generative Adversarial Minority Oversampling (GAMO) and resampling minority instances from a Conditional Generative Adversarial Network (cGAN) [19] are utilized.

DEAGO and cGAN are trained on data sampled from the minority class prior to training the classifier, and then are utilized to generated additional samples as a pre-processing step. GAMO learns a generator and discriminator for each

**Table 3.** Mean and standard deviation performances on MNIST Fashion (left) and CIFAR10 (right) level 1 long-tailed data. ReMix achieves the better performance than the more complex generative methods on MNIST Fashion. ReMix and cGAN achieve the top g-mean performance and ReMix is superior in terms of the balanced Brier score.

| | Mean GM | Std GM | Mean 1-BBS | Std 1-BBS | | Mean GM | Std GM | Mean 1-BBS | Std 1-BBS |
|---|---|---|---|---|---|---|---|---|---|
| ReMix$_{\alpha=0.5}$ | **0.787** | 0.015 | **0.853** | 0.005 | ReMix$_{\alpha=0.1}$ | **0.345** | 0.027 | **0.531** | 0.01 |
| GAMO | 0.515 | 0.298 | 0.815 | 0.015 | GAMO | 0.0 | 0.0 | 0.389 | 0.018 |
| cGAN | 0.743 | 0.023 | 0.827 | 0.006 | cGAN | 0.344 | 0.022 | 0.494 | 0.014 |
| DEAGO | 0.734 | 0.029 | 0.816 | 0.007 | DEAGO | 0.276 | 0.139 | 0.476 | 0.01 |
| Baseline | 0.738 | 0.027 | 0.825 | 0.007 | Baseline | 0.331 | 0.081 | 0.509 | 0.007 |

class, along with the multi-class classifier in parallel. Thus, a 10-class problem requires 10 generators and discriminators and one 10-class classifier. Training is conducted as an adversarial game that improves generation and classification of minority samples. After training, the multi-class classifier is used to classify the test set.

### 4.4 Metrics and Evaluation

We compare the imbalanced learning methods to the baseline classifier and one trained with Mixup. Classification performance is assessed with the g-mean (GM), a standard metric for imbalanced classification. It calculates the geometric mean of the true positive rate and true negative rate [5]. To be consistent with the work of [25,26], calibration is assessed with respect to the Brier score. We generalize the previous work as the arithmetic mean of the Brier score calculated independently for each class for multi-class settings and denote it the balanced Brier score (BBS):

$$BBS = \frac{\sum_j^K BS_j}{K},$$

where $K$ is the number of classes. This treats the calibration of each class as equally important, and is the calibration equivalent of the balanced per-class accuracy. Values closer to zero indicate better calibration. Our future work will evaluate the ECE metric for imbalance learning [13].

The results for each setup are recorded as the mean and standard deviation calculated across $10 \times 2$-fold cross validation. During model learning 30 percent of the training set is partitioned off for model validation. All experiments were performed with Tensorflow 2 on Ubuntu 18.04 desktop running a GeForce RTX 2080 Ti GPU[1].

---

[1] The code will be made available after publication.

# 5   Results

Table 3 present the mean and standard deviation of the g-mean and 1-BBS[2] results for the level 1, long-tailed MNIST Fashion and CIFAR10 datasets. The level 2 results are withheld for space considerations. ReMix produces better mean performance than the alternate methods at both imbalance levels for MNIST Fashion. On CIFAR10, both ReMix and cGAN are competitive with respect to the g-mean. According to the BBS, however, ReMix produces a model that is much more calibrated.

Table 4 summarizes the performance in terms of the rank of each training strategy across all IR[3]. Each cell shows the sum of ranks for the method, with lower being better. ReMix produces the best rank of GM and BBS for both the binary and multi-class data. Resampling with SMOTE produces the second best performance. It is worth emphasizing that relative to MixUp, ReMix produces a significant improvement in calibration and predictive performance.

**Table 4.** The sum of ranks on the binary and multi-class tabular datasets.

|  | Binary | | Multi-class | |
|---|---|---|---|---|
|  | GM | BBS | GM | BBS |
| Baseline | 65 | 63 | 31 | 38 |
| Cost Adjusted | 70 | 66 | 28 | 36 |
| SMOTE | 36 | 54 | 24 | 30 |
| MixUp | 77 | 63 | 40 | 22 |
| ReMix | **22** | **22** | **12** | **9** |

Figure 2 illustrates the performance of ReMix at different levels of imbalance. It shows the mean performance gain over the baseline deep NN achieved by ReMix and the comparison methods. The top row includes the GM gain, which is calculated as:

$$GM(f_{\text{alternative}}) - GM(f_{\text{baseline}}),$$

and the bottom row reports BBS gains, which is calculated as:

$$BBS(f_{\text{alternative}}) - BBS(f_{\text{baseline}}).$$

A large positive score indicates a greater improvement over the baseline deep NN. Each plot shows the gain for IRs 0.05, 0.025 and 0.01, along with the mean gain overall IRs (All).

The results show that ReMix produces a greater mean improvement than the alternative methods. The greatest improvements in GM are produced by ReMix on the binary data. Both ReMix and SMOTE (the second best method in terms of GM), see their GM gains decline with more extreme imbalance. As hypothesized, the BBS Gain indicates that ReMix produces a much larger improvement in calibration than the alternative methods on the binary and multi-class tabular data.

---

[2] We use 1-BBS so higher scores are better.
[3] Individual results for all datasets including means and standard deviation are included in the supplementary material.

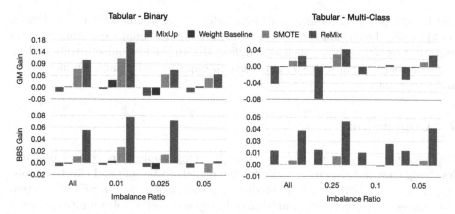

**Fig. 2.** Mean performance gains over the baseline on the tabular binary and multi-class classification data at different imbalance ratios.

# 6 Discussion

The reasonable range of the ReMix parameter is $0 < \alpha < 1$, where 0 produces instance replication and 1 produces uniform interpolation. Figure 3 shows the sensitivity of the resulting MNIST Fashion models on this range of parameters in terms of g-mean and balanced Brier score. The results indicate that ReMix is robust across the $0.1 \le \alpha \le 0.8$ range. From the imbalance Level 2 results, we see a slight preference for larger $\alpha$ values on more extreme levels of imbalance. We hypothesize that setting the $\alpha$ closer to 0.8 is helpful in these cases because it expands the minority space more, thereby reducing the predictive bias.

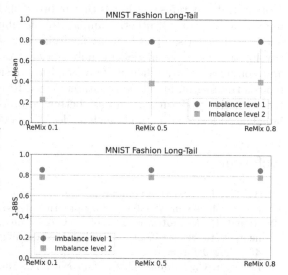

**Fig. 3.** Sensitivity of deep learning model to the $\alpha$ parameter in ReMix on MNIST Fashion.

A potential negative impact of large $\alpha$ values results from the risk of synthetic minority samples harmfully encroaching into other classes. This suggests a potential limitation of feature level mixing for imbalance classification. We note that image-specific data augmentations can be integrated into ReMix to improve performance on image data and avoid sample encroachment. The development and integration of automatic and data independent augmentation into ReMix could be very beneficial. If they can be made computationally efficient, the integration of recent methods, such as manifold MixUp [24] and out-of-manifold data augmentation [14], into ReMix may serve to improve performance on multi-class and image data.

# 7 Conclusion

Deep learning algorithms are increasingly being applied to critical domains involving class imbalance. In order for the models to be safe and effective tools, they must have both excellent predictive performance and be well calibrated. Nonetheless, the literature on imbalanced deep learning remains limited in scope and it is typically focused on optimized solutions for individual datasets and domains.

In this work, we analyze the efficacy of imbalance learning techniques for deep learning on long-tailed image data and imbalanced tabular data, and we propose the ReMix algorithm. It inflates the number of minority class samples in the mini-batches and expands the minority class space to reduce predictive bias. Moreover, it generates soft-labels that improve model calibration and reduce the impact of any noisy synthetic samples by distributing the probability mass in one-hot labels for samples that are far from their seed. ReMix is a general strategy for training deep models on imbalanced data that is simple to implement and computationally efficient. Our empirical results show that deep learning models training on imbalanced datasets using ReMix achieve equivalent or better predictive performance and are better calibrated, than models trained with imbalanced learning methods such as deep generation and resampling.

# References

1. Amodei, D., Olah, C., Steinhardt, J., Christiano, P., Schulman, J., Mané, D.: Concrete problems in AI safety. arXiv preprint arXiv:1606.06565 (2016)
2. Anand, R., Mehrotra, K.G., Mohan, C.K., Ranka, S.: An improved algorithm for neural network classification of imbalanced training sets. IEEE Trans. Neural Netw. **4**(6), 962–969 (1993)
3. Bellinger, C., Drummond, C., Japkowicz, N.: Manifold-based synthetic oversampling with manifold conformance estimation. Mach. Learn. **107**(3), 605–637 (2017). https://doi.org/10.1007/s10994-017-5670-4
4. Branco, P., Torgo, L., Ribeiro, R.P.: A survey of predictive modeling on imbalanced domains. ACM Comput. Surv. (CSUR) **49**(2), 1–50 (2016)
5. Buda, M., Maki, A., Mazurowski, M.A.: A systematic study of the class imbalance problem in convolutional neural networks. Neural Netw. **106**, 249–259 (2018)
6. Cao, K., Wei, C., Gaidon, A., Arechiga, N., Ma, T.: Learning imbalanced datasets with label-distribution-aware margin loss. In: Advances in Neural Information Processing Systems, pp. 1567–1578 (2019)
7. Chapelle, O., Weston, J., Bottou, L., Vapnik, V.: Vicinal risk minimization. In: Advances in Neural Information Processing Systems, pp. 416–422 (2001)

8. Chawla, N.V., Bowyer, K.W., Hall, L.O., Kegelmeyer, W.P.: SMOTE: synthetic minority over-sampling technique. J. Artif. Intell. Res. **16**, 321–357 (2002)
9. Cui, Y., Jia, M., Lin, T.Y., Song, Y., Belongie, S.: Class-balanced loss based on effective number of samples. In: Proceedings of the IEEE Conference on Computer Vision and Pattern Recognition, pp. 9268–9277 (2019)
10. Dai, W., Ng, K., Severson, K., Huang, W., Anderson, F., Stultz, C.: Generative oversampling with a contrastive variational autoencoder. In: 2019 IEEE International Conference on Data Mining (ICDM), pp. 101–109. IEEE (2019)
11. DeVries, T., Taylor, G.W.: Learning confidence for out-of-distribution detection in neural networks. arXiv preprint arXiv:1802.04865 (2018)
12. Dua, D., Graff, C.: UCI machine learning repository (2017). http://archive.ics.uci.edu/ml
13. Guo, C., Pleiss, G., Sun, Y., Weinberger, K.Q.: On calibration of modern neural networks. arXiv preprint arXiv:1706.04599 (2017)
14. Guo, H.: Nonlinear mixup: out-of-manifold data augmentation for text classification. In: AAAI, pp. 4044–4051 (2020)
15. Huang, C., Li, Y., Loy, C.C., Tang, X.: Learning deep representation for imbalanced classification. In: Proceedings of the IEEE Conference on Computer Vision and Pattern Recognition, pp. 5375–5384 (2016)
16. Johnson, J.M., Khoshgoftaar, T.M.: Survey on deep learning with class imbalance. J. Big Data **6**(1), 1–54 (2019). https://doi.org/10.1186/s40537-019-0192-5
17. Krawczyk, B., Bellinger, C., Corizzo, R., Japkowicz, N.: Undersampling with support vectors for multi-class imbalanced data classification. In: 2021 International Joint Conference on Neural Networks (IJCNN) (2021)
18. LeCun, Y., Bottou, L., Bengio, Y., Haffner, P.: Gradient-based learning applied to document recognition. Proc. IEEE **86**(11), 2278–2324 (1998)
19. Mirza, M., Osindero, S.: Conditional generative adversarial nets. arXiv preprint arXiv:1411.1784 (2014)
20. Mullick, S.S., Datta, S., Das, S.: Generative adversarial minority oversampling. In: Proceedings of the IEEE International Conference on Computer Vision, pp. 1695–1704 (2019)
21. Niculescu-Mizil, A., Caruana, R.: Obtaining calibrated probabilities from boosting. In: UAI, p. 413 (2005)
22. Rao, R.B., Krishnan, S., Niculescu, R.S.: Data mining for improved cardiac care. ACM SIGKDD Explor. Newsl. **8**(1), 3–10 (2006)
23. Sanz, J.A., Bernardo, D., Herrera, F., Bustince, H., Hagras, H.: A compact evolutionary interval-valued fuzzy rule-based classification system for the modeling and prediction of real-world financial applications with imbalanced data. IEEE Trans. Fuzzy Syst. **23**(4), 973–990 (2014)
24. Verma, V., et al.: Manifold mixup: better representations by interpolating hidden states. In: International Conference on Machine Learning, pp. 6438–6447 (2019)
25. Wallace, B.C., Dahabreh, I.J.: Class probability estimates are unreliable for imbalanced data (and how to fix them). In: 2012 IEEE 12th International Conference on Data Mining, pp. 695–704. IEEE (2012)
26. Wallace, B.C., Dahabreh, I.J.: Improving class probability estimates for imbalanced data. Knowl. Inf. Syst. **41**(1), 33–52 (2013). https://doi.org/10.1007/s10115-013-0670-6
27. Wang, Q., et al.: WGAN-based synthetic minority over-sampling technique: improving semantic fine-grained classification for lung nodules in CT images. IEEE Access **7**, 18450–18463 (2019)

28. Xiao, H., Rasul, K., Vollgraf, R.: Fashion-MNIST: a novel image dataset for bench-marking machine learning algorithms. arXiv preprint arXiv:1708.07747 (2017)
29. Zhang, H., Cisse, M., Dauphin, Y.N., Lopez-Paz, D.: mixup: beyond empirical risk minimization. arXiv preprint arXiv:1710.09412 (2017)

# Consensus Based Vertically Partitioned Multi-layer Perceptrons for Edge Computing

Haimonti Dutta[1](✉), Saurabh Amarnath Mahindre[2], and Nitin Nataraj[3]

[1] Department of Management Science and Systems, The State University of New York, Buffalo, NY 14260, USA
haimonti@buffalo.edu

[2] Institute for Computational and Data Sciences, The State University of New York, Buffalo, NY 14260, USA
smahindr@buffalo.edu

[3] o9 Solutions, Inc., 1501 Lyndon B. Johnson Freeway, Dallas, TX 75234, USA
nitin.nataraj@o9solutions.com

**Abstract.** Storing large volumes of data on distributed devices has become commonplace in recent years. Applications involving sensors, for example, capture data in different modalities including image, video, audio, GPS and others. Novel distributed algorithms are required to learn from this rich, multi-modal data. In this paper, we present an algorithm for learning *consensus based multi-layer perceptrons* on resource-constrained devices. Assuming nodes (devices) in the distributed system are arranged in a graph and contain vertically partitioned data and labels, the goal is to learn a *global* function that minimizes the loss. Each node learns a feed-forward multi-layer perceptron and obtains a loss on data stored locally. It then gossips with a neighbor, chosen uniformly at random, and exchanges information about the loss. The updated loss is used to run a back propagation algorithm and adjust local weights appropriately. This method enables nodes to learn the *global* function without exchange of data in the network. Empirical results reveal that the consensus algorithm converges to the centralized model and has performance comparable to centralized multi-layer perceptrons and tree-based algorithms including random forests and gradient boosted decision trees. Since it is completely decentralized, scalable with network size, can be used for binary and multi-class problems, not affected by feature overlap, and has good empirical convergence properties, it can be used for on-device machine learning.

**Keywords:** Multi-layer perceptron · Gossip · Consensus · Distributed learning

---

This work was done when the author was a student at the State University of New York at Buffalo.

C. Soares and L. Torgo (Eds.): DS 2021, LNAI 12986, pp. 253–267, 2021.
https://doi.org/10.1007/978-3-030-88942-5_20

# 1  Introduction

Distributed systems store, process and analyze large volumes of data [2] from the mega-scale cloud data-centers to resource constrained devices, such as the Internet of Things (IoT) and mobile devices. While the cloud can be used for executing large scale machine learning algorithms on big data, these algorithms exert severe demands in terms of energy usage, memory requirements and computing resources, limiting their adoption in resource constrained, network edge devices. The new breed of intelligent devices and high-stake applications (drones, augmented/virtual reality, autonomous systems, etc.), require distributed, *low-latency* and *reliable* machine learning at the wireless network edge. Thus computing services have now started to move from the cloud to the edge.

Deep learning-based intelligent services [25] and applications have become prevalent. However, their use in edge computing devices has been somewhat limited due to the following reasons: (a) Cost: Training and inference of deep learning models in the distributed infrastructures requires consumption of large amount of network bandwidth. (b) Latency: The access to data and services is generally not guaranteed and delay is not short enough for time-critical applications. (c) Reliability: Most distributed computing applications rely on wireless communications and backbone networks for connecting users to services, but intelligent services must be highly reliable, even when network connections are lost (d) Privacy: The data required for deep learning may involve private information, and privacy protocols need to be adhered. The current state of distributed deep learning systems on edge devices leaves much to be desired.

In this paper, we address this shortcoming by developing multi-layer perceptrons for *resource constrained* edge devices. When compute power is abundant and devices are not resource-constrained, deep neural networks can be trained using the DistBelief framework [8] with model parallelism within (via multi-threading) and across machines (via message passing). Aside from the fact that a parallel architecture has a single point of failure and therefore often unsuitable for adoption in resource constrained leaderless environments, synchronization requirements lends these algorithms even more unsuitable for use on edge devices. Our algorithm operates on *peer-to-peer computing* environments and as such interweaves local learning and label propagation [30]. Specifically, it optimizes a trade-off between smoothness of the model parameters over the network on the one hand and the model's local learning on the other. It has similarities to collaborative learning of personalized (peer-to-peer) models over networks [3] – however, unlike them, the work presented here learns the global function in the network, instead of solitary, local models.

Finally, it must be pointed out that this work explicitly considers *vertically* partitioned data or the setting in which features are distributed across nodes. Recent work on large scale distributed deep networks has primarily focused on horizontal partitions (for e.g. cross-data silo Federated Learning [15]) wherein all features are observed at the nodes [4,19] and a centralized parameter server updates models. Our work is closely related to the cross-silo Federated Learning [15,28] model, except that the single point of failure parameter server in those models is replaced with a peer-to-peer architecture. This seemingly minor change has far reaching implications – it removes the need for synchronization with the parameter server at every iteration of the algorithm. While peer-to-peer Federated Learning has been proposed in recent work [17,27] to the best of our knowledge its use in the design of deep learning algorithms in the cross-silo setting has not been explored.

This paper is organized as follows: Sect. 2 describes use cases for consensus based multi-layer perceptrons; Sect. 3 describes related work; Sect. 4 provides details of the algorithm and empirical results are presented in Sect. 5. Section 6 concludes the paper.

## 2   Use Cases for Learning Consensus Based Multi-layer Perceptrons

We motivate the need to develop consensus based multi-layer perceptrons by describing the following applications:

- **Medical Diagnosis:** Collaborations amongst health entities[1] on mobile devices [10] require examination of different modalities of patient data such as Electronic Health Records (EHR), imaging, pathology results, and genetic markers of a disease.
- **Drug Discovery:** The pharmaceutical industry requires platforms that enable drug discovery using private and competitive Drug Discovery related data and hundreds of TBs of image data[2].
- **Autonomous Vehicles:** Google, Uber, Tesla, and many automotive companies have developed autonomous driving systems. Applications (such as forward collision warning, blind spot, lane change warnings, and adaptive cruise control.) are time critical and require real time learning and updates from individual vehicles [21].
- **Home Sensing:** In home monitoring and sensing applications [13] non-intrusive load monitoring systems are used to study fluctuations in signals.
- **Manufacturing Operations:** which requires industrial data that is interoperable and scalable[3]. In applications of this genre, the sensors and IoT devices collect data at different time points often from different locations and these are then subjected to analysis.

## 3   Related Work

Scalable algorithms for deep learning have been explored in several papers in recent years. We discuss related work which make use of two different architectures: (a) Parallel – which ensures the presence of a master to control slave workers and (b) Distributed which is a fully decentralized, peer-to-peer architecture without the need for a master.

---

[1] https://featurecloud.eu/about/our-vision/.
[2] https://cordis.europa.eu/project/id/831472.
[3] https://musketeer.eu/project/.

**Parallel DNN Algorithms:** A large proportion of the research in this domain has focused on data parallelism and the ability to exploit compute power of multiple slave workers, with a single master controlling the execution of slaves. McDonald et al. [18] present two different strategies for parallel training of structured perceptrons and use them for named entity recognition and dependency parsing. TernGrad [26] uses ternary levels $\{-1, 0, +1\}$ to reduce overhead of gradient synchronization and communication. DoReFa-Net [31] train convolutional neural networks that have low bit width weights, activations and gradients. Seide et al. [22] show that it is possible to quantize gradients aggressively during training of deep neural networks using SGD making it feasible to use in data parallel fast processors such as GPUs. Quantized SGD (QSGD) [1] explores the trade-off between accuracy and gradient precision. A slightly different line of work [29] explores the utility of asynchronous Stochastic Gradient Descent algorithms suggesting that if the learning rate is modulated according to the gradient staleness, better theoretical guarantees for convergence can be established than the synchronous counterpart.

**Distributed DNN Algorithms:** In the fully decentralized setting, [14] present a consensus-based distributed SGD (CDSGD) algorithm for collaborative deep learning over fixed topology networks that enables data parallelization as well as decentralized computation. Sutton et al. [23] explore neural network architectures in which the structures of the models are partitioned prior to training. Partitioning of deep neural networks have also been studied in the context of distributed computing hierarchies such as the cloud, end and edge devices [24]. Gupta et al. [10] present an algorithm for training DNNs over multiple data sources. The research described above fundamentally differ from the material presented in this paper in that our consensus algorithm relies on both model and data partitioning to construct local multi-layer perceptron models which can independently learn global information.

## 4    Distributed Multi-layer Perceptrons

We present the consensus based multi-layer perceptron algorithm in this section.

In the distributed setting, let $M$ denote an $N \times n$ matrix with real-valued entries. This matrix represents a dataset of $N$ tuples of the form $x_i \in \mathbb{R}^n, 1 \leq i \leq N$. Each tuple has an associated label $y_i = \{+1, -1\}$. Assume this dataset has been *vertically*[4] distributed over $m$ nodes $S_1, S_2, \cdots, S_m$ such that node $S_i$ has a data set $M_i \subset M, M_i : N \times n_i$ and each $x_j \in M_i$ is in $\mathbb{R}^{n_i}, n_i \leq n$. Thus, $M = M_1 \cup M_2 \cup \cdots \cup M_m$ denotes the concatenation of the local data sets. The labels are shared across all the nodes. The goal is to learn a deep neural network on the global data set $M$, by learning local models[5] at the nodes, allowing

---

[4] This implies that all the nodes have access to all $N$ tuples but have limited number of features i.e. $n_i \leq n$.

[5] We assume that the models have the same structure i.e. the same number of input, hidden and output layers and connections.

exchange of information among them using a gossip based protocol [9,16] and updating the local models with new information obtained from neighbors. This ensures that there is no actual data transfer amongst nodes.

**Model of Distributed Computation.** The distributed algorithm evolves over discrete time with respect to a "global" clock[6]. Each node has access to a local clock or no clock at all. Furthermore, each node has its own memory and can perform local computation (such as estimating the local weight vector). It stores $f_i$, which is the estimated local function. Besides its own computation, nodes may receive messages from their neighbors which will help in evaluation of the next estimate for the local function.

**Communication Protocols.** Nodes $S_i$ are connected to one another via an underlying communication framework represented by a graph $G(V, E)$, such that each node $S_i \in \{S_1, S_2, \cdots, S_m\}$ is a vertex and an edge $e_{ij} \in E$ connects nodes $S_i$ and $S_j$. Communication delays on the edges are assumed to be zero.

**Distributed MultiLayer Perceptron (DMLP):** Assume that each node $S_t$ has a simple model of a fully connected multi-layer perceptron with Rectified Linear Unit (ReLU) activations for hidden layers and softmax for the output layer. The network is called $\mathcal{N}_t$. It has $L$ layers – the $0^{th}$ is the input layer, followed by $(L-1)$ hidden layers and the $L^{th}$ layer is the output layer. Let $r_i$ denote the number of units in the $i^{th}$ layer (note that $r_0 = n_i$ and $r_L = 1$).

**Feed-Forward Learning:** Let $\omega_{ij}^k$ denote the weight from $i^{th}$ node of $(k-1)^{th}$ layer to $j^{th}$ node of $k^{th}$ layer, $a_j^k$ is the weighted sum of inputs from the previous layer to the $j^{th}$ node of $k^{th}$ layer, $o_j^k$ is the output of $j^{th}$ node of $k^{th}$ layer, $b_j^k$ is the bias to $j^{th}$ node of $k^{th}$ layer. The feed-forward step for the first node of the first hidden layer can then be written as: $a_1^1 = b_1^1 + x_1^0 * \omega_{11}^1 + x_2^0 * \omega_{21}^1 + ... + x_{n_i 1}^0 * \omega_{n_i 1}^1$. The output $a_1^1$ is given by: $o_1^1 = \max(0, a_1^1)$. So, the output of $j^{th}$ node of $k^{th}$ layer is, $o_j^k = \max(0, a_j^k)$ where, $a_j^k = b_j^k + (\sum_{i=1}^{r_{k-1}} o_i^{k-1} * w_{ij}^k)$. The output from the network $\mathcal{N}_t$ is given by $\hat{y}_i^{\mathcal{N}_t} = softmax(a_1^L)$, where softmax refers to the normalized exponentiated function. The local loss at node $S_i$ is then given by $\mathcal{L}_t$.

**Gossip:** Node $S_t$ selects uniformly at random, a neighbor $S_u$ with whom it wishes to gossip. Both $S_t$ and $S_u$ have computed their local losses. When gossiping each node updates its current local loss with $\mathcal{L}_{gossip} = \frac{\mathcal{L}_t + \mathcal{L}_u}{2}$. This new loss is used for back propagation at both nodes $S_t$ and $S_u$.

---

[6] Existence of this clock is of interest only for theoretical analysis.

**Back Propagation:** The back propagation algorithm learns the weights for a multi-layer network, given a network with a fixed set of units and interconnections. It employs gradient descent to attempt to minimize the error between the network output values and the target values for these outputs. We use the new loss ($\mathcal{L}_{gossip}$) obtained after gossiping with a neighbor, in-place of the local loss ($\mathcal{L}_t$), for our back propagation phase. This modification helps the local node $S_t$ to incorporate information about the loss from its neighbor $S_u$ into its back propagation learning phase, thereby helping to minimize the *global* loss instead of the local loss. This is a crucial step in our algorithm. The local loss at node $S_t$ after gossip is then given by $\mathcal{L}_{gossip} = \frac{1}{2}\sum_{i=1}^{N}(y_i - \frac{\hat{y}_i^{N_t}+\hat{y}_i^{N_u}}{2})^2 = \frac{1}{2}(\mathbf{y} - \mathbf{y_{gossip}})^2; \mathbf{y_{gossip}} = \frac{\hat{y}_i^{N_t}+\hat{y}_i^{N_u}}{2}$ where the bold fonts are used to represent the loss vectors and the example assumes squared error[7]. Algorithm 1 presents the steps of the DMLP algorithm.

---

**Input**: $N \times n_i$ matrix at each node $S_i$, $G(V, E)$ which encapsulates the underlying communication framework, $T$ : no of iterations

**Output**: Each node $S_i$ has a multilayer perceptron network $\mathcal{N}_i$

**for** $t = 1$ *to* $T$ **do**

> (a) Node $S_i$ uses the network $\mathcal{N}_i$ for feedforward learning and locally estimates the loss on N instances;
> (b) Node $S_i$ gossips with its neighbors $S_j$ and obtains the loss from the neighbor;
> (c) **Gossip:** node $S_i$ averages the loss between $S_i$ and $S_j$ and sets this as the new loss;
> (d) Perform backpropagation on the current node and the neighbor node using the gossiped loss; Update the weight vectors in each layer using Stochastic Gradient Descent (SGD);
> (e) If there is no significant change in the local weight vectors, STOP ;

**end**

**Algorithm 1:** Distributed Multilayer Perceptron Learning (DMLP)

---

### 4.1  Discussion

Some interesting aspects of our algorithm are:

1. The algorithms presented in the above section are called anytime algorithms [32]. Anytime algorithms are those whose quality of results change gradually as computation time increases. At a given time a node may be interrupted to obtain an estimate of the performance.
2. Algorithm 1 works with different kinds of loss functions (such as cross-entropy, softmax) and activations (such as linear, tanh).

---

[7] Cross-entropy loss was used in empirical results.

3. The number of hidden layers of the multi-layer perceptron can be incremented as required by a node, without the need for any algorithmic changes.
4. It must be noted that the algorithm presented above does not depend on how the features are partitioned across the nodes. If a given node has more features than others, it would build a larger local model and have larger weight matrices as appropriate. Since only the loss is exchanged (and it is assumed that the number of instances at each node remains fixed), no algorithmic changes are required to adapt to a different split of the feature space. This is empirically verified in Sect. 5.

## 5    Empirical Results

The empirical results demonstrate the utility of the DMLP algorithm. We examine the following questions:

1. Is there empirical support for the conjecture that the performance of the distributed model is better than that of the centralized model?
2. Does the distributed model empirically converge to the centralized one?
3. How does the performance of the proposed method compare to feature subspace learning methods such as Random Forests [6] and tree boosting algorithms (such as XGBoost [7])?

The answers to the above questions are explored using the data sets shown in Table 1 [11].

**Table 1.** Characteristics of the datasets used for empirical analysis.

| Dataset | No. train | No. test | No. features |
|---|---|---|---|
| Arcene | 100 | 100 | 10000 |
| Dexter | 300 | 300 | 20000 |
| Dorothea Bal. | 156 | 68 | 100000 |
| Gisette | 6000 | 1000 | 5000 |
| Madelon | 2000 | 600 | 500 |
| HT Sensor | 14560 | 3640 | 10 |
| MNIST | 60000 | 10000 | 784 |

The experimental process is as follows:

1. The Peersim simulator [20] is used to construct a fully connected graph of 10 nodes. Each node can independently store vertically partitioned data.
2. The total number of features in the train data is split into 10 roughly equal parts. Each node is assigned the data with the corresponding split containing all the examples but only those features it has been assigned.

3. Each node builds a local neural network model. The local loss vector is generated.
4. Each node selects a neighbor uniformly at random according to the underlying distributed graph, and exchanges the local loss vector with its neighbor. The new loss vector is computed as the average of its own loss vector and that of the neighbor's.
5. Each node participates in back propagation using the new loss generated after gossiping with a neighbor.
6. The above process is repeated for several iterations until the nodes converge to a solution.

**Testing the Model:** Each node is provided with the test set having only those features that the node used to construct the local model. Therefore, each node can test its own performance. For experiments presented here, we construct the following *hypothetical* scenario: for each test sample, an average predicted probability is obtained across all nodes, and the distributed test AUC is then estimated. This is not a requirement of the algorithm, but it enables us to compare performance against benchmarks.

**Table 2.** Performance of the centralized (C) and distributed algorithms (D). The consensus multi-layer perceptron uses cross-entropy loss function, ReLU activation for the hidden layer, and softmax activation for the output layer. The results are averaged over three trials.

| Dataset | No. hidden neurons (C) | No. hidden neurons (D) | Learning rate | Centralized AUC | Distributed AUC | 95% C. I. | Cent. Itr. $(I_C)$ | Dist. Itr. $(I_D)$ |
|---|---|---|---|---|---|---|---|---|
| Arcene | 50 | 5 | 0.001 | $0.94 \pm 0.01$ | $0.94 \pm 0.01$ | $[0.90, 0.97]$ | $3000 \pm 1256$ | $6466 \pm 1087$ |
| Dexter | 15 | 2 | 0.05 | $0.67 \pm 0.07$ | $0.85 \pm 0.02$ | $[0.79, 0.86]$ | $171 \pm 27$ | $168 \pm 22$ |
| Dorothea Bal | 100 | 10 | 0.5 | $0.93 \pm 0.01$ | $0.92 \pm 0.01$ | $[0.87, 0.97]$ | $180 \pm 99$ | $130 \pm 28$ |
| Gisette | 200 | 20 | 0.001 | $0.98 \pm 0.00$ | $0.99 \pm 0.00$ | $[0.991, 0.997]$ | $6400 \pm 698$ | $7667 \pm 1497$ |
| Madelon | 50 | 5 | 0.005 | $0.62 \pm 0.01$ | $0.64 \pm 0.00$ | $[0.61, 0.68]$ | $18333 \pm 4497$ | $16411 \pm 2867$ |
| HT Sensor | 20 | 2 | 0.1 | $0.99 \pm 0.00$ | $0.99 \pm 0.01$ | $[0.92, 0.94]$ | $910 \pm 0$ | $1533 \pm 618$ |
| MNIST | 4000 | 400 | 0.0001 | $0.82 \pm 0.10$ | $0.91 \pm 0.02$ | $[0.91, 0.91]$ | $27333 \pm 8379$ | $45000 \pm 16309$ |

We measure the performance of the model by the area under the Receiver Operating Characteristic (ROC) curve [5] denoted by $\theta$. The centralized algorithm is executed by assuming that the entire dataset is available at a node. In the distributed setting, a neural network is employed at each node, and is fed partial data, partitioned in the feature space. The number of total hidden neurons is kept the same for both the centralized and distributed experiments. This implies that each distributed node has roughly $\frac{\text{No. of hidden neurons in cent. model}}{\text{No. of nodes}}$ hidden neurons in its model. We tune the model(s) in each experiment by selecting different parameters for learning rate, number of hidden neurons, number of hidden layers and activation functions.

**Table 3.** Comparison of the performance of the consensus algorithm to tree based algorithms Random Forest (RF) and XGBoost.

| Dataset | RF_AUC | XGBoost_AUC | Dist_AUC | Cent_AUC |
|---|---|---|---|---|
| Arcene | 0.79 | 0.84 | 0.94 | 0.94 |
| Dexter | 0.93 | 0.95 | 0.67 | 0.85 |
| Dorothea Bal. | 0.88 | 0.89 | 0.92 | 0.93 |
| Gisette | 0.99 | 0.99 | 0.99 | 0.98 |
| Madelon | 0.77 | 0.71 | 0.64 | 0.62 |
| HT | 1 | 0.99 | 0.99 | 0.99 |
| MNIST | 0.67 | 0.71 | 0.91 | 0.82 |

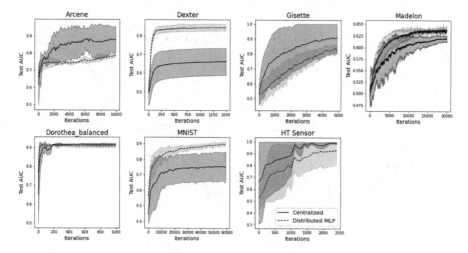

**Fig. 1.** AUC on the test sets for both centralized and distributed settings on all the datasets discussed above. For the distributed algorithm, test AUC results averaged over three random vertical feature splits without overlap are presented.

The steps outlined for the distributed algorithm above were repeated for three random feature splits and the test AUC averaged over the trials. We also compute the symmetric 95% confidence interval for distributed test AUC ($\theta_D$) and observe centralized test AUC ($\theta_C$) in relation to this interval. Figure 1 shows the AUC curves for all the datasets used in this study. The Standard Error ($SE$) for estimated area under the ROC curve in relation to the sample size ($n$) and $\theta_D$ can be computed as described in [12]:

$$SE = \sqrt{\frac{\theta_D(1-\theta_D)+(n-1)(Q_1+Q_2-2\theta_D^2)}{n^2}} \text{ where } Q_1 = \frac{\theta_D}{2-\theta_D}, Q_2 = \frac{2\theta_D^2}{1+\theta_D}.$$

Given SE, the symmetric 95% confidence interval ($CI$) is given by $\theta_D \pm 1.96(SE)$. The centralized algorithm and distributed algorithm can be deemed approximately comparable if $\theta_C$ lies within these bounds, i.e. if $\theta_D -$

**Table 4.** Performance of the centralized(C) and distributed algorithms(D) with 20% overlap of features. The consensus neural network uses cross-entropy loss function, ReLU activation for the hidden layer, and softmax activation for the output layer. The results are averaged over three trials.

| Dataset | Cent AUC | Dist. w/overlap AUC | 95% C. I. |
|---|---|---|---|
| Arcene | $0.94 \pm 0.01$ | $0.92 \pm 0.00$ | $[0.88, 0.96]$ |
| Dexter | $0.67 \pm 0.07$ | $0.90 \pm 0.02$ | $[0.87, 0.93]$ |
| Doro. Bal. | $0.93 \pm 0.01$ | $0.91 \pm 0.03$ | $[0.87, 0.97]$ |
| Gisette | $0.98 \pm 0.00$ | $0.99 \pm 0.00$ | $[0.98, 0.99]$ |
| Madelon | $0.62 \pm 0.01$ | $0.60 \pm 0.04$ | $[0.57, 0.63]$ |
| HT Sensor | $0.99 \pm 0.00$ | $0.99 \pm 0.01$ | $[0.99, 0.99]$ |
| MNIST | $0.82 \pm 0.10$ | $0.95 \pm 0.02$ | $[0.95, 0.95]$ |

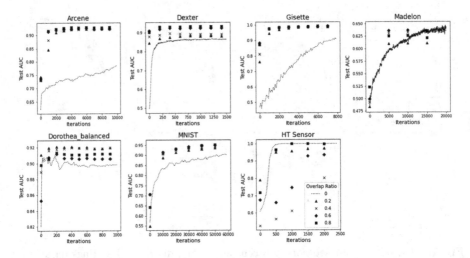

**Fig. 2.** AUC on the test sets for both centralized and distributed settings on all the datasets discussed above with varying degree of feature overlap. For the distributed algorithm, test accuracy results averaged over three random vertical feature splits with and without overlap are presented.

$1.96(SE) <= \theta_C <= \theta_D + 1.96(SE)$. In empirical studies (Table 2), it was found that the distributed algorithm obtains comparable or better test AUC scores than the centralized algorithm for all the datasets.

**Implementation Details:** An implementation of the DMLP Algorithm is available from: https://github.com/Saurabh7/consensus_based_dl.

## 5.1  Empirical Convergence of DMLP Algorithm

To test the convergence of the DMLP algorithm, we measure the difference in $L2$ norms of the normalized weight vectors in the centralized and distributed algorithms as they progress through the algorithm. Figure 3 shows that for all the data sets, this difference approaches zero, thereby supporting our conjecture that the distributed algorithm follows the behavior of the centralized one. This is a very important result, because even though the multi-layer perceptrons were trained on separate nodes with partial data, the global objective function was being minimized lending it useful for many applications on resource constrained devices. Furthermore, the anytime nature of the algorithm is also demonstrated by these experiments. After convergence, if new information arrives at a node it will continue to re-compute and update its local weights.

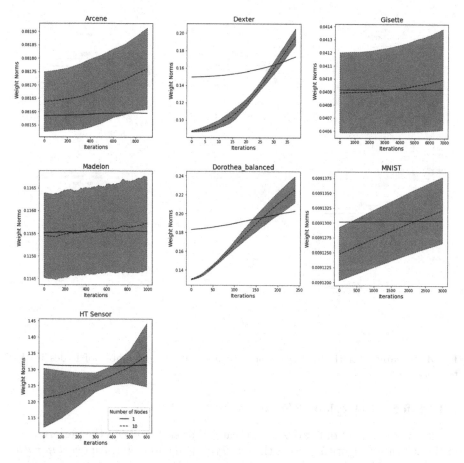

**Fig. 3.** Verification of the empirical convergence of the distributed algorithm. The distributed algorithm has 10 nodes.

## 5.2   Scalability of DMLP with Number of Nodes in the Network

We study the performance of the distributed algorithm when the number of nodes in the network is varied from between 10–100 and compare it to the centralized setting. The results (Fig. 4) show that an increase in network size does not significantly change the performance of the algorithm in all the datasets.

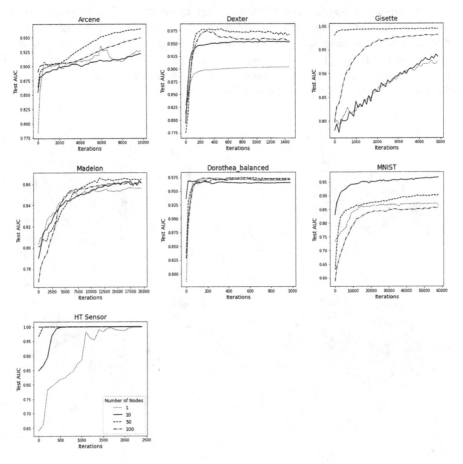

**Fig. 4.** Scalability of the DMLP algorithm in terms of the number of nodes in the network.

## 5.3   Effect of Overlap of Features

We study the impact of the overlap of features at each node on the performance of the consensus algorithm using the *overlap_ratio* parameter. An *overlap_ratio* of 0 indicates that the features present at one node are not present at any other node, i.e. the feature space is partitioned with mutual exclusivity. On the other hand, an *overlap_ratio* greater than 0, indicates that a subset of the feature space is shared among all nodes. When *overlap_ratio* is 1, all data is available at all nodes but model partition still exists. Table 4 presents the results of the experiments with *overlap_ratio* set to 0.2. Figure 2 shows the effect of variation of the *overlap_ratio* parameter on performance of the algorithm. In general, it is observed that when the *overlap_ratio* is incremented by a factor of 0.2, the AUC on the test set gradually improves. Our results reveal that in general, the overlap of features amongst nodes is beneficial and boosts the performance of the consensus algorithm. However, this behavior is not consistent for highly nonlinear datasets (such as Madelon) and those which have very large number of features (such as Dorothea) where-in the performance decreases as overlap increases and overfitting sets in.

### 5.4 Comparison with Feature Sub-space Learning Algorithms

Given that data partition at each node involves exploring a subset of the feature space, we compare the consensus algorithm to state-of-the-art tree-based algorithms which learn on feature subspaces (such as Random Forests and XGBoost). The results are presented in Table 3. We observe that the consensus algorithm has comparable performance to RF and XGBoost in all the datasets, except Dexter and Madelon – two particularly difficult datasets with no informative features [11].

## 6    Conclusion and Future Work

This paper presents the Distributed Multi-Layer Perceptron (DMLP) algorithm for learning consensus based vertically partitioned multi-layer perceptrons in resource constrained edge devices. The algorithm interweaves local learning with label propagation in the network. Each node constructs a local model by feed forward learning, exchanges losses with a randomly chosen neighbor, averages losses and uses this new loss for back propagation in the network. The iterative algorithm demonstrates good empirical convergence properties and can be used for both binary and multi-class classification problems. Empirical results on several real world data sets reveal that the DMLP algorithm has performance comparable to the centralized counterpart and tree-based learning algorithms. The performance of the algorithm is not affected by an increase in network size and therefore it can be used efficiently on edge devices in completely decentralized environments for on-device machine learning.

## References

1. Alistarh, D., Grubic, D., Li, J., Tomioka, R., Vojnovic, M.: QSGD: communication-efficient SGD via gradient quantization and encoding. In: Advances in Neural Information Processing Systems, pp. 1709–1720 (2017)
2. Bekkerman, R., Bilenko, M., Langford, J.: Scaling Up Machine Learning: Parallel and Distributed Approaches. Cambridge University Press, New York (2011)

3. Bellet, A., Guerraoui, R., Taziki, M., Tommasi, M.: Personalized and private peer-to-peer machine learning. In: International Conference on Artificial Intelligence and Statistics, AISTATS, vol. 84, pp. 473–481 (2018)
4. Blot, M., Picard, D., Thome, N., Cord, M.: Distributed optimization for deep learning with gossip exchange. Neurocomputing **330**, 287–296 (2019)
5. Bradley, A.P.: The use of the area under the roc curve in the evaluation of machine learning algorithms. Pattern Recogn. **30**(7), 1145–1159 (1997)
6. Breiman, L.: Random forests. Mach. Learn. **45**(1), 5–32 (2001)
7. Chen, T., Guestrin, C.: XGBoost: a scalable tree boosting system. In: Proceedings of the 22nd ACM SIGKDD International Conference on Knowledge Discovery and Data Mining (2016)
8. Dean, J., et al.: Large scale distributed deep networks. In: Proceedings of the 25th International Conference on Neural Information Processing Systems - Volume 1, pp. 1223–1231 (2012)
9. Demers, A., et al.: Epidemic algorithms for replicated database maintenance. In: ACM Symposium on Principles of Distributed Computing, pp. 1–12 (1987)
10. Gupta, O., Raskar, R.: Distributed learning of deep neural network over multiple agents. J. Netw. Comput. Appl. **116**, 1–8 (2018)
11. Guyon, I., Gunn, S., Hur, A.B., Dror, G.: Result analysis of the NIPS 2003 feature selection challenge. In: Proceedings of the 17th International Conference on Neural Information Processing Systems, NIPS 2004, pp. 545–552 (2004)
12. Hanley, J., Mcneil, B.: A method of comparing the areas under receiver operating characteristic curves derived from the same cases. Radiology **148**, 839–43 (1983)
13. Huerta, R., Mosqueiro, T., Fonollosa, J., Rulkov, N.F., Rodríguez-Luján, I.: Online decorrelation of humidity and temperature in chemical sensors for continuous monitoring. Chemom. Intell. Lab. Syst. **157**, 169–176 (2016)
14. Jiang, Z., Balu, A., Hegde, C., Sarkar, S.: Collaborative deep learning in fixed topology networks. In: Proceedings of the 31st International Conference on Neural Information Processing Systems, pp. 5906–5916 (2017)
15. Kairouz, P., et al.: Advances and open problems in federated learning. arXiv:abs/1912.04977 (2019)
16. Kempe, D., Dobra, A., Gehrke, J.: Gossip-based computation of aggregate information. IEEE Symposium on Foundations of Computer Science, pp. 482–491 (2003)
17. Lalitha, A., Shekhar, S., Javidi, T., Koushanfar, F.: Fully decentralized federated learning. In: Third Workshop of Bayesian Deep Learning (2018)
18. McDonald, R., Hall, K., Mann, G.: Distributed training strategies for the structured perceptron. In: Human Language Technologies: The 2010 Annual Conference of the North American Chapter of the Association for Computational Linguistics, HLT 2010, pp. 456–464 (2010)
19. McMahan, B., Moore, E., Ramage, D., Hampson, S., Agüera y Arcas, B.: Communication-efficient learning of deep networks from decentralized data. In: Proceedings of the International Conference on Artificial Intelligence and Statistics, pp. 1273–1282 (2017)
20. Montresor, A., Jelasity, M.: PeerSim: a scalable P2P simulator. In: Proceedings of the 9th International Conference on Peer-to-Peer (P2P 2009), pp. 99–100, September 2009
21. Provodin, A., et al.: Fast incremental learning for off-road robot navigation. CoRR abs/1606.08057 (2016)
22. Seide, F., Fu, H., Droppo, J., Li, G., Yu, D.: 1-bit stochastic gradient descent and application to data-parallel distributed training of speech DNNs. In: Interspeech 2014, September 2014

23. Sutton, D.P., Carlisle, M.C., Sarmiento, T.A., Baird, L.C.: Partitioned neural networks. In: Proceedings of the 2009 International Joint Conference on Neural Networks, IJCNN 2009, pp. 2870–2875 (2009)
24. Teerapittayanon, S., McDanel, B., Kung, H.T.: Distributed deep neural networks over the cloud, the edge and end devices. In: 37th IEEE International Conference on Distributed Computing Systems, ICDCS 2017, Atlanta, GA, USA, pp. 328–339 (2017)
25. Wang, X., Han, Y., Leung, V.C.M., Niyato, D., Yan, X., Chen, X.: Convergence of edge computing and deep learning: a comprehensive survey. IEEE Commun. Surv. Tutor. **22**(2), 869–904 (2020)
26. Wen, W., et al.: TernGrad: ternary gradients to reduce communication in distributed deep learning. In: Advances in Neural Information Processing Systems 30, pp. 1509–1519 (2017)
27. Wittkopp, T., Acker, A.: Decentralized federated learning preserves model and data privacy. CoRR abs/2102.00880 (2021)
28. Yang, Q., Liu, Y., Chen, T., Tong, Y.: Federated machine learning. ACM Trans. Intell. Syst. Technol. (TIST) **10**, 1–19 (2019)
29. Zhang, W., Gupta, S., Lian, X., Liu, J.: Staleness-aware Async-SGD for distributed deep learning. In: Proceedings of the Twenty-Fifth International Joint Conference on Artificial Intelligence, IJCAI 2016, pp. 2350–2356 (2016)
30. Zhou, D., Bousquet, O., Lal, T.N., Weston, J., Schölkopf, B.: Learning with local and global consistency. In: Proceedings of the 16th International Conference on Neural Information Processing Systems, NIPS 2003, pp. 321–328 (2003)
31. Zhou, S., Wu, Y., Ni, Z., Zhou, X., Wen, H., Zou, Y.: DoReFa-Net: training low bitwidth convolutional neural networks with low bitwidth gradients. arXiv preprint arXiv:1606.06160 (2016)
32. Zilberstein, S.: Operational rationality through compilation of anytime algorithms. Ph.D. thesis, Computer Science Division, University of California Berkeley (1993)

# Controlling BigGAN Image Generation
# with a Segmentation Network

Aman Jaiswal[✉], Harpreet Singh Sodhi[✉], Mohamed Muzamil H,
Rajveen Singh Chandhok, Sageev Oore, and Chandramouli Shama Sastry

Dalhousie University, Halifax, NS, Canada
{aman.jaiswal,harpreet,mohamed.muzamilh,rajveen,sageev,cssastry}@dal.ca

**Abstract.** GANS have been used for a variety of unconditional and conditional generation tasks; while class-conditional generation can be directly integrated into the training process, integrating more sophisticated conditioning signals within the training is not as straightforward. In this work, we consider the task of sampling from $P(X)$ such that the silhouette of (the subject of) $X$ matches the silhouette of (the subject of) a given image; that is, we not only specify *what* to generate, but we also control *where* to put it: more generally, we allow a mask (this is actually another image) to control the silhouette of the object to be generated. The mask is itself the result of a segmentation system applied to a user-provided image. To achieve this, we use pre-trained BigGAN and State-of-the-art segmentation models (e.g. DeepLabV3 and FCN) as follows: we first sample a random latent vector $z$ from the Gaussian Prior of BigGAN and then iteratively modify the latent vector until the silhouettes of $X = G(z)$ and the reference image match. While the Big-GAN is a class-conditional generative model trained on the 1000 classes of ImageNet, the segmentation models are trained on the 20 classes of the PASCAL VOC dataset; we choose the "Dog" and the "Cat" classes to demonstrate our controlled generation model.

**Keywords:** Generative model · Image segmentation · Computational creativity tools

## 1 Introduction

Generative adversarial networks (GANS) have been used for a variety of unconditional and conditional generation tasks; while unconditional generation involves learning and sampling from $P(X)$, conditional generation can be described as sampling from $P(X|f(X) = 1)$, where $f$ is a binary indicator function. Most commonly studied conditional generation are class-conditional generation wherein $f$ is a binary class-membership function. While class-conditional generation can be directly integrated into the training process, integrating more sophisticated indicator functions within the training is not as straightforward. Specifically, in this work, we aim learn to replace the silhouette of a subject in an image (in our

© Springer Nature Switzerland AG 2021
C. Soares and L. Torgo (Eds.): DS 2021, LNAI 12986, pp. 268–281, 2021.
https://doi.org/10.1007/978-3-030-88942-5_21

examples, an animal) with a different subject (e.g. a different animal) that still fits the exact same silhouette. Some generative models conditionally generate images by transforming vectors that lie in a large *latent* space. BigGAN, for example, has been trained to conditionally generate realistic images from any of the 1000 different categories of Imagenet, including various breeds of dogs and cats. While it is straightforward to sample from any of these categories, attributes like shape, size and posture cannot be directly manipulated to match our preference. These visible attributes are influenced by the choice of latent vector, but the nature of that influence is neither explicit, nor easily invertible, i.e. it is not clear how to choose a latent vector in order to achieve a desired visual attribute.

We introduce an iterative optimization-based approach to allow control over the silhouette of the image subject. We use a publicly-available pre-trained segmentation model to obtain a proxy for the silhouettes and the pre-trained Big-GAN generator to conditionally generate our desired subject. We compute the differences in both silhouettes and optimize to iteratively produce images that can match silhouette of the given subject. This is done by (locally) optimizing over the latent-space of GAN until the euclidean distance between the segmentation maps is minimized. Figure 1 shows an example of our final system's output as it iterates to find an image whose silhouette matches that of the source image (8a).

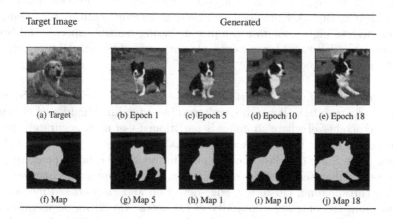

**Fig. 1.** Image (8a) is a provided source image. In this case, the source image happens to be itself a generated image (i.e. none of these images are photographs). Image (8b) shows a class-conditionally generated dog image for a random initial latent vector $z$, and (8c–8e) show the progression of images as we optimize $z$ through the latent space (described in Sect. 3) to arrive at an image (8e) of a different dog from the original source, but whose *silhouette* matches that of the source image. These images were found using the *ensemble model* as described in Sect. 4. Images (8f)–(8j) show the corresponding segmentation maps.

## 1.1    Background

Our system depends crucially on two types of models: a GAN-based generator, and segmentation model. We discuss each of these.

*Image Generation.* Generative adversarial networks (GANS) [6] use a neural network $(G)$ to transform a latent vector $z$ sampled from a prior distribution $p(z)$ to produce an output image $X = G(z)$. The generator network further comprises of intermediary layers $G_1...G_l$, where the first layer takes as input the latent vector to produce features tensors. The initial features are used by the next layer to produce higher abstraction of these features $y_i = G(y_{i-1})$. Lastly, The final layer is responsible for producing an output image $X = G_l(y_{l-1})$. Large scale class-conditional image synthesis [1] demonstrated that GANS could improve sample variety and fidelity from scaling up the number of parameters and batch size. BigGAN employs a shared class embedding $c$ that is linearly projected to every layer and uses skip connections from the noise vector $z$ to multiple layers of generator $y_i = G_i(y_{i-1}, z)$. This allows the latent space to directly influence features at different levels of hierarchy. This is done by splicing $z$ into one chunk per resolution and concatenating it with the shared class embedding $c$. They truncate the latent prior $N(0, I)$ during inference to improve sample quality. Although sampling from a truncated prior distribution during inference improves individual sample quality it also introduces undesirable saturation artifacts. In our framework, BigGAN provides a suitable generator because of its ability to generate diverse high resolution samples including multiple species and breeds of various animals (useful for our example purposes). The architecture details of BigGAN are described in Fig. 2.

*Image Segmentation.* An important part of our framework is realised using *Semantic segmentation.* It can be explained by extending the idea of classification to the pixel level where an image is partitioned—or more accurately, the set of *pixels* of an image is partitioned—such that each pixel in a partition belongs the same class. Since the class of every pixel in the image is being predicted, this task is commonly referred as *dense prediction*. Earlier approaches have relied on primitive thresholding, clustering, edge-detection and graph-based methods for segmentation. In contrast, a majority of the work [18] [2] for segmentation in deep learning builds on convolution neural networks (or CCNs) which helps by learning increasingly abstract feature representations. However, this approach introduces challenges like reduced resolution which may impede dense prediction tasks, where detailed spatial information is desired. State-of-the-art models like Deeplabv3 [3]use atrous convolution (also known as dilated convolutions) Fig. 3 to overcome this problem. For example, a kernel of size $K \times K$ with a dilation rate of $N$ will cover $(N-1) * K \times (N-1) * K$ pixels for an expansion of $(N-1) \times (N-1)$. This allows to control the resolution of features while preserving the number of parameters. The pre-trained semantic segmentation model from deeplabv3 provides segmentation images of 20 classes in the

| $z \in \mathbb{R}^{160} \sim \mathcal{N}(0, I)$ Embed$(y) \in \mathbb{R}^{128}$ |
| :---: |
| Linear $(20 + 128) \rightarrow 4 \times 4 \times 16ch$ |
| ResBlock up $16ch \rightarrow 16ch$ |
| ResBlock up $16ch \rightarrow 8ch$ |
| ResBlock up $8ch \rightarrow 8ch$ |
| ResBlock up $8ch \rightarrow 4ch$ |
| Non-Local Block $(64 \times 64)$ |
| ResBlock up $4ch \rightarrow 2ch$ |
| ResBlock up $2ch \rightarrow ch$ |
| ResBlock up $ch \rightarrow ch$ |
| BN, ReLU, $3 \times 3$ Conv $ch \rightarrow 3$ |
| Tanh |

(a) BigGAN 512 generator

| $z \in \mathbb{R}^{140} \sim \mathcal{N}(0, I)$ Embed$(y) \in \mathbb{R}^{128}$ |
| :---: |
| Linear $(20 + 128) \rightarrow 4 \times 4 \times 16ch$ |
| ResBlock up $16ch \rightarrow 16ch$ |
| ResBlock up $16ch \rightarrow 8ch$ |
| ResBlock up $8ch \rightarrow 8ch$ |
| ResBlock up $8ch \rightarrow 4ch$ |
| ResBlock up $4ch \rightarrow 2ch$ |
| Non-Local Block $(128 \times 128)$ |
| ResBlock up $2ch \rightarrow ch$ |
| BN, ReLU, $3 \times 3$ Conv $ch \rightarrow 3$ |
| Tanh |

(b) BigGAN 256 generator

**Fig. 2.** BigGAN generators architecture

PASCAL dataset including dogs and cats. This provides us the required segment masks to facilitate controlled generation from BigGAN.

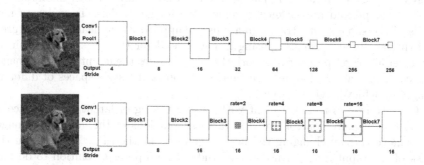

**Fig. 3.** Cascaded modules without and with atrous convolution [3]

## 2   Related Work

GAN frameworks [6], like BigGAN [1] and StyleGAN [10, 11] are powerful image synthesizers and have achieved impressive results in generation of variety of high quality images. Various improvements have been made to the original GAN model over the years, primarily to obtain higher quality images and more stable training, but most of those improved models still provide little direct control over the generated images other than selecting image classes or adjusting StyleGAN's style vector.

In studies like [5,9,15,16,20] there were attempts made to add control over the generated output images by focusing on supervised learning of latent directions. A few studies like [13,17,19] also provided useful control over spatial layout of the synthesized output images.

Our work focuses on exploring changes in the manifolds corresponding to the spatially localized region within the masked area of the image. We hope to discover smoothly varying sequences of latent vectors that lead to smooth transition of the generated "new subject" image (e.g. the new breed of dog) to exactly fit the mask corresponding to that of the target image (e.g. the silhouette of the dog in the provided source image).

A study by Yang et al. [21] have explored similar results by applying a rectangular mask over features of the image like eye or mouth regions and learning a function that can be applied over the latent vector that allowed targeted control over the appearance of feature within the rectangular mask. Another study by Srinivas et al. [8] shows that we can identify interpretable control over GAN generated image's pose, shape, facial and landscape attributes by applying principal component analysis (PCA) in latent space for StyleGAN, and feature space for BigGAN. Shen et al. [16] propose a framework called InterFaceGAN, to identify the semantics encoded in the latent space of well-trained face synthesis models and then utilize them for semantic face editing. A paper by Nguyen-Phuo et al. [14] proposes a novel method for the task of unsupervised learning of 3D representations from natural images. Their method enables direct manipulation of view, shape and appearance in generative image models. To generate new views of the same scene, transformations are applied to the learnt 3D features, and the results are visualised using a neural renderer that was jointly trained. Huang et al. [7] propose a framework that decomposes the latent space of images into content space and style space and recombines the style spaces of different images to achieve style transfer.

Our attempt is to experiment with careful tuning of latent vector space in order to gain more control over the targeted portions of the generated image. We achieve this by changing the latent vector to fit the targeted image in the mask of the input image. An another study [12] proposes a solution to do face swap by combining neural networks with simple pre- and post-processing steps. We achieved subject-swapping for animals by pre-processing the input and by defining a loss function which takes input from both an image segmentation model and BigGAN model.

## 3   Model

A conditional GAN is a latent generative model that maps a point $z$ in the latent space $Z$ to an image $G(z)$ that follows a lifelike distribution $R$. The likelihood of $G(z) \sim R$ is influenced by the selection of $z$. Empirical evidence [4,16] suggests that this mapping from $z \rightarrow G(z)$ is not always smooth and there are hidden but expressive transformations that remain to be explored. We propose a mask-guided image editing framework to swap a given subject in a given image (e.g. a dog) with another subject (e.g. a different dog!) using manifold transformation exploration. Our optimization framework requires four inputs: a source image $X_s$, a mask $M(X_s)$ (of the subject of interest) in the source image, a generated image $X_g$ and its corresponding subject mask $M(X_g)$. Note that the user only provides a single source image $X_s$; the other image $X_g$ is generated, and the masks of both images are obtained by the resnet-based semantic segmentation models. The framework $F(X_s, z_g)$, where $X_s$ being the target image and $z_g$ being the input latent vector for image to be generated optimises $z$ to discover a meaningful transformation that can overlap the subject in the generated image $X_g$ with that in $X_s$. The source image $X_s$ can also come from another class of the generator. The optimization based exploration progresses using $L2$ loss between the source image segmentation map $M(X_s)$ and generated image segmentation map $M(X_g)$.

Our optimization framework is described in Fig. 4. The segmentation model is used to get the segmentation maps for both the target image and the BigGAN generated image. Mean squared error is computed between the segmentation maps of the target image and the BigGAN generated image. The computed loss is then back propagated through the model to the input latent vector $z$ of the generator. This vector, in turn, is optimised to minimize the MSE between these maps, and thus incrementally generate images that can fit within the segmentation map of the target image. We used the BigGAN generator due to both availability of pretrained parameters[1] and its ability to generate diverse samples.

## 4    Experimental Results

We initially tested this framework by performing small transformations, such as translations and rotation on a generated image, where we had access to the latent vector $z_s$ used to generate the source image. This allows to assess model's ability to find the transformed vector from a good initialization point. This can be done by using the same BigGAN generated image as source and target image, where a known and controlled transformation has been applied to the source in order to generate the target. Figure 5 shows the results of this test.

The experiment in Appendix C (Fig. 12) shows that the segmentation model struggles to segment the generated image when it is undergoing transitions. The segmentation part of the model is regularised by adding another, second, segmentation model into the pipeline as shown in Fig. 7. FCN ResNet101 is selected for supplementation because it has a global pixel-wise accuracy of 91.9% on COCO val2017 dataset and also shares the same architectural backbone as DeepLabv3.

---

[1] https://tfhub.dev/s?network-architecture=BIGGAN,BIGGAN-deep&publisher=deepmind.

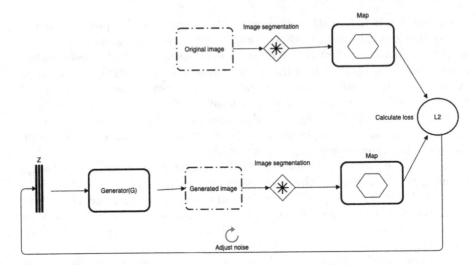

**Fig. 4.** Optimizing in $Z$: The top "row" of this figure stays fixed during optimization: given input image $X_s$ is passed through a segmentation model to get a segmentation map, resulting in a mask $M(X_s)$. This is the source mask. In the bottom "row", the latent variable $z$ is optimized using the $L2$ loss between the target map $M(X_s)$ and segmentation map $M(X_g)$ of the generated image $X_g$. The generated image, $X(g)$, is a itself generated based on the latent variable, i.e. $X_g = G(z)$. This allows us to incrementally update $z$ until we are able to generate an image $G(z)$ such that its mask is very close to that of the source image, i.e. $M(G(z)) \approx M(X_s)$.

The average segmentation map is generated by computing the weighted average of the two maps (obtained by DeeplabV3 and FCN Resnet101). MSE loss is computed on the average maps of both the generated and target image. The model can benefit from averaging due to partial independent errors of the individual models. Appendix B contains further details about implementation. Another ensemble method which was also implemented but did not produce desirable results is discussed in Appendix A.

## 5    Discussion

We observed that when the generator did not yield a high fidelity initial image the model finds it difficult to converge and find a generation that can fit the target map. Also, we found that certain classes used by the generator seemed to allow better convergence than others, this may be attributed to the biases of the BigGAN generator. We experimented with both the original proposed pipeline and the extended ensemble version. The results shown in Fig. 8 are generated using a single segmentation model while computing MSE on the target and generated segmentation maps. The results in figure are generated using an ensemble of segmentation modules as illustrated in Fig. 1.

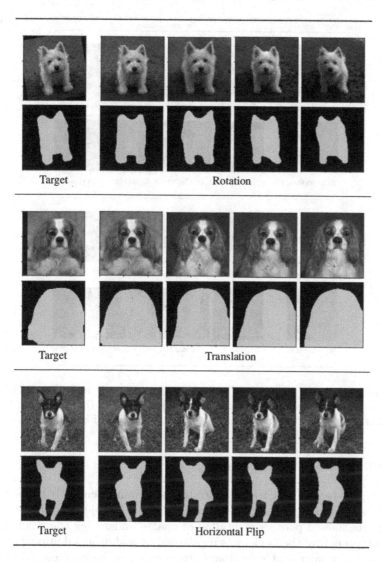

**Fig. 5.** Transformed generated image used as target. The target image is shown in 8a. The generated images are results from Epoch 1, 5, 10, 20 respectively. The rotation target is rotated 10 degrees to the left and the translation target is translated 10 pixels to the right

## Class Experiments

The multi-task nature of the models used in the pipeline allows for generation of variety of animals. The segmentation models used are capable of segmenting birds and cats. Figure 6 demonstrate the ability of the model to fit different birds and cats. We can see in this example that while the cat image had a very well-matched silhouette, the bird silhouette did not quite converge.

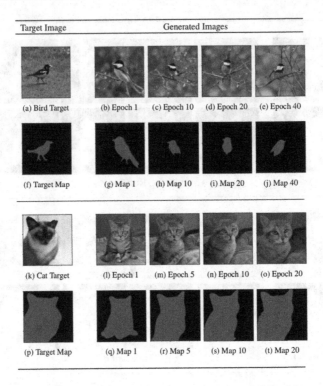

**Fig. 6.** *Different classes*, Figures (b)–(e) shows the generated bird images with their corresponding segmentation maps from Figure (g)–(j). Figures (l)–(o) shows generated cat images with their corresponding segmentation maps from Figure (q)–(t).

## 6   Conclusion

In this paper, we were able to demonstrate that two independently trained modules when stacked together can achieve the task of subject swapping. Initial experiments showed poor segmentation of images undergoing transition, so the segmentation part of the model was regularised by adding an additional segmentation model in an ensemble fashion. Future work may include using a discriminator to further regularize the model to provide more gradient feedback.

## A   Alternative Ensemble methods

The segmentation models used share similar architecture(Resnet101) and training dataset. Although the range of the logits vary from network to network, we could not find any evidence that computing an average across the logits produced by different segmentation modules should not necessarily produce good results. Therefore, we tried averaging the logits and then applying soft-max on the channel dimension before computing BCE Loss. The results of the average segmentation are shown in (Fig. 9). We also tried a method where we average

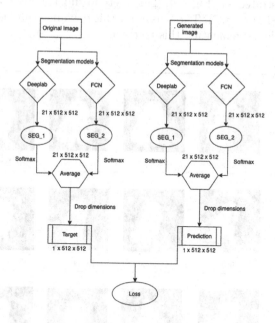

**Fig. 7.** Ensemble of two segmentation models

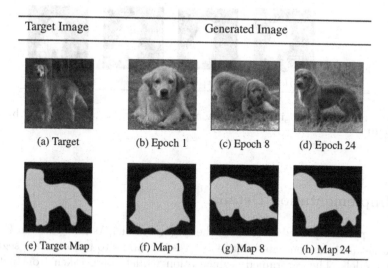

**Fig. 8.** *Shifting Dog Face*, target image is shown in (a), generated images are shown in (b), (c), (d) and their corresponding segmentation maps in (f), (g), (h). The face of the dog shifts from the right-side towards left-side.

the losses as illustrated in (Fig. 10). This method did not work as well as the method illustrated in (Fig. 7), The reason for this deviancy can be the BCE loss that we used while implementing this method.

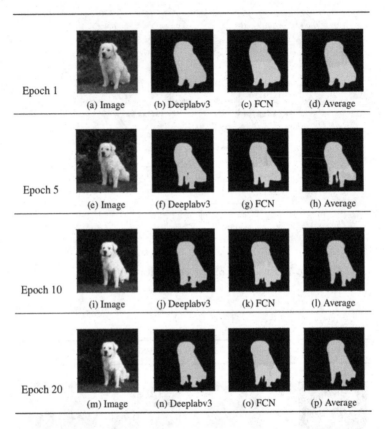

**Fig. 9.** Average over logits of two segmentation models, a) Deeplabv3 b) FCN ResNet101

## B    Implementation Details

We use a pytorch ported version[2] of the original model(As illustrated in Fig. 7). The target and the generated image are used as inputs to two separate segmentation models. The pretrained segmentation models were taken from pytorch hub[3],[4].

---

[2] https://github.com/ivclab/BIGGAN-Generator-Pretrained-Pytorch.
[3] https://pytorch.org/hub/pytorch_vision_fcn_resnet101/.
[4] https://pytorch.org/hub/pytorch_vision_deeplabv3_resnet101/.

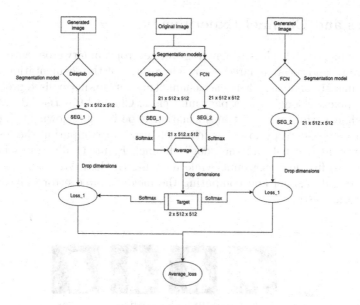

**Fig. 10.** Average losses

We use Adam optimzer with a learning rate of $1e - 1$ and beta values of 0.5–0.99. The model is run for a maximum of 25 epochs. The segmentation models expects the RGB channel to have the corresponding Mean($\mu$) = $[0.485, 0.456, 0.406]$ and Variance ($\sigma$) = $[0.229, 0.224, 0.225]$ values, This is done explicitly for every generated image. *Mean squared error* is used for computing the loss over the "Dog" channel of the two segmentation maps. Weighted average with the ratio 0.6 : 0.4 is used for the segmentation models because the DeepLabv3 segmentation model works better than FCN resnet101 segmentation model.

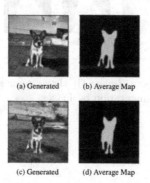

(a) Generated        (b) Average Map

(c) Generated        (d) Average Map

**Fig. 11.** *Background change.* The model changes the background owing to inclusion of background channel in loss computation

## C    Loss and Channel Experiments

During training, we tried losses including cross-entropy, binary cross-entropy, soft cross-entropy, and mean squared error. The results of the segmentation models used contains 21 channels, where each channel outputs un-normalised probability values for pixels belonging to a particular class. Channel 0 is the "Background class". While performing the transformation experiments shown in Fig. 5, we used binary cross-entropy loss. The channels used for computing the loss were the background channel and the "dog" channel. Figure 11 illustrates how the model tries to fit the background while reducing the loss. We found excluding the background channel and computing the mean squared error only on 'dog' channel works best.

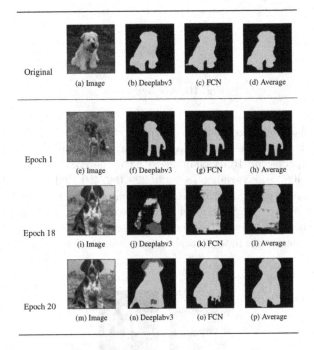

**Fig. 12.** *Ensemble Segmentation.* Image (j) shows poor segmentation by DeeplabV3 and Image (o) shows poor segmentation by FCN.

## References

1. Brock, A., Donahue, J., Simonyan, K.: Large scale GAN training for high fidelity natural image synthesis. In: 7th International Conference on Learning Representations, ICLR 2019, New Orleans, LA, USA, 6–9 May 2019. OpenReview.net (2019). https://openreview.net/forum?id=B1xsqj09Fm

2. Chen, L.C., Papandreou, G., Kokkinos, I., Murphy, K., Yuille, A.L.: Semantic image segmentation with deep convolutional nets and fully connected CRFs (2016)
3. Chen, L., Papandreou, G., Schroff, F., Adam, H.: Rethinking atrous convolution for semantic image segmentation. CoRR abs/1706.05587 (2017). http://arxiv.org/abs/1706.05587
4. Creswell, A., Bharath, A.A.: Inverting the generator of a generative adversarial network (ii) (2018)
5. Goetschalckx, L., Andonian, A., Oliva, A., Isola, P.: GANalyze: toward visual definitions of cognitive image properties (2019)
6. Goodfellow, I.J., et al.: Generative adversarial networks. Commun. ACM **63**(11), 139–144 (2020). https://doi.org/10.1145/3422622
7. Huang, X., Liu, M.Y., Belongie, S., Kautz, J.: Multimodal unsupervised image-to-image translation (2018)
8. Härkönen, E., Hertzmann, A., Lehtinen, J., Paris, S.: GANspace: Discovering interpretable GAN controls (2020)
9. Jahanian, A., Chai, L., Isola, P.: On the "steerability" of generative adversarial networks (2020)
10. Karras, T., Laine, S., Aila, T.: A style-based generator architecture for generative adversarial networks (2019)
11. Karras, T., Laine, S., Aittala, M., Hellsten, J., Lehtinen, J., Aila, T.: Analyzing and improving the image quality of StyleGAN (2020)
12. Korshunova, I., Shi, W., Dambre, J., Theis, L.: Fast face-swap using convolutional neural networks (2017)
13. Kulkarni, T.D., Whitney, W., Kohli, P., Tenenbaum, J.B.: Deep convolutional inverse graphics network (2015)
14. Nguyen-Phuoc, T., Li, C., Theis, L., Richardt, C., Yang, Y.L.: HoloGAN: unsupervised learning of 3D representations from natural images (2019)
15. Plumerault, A., Borgne, H.L., Hudelot, C.: Controlling generative models with continuous factors of variations (2020)
16. Shen, Y., Gu, J., Tang, X., Zhou, B.: Interpreting the latent space of GANs for semantic face editing (2020)
17. Singh, K.K., Ojha, U., Lee, Y.J.: FineGAN: unsupervised hierarchical disentanglement for fine-grained object generation and discovery (2019)
18. Srinivas, S., Sarvadevabhatla, R.K., Mopuri, K.R., Prabhu, N., Kruthiventi, S.S.S., Babu, R.V.: A taxonomy of deep convolutional neural nets for computer vision. Front. Robot. AI **2** (2016). https://doi.org/10.3389/frobt.2015.00036
19. Tran, L., Yin, X., Liu, X.: Disentangled representation learning GAN for pose-invariant face recognition. In: 2017 IEEE Conference on Computer Vision and Pattern Recognition (CVPR), pp. 1283–1292 (2017). https://doi.org/10.1109/CVPR.2017.141
20. Yang, C., Shen, Y., Zhou, B.: Semantic hierarchy emerges in deep generative representations for scene synthesis (2020)
21. Yang, M., Rokeby, D., Snelgrove, X.: Mask-guided discovery of semantic manifolds in generative models (2021)

# GANs for Tabular Healthcare Data Generation: A Review on Utility and Privacy

João Coutinho-Almeida[1,2](✉) ⓘ, Pedro Pereira Rodrigues[1,2] ⓘ,
and Ricardo João Cruz-Correia[1,2] ⓘ

[1] CINTESIS - Centre for Health Technologies and Services Research,
University of Porto, Porto, Portugal
[2] MEDCIDS – Faculty of Medicine, University of Porto, Porto, Portugal

**Abstract.** Data is a major asset in today's healthcare scenery. Hospitals are one of the primary producers of healthcare-related data and the value this data can provide is enormous. However, to use this to improve healthcare practice and push science forward, it is necessary to safeguard the patient's privacy and the ethical use of the data. The ethical and legal requirements are vast and complex. Synthetic data appears as a tool to overcome these hurdles and provide fast and reliable access to data without compromising utility nor privacy. Even though Generative Adversarial Networks (GANs) are receiving a lot of attention lately, the application of most common models and architectures are not suited to tabular data – the most prevalent healthcare-related data. This study surveys the current GAN implementations tailored to this scenario. The analysis was focused mainly on the models employed, datasets used, and metrics reported regarding the quality of the generated data in terms of utility, privacy and how they compare among themselves. We aim to help institutions and investigators get a grasp of the tools to facilitate access to healthcare data, as well as recommendations for testing data synthesizers with privacy concerns.

**Keywords:** Synthetic data · Generative adversarial networks · Privacy

## 1 Introduction

With the growing technological advances, the quantity of healthcare-related data produced around the world increased exponentially [22,27]. Consequently, the potential for harvesting this data also increases. The value locked within this data could help provide better healthcare with new information about diseases, drugs, and preventive therapies. It can also help create better health information systems, meaning an overall better clinical practice [18]. But for this to happen, data must reach capable hands at the right time. But the release of clinical data has several barriers attached and rightly so. The leakage of patient's privacy can break the confidence of the population in the healthcare professionals and

C. Soares and L. Torgo (Eds.): DS 2021, LNAI 12986, pp. 282–291, 2021.
https://doi.org/10.1007/978-3-030-88942-5_22

institutions. Patient safety and privacy should be kept at all costs. However, the current mechanisms for privacy maintenance are very long, bureaucratic and time-consuming, nationally [23], and internationally [37]. The current scenario and general methods for privacy safeguards are related to pseudo-anonymisation techniques. The removal of certain attributes, identifier modification, code grouping, or discretization are some methodologies. But not even these are totally safe [24]. Synthetic data appear as an alternative for clinical data sharing, promising great data utility with minimal privacy concerns. Synthetic data is data that is generated automatically through programmatic processes. This is especially impactful for the case at hand since synthetic data has no explicit connection with the original data. There are several mechanisms for data synthesis like postulated by [25], there are process-driven methods and data-driven methods. Process-driven methods generate data through pre-determined models inputted into the generator. Data-driven methods produce new data based on inputted source data. With this, it is possible to create new patient data that has no relation to reality while providing the same statistical relations between variables. This provides the basis for quality clinical research on top of this new data. Even though these techniques are still new and in rapid development, the results seem interesting [25], but not without questions and doubts [40]. Creating a thorough survey based on the generation of synthetic data is seldom a simple task when compared to other surveys since synthetic data is present across several domains and has several uses, like software testing, assessing methods, or generating hypotheses. Moreover, synthesis has the double meaning of summing up information and generating something, easily wielding hundreds of results per query. Finally, trying to filter algorithms aimed at tabular data is also burdensome, since not always it is easy to discriminate input types. These factors make the survey interesting to focus on the state-of-the-art mechanisms of generating tabular data.

## 2   Theoretical Background

### 2.1   GANs

First introduced just over seven years ago, Generative Adversarial Networks (GANs) [26] have been under the scope and have been proven very good for generating complex data. Images, text, video have been successfully generated with very good performances. The original architecture is based on two artificial neural networks trained simultaneously in a competitive manner. One of them, the generator, has the objective of generating the most realistic possible data, while the second network – the discriminator, has the opposite aim of aiming to distinguish the realistic data from the synthetic data the best it can. So, the elegance of this architecture is that each network tries to make the other perform better every time. The GAN architecture is shown in Fig. 1.

The generator is represented by $G_\theta$ where the parameter $\theta$ represents the weights of the neural network. It takes as input, a Gaussian random variable, and outputs $G_\theta(Z)$. Distribution of $G_\theta(Z)$ is denoted by $P_\theta$. The goal of the generator

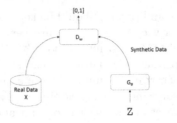

**Fig. 1.** Generative Adversarial Network framework

is to choose $\theta$ such that the output $G_\theta(Z)$ has a distribution close to the real data. The discriminator is represented by $D_\omega$, parametrised by weights $\omega$. The goal of the discriminator is to assign 1 to the samples from the real distribution $P_X$ and 0 to the generated samples ($P_\theta$). So, GANs can be mathematically represented by a minimax game identified by:

$$\min_G \max_D E[log(D_\omega(X)) + log(1 - D_\omega(G_\theta(Z)))] \tag{1}$$

So, G must minimise this equation and D must maximise it, each one tweaking the weights of its network ($\theta$ and $\omega$) to do so. This is the loss function on the initial GAN architecture. After the classification of D, the G is trained again with the error signal from D through backpropagation. This equation is the log of the probability of D predicting that the real data is genuine and the log probability of D classifying synthetic data as not genuine. The equation is essentially the same as minimising the *Jensen-Shannon divergence* [26]:

$$\min_G JS(P_x||P_\theta) \tag{2}$$

Where the JS means the *Jensen-Shannon divergence* between the probability of the real data and the probability of the generated data. The JS divergence provides a measure of distance between two probability distributions. Therefore, the minimisation over $\theta$ means, choosing the $P_\theta$ that is closest to the target distribution $P_X$ in the JS divergence distance. Despite the significant results provided by GANs with continuous real values, categorical values still seem to be a problem for this approach [30], since it is not directly applicable for calculating the gradients of latent categorical variables in order to train these networks through backpropagation. This happens since the output of the generator, even though can be transformed into a multinomial distribution with a softmax layer, sampling from it is not a differentiable operation, limiting the backpropagation process of the GAN.

## 3   Methods

This search was made during December 2020 and January 2021. It was made on "Web of Science", IEEE, PubMed, Arxiv and finally GitHub. The terms searched were related to GANs, synthetic data generation, electronic health records, patient data, or tabular data. Applications of GANs to non-tabular data were filtered, like image, sound, video, or graphs. Time series and text data were also removed since the methodology for synthesising this type of data has specific functions related to the nature of the data. The filter for date was after 2014 since GANs were introduced at that time. The queries used were similar to the one below, adapted for the search mechanics for each website.

("generation" OR "creation" OR "synthesis" OR "synthesizing" OR "generating" OR "creating") AND ("synthetic data" OR "synthetic patient" OR "synthetic electronic health record" OR "synthetic EHR" OR "realistic patient data" OR "realistic health record" OR ("synthetic" AND "privacy" AND "utility")) AND ("GAN" OR "Generative Adversarial Network")

From the total articles found (1165) with all the queries, 100 articles were chosen for full text and in the end, 22 papers with GAN implementations that were tested on tabular data were selected.

## 4   Results

The selected papers ranged from 2017 to 2020. Being that 2 are from 2017, 4 from 2018, 8 from 2019 and 8 from 2020. All authors showed original GAN implementations, apart from 2 papers. Beaulieu-Jones et al. [19] used a GAN

Table 1. Summary of the articles selected.

|    | Year | Acronym | Article | Metric | Code |
|----|------|---------|---------|--------|------|
| 1  | 2017 | medGAN | [22] | Utility, Privacy, Clinical | [7] |
| 2  | 2017 | POSTER | [33] | Utility, Privacy | [10] |
| 3  | 2018 | table-GAN | [38] | Utility, Privacy | [14] |
| 4  | 2018 | dp-GAN | [46] | Utility, Privacy | [3] |
| 5  | 2018 | mc-medGAN | [21] | Utility | [6] |
| 6  | 2018 | TGAN | [48] | Utility | [15] |
| 7  | 2019 | PATE-GAN | [29] | Utility, Privacy | – |
| 8  | 2019 | SPRINT-GAN | [19] | Utility, Privacy, Clinical | [13] |
| 9  | 2019 | GAN-based | [32] | Utility, Privacy | – |
| 10 | 2019 | CTGAN | [47] | Utility | [2] |
| 11 | 2019 | WGAN-DP | [20] | Utility, Privacy | [16] |
| 12 | 2019 | PPGAN | [31] | Utility, Privacy | [11] |
| 13 | 2019 | medBGAN | [18] | Utility | – |
| 14 | 2019 | medWGAN | [17] | Utility | [8] |
| 15 | 2020 | ADS-GAN | [50] | Utility, Privacy | – |
| 16 | 2020 | corGAN | [42] | Utility, Privacy | [1] |
| 17 | 2020 | CGAN | [44] | Utility | – |
| 18 | 2020 | DPAutoGAN | [41] | Utility, Privacy | [4] |
| 19 | 2020 | GAN Boosting | [35] | Utility, Privacy | [9] |
| 20 | 2020 | RDP-CGAN | [43] | Utility, Privacy | [5] |
| 21 | 2020 | WCGAN-GP | [45] | Utility, Privacy | – |
| 22 | 2020 | SMOOTH-GAN | [39] | Utility | [12] |

architecture that was originally published with usage on image datasets [36]. Additionally, Vega-Marquez et al. [44] used an already known implementation of conditional GANs [34]. We classified papers regarding 3 metrics: utility, privacy and clinical. For utility, we looked for methods for measuring the generated data's quality. As for privacy, we aimed for some mechanism for measuring the privacy loss of the new data. Concerning clinical metrics, any kind of evaluation from healthcare professionals was considered. This can be seen in Table 1.

The metrics the authors used are exhibited in Table 2. Regarding privacy, 15 papers assessed it or included some kind of mechanism to improve data protection. The most common was including Differential Privacy (DP) in the generation process. Other mechanisms for measuring privacy loss were Membership Inference (Member. Inf.), Attributes Disclosure (Attrib. Disc.), Euclidean distance (Eucl.), record-linkage (R. Linkage) and Nearest Neighbours (KNN). As for utility, all papers assessed it. There were 3 major areas of utility assessment: Dimension-wise (DW) probability, cross-testing, and distance metrics. The most basic one was dimension-wise probability, which is important for making sanity checks for the generated data, comparing the distributions of each column between real and synthetic. In this category we can find Bernoulli (Bern.), cumulative distributions (Cumul. Dist.), Pearson correlation (Pearson) and Spearman correlation (Spearman), correlation coefficients (CCS), chi-squared test ($\chi^2$), *Kolmogorov-Smirnov* (KS) or Correlation Matrices (Corre. Mat.). Cross-testing was about training machine-learning algorithms with both datasets in order to compare the results. The key factor is generating a synthetic dataset based on the training set and then train models on the original train set and the generated dataset. Then the models are compared regarding their predictive capability on the (real) test set. This was a way of assessing if the generator models were capturing inter-variable relationships. The authors applied different metrics from AUC, F1, AUPRC, Accuracy (Acc.) to Mean Relative Error (MRE). Finally, there was also the application of distance metrics, for measuring the difference between column distribution in both datasets. Jensen Shannon divergence (JSD), Wasserstein Distance (WD), Bhattacharyya Distance (BD) or Generate Scores (GS) that was a metric implemented by the authors of [31] that creates a metric based on the sum of the mean of *kullblack-leibler* distance of all columns. Other less used methods were Principal Component Analysis (PCA), Also, propensity score mean squared error ratio (pMSE). NMI (Normalised Mutual Information), which is the ability to capture correlations between columns by computing the pairwise mutual information and MMD (Maximum Mean Discrepancy), which is similar to distance metrics were also used. Regarding datasets utilised, the most used was MIMIC-III [28] (9 times). The papers used 27 different datasets, being 16 healthcare-related and 11 non-healthcare related. Finally, regarding clinical evaluation, only two papers assessed it, like it is possible to see in Table 1. Both had a group of clinicians assessing a sample of both real and synthetic information and evaluating from 0 to 10, where 10 is most realistic. One major point preventing a larger comparison is that despite some papers using the same dataset and same methodologies, the presented values are different, making it difficult for a clear comparison of results. One example is dimension-wise prediction with F1 score for MIMIC-III. CorGAN presents the mean difference between the two classifications (real on real and synthetic on synthetic), while medBGAN presents the correlation coefficients of the two, and medGAN only presents the visual comparisons. Regarding code availability, 16 papers had the code publicly available in some form. As of January 2021, papers pointed in Table 1 have public code.

## 5 Implications for Future Research

From the work done on this paper, it is clear that synthetic data generation is a growing field. The increasing number of papers through the years as the growing quality in the mechanisms of generating data and assessing its quality are a clear proof. It also became apparent that privacy and utility in synthetic data represent a delicate balance. The very same definition of differential privacy represents it. The compromise between privacy and utility is real and should be taken into account when creating privacy demanding datasets. Creating statistically good tabular datasets is already possible, but that task becomes increasingly difficult if privacy concerns are added. However, privacy is also a complex subject, and the context of the setting is important for privacy assessment, which explains the different approaches for evaluating privacy protection of synthetic data. From this review, we believe that a proper evaluation of synthetic data generators in the healthcare setting with privacy concerns should at least include utility and privacy evaluations. For utility, we believe that evaluating column-wise is a nice first check but insufficient alone. For table-wise, since there is not a fundamental metric for assessing the inter-column correlations between mixed-type variables, cross-testing is the best next thing. Distance metrics are a nice to have and seem to have the potential for creating a table-wise metric [49], so presenting them is important. Second, for privacy evaluation, we believe that Differential Privacy in itself is not a guarantee of protection for real patients. More research and depth should be employed when presenting results for such generators; record-linkage and attribute disclosure can provide extra guarantees. Thirdly, a clinical evaluation should be done as well to understand if the synthetic patients are a reality in the clinical setting. Since the correlations could be correct but clinically (or biologically) they might not make sense. Finally, in the scope of this paper, only GANs were assessed, but there are more mechanisms for generating data and could be interesting to assess how all of them perform on the same datasets. There are other methods for handling the mixed data types that regularly appear in clinical settings, like Variational Autoencoders, Gaussian Mixtures, Bayesian Networks, and imputation mechanisms, making them excellent candidates for this assessment.

**Table 2.** Metrics utilised for evaluation

| Acronym | Utility | Privacy |
|---|---|---|
| medGAN | 1. Bern. 2. Pred F1 | 1. Attrib. disc. 2. Memb. inf. 3. KNN |
| POSTER | 1. Pred Acc. 2. Corre. Mat. 3. BD | DP |
| table-GAN | 1. Cumul. Dist. 2. Pred F1\|MRE | 1. Eucl. 2. Member. inf. |
| dp-GAN | 1. Pred AUC 2. Bern. | DP |
| mc-medGAN | 1. Pred F1\|AUC 2. Bern. 3. ME F1\|Acc | – |
| TGAN | 1. KNN 2. NMI 3. Pred F1 | – |
| PATE-GAN | 1. Pred AUC\|AUPRC | DP |
| SPRINT-GAN | 1. Pred AUC 2. Corre. Mat. | DP |
| GAN-based | 1. Pred Acc. 2. Corre. Mat. | 1. Hit. Rate 2. R. Linkage 3. Eucl. |
| CTGAN | 1. Pred F1\|R2\|Acc. | – |
| WGAN-DP | 1. Corre. Mat. 2. PCA 3. Pearson RMSE 4. Pred F1\|RMSE\|1-MAPE(F1) | 1. Eucl. 2. Dupl. 3. DP |
| PPGAN | 1. GS | DP |
| medBGAN | 1. Assoc. Rul. 2. CCS Pred F1 3. KS | – |
| medWGAN | 1. Assoc. Rul. 2. CCS Pred F1 3. KS | – |
| ADS-GAN | 1. $\chi^2$ 2. JSD 3. WD 4. t-test 5. Pred AUROC 6. Corre. Mat. | DP |
| CorGAN | 1. Pred F1 2. Bern. | Member. Inf. |
| CGAN | 1. Pearson 2. Spearman 3. Pred F1\|AUC\|Acc | – |
| DPAutoGAN | 1. Pred AUROC\|R2 2. Bern. | DP |
| GAN Boosting | 1. pRMSE 2. Pred AUROC\|AUPRC\|Acc. | DP |
| RDP-CGAN | 1. Pred F1\|AUROC\|AUPRC 2. MMD | DP |
| WCGAN-GP | 1. Corre. Mat. 2. Pred F1 | 1. Dupl. 2. Eucl. |
| SMOOTH-GAN | 1. DW MAE 2. Pearson 3. Pred AUROC\|AUPRC | – |

# 6   Summary

In this paper, we had the opportunity of surveying the current framework for generating tabular data using GANs and which ones were already tested in the healthcare setting. We summarised the utility and privacy metrics employed, and the datasets used to measure them. We analysed the code availability and made suggestions for further work on cataloging, comparing, and assessing synthetic health data generators. A survey with a global benchmark of methodologies, despite being arduous, could yield great results for the community and take the aim of this paper further.

**Acknowledgments.** This work has been done under the scope of - and funded by - the Ph.D. Program in Health Data Science of the Faculty of Medicine of the University of Porto, Portugal - heads.med.up.pt.

# References

1. corGAN Repository. https://github.com/astorfi/cor-gan
2. CTGAN Repository. https://github.com/sdv-dev/CTGAN
3. dp-GAN Repository. https://github.com/illidanlab/dpgan
4. DPAutoGAN Repository. https://github.com/DPautoGAN/DPautoGAN

5. DRP-CGAN Repository. https://github.com/astorfi/differentially-private-cgan
6. mc-medGAN Repository. https://github.com/rcamino/multi-categorical-gans
7. medGAN Repository. https://github.com/mp2893/medgan
8. medWGAN Repository. https://github.com/baowaly/SynthEHR
9. Post-GAN Boosting Repository. https://github.com/mneunhoe/post-gan-boosting
10. POSTER Repository. https://goo.gl/94qyQz
11. PPGAN Repository. https://github.com/niklausliu/PPGANs-Privacy-preserving-GANs
12. SMOOTH-GAN Repository. https://github.com/anuragdutt/synthehr_medgan
13. SPRINT-GAN Repository. https://github.com/greenelab/SPRINT_gan
14. table-GAN Repository. https://github.com/mahmoodm2/tableGAN
15. TGAN Repository. https://github.com/sdv-dev/TGAN
16. WGAN-DP Repository. https://github.com/Baukebrenninkmeijer/On-the-Generation-and-Evaluation-of-Synthetic-Tabular-Data-using-GANs
17. Baowaly, M.K., Lin, C.C., Liu, C.L., Chen, K.T.: Synthesizing electronic health records using improved generative adversarial networks. J. Am. Med. Inform. Assoc. 26(3), 228–241 (2019). https://doi.org/10.1093/jamia/ocy142
18. Baowaly, M.K., Liu, C.L., Chen, K.T.: Realistic data synthesis using enhanced generative adversarial networks. In: 2019 IEEE 2nd International Conference on Artificial Intelligence and Knowledge Engineering (AIKE), pp. 289–292. IEEE; IEEE Computer Society (2019). https://doi.org/10.1109/AIKE.2019.00057
19. Beaulieu-Jones, B.K., et al.: Privacy-preserving generative deep neural networks support clinical data sharing. Circ. Cardiovasc. Qual. Outcomes 12(7), 139–148 (2019). https://doi.org/10.1161/CIRCOUTCOMES.118.005122, https://www.ahajournals.org/doi/10.1161/CIRCOUTCOMES
20. Brenninkmeijer, B.: On the generation and evaluation of tabular data using GANs. Ph.D. thesis (2019)
21. Camino, R., Hammerschmidt, C., State, R.: Generating multi-categorical samples with generative adversarial networks (2018). arXiv:1807.01202
22. Choi, E., Biswal, S., Malin, B., Duke, J., Stewart, W.F., Sun, J.: Generating multi-label discrete patient records using generative adversarial networks, vol. 68, pp. 1–20 (2017). arXiv:1703.06490
23. Comissão Nacional Proteção de dados: Princípios aplicáveis aos tratamentos de dados efetuados no âmbito da investigação clínica (2015)
24. El Emam, K., Jonker, E., Arbuckle, L., Malin, B.: A systematic review of re-identification attacks on health data. PLOS ONE 6(12), e0126772 (2011). https://doi.org/10.1371/journal.pone.0028071
25. Goncalves, A., Ray, P., Soper, B., Stevens, J., Coyle, L., Sales, A.P.: Generation and evaluation of synthetic patient data. BMC Med. Res. Methodol. 20(1), 1–40 (2020). https://doi.org/10.1186/s12874-020-00977-1
26. Goodfellow, I.J., et al.: Generative adversarial networks. Commun. ACM 63(11), 139–144 (2014). https://doi.org/10.1145/3422622, arXiv:1406.2661
27. Henry, J., Pylypchuk, Y., Searcy, T., Patel, V.: Adoption of electronic health record systems among U.S. Non-Federal Acute Care Hospitals: 2008–2015. Technical report (2016). https://dashboard.healthit.gov/evaluations/data-briefs/non-federal-acute-care-hospital-ehr-adoption-2008-2015.php
28. Johnson, A.E., et al.: MIMIC-III, a freely accessible critical care database. Sci. Data 3, 160035 (2016)

29. Jordon, J., Yoon, J., Van Der Schaar, M.: PATE-GaN: generating synthetic data with differential privacy guarantees. In: 7th International Conference on Learning Representations, ICLR 2019, pp. 1–21 (2019)

30. Kusner, M.J., Hernández-Lobato, J.M.: GANS for sequences of discrete elements with the Gumbel-softmax distribution, pp. 1–6 (2016). arXiv:1611.04051

31. Liu, Y., Peng, J., Yu, J.J., Wu, Y.: PPGAN: privacy-preserving generative adversarial network. In: Proceedings of the International Conference on Parallel and Distributed Systems, ICPADS 2019, December 2019, pp. 985–989 (2019). ISBN 9781728125831. https://doi.org/10.1109/ICPADS47876.2019.00150, arXiv:1910.02007v1

32. Lu, P.H., Wang, P.C., Yu, C.M.: Empirical evaluation on synthetic data generation with generative adversarial network. In: Proceedings of the 9th International Conference on Web Intelligence, Mining and Semantics, WIMS 2019 (2019). https://doi.org/10.1145/3326467.3326474

33. Lu, P.H., Yu, C.M.: POSTER: a unified framework of differentially private synthetic data release with generative adversarial network. In: Proceedings of the 2017 ACM SIGSAC Conference on Computer and Communications Security, CCS 2017, pp. 2547–2549. ACM SIGSAC; Association of Computer Machinery; AT & T Business; Baidu; NSF; CISCO; Internet Finance Authenticat Alliance; Samsung; University of Texas Dallas; Google; IBM Res; Paloalto Networks; Visa Res; Army Res Off; Nasher Sculpture Ctr (2017). https://doi.org/10.1145/3133956.3138823

34. Mirza, M., Osindero, S.: Conditional generative adversarial nets, pp. 1–7 (2014). arXiv:1411.1784

35. Neunhoeffer, M., Wu, Z.S., Dwork, C.: Private post-GAN boosting (2020)

36. Odena, A., Olah, C., Shlens, J.: Conditional image synthesis with auxiliary classifier GANs. In: 34th International Conference on Machine Learning, ICML 2017, vol. 6, pp. 4043–4055 (2017). ISBN 9781510855144. arXiv:1610.09585

37. Office for Civil Rights.: Guidance Regarding Methods for De-identification of Protected Health Information in Accordance with the Health Insurance Portability and Accountability Act (HIPAA) Privacy Rule. U.S. Department of Health and Human Services, 20 November 2013 (2013). https://www.hhs.gov/hipaa/for-professionals/privacy/special-%20topics/de-identification/index.html

38. Park, N., Mohammadi, M., Gorde, K., Jajodia, S., Park, H., Kim, Y.: Data synthesis based on generative adversarial networks. Proc. VLDB Endow. 11(10), 1071–1083 (2018). https://doi.org/10.14778/3231751.3231757, arXiv:1806.03384

39. Rashidian, S., et al.: SMOOTH-GAN: towards sharp and smooth synthetic EHR data generation. In: Michalowski, M., Moskovitch, R. (eds.) AIME 2020. LNCS (LNAI), vol. 12299, pp. 37–48. Springer, Cham (2020). https://doi.org/10.1007/978-3-030-59137-3_4

40. Stadler, T., Oprisanu, B., Troncoso, C.: Synthetic data - a privacy mirage. arXiv arXiv:2011.07018 (2020)

41. Tantipongpipat, U., Waites, C., Boob, D., Siva, A.A., Cummings, R.: Differentially private synthetic mixed-type data generation for unsupervised learning. arXiv arXiv:cs.LG/1912.03250 (2020)

42. Torfi, A., Fox, E.A.: CorGAN: Correlation-capturing convolutional generative adversarial networks for generating synthetic healthcare records. arXiv arXiv:2001.09346 (2020)

43. Torfi, A., Fox, E.A., Reddy, C.K.: Differentially private synthetic medical data generation using convolutional GANs. arXiv arXiv:2012.11774 [cs] (December 2020). https://web.archive.org/web/20210618105126/

44. Vega-Marquez, B., Rubio-Escudero, C., Riquelme, J.C., Nepomuceno-Chamorro, I.: Creation of synthetic data with conditional generative adversarial networks. In: 14th International Conference on Soft Computing Models in Industrial and Environmental Applications, SOCO 2019, vol. 950, pp. 231–240. Startup Ole; IEEE SMC Spanish Chapter (2020). https://doi.org/10.1007/978-3-030-20055-8_22
45. Walia, M., Tierney, B., McKeever, S.: Synthesising tabular data using Wasserstein conditional GANs with gradient penalty, p. 13 (2020)
46. Xie, L., Lin, K., Wang, S., Wang, F., Zhou, J.: Differentially private generative adversarial network. arXiv arXiv:1802.06739 (2018). ISBN 1234567245
47. Xu, L., Skoularidou, M., Cuesta-Infante, A., Veeramachaneni, K.: Modeling tabular data using conditional GAN. arXiv arXiv:1907.00503 32(NeurIPS) (2019)
48. Xu, L., Veeramachaneni, K.: Synthesizing tabular data using generative adversarial networks. arXiv arXiv:1811.11264 (November 2018)
49. Xu, Q., et al.: An empirical study on evaluation metrics of generative adversarial networks. arXiv:1806.07755 [cs, stat] (August 2018). https://web.archive.org/web/20200604163128/
50. Yoon, J., Drumright, L.N., van der Schaar, M.: Anonymization through data synthesis using generative adversarial networks (ADS-GAN). IEEE J. Biomed. Health Inform. **24**(8), 2378–2388 (2020). https://doi.org/10.1109/JBHI.2020.2980262

# Preferences and Recommender Systems

# An Ensemble Hypergraph Learning Framework for Recommendation

Alireza Gharahighehi[1,2](✉) (iD), Celine Vens[1,2], and Konstantinos Pliakos[3]

[1] Itec, imec Research Group at KU Leuven, Kortrijk, Belgium
[2] Department of Public Health and Primary Care, KU Leuven, Campus KULAK, Kortrijk, Belgium
[3] Department of Management, Strategy and Innovation, KU Leuven, Leuven, Belgium
{alireza.gharahighehi,celine.vens,konstantinos.pliakos}@kuleuven.be

**Abstract.** Recommender systems are designed to predict user preferences over collections of items. These systems process users' previous interactions to decide which items should be ranked higher to satisfy their desires. An ensemble recommender system can achieve great recommendation performance by effectively combining the decisions generated by individual models. In this paper, we propose a novel ensemble recommender system that combines predictions made by different models into a unified hypergraph ranking framework. This is the first time that hypergraph ranking has been employed to model an ensemble of recommender systems. Hypergraphs are generalizations of graphs where multiple vertices can be connected via hyperedges, efficiently modeling high-order relations. We perform experiments using four datasets from the fields of movie, music and news media recommendation. The obtained results show that the ensemble hypergraph ranking method generates more accurate recommendations compared to the individual models and a weighted hybrid approach.

**Keywords:** Recommender systems · Hypergraph learning · Ensemble methods

## 1 Introduction

Nowadays, people use digital services more and more to fulfill their needs. The owners of these services monitor users' behavior and utilize users' interactions with provided items, such as movies, songs, commercial products, to predict users' preferences. This enables the personalization of digital services and the rise of effective recommender systems (RSs) which learn from users' preferences and provide them with accurate recommendations. Generally, there are two main categories in RSs: content-based filtering and collaborative filtering approaches. Content-based RSs use the features that describe the items for computing similarities between the items and the user interaction profile. Next, they recommend items that are more similar to this user profile. Upon a recommendation query

© Springer Nature Switzerland AG 2021
C. Soares and L. Torgo (Eds.): DS 2021, LNAI 12986, pp. 295–304, 2021.
https://doi.org/10.1007/978-3-030-88942-5_23

for a target user, these RSs do not consider the interactions of the other users in generating the recommendation list. In contrast to that, collaborative filtering approaches infer the users' preferences by processing the collaborative information between users or items. In many applications, collaborative filtering RSs generate more accurate [1] and less obvious [15] recommendations compared to content-based approaches.

Each type of RS processes the information based on different assumptions to decide which items should be ranked higher among many available ones. For instance, memory-based collaborative filtering approaches (user-based and item-based) assume that users (items) with similar interactions have similar interests. Therefore, these approaches form neighborhoods to generate recommendations. Model-based collaborative filtering approaches assume that users and items can be represented in a common feature space and they use different learning methods to learn these latent features. While these approaches might vary in prediction power, they convey relevant information from different perspectives, following practically different learning strategies for the same recommendation task. Ensemble methods include multiple learning methods and integrate their predictive power into a single system, achieving superior predictive performance to individual models. Examples of ensembles in machine learning are bagging and boosting. In recommendation tasks a hybrid RS can be applied to exploit several data sources or the prediction power of different RSs to generate more relevant recommendations. An ensemble RS is a hybrid model that employs the ranking lists of multiple RSs to decide which items should be recommended to each user [2].

In this paper we propose an ensemble hypergraph learning framework for recommendation. This way we integrate the predictive power of several models into a unified RS powered by hypergraph ranking. Unlike regular graphs, where edges connect pairs of nodes, in hypergraphs multiple nodes can be connected via hyperedges. These higher order relations in hyperedges empower hypergraphs to cast more reliable information in the model [21]. Furthermore, hypergraph learning can inherently model the complex relations between different types of entities in a unified framework. It is therefore a deliberate choice for the construction of an ensemble of individual RSs driven by different types of information. Moreover, as was shown in [12], hypergraph ranking-based methods can mitigate popularity bias, enhance fairness and coverage as well as act as innate multi-stakeholder RSs. The main contribution of this paper is to construct a hypergraph as an ensemble framework for recommendation tasks. Despite its capability to stack multiple connections in a unified model, to the best of our knowledge hypergraphs have not been employed to form ensembles of RSs.

The structure of this paper is as follows: Studies about applications of hypergraph learning in RSs are presented in Sect. 2. Next, in Sect. 3, we show how a unified hypergraph can be formed as an RS (Sect. 3.1) and how it can formulate an ensemble of RSs (Sect. 3.2). In Sect. 4, four recommendation datasets are described and the experimental setup in designing and testing the proposed model is described. Next, the obtained results of comparing the proposed ensemble model against other methods on these four datasets are presented and discussed in Sect. 5. Finally, we draw conclusions and outline some directions for future research in Sect. 6.

## 2  Related Work

Hypergraph learning has been applied to generate recommendation lists in several applications. For instance in the music domain, Bu et al. [3] used hypergraph learning to recommend music tracks where the relations between users, tracks, albums and artists were modeled using a unified hypergraph. Hypergraph ranking has been also used in news recommendation tasks [12,14]. News usually contains very rich features such as text, tags and named entities. Therefore, hypergraph learning can effectively model the relations between these entities. Moreover, Pliakos et al. [19] used hypergraph ranking for a tag recommendation task. They built a hypergraph ranking model to capture the complex relations between different entities in the system, such as users, images, tags, and geo-tags. Hypergraph-based RSs have been also used in e-commerce applications [16,22]. For instance in [16], a multipartite hypergraph is used to model the relations between users, restaurants and attributes in a multi-objective setting. In such applications, item attributes and sequences of user-item interactions are effectively modeled in hypergraphs.

Hypergraph learning has been employed to address various issues in RSs. A hypergraph can model the relations between different types of stakeholders and objects and therefore, it can be intrinsically used as a multi-stakeholder RS [11]. Additionally, it can be used to burst the filter bubble around the user by querying a more diverse recommendation list based on the user history [12,14]. Moreover, hypergraph learning has been used to address fairness [12], the cold-start problem [24] as well as context-awareness [23] in recommendation tasks.

An ensemble RS is a type of hybrid RSs that integrates the recommendations of multiple individual RSs. Aggarwal [2] categorized hybrid RSs to monolithic, ensembles, and mixed RSs. Burke et al. [4] provided another categorization where hybrid models are categorized into weighted, switching, cascade, feature augmentation, feature combination, meta-level and mixed RSs. A weighted hybrid RS uses the weighted average of the scores from individual RSs to generate the recommendation list. For instance, Do et al. [6] applied a weighted hybrid RS based on collaborative and content-based filtering approaches on *Movielens* dataset and showed that it is more effective compared to the individual collaborative and content-based RSs. Here, we employ a unified hypergraph as an ensemble RS. Although hypergraph learning is very promising and effective in addressing many problems in RSs, to the best of our knowledge, it has never been studied as an ensemble RS.

## 3   Methodology

### 3.1   Hypergraphs as Recommender Systems

Hereafter, uppercase bold letters are used for matrices, lowercase bold letters represent vectors, uppercase non-bold letters are used for sets and lowercase non-bold letters represent constants. The element in $i^{th}$ row and $j^{th}$ column of matrix $\mathbf{X}$ is denoted as $\mathbf{X}(i,j)$.

A hypergraph consists of a set of nodes (vertices) $N : \{n_1, n_2 \cdots, n_{|N|}\}$ and a set of hyperedges $E : \{e_1, e_2 \cdots, e_{|E|}\}$ that connect the nodes. Each hyperedge can connect multiple nodes in the hypergraph. Based on the application, different types of hyperedges can be defined that capture different forms/sources of information. We define these hyperedge types in Sect. 3.2. In a typical collaborative filtering setting there are two types of entities in a hypergraph: users $U : \{u_1, u_2 \cdots, u_{|U|}\}$ and items $I : \{i_1, i_2 \cdots, i_{|I|}\}$. Therefore, the set of nodes $N$ in a hypergraph is formed based on users and items ($N : \{U \cup I\}$).

Let $\mathbf{H}$ of size $|N| \times |E|$ be the incidence matrix of the hypergraph, where $H(n, e) = 1$, if node $n$ is in hyperedge $e$ and *zero* otherwise. Based on $\mathbf{H}$, the symmetric matrix $\mathbf{A}$ can be formed using Eq. 1:

$$\mathbf{A} = \mathbf{D_n}^{-1/2}\mathbf{HWD_e}^{-1}\mathbf{H}^T\mathbf{D_n}^{-1/2} \tag{1}$$

where $\mathbf{D}_n$ and $\mathbf{D}_e$ are the diagonal matrices that contain the node and hyperedge degrees and $\mathbf{W}$ is the diagonal hyperedge weight matrix (here $\mathbf{W} = \mathbf{I}$). Each element $\mathbf{A}(i,j)$ reflects the relatedness between nodes $i$ and $j$. Higher values indicate stronger relations between the corresponding nodes. Then, the recommendation problem is formulated as finding a ranking (score) vector $\mathbf{f} \in \mathbb{R}^{|N|}$ that minimizes the following loss function [3]:

$$Q(\mathbf{f}) = \frac{1}{2}\mathbf{f}^T\mathbf{L}\mathbf{f} + \vartheta\|\mathbf{f} - \mathbf{y}\|_2^2 \tag{2}$$

where $\mathbf{L}$ is the hypergraph Laplacian matrix (i.e. $\mathbf{L} = \mathbf{I} - \mathbf{A}$), $\vartheta$ is a regularizing parameter and $\mathbf{y} \in \mathbb{R}^{|N|}$ is the query vector. Every item of the ranking vector $\mathbf{f}$ or query vector $\mathbf{y}$ corresponds to a node. Typically, to generate the recommendation list for user $u$ in a regular recommendation task, one can query the hypergraph for user $u$ by setting the corresponding value in the query vector to *one* ($\mathbf{y}(u) = 1$) and all the other values that correspond to other nodes to *zero*. By solving the optimization problem in Eq. 2, the optimal score (ranking) vector can be calculated using Eq. 3:

$$\mathbf{f}^* = \frac{\vartheta}{1 + \vartheta}\Big(\mathbf{I} - \frac{1}{1 + \vartheta}\mathbf{A}\Big)^{-1}\mathbf{y}. \tag{3}$$

Finally, the top k items that have the highest scores in $\mathbf{f}^*$ are recommended to the user $u$.

## 3.2   An Ensemble Hypergraph-Based Recommender System

An ensemble RS[1] utilizes the decisions of multiple individual RSs to decide which items should be ranked higher in the final recommendation lists. Let $M : \{m_1, m_2 \cdots, m_{|M|}\}$ be the set of individual methods that we want to incorporate in our ensemble RS. Each of these individual methods $m_i$ can generate its own top $k$ rankings $\mathbf{R}_i \in \mathbb{R}^{|U| \times k}$ where each row in $\mathbf{R}_i$ is the top $k$ ranked items for the corresponding user. Then, based on the recommendation lists of each RS, hyperedges are formed to connect users to their top $k$ recommendations.

As is mentioned previously, the hypergraph consists of multiple types of hyperedges. We consider three types of hyperedges, which are defined in Table 1. The $E_{UI}$ hyperedges connect the users with the items that they have interacted with. To make the relations between users with similar tastes more explicit, the $E_{UU}$ hyperedges connect users to their $k$ nearest neighbors. To find these neighbors we use the user-item interaction matrix $\mathbf{Z}$, where $\mathbf{Z}(i, j) \in \{0, 1\}$. The $k$ nearest neighbors of user $u$ are users that have the highest cosine similarity with $u^{th}$ row of matrix $\mathbf{Z}$. The $E_M$ hyperedges are considered to integrate the recommendations of multiple RSs in the hypergraph. These RSs can be from different families such as collaborative filtering or content-based approaches. The fact that recommendations from any type of RS can be directly modeled as hyperedges in our system is a vital advantage of the proposed method.

We constructed the $E_M$ hyperedge set using two well-established and powerful matrix completion-based recommendation methods, namely Bayesian Personalized Ranking (BPR) [20] and Weighted Regularized Matrix Factorization (WRMF) [13,18]. *BPR* is a learning-to-rank matrix completion approach which uses user-specific relative preferences between observed and unobserved items to learn items' and users' low rank matrices. *WRMF* is a matrix factorization approach for implicit feedback datasets that uses the alternating-least-squares optimization process to learn items and users' parameters.

**Table 1.** Hyperedge definitions

| Hyperedge | Definition | # of hyperedges |
|---|---|---|
| $E_{UI}$ | Each user is connected to the items that the user has interacted with | $|U|$ |
| $E_{UU}$ | Each user is connected to the k most similar users | $|U|$ |
| $E_M$ | Each user is connected to top k recommended items by a RS | $|M| \times |U|$ |

The hypergraph and its incidence matrix $\mathbf{H}$ are constructed using the hyperedge sets of Table 1. Following that, the affinity matrix $\mathbf{A}$ is computed and the

---

[1] The source code is available at https://github.com/alirezagharahi/ensemble_hypergraph.

recommendation task is addressed as was described in Sect. 3.1. For the sake of simplicity we consider equal weights for recommendations of different models and also similar weights for items with different rankings in top $k$ recommendation lists and leave the weight optimization as future work.

## 4   Experimental Setup

To evaluate the performance of the proposed approach we use four datasets from news, music and movie application domains. These datasets are described in Table 2. AOTM is a publicly available dataset collected from the Art-of-the-Mix platform that is based on user playlists [17]. Movielens[2] is a publicly available movie rating dataset [5]. As we only encode interactions in the hypergraph for this dataset we transform ratings to binary feedback. Globo[3] and Roularta[4] are news datasets that contain readers' interactions with news articles.

**Table 2.** Datasets descriptions

|           | AOTM        | Movielens | Globo        | Roularta     |
|-----------|-------------|-----------|--------------|--------------|
| Item type | Music track | Movie     | News article | News article |
| # users   | 1,605       | 1,573     | 3,903        | 5.082        |
| # items   | 2,199       | 2,053     | 1,246        | 2,739        |
| Sparsity  | 3.8%        | 19.9%     | 5.7%         | 8.5%         |

In our experiments we consider the following five approaches[5]:

- **BPR:** Bayesian Personalized Ranking (BPR) [20] is a learning-to-rank matrix completion approach as presented in the previous section.
- **WRMF:** Weighted Regularized Matrix Factorization (WRMF) [13,18] is a MF approach using the alternating-least-squares optimization process to learn items and users' parameters as presented in the previous section.
- **Hybrid:** A weighted hybrid model that uses scores of *BPR* and *WRMF* and then considers the weighted average of these scores to generate the final ranking lists.
- **H:** A hypergraph-based RS explained in Sect. 3.1 that only contains the hyperedge types of $E_{UI}$ and $E_{UU}$ from Table 1.
- **H$_{Ens}$:** The proposed hypergraph-based ensemble RS explained in Sect. 3.2.

---

[2] http://www.grouplens.org.
[3] http://www.globo.com.
[4] http://www.roularta.be.
[5] For BPR and WRMF we used implicit library (https://implicit.readthedocs.io/en/latest/index.html).

To validate the performance of the proposed method against the compared methods we randomly hide *ten* interactions of each user from training and then measure the ability of the methods in predicting these hidden interactions[6]. We use *precision@10* to measure the accuracy of predictions. *Precision* is a standard information retrieval accuracy measure that reflects the proportion of relevant items in the recommendation list. As the number of relevant items and length of the recommendation lists are the same (10 items), *precision, recall* and *F1-score* are all the same. Therefore, we only report *precision* in this paper. The compared methods have some hyperparameters to be tuned. *BPR* and *WRMF* have number of latent features, number of iterations, regularizing parameter and learning rate, *Hybrid* model has a hybridization weight and $H$ as well as $H_{Ens}$ have a regularizer as a hyperparameter. To tune these hyperparamters we form a validation set for each dataset by randomly drawing *five* interactions of each user from the training set as the validation set. The final tuned hyperparameter values are based on *precision@10* and are reported in Table 3.

**Table 3.** Hyperparameters

|  |  | Range | AOTM | Movielens | Globo | Roularta |
|---|---|---|---|---|---|---|
| BPR | # iterations | [1000, 2000] | 1645 | 1984 | 1598 | 1984 |
|  | # latent features | [100, 250] | 129 | 500 | 168 | 129 |
|  | regularizing parameter | [0.01, 0.05] | 0.0194 | 0.0412 | 0.0374 | 0.0412 |
|  | learning rate | [0.001, 0.07] | 0.0284 | 0.0092 | 0.0174 | 0.0092 |
| WRMF | # iterations | [1000, 2000] | 1276 | 1393 | 1129 | 1288 |
|  | # latent features | [100, 250] | 201 | 107 | 152 | 109 |
|  | regularizing parameter | [0.01, 0.05] | 0.0374 | 0.0225 | 0.0432 | 0.0315 |
| Hybrid | Hybridization weight | [0.01, 0.99] | 0.6664 | 0.6664 | 0.3986 | 0.2701 |
| H | Regularizing parameter | [0.01, 0.99] | 0.2414 | 0.2414 | 0.0656 | 0.0616 |
| $H_{Ens}$ | Regularizing parameter | [0.01, 0.99] | 0.4554 | 0.4554 | 0.8301 | 0.6325 |

## 5    Results and Discussion

The results of the proposed hypergraph-based ensemble RS and the selected approaches on the four datasets are reported in Table 4. The reported values are in terms of average *precision@10* of the recommendation lists generated by the compared approaches. As is shown in Table 4, the proposed hypergraph-based ensemble RS ($H_{Ens}$) has superior predictive performance compared to all the competitor approaches including the hybrid model in all datasets. The competitor methods have different performance rankings in the four datasets. Each of these methods processes the information based on different assumptions and learning approaches. The effectiveness of these assumptions and learning approaches differs across different applications. For instance, the pair-wise

---

[6] Users with few interactions are omitted from experiments.

learning-to-rank approach in *BPR* is more effective in *Globo* dataset compared to the point-wise error minimization approach in *WRMF*, while this does not hold in the other datasets. An ensemble RS exploits the combined predictive power of the individual methods. It considers all assumptions and decisions of various independent RSs and achieves overall superior performance regardless of the application domain of the recommendation task.

**Table 4.** Results (*precision@10*)

|         | AOTM   | Movielens | Globo  | Roularta |
|---------|--------|-----------|--------|----------|
| BPR     | 0.0373 | 0.1716    | 0.0937 | 0.0704   |
| WRMF    | 0.0402 | 0.1718    | 0.0921 | 0.0764   |
| H       | 0.0338 | 0.1503    | 0.1125 | 0.0657   |
| Hybrid  | 0.0388 | 0.1828    | 0.0979 | 0.0769   |
| $H_{Ens}$ | **0.0412** | **0.1860** | **0.1140** | **0.0773** |

In this study we keep the experiments simple by only using the collaborative information, i.e. user-item interactions, to make them applicable on available datasets and various application fields (i.e. movies, music, news). Nevertheless, in cases where side information is available for users or items, content-based approaches can be included in the ensemble RS. Hypergraph learning has the natural capability of modeling the complex relations between different types of entities in a unified hypergraph and therefore is a deliberate choice to construct an ensemble of RSs with different types of information.

## 6     Conclusion

We proposed a new ensemble hypergraph learning-based RS. A unified hypergraph can integrate multiple connections between entities (here users and items) and therefore can combine the predictive power of various individual RSs boosting the precision of final recommendation lists. We empirically tested this method on four datasets from different application domains, such as news, music, and movies. The obtained results showed that the hypergraph-based ensemble RS achieves superior performance compared to all the individual models, as well as compared to a hybrid approach that averages individual scores to produce final rankings, in all datasets.

For future work we outline the following directions:

– **Weight optimization:** For the sake of simplicity we considered equal weights for individual RSs in the hypergraph-based ensemble RS and also similar weights for items in different rankings. These weights can be optimized to achieve even better performance.

- **Beyond accuracy evaluation:** In this paper we only used user-item inter-actions. Future approaches could include additional information and relevant stakeholders so that fairness [12] and diversity [9] are also taken into account.
- **Consumption level:** We only captured the binary feedback between users and items. In real applications usually the user feedback is graded [7] which shows to what extend the user is interested in the item. This graded feedback could be reflected in the hypergraph to model user preferences more precisely.
- **Long-term vs short-term preferences:** In some applications such as news [8] and music [10] recommendation tasks, users' short-term preferences play important roles. Session-based RSs have been used to model such user short-term preferences. An ensemble RS could include models for both long-term and short-term preferences.

**Acknowledgments.** This work was executed within the imec.icon project NewsButler, a research project bringing together academic researchers (KU Leuven, VUB) and industry partners (Roularta Media Group, Bothrs, ML6). The NewsButler project is co-financed by imec and receives project support from Flanders Innovation & Entrepreneurship (project nr. HBC.2017.0628). The authors also acknowledge support from the Flemish Government (AI Research Program).

# References

1. Adomavicius, G., Tuzhilin, A.: Toward the next generation of recommender systems: a survey of the state-of-the-art and possible extensions. IEEE Trans. Knowl. Data Eng. **17**(6), 734–749 (2005)
2. Aggarwal, C.C.: Ensemble-based and hybrid recommender systems. In: Recommender Systems, pp. 199–224. Springer, Cham (2016). https://doi.org/10.1007/978-3-319-29659-3_6
3. Bu, J., Tan, S., Chen, C., Wang, C., Wu, H., Zhang, L., He, X.: Music recommendation by unified hypergraph: combining social media information and music content. In: Proceedings of the 18th ACM international conference on Multimedia, pp. 391–400 (2010)
4. Burke, R.: Hybrid recommender systems: survey and experiments. User Model. User Adapt. Interact. **12**(4), 331–370 (2002)
5. Cantador, I., Brusilovsky, P., Kuflik, T.: 2nd workshop on information heterogeneity and fusion in recommender systems (HetRec 2011). In: Proceedings of the 5th ACM conference on Recommender systems. RecSys 2011. ACM, New York (2011)
6. Do, H., Le, T., Yoon, B.: Dynamic weighted hybrid recommender systems. In: 2020 22nd International Conference on Advanced Communication Technology (ICACT), pp. 644–650 (2020). https://doi.org/10.23919/ICACT48636.2020.9061465
7. Gharahighehi, A., Vens, C.: Extended Bayesian personalized ranking based on consumption behavior. In: Bogaerts, B., Bontempi, G., Geurts, P., Harley, N., Lebichot, B., Lenaerts, T., Louppe, G. (eds.) BNAIC/BENELEARN -2019. CCIS, vol. 1196, pp. 152–164. Springer, Cham (2020). https://doi.org/10.1007/978-3-030-65154-1_9
8. Gharahighehi, A., Vens, C.: Making session-based news recommenders diversity-aware. In: Proceedings of the Workshop on Online Misinformation- and Harm-Aware Recommender Systems (2020)

9. Gharahighehi, A., Vens, C.: Diversification in session-based news recommender systems. Pers. Ubiquit. Comput., 1–11 (2021). https://doi.org/10.1007/s00779-021-01606-4

10. Gharahighehi, A., Vens, C.: Personalizing diversity versus accuracy in session-based recommender systems. SN Comput. Sci. **2**(1), 1–12 (2021). https://doi.org/10.1007/s42979-020-00399-2

11. Gharahighehi, A., Vens, C., Pliakos, K.: Multi-stakeholder news recommendation using hypergraph learning. In: Koprinska, I., et al. (eds.) ECML PKDD 2020. CCIS, vol. 1323, pp. 531–535. Springer, Cham (2020). https://doi.org/10.1007/978-3-030-65965-3_36

12. Gharahighehi, A., Vens, C., Pliakos, K.: Fair multi-stakeholder news recommender system with hypergraph ranking. Info. Process. Manag. **58**(5), 102663 (2021) https://doi.org/10.1016/j.ipm.2021.102663.    https://www.sciencedirect.com/science/article/pii/S0306457321001515

13. Hu, Y., Koren, Y., Volinsky, C.: Collaborative filtering for implicit feedback datasets. In: 2008 Eighth IEEE International Conference on Data Mining, pp. 263–272. IEEE (2008)

14. Li, L., Li, T.: News recommendation via hypergraph learning: encapsulation of user behavior and news content. In: Proceedings of the Sixth ACM International Conference on Web Search and Data Mining, pp. 305–314 (2013)

15. Lops, P., de Gemmis, M., Semeraro, G.: Content-based recommender systems: state of the art and trends. In: Ricci, F., Rokach, L., Shapira, B., Kantor, P.B. (eds.) Recommender Systems Handbook, pp. 73–105. Springer, Boston (2011). https://doi.org/10.1007/978-0-387-85820-3_3

16. Mao, M., Lu, J., Han, J., Zhang, G.: Multiobjective e-commerce recommendations based on hypergraph ranking. Inf. Sci. **471**, 269–287 (2019). https://doi.org/10.1016/j.ins.2018.07.029

17. McFee, B., Lanckriet, G.R.: Hypergraph models of playlist dialects. In: ISMIR, vol. 12, pp. 343–348. Citeseer (2012)

18. Pan, R., et al.: One-class collaborative filtering. In: 2008 Eighth IEEE International Conference on Data Mining, pp. 502–511. IEEE (2008)

19. Pliakos, K., Kotropoulos, C.: Simultaneous image tagging and geo-location prediction within hypergraph ranking framework. In: 2014 IEEE International Conference on Acoustics, Speech and Signal Processing (ICASSP), pp. 6894–6898. IEEE (2014)

20. Rendle, S., Freudenthaler, C., Gantner, Z., Schmidt-Thieme, L.: BPR: Bayesian personalized ranking from implicit feedback. arXiv preprint arXiv:1205.2618 (2012)

21. Vinayak, R.K., Hassibi, B.: Crowdsourced clustering: querying edges vs triangles. In: Advances in Neural Information Processing Systems, pp. 1316–1324 (2016)

22. Wang, J., Ding, K., Hong, L., Liu, H., Caverlee, J.: Next-item recommendation with sequential hypergraphs. In: Proceedings of the 43rd International ACM SIGIR Conference on Research and Development in Information Retrieval, SIGIR 2020, pp. 1101–1110. Association for Computing Machinery, New York (2020). https://doi.org/10.1145/3397271.3401133

23. Yu, C.A., Tai, C.L., Chan, T.S., Yang, Y.H.: Modeling multi-way relations with hypergraph embedding. In: Proceedings of the 27th ACM International Conference on Information and Knowledge Management, pp. 1707–1710 (2018)

24. Zheng, X., Luo, Y., Sun, L., Ding, X., Zhang, J.: A novel social network hybrid recommender system based on hypergraph topologic structure. World Wide Web **21**(4), 985–1013 (2017). https://doi.org/10.1007/s11280-017-0494-5

# KATRec: Knowledge Aware aTtentive Sequential Recommendations

Mehrnaz Amjadi, Seyed Danial Mohseni Taheri[✉], and Theja Tulabandhula

University of Illinois at Chicago, Chicago, IL 60607, USA
{mamjad2,smohse3,theja}@uic.edu

**Abstract.** Sequential recommendation systems model dynamic preferences of users based on their historical interactions with platforms. Despite recent progress, modeling short-term and long-term behavior of users in such systems is nontrivial and challenging. To address this, we present a solution enhanced by a knowledge graph called KATRec (Knowledge Aware aTtentive sequential Recommendations). KATRec learns the short and long-term interests of users by modeling their sequence of interacted items and leveraging pre-existing side information through a knowledge graph attention network. Our novel knowledge graph-enhanced sequential recommender contains item multi-relations at the entity-level and users' dynamic sequences at the item-level. KATRec improves item representation learning by considering higher-order connections and incorporating them in user preference representation while recommending the next item. Experiments on three public datasets show that KATRec outperforms state-of-the-art recommendation models and demonstrates the importance of modeling both temporal and side information to achieve high-quality recommendations.

**Keywords:** Sequential recommendations · Attention mechanism · Bidirectional transformers · Knowledge graph

## 1 Introduction

With the exploding growth of online platforms in recent years, recommendation systems [11] have become an essential component in elevating user engagement levels and thus have taken a central role in business success. Many services leverage historical data of users and their interactions with their service (e.g., an app or a website) to personalize recommendations. Such recommendation systems are increasingly popular in various domains including: e-commerce, social media, search engines, content portals, and online publishing platforms.

In this work, we focus on exploiting the sequential behavior of users in order to predict their upcoming interactions. Existing sequential recommender designs, e.g., Markov chains [4,14,15], recurrent neural networks (RNN) [3,7], graph convolutional neural networks (GCN) [20,23], and self-attention based models [9,17] (to name a few), primarily focus on various ways to model such historical data. However, these sequential models tend to disregard the relationship between

© Springer Nature Switzerland AG 2021
C. Soares and L. Torgo (Eds.): DS 2021, LNAI 12986, pp. 305–320, 2021.
https://doi.org/10.1007/978-3-030-88942-5_24

items. In particular, platforms usually have access to two types of information that can be valuable for recommendations: (i) interactions of users and the service, which may evolve over time, and (ii) side information about users, items, and other auxiliary components. Recommender systems can generate more relevant content by taking advantage of item relations that are hard to elicit from interaction sequences of users. Such side information about the items can be based on higher-order item-entity connections and co-occurrence patterns, which can provide implicit information about related items (for instance, a mouse and a laptop).

Efficient exploitation of both temporal data of users and side information is the primary gap this paper attempts to fill. While there are many well-performing solutions in the literature, they do not effectively capture both types of information. For instance, models such as BERT4Rec, SASRec, and GRU4Rec [7,9,17] heavily focus on the temporal aspect by encoding user behavior sequences. Another stream of literature focuses on graph structure to capture item relations and side information, for example: TransE [1], TransH [21], and TransR [12]. In this work, we build on both these prior works and provide a novel way to integrate them. We follow this up by systematically exploring the importance of capturing both types of information on recommendation quality.

To capture the short-term preferences, long-term interests, and item-item relations, we propose the Knowledge Aware aTtentive Sequential Recommendations (KATRec) system. Our recommendation system consists of two modules: (i) a bidirectional transformer, which captures sequential interests by considering the inter-dependencies among items at any temporal distance, and (ii) a knowledge graph attention network that models higher-order user-item and item-item relations. The user-item relations are based on the interactions of users with items, e.g., click, purchase, view, etc., and capture *collaborative information*. The item-item relations capture the *semantic relatedness* among items based on their shared entities (e.g., movies with the same actor or genre). The importance of capturing such information while making sequential recommendations has been previously discussed in [8,13,22] to name a few. In addition to first-order relations, sequential recommendations generated by KATRec can use higher-order connections and relations among items (e.g., an individual can be an actor in a movie and a producer in another). Figure 1 illustrates two sequences of user interactions. A traditional sequential recommendation system models these two users differently, as they interact with different movies, although these users show similar interests to movies with shared entities (genre, actor, producer, etc.). Therefore, incorporating information using a knowledge graph can enhance their representations, and consequently improve recommendation performance.

This work proposes a novel deep neural network architecture incorporating both sequential behavior of users and side information about items. Our proposed structure captures the temporal information using a sequential attention mechanism and spatial information via a knowledge graph attention mechanism. To summarize, the key contributions of this work are:

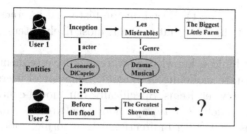

**Fig. 1.** Sequential interactions of two users with the systems. While users interact with different items, their collaborative signal can be detected via shared entities.

- KATRec builds on a knowledge graph neural network that captures multi-relationships between items by tying them together using the underlying entities. This allows for better representations of items and enhances the recommendation performance.
- KATRec models the short-term and long-term user preferences by adaptively aggregating dynamic interactions and item-item multi-relations through a gating mechanism. This mechanism can significantly alleviate the sparsity in both user sequences and item relationships.
- KATRec captures co-occurrence information and collaborative signals by leveraging attention mechanisms in the knowledge graph, which ultimately impact the attention weights in the bidirectional transformer module and the overall recommendations.
- We conduct experiments to evaluate the impact of different components on the performance of KATRec, and show that it outperforms state-of-the-art baselines on three public datasets.

The paper is organized as follows: in Sect. 2, we define the problem and present the new architecture KATRec. We conduct a detailed comparison of the performance of KATRec with multiple competitive baselines in Sect. 3. Section 4 concludes with some pointers to future directions.

## 2   Method

Our goal is to provide a personalized next item recommendation for users based on their history and higher-order relations between items. In this section, we first state a formal definition of the problem, and then we elaborate on different parts of our proposed solution.

### 2.1   Problem Definition and Solution Overview

Given a set of users $\mathcal{U}$ and items $\mathcal{I}$, we have a sequence of items $\mathcal{S}^u = \{S_1^u, \cdots, S_T^u\}$ that user $u$ has interacted with over $T$ time steps ($T = |\mathcal{S}^u|$). Also, we have access to side information related to items (e.g., actors, directors, and genre as shown in Fig. 1). Based on the historical interaction sequence $\mathcal{S}^u$, our goal is to predict the item that user $u$ is most likely interested in at the next time step $T + 1$.

In KATRec, we build a knowledge-aware attentive sequential recommendation system to facilitate modeling of dynamic behavior of users while capturing the multi-relations between items. Specifically, our model contains two modules as shown in Fig. 2, namely: (a) a graph neural network $\mathcal{G}$ which captures item level multi-relations, and (b) a bidirectional transformer module that incorporates item embeddings from the knowledge graph network into the representations of dynamic preferences exhibited by the users. In the following, we discuss the details of each.

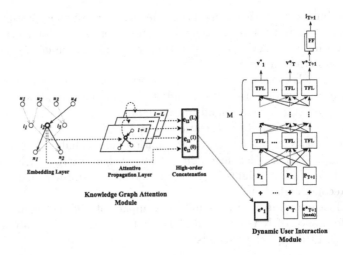

**Fig. 2.** Illustration of different components of the knowledge graph attention module (left) and the dynamic user interaction module (right). First, we learn the initial item embeddings using the knowledge graph attention module. Then, we feed these item embeddings and the user's interaction sequence $\mathcal{S}^u$ through $M$ layers of the bidirectional transformer (TFL) to obtain the final item embeddings that are then used for next item recommendation.

## 2.2   Knowledge Graph Module with Attention

This module encodes items' metadata as a unified graph to exploit the higher-order connectivity between items. The graph $\mathcal{G} = (\mathcal{E}, \mathcal{R})$, where $\mathcal{I} \subset \mathcal{E}$, incorporates entities as nodes and relationships as the edges. For instance, entities in a movie dataset include the items and their side information e.g., genres, producer, actor, etc. We use entities as building blocks that capture connections between items, and we focus on paths that start and end with items. Formally, the $K$-order connectivity between items is a path that captures a higher-order relationship between items $(i, j)$ as: $i \xrightarrow{r_1} n_1 \xrightarrow{r_2} n_2 \xrightarrow{r_3} \cdots \xrightarrow{r_K} j$, where $(i, j) \in \mathcal{I}$, entity $n_k \in \mathcal{E}$, and relation $r_k \in \mathcal{R}$ for $k \in \{1, 2, \cdots, K\}$. In addition to item and entities in the paths, we also model the joint occurrence of commonly related items using a collaborative knowledge graph by adding users as nodes in the graph. In particular, we integrate the item-user relations $\mathcal{R}_{\mathcal{U}}$ into the knowledge graph, so in the resulting graph, the interactions of users with items are also captured. We can represent the graph by a pair of nodes and their relation as $\mathcal{G} = \{(h, r, t) | h, t \in \mathcal{E}', r \in \mathcal{R}'\}$, where $\mathcal{E}' \subset \mathcal{E} \cup \mathcal{U}$ and $\mathcal{R}' \subset \mathcal{R} \cup \mathcal{R}_{\mathcal{U}}$. Figure 3 illustrates a collaborative knowledge graph where movie $i$ and movie $j$ are second order neighbors related by entity $n_1$ in the path $i \xrightarrow{r_4} n_1 \xrightarrow{r_5} j$ or related by user $u_2$ in the path $i \xrightarrow{r_3} u_2 \xrightarrow{r_2} j$.

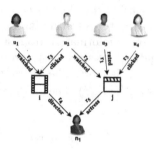

**Fig. 3.** Collaborative knowledge graph relates items $i$ and $j$ through user $u_2$ and entity $n_1$ with different types of relations.

In order to encode these relations as item embeddings, we use the TransR method [12]. TransR learns the embedding of each node and relation via the translation principle: $W^r e_h + e_r \approx W^r e_t$ ($e_h, e_t \in \mathbb{R}^d$, $e_r \in \mathbb{R}^k$, and $W^r \in \mathbb{R}^{k \times d}$), if triplet $(h, r, t)$ exists in the knowledge graph. For each triplet $(h, r, t)$ the dissimilarity score is computed using:

$$s(h, r, t) = \|W^r e_h + e_r - W^r e_t\|_2^2,$$

The lower the score of $s(h, r, t)$ the more likely is the triplet in the KG. Following [20], we capture each relation's importance by generating attentive weights between a node and its higher-order neighbors. So, for each node $h$, we initially consider all nodes that have the first-order relation with it, i.e. $\mathcal{N}_h = \{(h', r, t) | (h', r, t) \in \mathcal{G} \text{ with } h' = h\}$.

The first order connectivity embedding of head node $h$ is defined as the linear combination of the embeddings of its neighbors (ego network) $e_{\mathcal{N}_h} = \sum_{(h,r,t) \in \mathcal{N}_h} \pi(h, r, t) e_t$. The attention factor $\pi(h, r, t)$ controls how much information from different tails can be propagated to head $h$ based on specific relations, and can be computed as $\pi(h, r, t) = (W^r e_t)^\top \tanh(W^r e_h + e_r)$.

This attention mechanism will propagate more information from closer entities in the relationship space. Then, we normalize the coefficients across all $h$'s first-order relations using a softmax function $\pi'(h, r, t) = \frac{\exp(\pi(h,r,t))}{\sum_{(h,r',t') \in \mathcal{N}_h} \exp(\pi(h,r',t'))}$.

We update the node's representation by aggregating its representation and its ego network/connectivity representation using the relation: $e_h^{(1)} = f_{\mathcal{G}}^{(1)}(e_h, e_{\mathcal{N}_h})$, where aggregator $f_{\mathcal{G}}^{(1)}$ is defined as follows:

$$f_{\mathcal{G}}^{(1)} = \sigma\left(W_1^{(1)}(e_h + e_{\mathcal{N}_h})\right) + \sigma\left(W_2^{(1)}(e_h \odot e_{\mathcal{N}_h})\right).$$

Here $\sigma$ is the Leaky ReLU activation function and $W_1^{(1)}, W_2^{(1)} \in \mathbb{R}^{d^{(1)}} \times \mathbb{R}^d$ are trainable weight matrices. We follow a similar intuition for residual connections by aggregating information through the sum of two representations $e_h$ and $e_{\mathcal{N}_h}$ and retain a copy of input by using the identity transformation. Note that $\odot$ is the element-wise product that captures feature interaction between $e_h$ and $e_{\mathcal{N}_h}$, and ensures richer propagation of information from similar nodes.

For higher-order propagation, we stack more propagation layers to cascade information from higher-order neighbors. The $l$-th step node representation can be formulated as $e_h^{(l)} = f_{\mathcal{G}}^{(l)}(e_h^{(l-1)}, e_{\mathcal{N}_h}^{(l-1)})$. The information cascaded from $l-1$-th ego network is defined as:

$$e_{\mathcal{N}_h}^{(l-1)} = \sum_{(h,r,t)\in\mathcal{N}_h} \pi'(h, r, t)e_t^{(l-1)}.$$

Using this embedding propagation mechanism to stack $L$ layers, the higher-order connectivities can be captured in the node representation. Finally, we concatenate these representations into one vector to get the final representation of the node, $e_h^* = [e_h^{(0)} \| \cdots \| e_h^{(L)}] \in \mathbb{R}^q$, where $e_h^{(0)} := e_h$ and $q = d + d^{(1)} + \cdots + d^{(L)}$. The different layers of knowledge graph attention module described above are shown in Fig. 2 (left).

## 2.3  Dynamic User Interaction Module

To capture the sequential patterns among successive items that a user has interacted with, we use the Bidirectional Encoder Representations from Transformers (BERT) architecture [2,19]. In our context, BERT uses the historical item sequence $\mathcal{S}^u$ corresponding to user $u$ and aims to predict the item that the user is interested in the next time step $T + 1$. BERT models $M$ bidirectional transformer layers and revises each item's representation at each layer by exchanging information across all positions at the previous layer. Our key contribution is that we embed the higher-order connectivity of the item relations discussed in Subsect. 2.2 into the BERT module to capture user preferences under a more informative context. Below, we briefly discuss the self-attention structure used in the BERT module of KATRec.

**Embedding Layer:** Since the self-attention mechanism doesn't include any recurrent or convolutional blocks, it cannot be aware of items' position embeddings. So, we incorporate a pre-determined positional embedding $P \in \mathbb{R}^{T \times q}$

into the input embedding, $v_i^{(0)} = e_i^* + p_i$. In this equation, $v_i^{(0)}$ computes the input representation of items at each position index $i = 1, \cdots, T$. Concatenating embedding of $T$ items in a user sequence, $v_i^{(0)} \in \mathbb{R}^q$, results in $V^{(0)} \in \mathbb{R}^{T \times q}$. Next, this positional embedding matrix is provided as an input to the first transformer layer. The transformer layer contains two sublayers, a multi-head self-attention sublayer, and a position-wise feed-forward network. In the following, we describe each of these sublayers briefly.

**Transformer Layer:** The attention mechanism helps the model capture dependencies between each pair of items at any distance in the input sequence, across multiple subspace representations simultaneously. To learn information in different representation subspaces, we use multi-head attention [2]. First, we linearly project $V^{(0)}$ into $k$ subspaces using different learnable projections and then apply attention function on each in parallel to create $k$ heads $H_1, \cdots, H_k$ as follows:

$$H_j = \text{Attention}(V^{(0)} W_j^{\mathcal{Q}}, V^{(0)} W_j^{\mathcal{K}}, V^{(0)} W_j^{\mathcal{V}}),$$

where $W_j^{\mathcal{Q}}$, $W_j^{\mathcal{K}}$ and $W_j^{\mathcal{V}}$ are all $\mathbb{R}^{q \times q/k}$ learnable projection matrices corresponding to head index $j = 1, \cdots, k$. The attention function is a scaled dot-product defined as $\text{Attention}(\mathcal{Q}, \mathcal{K}, \mathcal{V}) = \text{softmax}\left(\frac{\mathcal{Q}\mathcal{K}^\top}{\sqrt{q/k}}\right)\mathcal{V}$. We concatenate these $k$ heads and then project them. We input the multi-head attention output, $\text{MH}(V^{(0)})$, to a feed-forward sublayer at each position $i = 1, \cdots, T$, $\text{PFF}(V^{(0)}) = [FF(v_1^{(0)})^\top \parallel \cdots \parallel FF(v_T^{(0)})^\top]$. Other operations including the dropout, residual connections, and layer normalization (LN) are kept similar to the original BERT architecture for language modeling.

**Output Layer:** Capturing pairwise item relations plays an important role in the effectiveness of the recommendation systems and also allows some degree of interpretability. To include item-item relations in item embeddings, we concatenate item embeddings learned by the sequential module and the knowledge graph.

$$\hat{E} = \sigma((V^* \parallel E^*)\hat{W} + \hat{b}),$$

where $\hat{E} \in \mathbb{R}^{|\mathcal{I}| \times q}$ is the set of final KATRec embeddings for item set $\mathcal{I}$ and $\hat{W} \in \mathbb{R}^{2q \times q}$ and $\hat{b} \in \mathbb{R}^q$ are learnable parameters. $V^*$ is the embedding matrix of items learned by sequential module, and $E^*$ is the items embedding table learned by the knowledge graph. Finally, the next item at time $T + 1$ is predicted by:

$$P(\mathcal{I}) = \text{softmax}(\text{GELU}(v_{T+1}^{(M)} W^P + b^P)\hat{E}^\top + b^O),$$

where $W^P, b^P$, and $b^O$ are learnable parameters, $\hat{E} \in \mathbb{R}^{|\mathcal{I}| \times q}$ is the embedding matrix for item set $\mathcal{I}$, and $P(\cdot)$ is the KATRec model's predicted distribution over the target items. $v_{T+1}^{(M)}$ is the hidden state of position $T + 1$ after $M$ transformer layers, denoted by $v_{T+1}^*$ in Fig. 2.

Existing approaches provide personalized recommendation by modeling the user embedding either explicitly [11,15,18] based on users' previous actions, or implicitly [6,9,17] based on embeddings of the sequence of visited items by a user. KATRec belongs to the latter category as we predict the next item at time step $T + 1$ by considering the hidden state embedding $v_{T+1}^{(M)}$.

As an aside, we also considered explicit user behavior by incorporating user embeddings learned from the knowledge graph into the user's hidden state via concatenation: $[e_u^* \parallel v_{T+1}^{(M)}]$, where $e_u^*$ is the embedding of user $u$ in the knowledge graph. Although this concatenation seems promising, we empirically did not observe improvement in the model's performance. This could be potentially because the model learned users' embedding very well by considering the sequence of interacted items.

## 2.4 Optimization

The loss function of KATRec contains the TransR objective along with a regularizer. In particular, we use the TransR to train the entity embeddings. The objective function can be minimized by discriminating between valid and invalid triplets in the collaborative knowledge graph:

$$\mathcal{L_G} = \sum_{(h,r,t,t')} -\ln \sigma\Big(s(h,r,t') - s(h,r,t)\Big) + \lambda\|\Theta\|_2^2, \tag{1}$$

where the sum is over all valid $(h, r, t) \in \mathcal{G}$ and invalid $(h, r, t') \notin \mathcal{G}$ triplets in the knowledge graph $\mathcal{G}$, $\sigma(\cdot)$ is the sigmoid function, and $\lambda\|\Theta\|_2^2$ represents $\ell_2$ regularization. Similar to [2], we implement the *Cloze task* training approach in addition to pairwise ranking loss in Eq. (1). This allows us to learn the parameters of the encoders in the transformer layers. In this approach, we randomly mask a portion of items in the input sequence $\mathcal{S}_u$ and try to predict them. The loss for each masked input $S_u'$ is given by:

$$\mathcal{L_S} = \frac{1}{|\mathcal{S}_u^m|} \sum_{i_m \in \mathcal{S}_u^m} -\log P(i_m = i_m^\star | \mathcal{S}_u'),$$

where $\mathcal{S}_u^m$ is the set of randomly masked items, $i_m^\star$ is the true item corresponding to the masked item $i_m$, and $P(.)$ is the predicted probability mass function over the target item. These two losses are jointly minimized over their respective parameters using standard first-order approaches (see the next section for details).

## 3    Experiments

We evaluate our model on three real-world datasets, which are different in domains and have varying levels of sparsity[1]. We aim to answer the following questions in this section:

---

[1] Code is available at https://github.com/DanialTaheri/KATRec.

Q1: How does KATRec perform compared to the current state-of-the-art sequential recommendation methods? Q2: How do different components of the model (viz., knowledge graph based attention mechanism, information aggregation, and pre-training) affect the performance of KATRec? Q3: How KATRec's performance change in different settings and datasets?

Following prior work [9,17], we convert all numeric ratings to positive interaction with a value of 1, which indicates that the user has interacted with the item. Then, we sort each user's interactions by timestamp to build her interaction sequence. Similar to prior works, we split the user sequences into three parts. The test dataset includes the most recent item each user has interacted with ($S^u_{T+1}$), the validation dataset consists of the second most recent item interacted by each user ($S^u_T$), and the remaining items in the sequence belong to the training data. To construct knowledge graph aware attention, we first build the item knowledge graph for each dataset. We follow [20,23] to capture knowledge graph triplets by mapping items into *freebase* entities. We include triplets with one-hop and two-hop neighbor entities and filter out entities with less than ten occurrences, and relations less than fifty occurrences. For $\mathcal{L}_G$ in Eq. (1), we pair each observed triplet with a broken (unobserved) triplet.

### 3.1  Datasets Description

We consider three datasets with different levels of sparsity. For most of the experiments, we only include users and items with at least ten interactions to ensure data quality [20]. The statistics of the datasets described below are presented in Table 1 for ease of reference:

**Table 1.** Statistics of datasets

| Datsets | Users | Items | Interactions | Entities | Relations | Triplets | Density |
|---|---|---|---|---|---|---|---|
| Amazon-book | 70679 | 24915 | 846434 | 88572 | 39 | 2557746 | 0.048% |
| LastFM | 23566 | 48123 | 8057269 | 58266 | 9 | 464567 | 0.7105% |
| Yelp2018 | 45919 | 45538 | 1185068 | 90961 | 42 | 1853704 | 0.057% |

Amazon-book: Amazon review data is one of the popular datasets in the recommendation systems literature [5]. The data has been categorized based on different product categories, and in this paper, we focus on the book category.

LastFM: This is a dataset about music listening patterns collected from the Last.fm online music platform [16]. In this dataset, tracks are viewed as items, and we consider a subset of the dataset from January 2014 to August 2014.

Yelp2018: This dataset is adopted from the Yelp 2018 recommendation system challenge. In this dataset, local restaurants and bars are represented as items.

We calculate each dataset's density based on the number of interactions, users, and items as $\frac{|Interactions|}{|Users| \cdot |Items|}$. Therefore, larger values in the density column represent datasets with more interactions (per user and item). Table 1 lists specific properties and the density of each dataset. It also highlights that LastFM has a substantially lesser number of relations and is significantly denser than others.

## 3.2    Experimental Settings

**Evaluation Metrics:** We use two common top-K metrics to evaluate the performance of our model. Hit@K and NDCG@K count the fraction of the time the ground truth item is among the top $K$ recommendations, without and with defining a position-aware weight respectively. Mean Average Precision (MAP) is also a ranked precision metric over all users that emphasizes correct predictions at the top of the list with a position-aware weight. Following the works of [9,17], we randomly sample 100 negative items for each user besides the ground-truth item. We report the average of the metrics Hit@K, NDCG@K over all users.

**Baselines:** To compare the performance of our model with others, we consider the following competitive baselines:

- GRU4Rec [7]: It implements a session-based recommendation model based on RNNs. We consider each user's sequence as a session.
- GRU4Rec$^{++}$ [6]: It modifies the way GRU4Rec is optimized by implementing a new loss function and a new sampling approach.
- SASRec [9]: It uses a self-attention mechanism with a left-to-right Transformer to improve the capturing of useful patterns in user sequences.
- BERT4Rec [17]: This is a recent state-of-the-art sequential recommendation model that adapts the bidirectional Transformers language model architecture to learn the temporal behavior of users.

**Parameters for Models:** We implement KATRec with Tensorflow (version 2.2.0). All parameters are initialized using a truncated normal distribution in the interval $[-0.02, 0.02]$. We use the Adam optimiser [10] with learning rate $10^{-4}$ that decays linearly, $\beta_1 = 0.9$, $\beta_2 = 0.999$, and a weight decay of 0.01. We fix the maximum sequence length proportional to the average sequence length in the dataset, i.e., $50, 50$, and 200 for Amazon-book, Yelp, and LastFM respectively. We set the dimension of the hidden fully connected layers of KATRec to be 128. We propagate neighbors' information up to three levels into each entity's embedding with hidden dimensions 32, 16, and 16. Finally, we set the embedding of entities in the knowledge graph to be 64.

We search for hyperparameters to select the best parameters for different baselines. These include changing embedding size from $\{8, 16, 32, 64, 128\}$ and the regularization hyperparameter $\lambda$ across $\{0.1, 0.05, 0.01, 0.005, 0.001, 0.0001\}$. We use the optimization schemes and parameters suggested by the authors whenever possible. The models are trained on a single GeForce GTX 1080Ti GPU.

**Performance Comparison:** Table 2 shows the recommendation performance of KATRec and baselines. We do not include Hit@1 since Hit@1 and NDCG@1 are equivalent. Since we have a single ground-truth, Hit@K is equivalent to Recall@K, and it is proportional to Precision@K. *We observe that KATRec provides improved relative recommendation performance over all alternatives by 6.29% and 3.82% in Hit and 7.15% and 4.64% in NDCG on average, respectively on the Amazon-book and Yelp2018 datasets.*

**Table 2.** KATRec versus baselines over three datasets. The improvement percentage compares KATRec versus the next-best alternative.

| Datasets | Metrics | GRU | $GRU^{++}$ | SASRec | BERT | KATRec | Improv |
|----------|---------|-----|-----------|--------|------|--------|--------|
| Amazon | NDCG@1 | 0.3485 | 0.3464 | 0.3749 | <u>0.4344</u> | **0.4706** | 8.33% |
| | NDCG@5 | 0.4404 | 0.4358 | 0.5267 | <u>0.5715</u> | **0.6110** | 6.91% |
| | NDCG@10 | 0.4598 | 0.4574 | 0.5600 | <u>0.6022</u> | **0.6401** | 6.2% |
| | Hit@5 | 0.5202 | 0.5148 | 0.6594 | <u>0.6910</u> | **0.7321** | 5.94% |
| | Hit@10 | 0.58 | 0.5814 | 0.7621 | <u>0.7856</u> | **0.8217** | 4.6% |
| | MAP | 0.42 | 0.4259 | 0.5065 | <u>0.5539</u> | **0.5907** | 6.64% |
| LastFM | NDCG@1 | 0.3646 | 0.3523 | <u>0.6771</u> | 0.6339 | **0.6931** | 2.36% |
| | NDCG@5 | 0.4648 | 0.4448 | **0.7765** | 0.7606 | <u>0.7725</u> | −0.51% |
| | NDCG@10 | 0.4881 | 0.4674 | **0.7930** | 0.7786 | <u>0.7911</u> | −0.24% |
| | Hit@5 | 0.5531 | 0.5263 | **0.8600** | 0.8281 | <u>0.8426</u> | −2.06% |
| | Hit@10 | 0.6249 | 0.5958 | **0.9105** | 0.8836 | <u>0.9001</u> | −1.15% |
| | MAP | 0.4577 | 0.4357 | <u>0.7598</u> | 0.7509 | **0.7618** | 0.26% |
| Yelp2018 | NDCG@1 | 0.3946 | 0.4148 | 0.3723 | <u>0.4149</u> | **0.4405** | 6.17% |
| | NDCG@5 | 0.5041 | 0.5143 | 0.5703 | <u>0.6039</u> | **0.629** | 4.15% |
| | NDCG@10 | 0.5278 | 0.5395 | 0.6068 | <u>0.6400</u> | **0.663** | 3.6% |
| | Hit@5 | 0.5991 | 0.6021 | 0.7434 | <u>0.7690</u> | **0.7927** | 3.08% |
| | Hit@10 | 0.6721 | 0.68 | 0.8551 | <u>0.8796</u> | **0.899** | 2.2% |
| | MAP | 0.49 | 0.515 | 0.5351 | <u>0.5706</u> | **0.5946** | 4.21% |

In particular, KATRec consistently outperforms BERT4Rec on Amazon-book and Yelp2018 datasets, which shows the importance of modeling item-item and user-item relations. KATRec achieves a considerable performance improvement in Amazon-book, while the improvement in Yelp is relatively small. This observation can be attributed to the difference in the sparsity of these two datasets. The importance of the impact of data density and number of relations on KATRec's performance is highlighted explicitly with the LastFM dataset, which we discuss in further detail in the ablation study that follows.

**Ablation Study:** In this section, we analyze variants of KATRec to understand the impact of different components on model performance. The variations are as follows: (1) *No Attention*: we remove the attention mechanism in the knowledge graph and allocate equal weights to each entity's neighbors. (2) *Level-1*: we decrease the level of information that can propagate from the neighbors to a node, and study the impact of only using immediate neighbors to improve node embeddings. (3) *Connection*: While there exists a connection between two encoder modules in our model, we consider the setting where both modules train independently using learnable embeddings. (4) *No Pretraining*: We forgo the pre-trained embeddings of entities in the knowledge graph. (5) *Concat*: We remove a part of our model that deals with item co-occurrence, and only consider the item embedding vector that results from the sequential module.

Results are shown in Table 3. We observe that incorporating the side information in the bidirectional encoder module during the training results in better parameter learning as shown in column *Connection*. Furthermore, the results of making the *NoPretrain* choice in the knowledge graph show that incorporating pre-trained embeddings increases the performance of the sequential recommendation model. The attention mechanism between neighbors in the knowledge graph increases the recommendation's performance. However, this increase is not substantial. We also observe that multiple layer information propagation in an item's embedding plays an important role in making next item recommendations. Also, as expected, propagating the first layer's information in the knowledge graph has a higher impact, and this impact decreases when we incorporate higher level of connections. Finally, the results of the *Concat* choice shows that incorporating non-linearity in the final layer is beneficial for learning item embeddings, especially by combining features learned through the knowledge graph and the bidirectional encoder.

**Table 3.** Ablation study of design choices in KATRec using three datasets.

| Datasets | Metrics | KATRec | NoAtten. | Level-1 | Connect. | NoPretrain | Concat. |
|---|---|---|---|---|---|---|---|
| Amazon-book | NDCG@10 | 0.6401 | 0.6371 | 0.6386 | 0.621 | 0.6306 | 0.6318 |
|  | Hit@10 | 0.8217 | 0.8178 | 0.8195 | 0.801 | 0.8092 | 0.8142 |
| LastFM | NDCG@10 | 0.7911 | 0.7836 | 0.7853 | 0.763 | 0.7587 | 0.7855 |
|  | HIT@10 | 0.9001 | 0.8967 | 0.8957 | 0.8796 | 0.8752 | 0.8908 |
| Yelp2018 | NDCG@10 | 0.663 | 0.6567 | 0.6546 | 0.6458 | 0.6359 | 0.6515 |
|  | HIT@10 | 0.899 | 0.8954 | 0.8929 | 0.885 | 0.8696 | 0.8881 |

We also study the recommendation performance of KATRec for users with different sequence lengths and compare it with competitive baselines. The intuition here is that the use of side information can compensate for potentially sparse user item interactions. In Fig. 4, we illustrate the percentage of users with varying sizes of item sequences associated with them in each dataset, and report each method's performance for each of the resulting user groups. KATRec outperforms two other competitive baselines for all user groups for Yelp2018 and Amazon-book datasets. Results for LastFM show that KATRec provides a reasonably robust recommendation performance across user groups, while the performance of the baselines is more sensitive to the user sequence length. This robustness of KATRec can be attributed to the item-item and user-item relationships that are explicitly learned. Our model performs relatively better for users with small sequence lengths. However, SASRec marginally outperforms KATRec for users with long sequence lengths. This can be attributed to the high number of interactions and low number of relations in the LastFM dataset, which indicates that the impact of side information on dense datasets may not be significant enough.

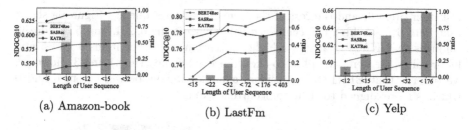

(a) Amazon-book

(b) LastFm

(c) Yelp

**Fig. 4.** Performance comparison of models across users with different sequence lengths on the Amazon-book, LastFM and Yelp2018 datasets.

**Visualizing Attention Weights:** This section visualizes the attention weights related to items and positions to find a meaningful pattern and discuss their differences with the BERT model's attention weights. Figure 5 shows the heatmaps of the average attention weights on the last 15 items of the sequences in the test dataset of Yelp2018. In order to calculate the accurate average weights, we do not incorporate weights of padded items in sequences shorter than 15 items. Comparing heatmaps (a) and (b) shows the impact of positional embeddings (PE). In particular, heatmap (a) illustrates how positional embeddings results in items attending more on recent items. Heatmaps (a) and (c) points out how items in various heads and layers focus on different parts of the sequence at both the right and left sides. To compare attentions in BERT4Rec and KATRec, we analyze weights in the final layer as it is directly connected to the output layer and plays an important role in the prediction (heatmaps (d) and (c)). The comparison indicates that while BERT4Rec inclines to focus more on the recent items due to the sparsity of the dataset, KATRec tends to attend on less recent items due to incorporating side information through the knowledge graph. This behavior is similar to the attention weights of self-attention blocks in dense datasets in [9].

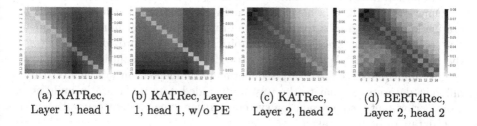

(a) KATRec, Layer 1, head 1

(b) KATRec, Layer 1, head 1, w/o PE

(c) KATRec, Layer 2, head 2

(d) BERT4Rec, Layer 2, head 2

**Fig. 5.** Yelp average attention weights on positions (x-axis) at time (y-axis)

**Attention Weight Case Study.** Figure 6a illustrates the co-occurrence ratio between six items in user sequences, computed by the average number of times that a pair of items appeared simultaneously in users' sequence. Figure 6b compares weights of the final attention layer in KATRec and Bert4Rec between the last item (item 6) and the rest of the items in a user sequence. Figure 6a shows that item 6 has a high co-occurrence value with items $0, 2, 4$, and $5$. Figure 6b confirms that KATRec places a higher attention values for items 4 and 5 and lower values for items with low co-occurrence value, i.e., time 1 and 3. However, we observe that BERT4Rec considers a higher weight between item 6 and 0 which is more aligned to their co-occurrence value.

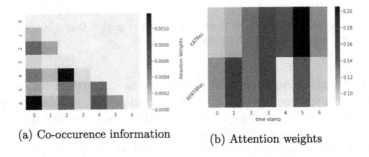

(a) Co-occurence information          (b) Attention weights

**Fig. 6.** Attention heatmap comparison of a random user in Yelp2018 dataset. 6a shows the co-occurrence ratio, which is the co-occurrence frequency of each pair of six items among all users' sequences. 6b compares these items' attention weight at the last position in KATRec and BERT4Rec.

## 4    Conclusion and Future Work

Designing robust deep neural network architectures that produce quality recommendations is challenging for several reasons. A couple of these challenges were addressed in this work, namely leveraging of side information and getting around data sparsity. In particular, we proposed incorporating item side information to alleviate both these shortcomings while making recommendations. This information is readily available in many real-world applications.

Our work introduces a novel neural network structure that leverages collaborative knowledge graphs to improve the representations of items in a sequential recommendation system setup. Empirical results are provided to illustrate the benefit via multiple evaluation metrics: the proposed solution is compared against multiple state-of-the-art sequential recommendation systems on three different datasets. Similar to the way we included item metadata in building a more performant recommendation system, further research in incorporating user metadata can be undertaken.

## References

1. Bordes, A., Usunier, N., Garcia-Duran, A., Weston, J., Yakhnenko, O.: Translating embeddings for modeling multi-relational data. In: Advances in Neural Information Processing Systems, pp. 2787–2795 (2013)

2. Devlin, J., Chang, M.W., Lee, K., Toutanova, K.: BERT: pre-training of deep bidirectional transformers for language understanding. arXiv preprint arXiv:1810.04805 (2018)

3. Donkers, T., Loepp, B., Ziegler, J.: Sequential user-based recurrent neural network recommendations. In: Proceedings of the Eleventh ACM Conference on Recommender Systems, pp. 152–160 (2017)

4. He, R., McAuley, J.: Fusing similarity models with Markov chains for sparse sequential recommendation. In: 2016 IEEE 16th International Conference on Data Mining (ICDM), pp. 191–200. IEEE (2016)

5. He, R., McAuley, J.: Ups and downs: modeling the visual evolution of fashion trends with one-class collaborative filtering. In: Proceedings of the 25th International Conference on World Wide Web, pp. 507–517 (2016)

6. Hidasi, B., Karatzoglou, A.: Recurrent neural networks with top-k gains for session-based recommendations, pp. 843–852 (2018). https://doi.org/10.1145/3269206.3271761

7. Hidasi, B., Karatzoglou, A., Baltrunas, L., Tikk, D.: Session-based recommendations with recurrent neural networks. arXiv preprint arXiv:1511.06939 (2015)

8. Ji, M., Joo, W., Song, K., Kim, Y.Y., Moon, I.C.: Sequential recommendation with relation-aware kernelized self-attention. arXiv preprint arXiv:1911.06478 (2019)

9. Kang, W.C., McAuley, J.: Self-attentive sequential recommendation. In: 2018 IEEE International Conference on Data Mining (ICDM), pp. 197–206. IEEE (2018)

10. Kingma, D.P., Ba, J.: Adam: a method for stochastic optimization. arXiv preprint arXiv:1412.6980 (2014)

11. Koren, Y.: Factorization meets the neighborhood: a multifaceted collaborative filtering model. In: Proceedings of the 14th ACM SIGKDD International Conference on Knowledge Discovery and Data Mining, pp. 426–434 (2008)

12. Lin, Y., Liu, Z., Sun, M., Liu, Y., Zhu, X.: Learning entity and relation embeddings for knowledge graph completion. In: Twenty-Ninth AAAI Conference on Artificial Intelligence (2015)

13. Ma, C., Ma, L., Zhang, Y., Sun, J., Liu, X., Coates, M.: Memory augmented graph neural networks for sequential recommendation. arXiv preprint arXiv:1912.11730 (2019)

14. Rendle, S., Freudenthaler, C., Gantner, Z., Schmidt-Thieme, L.: BPR: Bayesian personalized ranking from implicit feedback. arXiv preprint arXiv:1205.2618 (2012)

15. Rendle, S., Freudenthaler, C., Schmidt-Thieme, L.: Factorizing personalized Markov chains for next-basket recommendation. In: Proceedings of the 19th International Conference on World Wide Web, pp. 811–820 (2010)

16. Schedl, M.: The LFM-1b dataset for music retrieval and recommendation. In: Proceedings of the 2016 ACM on International Conference on Multimedia Retrieval, pp. 103–110 (2016)

17. Sun, F., et al.: BERT4Rec: sequential recommendation with bidirectional encoder representations from transformer. In: Proceedings of the 28th ACM International Conference on Information and Knowledge Management, pp. 1441–1450 (2019)

18. Tang, J., Wang, K.: Personalized top-n sequential recommendation via convolutional sequence embedding. In: Proceedings of the Eleventh ACM International Conference on Web Search and Data Mining, pp. 565–573 (2018)

19. Vaswani, A., et al.: Attention is all you need. In: Advances in Neural Information Processing Systems, pp. 5998–6008 (2017)

20. Wang, X., He, X., Cao, Y., Liu, M., Chua, T.: KGAT: knowledge graph attention network for recommendation. In: KDD, pp. 950–958 (2019)

21. Wang, Z., Zhang, J., Feng, J., Chen, Z.: Knowledge graph embedding by translating on hyperplanes. In: Twenty-Eighth AAAI Conference on Artificial Intelligence (2014)
22. Zhang, Y., He, Y., Wang, J., Caverlee, J.: Adaptive hierarchical translation-based sequential recommendation. In: Proceedings of The Web Conference 2020, pp. 2984–2990 (2020)
23. Zhao, W.X., et al.: KB4Rec: a data set for linking knowledge bases with recommender systems. Data Intell. 1(2), 121–136 (2019)

# Representation Learning and Feature Selection

# Elliptical Ordinal Embedding

Aïssatou Diallo[1]([✉]) and Johannes Fürnkranz[2]

[1] Research Training Group AIPHES, Technische Universität Darmstadt, Darmstadt, Germany
diallo@aiphes.tu-darmsatdt.de
[2] Computational Data Analytics Group, JKU Linz, Linz, Austria
juffi@faw.jku.at

**Abstract.** Ordinal embedding aims at finding a low dimensional representation of objects from a set of constraints of the form "item $j$ is closer to item $i$ than item $k$". Typically, each object is mapped onto a point vector in a low dimensional metric space. We argue that mapping to a density instead of a point vector provides some interesting advantages, including an inherent reflection of the uncertainty about the representation itself and its relative location in the space. Indeed, in this paper, we propose to embed each object as a Gaussian distribution. We investigate the ability of these embeddings to capture the underlying structure of the data while satisfying the constraints, and explore properties of the representation. Experiments on synthetic and real-world datasets showcase the advantages of our approach. In addition, we illustrate the merit of modelling uncertainty, which enriches the visual perception of the mapped objects in the space.

**Keywords:** Ordinal embedding · Representation learning

## 1 Introduction

A crucial problem in machine learning is the assessment of similarities between data instances. In fact, multiple tasks depend on such an ability. For example, in clustering, similar items should be grouped together, or in classification, where similar items should be assigned similar labels. In general, one expects to be given a collection of data instances and a similarity function that allows determining how similar objects are to each other. Yet, it is not always straight-forward to define such a similarity function for a given data representation. Thus, recent works in machine learning focus on a scenario in which the learner is only given relative comparisons between data instances [1,10]. Instead of directly querying the degree of similarity between items on an absolute scale, it has been shown that eliciting *ordinal feedback* from subjects in the form of "item $i$ is more similar to item $j$ than to item $k$" is a more reliable form of supervision, especially when the feedback is subjective [13]. The problem of interest is to learn representations in a low-dimensional metric space such that the relative distances of the representation satisfy a set of ordinal triplet constraints of the above type. This

© Springer Nature Switzerland AG 2021
C. Soares and L. Torgo (Eds.): DS 2021, LNAI 12986, pp. 323–333, 2021.
https://doi.org/10.1007/978-3-030-88942-5_25

problem is known as *ordinal embedding*. It dates back to the classic non-metric multidimensional scaling approach, but interest in the problem has renewed in recent years. The main expected result of this task is a faithful geometric representation that allows to easily visualize similarities between data instances.

We argue that the classical representation of items as points does not allow to capture the inherent noise of the ordinal feedback and the resulting uncertainty of the representation. Consider the case of an object for which the different triplets are conflicting with each other or even revealing contradicting underlying patterns. Such a discrepancy should be reflected and possibly visually expressed by the learnt embedding. As a remedy, we propose to embed items as probability distributions in $\mathbb{R}^d$, with a location and a scale parameter, which captures the uncertainty of the location. In particular, we focus on Gaussian distributions which enjoy some desirable properties.

The paper is organized as follows: Sect. 2 formally introduce the problem of ordinal embedding, Sect. 3 presents our approach for elliptical ordinal embedding. Finally, Sect. 4 illustrates our empirical studies and analyze the results.

## 2    Problem Statement

In this section, we formally state the ordinal embedding problem and establish the notation, for which we follow [11]. $\|\cdot\|$ denotes the $\ell_2$ norm. $\mathcal{S}_+^d$ is the set of all positive definite matrices. In the scope of this work, we only focus on Gaussian distributions which belong to the family of parametrized probability distributions $z_{h,\mathbf{a},\mathbf{A}}$ having a location vector $\mathbf{a} \in \mathbb{R}^d$ which represents the shift of the distribution, a scale parameter $\mathbf{A} \in \mathcal{S}_+^d$, which represents the statistical dispersion of the distribution, and a characteristic generator function $h$. Specifically, for Gaussian distributions, the scale parameter coincides with the covariance matrix $var(z_{h,\mathbf{a},\mathbf{A}}) = \mathbf{A}$. From now on, we denote Gaussian distributions (or embeddings) as $z_{(h,\mathbf{a},\mathbf{A})} = \mathcal{N}(\mathbf{a}, \mathbf{A})$.

Consider $n$ items in an abstract space $\mathcal{X}$, which we represent by their indices $[n] = 1, ..., n$. It is worth mentioning that no explicit representation of the items is available so it is not possible to analytically express the dissimilarity between the items. We assume a latent underlying dissimilarity (or similarity) function $\delta : \mathcal{X} \times \mathcal{X} \to \mathbb{R}_{\geq 0}$. Let $\mathcal{T} := \{\langle i, j, k \rangle : 1 \leq i \neq j \neq k \leq n\}$ be a set of unique triplets of elements in $\mathcal{X}$. We further have access to an oracle $\mathcal{O}$ which indicates whether the inequality $\delta(i, j) < \delta(i, k)$ holds or not:

$$\mathcal{O}(\langle i, j, k \rangle) = \begin{cases} +1 \text{ if } \delta(i, j) < \delta(i, k) \\ -1 \text{ if } \delta(i, j) > \delta(i, k) \end{cases} \tag{1}$$

Note that at this stage, we do not require the latent function $\delta$ to be a metric. Together, $\mathcal{T}$ and $\mathcal{O}$ represent the observed *ordinal constraints* on distances. We can now formally define the problem as follows:

**Definition 1 (Ordinal Embedding).** *Consider $n$ vector points $\mathbf{X} = (\mathbf{x_1}, \mathbf{x_2}, ..., \mathbf{x_n})$ in a $d$-dimensional Euclidean space $\mathcal{X}$. Given a set of triplets*

$\mathcal{T} \subset \mathcal{X}^3$ and an oracle $\mathcal{O} : \mathcal{X}^3 \to \{-1, 1\}$, the ordinal embedding problem consists of recovering $\mathbf{X}$ given $\mathcal{O}$ and $\mathcal{T}$.

## 3    Elliptical Ordinal Embedding

We propose to learn probabilistic embeddings in lieu of the conventional Euclidean embeddings, taking advantage of the fact that vectors can be considered as an extreme case of probability measures, namely a Dirac [11]. For this purpose, we focus on the family of elliptical distributions, more precisely Gaussian distributions, which enjoy many advantages. Our goal is to extend the ordinal embedding problem, from Definition 1 for point embeddings, to Gaussian embeddings. Hence the considered problem becomes:

**Definition 2 (Probabilistic Ordinal Embedding).** *Suppose $\mathcal{T} \subset \mathcal{X}^3$ is a set of triplets over $\mathcal{X}$ and $\mathcal{O} : \mathcal{X}^3 \to \{-1, 1\}$ is an oracle as defined in (1). Let $\mathbf{Z} = \{\mathbf{z}_1, \dots, \mathbf{z}_n\}$ the desired probabilistic embedding, where each of the original points $\mathbf{x}_i$ is mapped to probability distribution parametrized by $\mathbf{z}_i$.* Probabilistic ordinal embedding *is the problem of obtaining $\mathbf{Z}$ from ordinal constraints $\mathcal{T}$ and $\mathcal{O}$ and a distance measure $d$ such that $\mathrm{sgn}(d(\mathbf{z}_i, \mathbf{z}_j) - d(\mathbf{z}_i, \mathbf{z}_k)) = O(\langle i, j, k \rangle)$, for $\langle i, j, k \rangle \in \mathcal{T}$.*

This definition requires a distance measure $d$ between distributions. For this purpose, we selected the *Wasserstein distance* [12] which has been previously used as a loss function for supervised learning [5] and in several applications.

**The 2-Wasserstein Distance.** In Optimal Transport (OT) theory, the Wasserstein or Kantorovich–Rubinstein metric is a distance function defined between probability distributions (measures) on a given metric space $M$. The squared Wasserstein metric for two arbitrary probability measures $\mu, \nu \in \mathcal{P}(\mathbb{R}^d)$ is defined as: $W_2^2(\mu, \nu) \overset{\text{def}}{=} \inf_{X \sim \mu, Y \sim \nu} \mathbb{E}\|X - Y\|^2$. In the general case, it is difficult to find analytical solutions for the Wasserstein distance. However, a closed form solution exists in the case of Gaussian distributions. Let $\alpha \overset{\text{def}}{=} \mathcal{N}(\mathbf{a}, \mathbf{A})$ and $\beta \overset{\text{def}}{=} \mathcal{N}(\mathbf{b}, \mathbf{B})$, where $\mathbf{a}, \mathbf{b} \in \mathbb{R}^d$ and $\mathbf{A}, \mathbf{B} \in \mathcal{S}_+^d$ are positive semi-definite. When $\mathbf{A} = \mathrm{diag}\ \mathbf{d_A}$ and $\mathbf{B} = \mathrm{diag}\ \mathbf{d_B}$ are diagonal, $W_2^2$ simplifies to the sum of two terms:

$$W_2^2(\alpha, \beta) = \|\mathbf{a} - \mathbf{b}\|^2 + \mathfrak{h}^2(\mathbf{d_A}, \mathbf{d_B}) \tag{2}$$

where $\mathfrak{h}^2(\mathbf{d_A}, \mathbf{d_B}) \overset{\text{def}}{=} \|\sqrt{\mathbf{d_A}} - \sqrt{\mathbf{d_B}}\|^2$ is the *squared Hellinger distance* [2] between the diagonal $\mathbf{d_A}$ and $\mathbf{d_B}$.

**Learning Problem.** As mentioned earlier, the goal is to learn a function that maps each item to a $d$-dimensional Gaussian embeddings in $\mathbb{R}^d$ such that the 2-Wasserstein distances between the embeddings satisfy as many triplets as possible. Each Gaussian embedding is denoted as $\mathbf{z}_{\mu, \Sigma}$, which for the sake of compactness, we abbreviate in $\mathbf{z}$. Let $E_{ij}$ be the energy function between two items

$(i, j)$ [9] which characterizes our energy-based learning approach. In particular, we set $E_{ij} = W_2^2(\mathbf{z}_i, \mathbf{z}_j)$. Finally, the corresponding optimization problem is the following: $\max_{\mathbf{z}_1,\ldots\mathbf{z}_n \in \mathbb{R}^d} \sum_{t=(i,j,k) \in \mathcal{T}} \mathcal{O}(t) \cdot \mathrm{sgn}(E_{ij} - E_{ik})$ which is discrete, non-convex and NP-hard. For these reasons, a relaxation of this optimization problem is needed. We make the choice of using the hinge loss $\mathcal{L}((t = \langle i, j, k \rangle, \mathcal{O}(t))$, a well established loss function in contrastive metric learning, as a convex surrogate:

$$\mathcal{L} = \sum_{t=\langle i,j,k \rangle \in \mathcal{T}} \max(1 - \mathcal{O}(t) \cdot (E_{ij} - E_{ik}), 0) \tag{3}$$

The empirical performance of embedding methods is evaluated by the *empirical error*: $Err = \frac{1}{|\mathcal{T}'|} \sum_{\langle i,j,k \rangle \in \mathcal{T}} \mathbb{1}[(y \cdot \mathrm{sgn}(E_{ij} - E_{ik})) = 1]$.

**Complexity.** The training complexity is linear to the size of $\mathcal{T}$, which is the set of all triplets and bounded by $\mathcal{O}(n^3)$. However, a well chosen sampling strategy may decrease this bound. It has been shown by [7] that the minimum number of triplets to recover the ordinal embedding is $\Omega(nd \log n)$ in $\mathbb{R}^d$. We adapt this result to the setting in which the parameters to be learnt are a mean vector in $\mathbb{R}^d$ and a covariance matrix $\mathcal{S}_+^d$. Hence, the dimensionality can be considered to be $d' = d + d^2$ and $\mathcal{O}(d^2)$. Thus, the new recovered lower bound for the triplets becomes $\Omega(d^2 n \log n)$, which is still polynomial in $d$ and $\mathcal{O}(n \log n)$. Since ordinal embeddings typically map into a low-dimensional space, this is not a drastic loss in efficiency.

**Architecture and Hyperparameters.** We noticed that EIOE is not very sensitive to the number and size of hidden layers. For such, we chose a sufficiently large hidden size, specifically $h_{dim} = 50$. To embed the item $i$:

$$\mathbf{h}_i = \mathrm{relu}(\mathbf{x}_i \mathbf{W}_i + \mathbf{b}) \quad \mu_i = \mathbf{h}_i \mathbf{W}_\mu + \mathbf{b}_\mu \quad \sigma_i = \exp(\mathbf{h}_i \mathbf{W}_\Sigma + \mathbf{b}_\Sigma)$$

where $\mathbf{x}_i$ is a random sample from $\mathcal{N}(0, I_h)$ and relu is the rectifier linear unit. We apply the exponential function to make sure that $\sigma_i$ is positive (and $\Sigma_i$ is positive definite). Weight matrices $\mathbf{W}_\mu$, $\mathbf{W}_\Sigma$ and $\mathbf{W}$ are initialized with Xavier initialization. As stated earlier, we do not regularize the norm of the mean vectors but we bound the values of the covariance matrices with $C = \log(100)$. However, we observe that this additional precaution is not needed unless the number of contradicting triplets is too large. This is due to the self-regularizing nature of the Wasserstein distance and it was confirmed by our experiments in which the average value of the variance is far from that bound for reasonable levels of noise. All parameters are optimized using Adam, with a fixed learning rate of 0.01 and a learning rate decay of $10^{-5}$.

## 4   Experiments

In this section, we evaluate our method in two settings. First, we perform experiments on synthetic datasets in order to gain some insight regarding our approach.

We then apply our approach to real datasets in order to assess the performance of our model in real cases. When the ground truth is available, an adequate error metric to assess the embedding quality is the Procrustes distance [4] for which we provided an extension defined in Eq. (4). Additional results and figures are presented in the extended version of the paper [3].

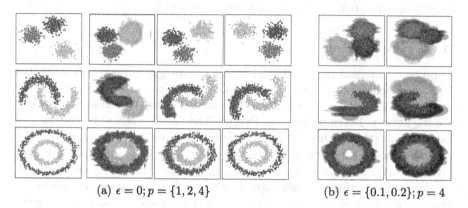

(a) $\epsilon = 0; p = \{1, 2, 4\}$        (b) $\epsilon = \{0.1, 0.2\}; p = 4$

**Fig. 1.** ElOE embeddings for synthetic experiments. From left to right, columns represent ground-truths, elliptical embeddings in nose-free setting ($\epsilon = 0$) for increasing numbers of triplets $pd^2 n \log n$ for $p = \{1, 2, 4\}$ (col. 2–4) and increasing noise for $p = 4$ and $\epsilon = \{0.1, 0.2\}$ (col. 5–6). Color indicates labels not used for training. (Color figure online)

**Definition 3 (Procrustes Distance between distributions).** *Given two finite sequences* $X = (x_i)_{i=1}^n$, $X' = (x'_{i, \mu, \Sigma})_{i=1}^n$ *in* $\mathbb{R}^d$ *of equal length with centroids in* $\bar{x}$, $\bar{x}'$ *and centroids sizes* $S_X$, $S_{X'}$ [1], *respectively, the Procrustes distance* $d_P^{\star}(X, X')$ *between* $X$ *and* $X'$ *is defined as:*

$$d_P^{\star}(X, X') = \inf_{R \in \mathcal{R}} \left( \sum_{i=1}^n \left\| \frac{R x_i}{S_X} - \frac{\mu_i}{S_{X'}} \right\|^2 + \frac{\mathrm{Tr}(\Sigma_i)}{S_{X'}^2} \right)^{\frac{1}{2}} \tag{4}$$

**Visualization of Embeddings Using Ellipses.** The most significant difference between our distribution-based approach and the point-based embeddings is the variance. In particular, the variance has the purpose of reflecting the uncertainty. It does so by enriching the scope of the embeddings and by providing the possibility of continuously representing a discrete object in the metric space. In most cases related to multidimensional scaling, the output dimensionality of the representations is low, thus the learned embeddings can be visualized as they

---

[1] Let us define the centroid $\bar{x}$ as $\bar{x} = \frac{1}{n} \sum_{i=1}^n x_i$, then the centroid size $S_X$ is $S_X = (\frac{1}{n} \sum_{i=1}^n (\bar{x} - x_i)^2)^{1/2}$, provided we ignore the trivial case in which all points coincide.

are. We argue that mapping objects into ellipses on a plane allows to better observe the relationship between objects visually. [11] state that visualizing the variances as they are is not natural to the human eye, and they instead favor a representation by the precision matrix rather than the covariance matrix. On the contrary, we believe that in this context a visualization based on the variance is preferable when the focus is on illustrating the spread around the location rather than the distance between the embeddings themselves.

### 4.1   Reconstruction

We present empirical results that aim to evaluate the reconstruction abilities of the proposed approach. For this, we follow the same experimental setting of [6]. More specifically, we use three 2-dimensional synthetic datasets generated with the `scikit-learn` package[2] in Python. The datasets are: a) two interleaved moons, b) a mixture of three Gaussians $\mathcal{N}(\mu, \frac{1}{\sqrt{2}}I_2)$ and c) two concentric circles. For each dataset, $n = 1000$ points are generated. The label information is used only for visualization purposes. We generate $|\mathcal{T}|$ random triplets sampled from a uniform distribution. To simulate the ordinal feedback from the oracle, we compute the difference of the squared $\ell_2$ norm between the points for a given triplet. The total number $|\mathcal{T}|$ is set to be $pd^2 n \log n$ for $p = \{1, 2, 4\}$. To evaluate the performance, we compute the triplet error as well as the Procrustes distance shown in (4).

**Noise-Free Setting.** In this series of experiments, we aim at investigating the influence of the number of triplets on the reconstruction ability, specifically on the variance of the elliptical embeddings. We first test in a noise-free setting. Figure 1(a) depicts the original datasets (to the left) and the learned embeddings for different values of $T$. From left to right, the number of used triplets $|\mathcal{T}|$ increases with $|\mathcal{T}| = pd^2 n \log n$, where $p \in \{1, 2, 4\}$. For all three datasets, we observe that the reconstruction abilities w.r.t the location point improves when the number of triplets increases. Furthermore, we observe that on average the variance decreases with increasing $|\mathcal{T}|$, which confirms that the uncertainty about a point's location decreases when more exact comparisons are available. For example, for $p = 4$, the average area of the ellipses is minimal. We can also observe that the variance enriches the visual representation. A point-vector representation may be misleading because when the algorithm is given few triplets, it has also to satisfy fewer constraints which means that the overall degree of freedom for selecting the individual points is greater. However a point-based visualization does not appreciate this fact.

**Noisy Setting.** Our next goal was to investigate the influence of noisy or erroneous triplets on the behaviour of the variance. We follow the procedure described above, but simulated noise by randomly swapping the assessment of

---

[2] https://scikit-learn.org/.

the oracle with a probability of $\epsilon = \{0.1, 0.2\}$. Figure 1(b) shows the results obtained. We notice that when the proportion of erroneous triplets increases (from left to right), the variance on average increases for all triplets. Additionally, in order to quantitatively estimate the performance of our approach we measure the Procrustes distance $d_P^*$ Equation (4) with respect to the ground truths. As a baseline, we compare our model to STE [14]. For 10 rounds, we compute $d_P^*$ e Equation (4) w.r.t $\epsilon$. We notice that generally, EIOE recovers better the density estimate even considering the variance of the ellipses. Figure 2 illustrates the relation between Procrustes distance and empirical triplet error and the number of triplets, for STE and EIOE. We notice that overall, EIOE performs better than the baseline embeddings.

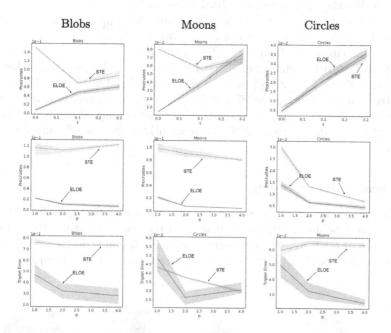

**Fig. 2.** 1st row: Procrustes distance vs $\epsilon$. 2nd row: Procrustes distance vs $p$. 3rd row: Empirical error vs $p$.

### 4.2 Ordinal Embedding

**Food Dataset.** We evaluate our method on the Food relative similarity dataset [15], humans were presented images of dishes and asked to compare similar dishes based on their taste. A good embedding method should show clusters of dishes of the same type. We compute 2d embeddings of the food images based on the available unique triplets of 100 images with $|\mathcal{T}| = 190376$ and 9349 pairs of contradicting triplets. We compare our embeddings to STE and we observe that our embeddings closely match the one produced by STE, which is the reference model used by the authors of the dataset. For lack of space, these are available at [3]. Note that for this dataset, no ground truth is available and hence there is no way other than visual inspection for evaluating our results.

**MNIST Dataset.** On this dataset, we reproduce the experiment conducted in [8]. For $n = 500, 1000$ and 2500, we uniformly chose $n$ MNIST digits randomly and we generate $200n \log n$ triplets comparisons based on the Euclidean distances between the digits. Each comparison is incorrect with probability $\epsilon = 0.15$. We then generate an ordinal embedding with $d = 5$ and compute a k-means clustering on the obtained embeddings. Section 4.2 compares the purity of the clusters obtained with STE and EIOE

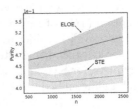

**Fig. 3.** Purity for MNIST dataset.

embeddings. Purity is computed as $\text{purity}(\Omega, \mathcal{C}) = n^{-1} \sum_k |w_k \cap c_j|$, where the clusters are $\Omega = \{w_1, \ldots, w_k\}$ and the classes are $\mathcal{C} = \{c_1, \ldots, c_j\}$. High purity is better. We concatenate the diagonal of the covariance matrix and the mean vector for each embedding for evaluating our embeddings. We observe that the purity of the clustering from EIOE is consistently higher for all values of $n$ considered.

### 4.3 Semantic Embedding

In this section, we intend to embed a real-world full or partial ordinal relation between data points, in this case images. In particular we study the following three types of relations, where the given ordinal relation is derived from various label structures of the objects, such as a linear or a hierarchical order.

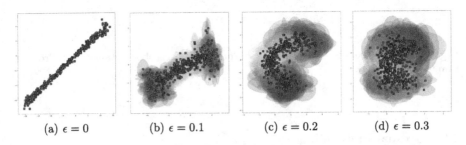

| (a) $\epsilon = 0$ | (b) $\epsilon = 0.1$ | (c) $\epsilon = 0.2$ | (d) $\epsilon = 0.3$ |

**Fig. 4.** Linear order in the MNIST dataset. Color indicates the label of the handwritten digits to better appreciate the linearity. (Color figure online)

Intuitively, we want all nodes that belong to the 1-hop neighborhood of item $i$ to be closer to $i$ in their embedding, compared to the nodes in the 2-hop neighborhood, which in turn will be closer than the items in the 3-hop neighborhood and so on. Moreover, we need to adapt the sampling strategy to deal with this task because uniformly sampling triplets leads to oversampling more frequent high-degree nodes. Thus, we use the following strategy: we first sample a node $i$, then we sample a node from each of its neighborhoods, and randomly choose one of those triplets.

This simple experiment aims at verifying whether our approach is able to capture the structural information when the underlying ordinal feedback derived from labels is a linear order with MNIST dataset. In this case, we use the same sampling strategy described in Sect. 4.1. We train our model with $200n \log n$ triplets and we sample 500 digits for visualization. We perturbed a subset of the available triplets defined by $\epsilon = \{0, 0.1, 0.2, 0.3\}$. In the noise-free case (Fig. 4(a)) we obtain a linear relation in which the embedding have very small variances. When perturbing 10% of the triplets, as shown in Fig. 4(b) the linear order is maintained but we can observe an average increase in the variance of the embeddings. Finally, for $\epsilon = 0.3$ (Fig. 4(d)), the proportion of noisy triplets is so high that even the linear order is perturbed. Nevertheless, the clusters defined by the classes of the embeddings are still easily identified.

**Hierarchical Relation.** This experiment was conducted on CIFAR100, a multi-class image dataset where each of the 60000 images has two different levels of labels, a super-class and a fine-class label There are 20 super-classes, each of which has 5 labels. The graph structures can be seen in [3]. We sample $n = 5000$ images to create $2nd^2 \log n$ triplets. The triplet score is computed with the methodology described earlier, through the shortest path distance between nodes.

To quantitatively assess the meaningfulness of the embeddings, we report the area under the ROC curve (AUC) and the average precision (AP) of randomly sampled triplets. We compare our results to [6] which we re-implemented with the same hidden size of EIOE and STE. It is worth noticing that this method can also be seen as the producing distributional vectors with null variance. The score considered for EIOE is $E_{ij}$,

**Table 1.** Link prediction scores.

|       | CIFAR |      | VOC  |      |
|-------|-------|------|------|------|
|       | AUC   | AP   | AUC  | AP   |
| STE   | 0.47  | 0.53 | 0.54 | 0.56 |
| [6]   | 0.88  | 0.90 | 0.93 | 0.94 |
| ELOE  | 0.89  | 0.92 | 0.95 | 0.95 |

$\|\mathbf{x}_i - \mathbf{x}_j\|^2$ for STE and [6], where $\mathbf{x}_i$ is the embedding of item $i$. Results are reported in Table 1. We see that EIOE embeddings satisfy more triplets than STE embeddings.

**Multilabel Distance.** Finally, we looked at the PASCAL VOC multi-label dataset, where each image can be assigned to multiple labels. Here $n = 5000$ and $p = 2$. The same concerns with respect to the sampling strategy occur in this case as well, and we apply the same methodology described earlier. In this case, an image node can be connected to multiple node classes. The obtained results showed in Table 1 confirm our intuition, nodes with less diverse neighborhoods have a lower variance, hence less uncertainty compared to nodes that belongs to multiple classes. In fact, the inclusion in multiple classes makes the embedding location less certain.

## 5    Conclusion

We have proposed to generalize the ordinal embedding problem by mapping objects in the space of Gaussian distributions endowed with the Wasserstein distance. This is based on the generalization of point embeddings in $\mathbb{R}^d$ to distributions. Each embedding is described by a location parameter $\mu$ and a scale parameter $\Sigma$, visualized as ellipses. We argue that this allows to more informative perceptual embeddings by representing uncertainty of the representation. In a number of experiments on different datasets we demonstrate the validity of our approach. We show that the proposed framework is robust and beneficial when the triplet comparisons are noisy. Overall, with our proposed approach we are able to obtain valid embedding that can be used for downstream tasks. As future work we aim to study other distributions beyond Gaussian for the problem of ordinal embedding.

**Acknowledgements.** This work has been supported by the German Research Foundation as part of the Research Training Group Adaptive Preparation of Information from Heterogeneous Sources (AIPHES) under grant No. GRK 1994/1.

## References

1. Agarwal, S., Wills, J., Cayton, L., Lanckriet, G., Kriegman, D., Belongie, S.: Generalized non-metric multidimensional scaling. In: Proceedings of the 11th International Conference on Artificial Intelligence and Statistics (AISTATS), pp. 11–18 (2007)
2. Beran, R.: Minimum Hellinger distance estimates for parametric models. Ann. Stat. **5**(3), 445–463 (1977)
3. Diallo, A., Fürnkranz, J.: Elliptical ordinal embedding. arXiv preprint (2021)
4. Dryden, I.L., Mardia, K.V.: Statistical shape analysis with applications in R. Wiley Series in Probability and Statistics, vol. 995. Wiley, New York (2016)
5. Frogner, C., Zhang, C., Mobahi, H., Araya, M., Poggio, T.A.: Learning with a Wasserstein loss. In: Advances in Neural Information Processing Systems 28 (2015)
6. Haghiri, S., Vankadara, L.C., von Luxburg, U.: Large scale representation learning from triplet comparisons. arXiv preprint arXiv:1912.01666 (2019)
7. Jamieson, K, G., Nowak, R, D.: Low-dimensional embedding using adaptively selected ordinal data. In: Proceedings of the 49th Annual Allerton Conference on Communication, Control, and Computing. IEEE (2011)
8. Kleindessner, M., von Luxburg, U.: Kernel functions based on triplet comparisons. In: Advances in Neural Information Processing Systems 30 (2017)
9. LeCun, Y., Chopra, S., Hadsell, R., Ranzato, M., Huang, F.: A tutorial on energy-based learning. Predict. Struct. Data **1**(0) (2006)
10. McFee, B.: More like this: machine learning approaches to music similarity. Ph.D. thesis, University of California, San Diego (2012)

11. Muzellec, B., Cuturi, M.: Generalizing point embeddings using the Wasserstein space of elliptical distributions. In: Advances in Neural Information Processing System 31, pp. 10258–10269 (2018)
12. Olkin, I., Pukelsheim, F.: The distance between two random vectors with given dispersion matrices. Linear Algebra Appl. **48**, 257–263 (1982)
13. Stewart, N., Brown, G.D., Chater, N.: Absolute identification by relative judgment. Psychol. Rev. **112**(4), 881 (2005)
14. Van Der Maaten, L., Weinberger, K.: Stochastic triplet embedding. In: Proceedings of the IEEE International Workshop on Machine Learning for Signal Processing, pp. 1–6. IEEE (2012)
15. Wilber, M.J., Kwak, I.S., Belongie, S.J.: Cost-effective hits for relative similarity comparisons. In: Proceedings of the 2nd AAAI conference on Human Computation and Crowdsourcing (HCOMP) (2014)

# Unsupervised Feature Ranking
# via Attribute Networks

Urh Primožič[1] , Blaž Škrlj[1,2] , Sašo Džeroski[1,2] , and Matej Petković[1(✉)]

[1] Jozef Stefan Institute, Jamova 39, 1000 Ljubljana, Slovenia
urh.primozic@student.fmf.uni-lj.si,
{blaz.skrlj,saso.dzeroski,matej.petkovic}@ijs.si
[2] Jotef Stefan Postgraduate School, Jamova 39, 1000 Ljubljana, Slovenia

**Abstract.** The need for learning from unlabeled data is increasing in contemporary machine learning. Methods for unsupervised feature ranking, which identify the most important features in such data are thus gaining attention, and so are their applications in studying high throughput biological experiments or user bases for recommender systems. We propose FRANe (Feature Ranking via Attribute Networks), an unsupervised algorithm capable of finding key features in given unlabeled data set. FRANe is based on ideas from network reconstruction and network analysis. FRANe performs better than state-of-the-art competitors, as we empirically demonstrate on a large collection of benchmarks. Moreover, we provide the time complexity analysis of FRANe further demonstrating its scalability. Finally, FRANe offers as the result the interpretable relational structures used to derive the feature importances.

**Keywords:** Feature ranking · Feature selection · Unsupervised learning · Attribute networks · PageRank

## 1 Introduction

Increasing amounts of high-dimensional data, in fields such as molecular and systems' biology, require development of fast and scalable feature ranking algorithms [14]. By being able to prioritize the feature space with respect to a given target, feature ranking algorithms already offer, e.g., novel biomarker candidates. However, the amount of available labeled data is potentially much smaller when compared to the amount of unlabeled data, which remains largely unexploited. In response, *unsupervised* feature ranking algorithms (that operate only on unlabeled data) are actively developed.

We propose FRANe, a **F**eature **R**anking approach based on **A**ttribute **Ne**tworks), schematically shown in Fig. 1. FRANe achieves state-of-the-art performance by exploiting data-derived relations between the features (which form an undirected weighted graph). The contributions of this work are manifold, and can be summarized as follows:

Supported by the Slovenian Research Agency (grant P2-0103 and a young researcher grant), and European Commission (grant 952215).

C. Soares and L. Torgo (Eds.): DS 2021, LNAI 12986, pp. 334–343, 2021.
https://doi.org/10.1007/978-3-030-88942-5_26

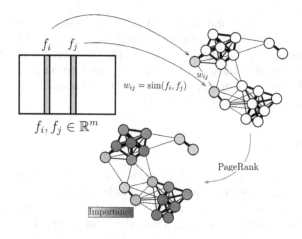

$$w_{ij} = \mathrm{sim}(f_i, f_j)$$

**Fig. 1.** Overview of the FRANe method.

1. We propose FRANe, a fast algorithm for unsupervised feature ranking based on reconstructing attribute networks and subsequent node ranking.
2. We demonstrate the algorithm's state-of-the-art performance on 26 datasets, validating our claims via Bayesian and classical performance analysis.
3. We present an extensive theoretical analysis of the proposed algorithm.
4. We offer an implementation of FRANe as a simple-to-use, freely available Python library, which also includes other baseline approaches.

The remainder of this work is structured as follows. In Sect. 2, we discuss the related work that has led us to propose FRANe. We describe the proposed method in Sect. 3. Next, we discuss the experimental setup (Sect. 4), followed by our results (Sect. 5) and conclusions (Sect. 6).

## 2   Related Work

Unsupervised feature ranking is a relatively new research endeavor. An overview of unsupervised ranking algorithms [13] was published only recently. Some of the currently well-established methods for unsupervised feature ranking include: **Laplace** [8], **MCFS**, and **NDFS**. All of them construct a network of instances by employing an instance similarity measure. Finally, recent work – awarded the best paper award at ECML PKDD 2019 – uses autoencoder [7]: the **AgnoS-S** algorithm gives feature ranking scores as a parameter vector at the early stages of a neural network, which learns to reconstruct the input space and assigns each input variable a score as a side-result.

Apart from the unsupervised feature ranking literature, we also draw inspiration from the literature on *network reconstruction* and its applications in gene expression analysis [5,9,12]. Network reconstruction derives a network from a tabular data set, so that relations between instances (rows) or features (columns) are identified, maintained, and used for a given down-stream task. Once a tabular data set is converted to a network (graph), various centrality measures can be used to determine the centrality of the nodes in the network. Our method uses PageRank centrality measure [11] and its generalization to weighted graphs. While we use it in the unsupervised fashion (somewhat similarly to [16]), it can be also used in the supervised scenario [1].

## 3    Method

Real data often consists of groups of similar features. Intuitively, each such group has a **representative feature** that is most similar to all others. This feature can be expected to predict the values of the other features in the group reasonably well, making the others **redundant**. Thus, the most central features are potentially good candidates for a set of features that a feature selection algorithm would return. When the number of features in the data goes into thousands and more, it is expected that many of them are effectively random noise or completely redundant. The corresponding noisy weights could prevent discovering the wanted centrality values: We therefore introduce a **minimal weight threshold** and only connect the features that are similar enough.

It is not clear in advance which threshold value is the best. Therefore, we try out a set of candidate thresholds, following geometric threshold progression and ranging from the minimal to the maximal edge weight. We calculate the centrality (feature importance) values from the corresponding graphs, and obtain a set of feature rankings. Among those, we choose the one that maximizes the heuristic that is based on the intuition that the feature importance values in a good ranking have a large spread. Rankings obtained with low thresholds are expected to be similar, whereas small increases of high thresholds can cause large changes in the rankings. Sets of candidate thresholds with few low thresholds and many large ones, e.g., the geometric threshold sequence, are expected to give good results.

### 3.1    Algorithm

Let $X = [x_{i,j}]_{i,j} \in \mathbb{R}^{m \times n}$ be a data set, where $m$ is the number of examples and $n$ is the number of features. The $i$-th example (row in the matrix $X$), $1 \leq i \leq m$, is given as $\boldsymbol{x}_i = [x_{i,1}, \dots, x_{i,n}]$. The $j$-th feature, $1 \leq j \leq n$, is given as a feature vector (column in the matrix $X$) $\boldsymbol{f}_j = [x_{1,j}, \dots, x_{m,j}]^T$.

The computation of FRANe is given in Algorithm 1. At input, it takes the (training) data $X$, a minimal edge threshold and the number of iterations $I$. First, it computes the feature similarly matrix $W = [w_{j,k}] \in \mathbb{R}^{n \times n}$. It then computes the geometric sequence $\boldsymbol{T}$ of (edge-weight) thresholds as follows. First, we define the set of similarities between different features $W' = \{w_{j,k} \mid j \neq k\}$, together with $M' = \max(W')$ and $m' = \min(W')$. Then, the dissimilarity values $D = \{M' - w \mid w \in W' \wedge w < M'\}$ are computed. Finally, the thresholds $t_i \in \boldsymbol{T}$

are defined as $T = [t_1, \ldots, t_I]$ where

$$t_i = M' - \min(D) \cdot \left( \frac{\max(D)}{\min(D)} \right)^{(i-1)/(I-1)} \tag{1}$$

The temporary resort to dissimilarities is necessary, because we want to analyze the region of larger similarities more thoroughly. For every threshold $t_i \in T$, we build a weighted graph $G(t_i)$ with $n$ vertices that correspond to features. An edge with the weight $w_{j,k}$ between $f_j$ and $f_k$ exists in $G(t_i)$, if $w_{j,k} \geq t_i$. To avoid too sparse graphs, we consider only those, for which the average degree $\bar{e} = |\{w_{j,k}|j < k \wedge w_{j,k} \geq t_i\}|/n$ exceeds $\bar{e}_{\min} = 1$.

We run a PageRank on the graph $G(t_i)$, which returns a possible ranking $r(t_i) = [\mathrm{PR}(f_1), \ldots, \mathrm{PR}(f_n)]$, where $\mathrm{PR}(f_j)$ is the PageRank importance $\mathrm{PR}(j)$ of the node of feature $f_j$ in $G(t_i)$, as defined in [1,11]. After iterating through all thresholds, calculating the rankings $r(t_i)$ for each $t_i \in T$, we pick as output the ranking with the highest value of the ranking quality heuristic RQH, where

$$\mathrm{RQH}(r) = \frac{\text{second largest score in } r}{\text{second smallest score in } r}. \tag{2}$$

The second largest and smallest scores are taken for stability reasons as the medians of the three largest and smallest scores, respectively.

---

**Algorithm 1:** FRANe($X$, $\bar{e}_{\min}$, $I$)

| | |
|---|---|
| 1 $W = $ compute $[w_{j,k}]_{j,k=1}^n = [PearsonCorr(f_j, f_k) + 1]_{j,k=1}^n$ | // $w_{jk} \geq 0$ |
| 2 $S = []$ | // candidate rankings |
| 3 $T = $ list of $I$ thresholds $t_i$ | // Eq. (1) |
| 4 **for** $t_i \in T$ **do** | |
| 5     $\bar{e} = |\{w_{j,k}|j, k \wedge w_{j,k} \geq t_i\}|/n$ | // Avoid sparse graphs |
| 6     **if** $\bar{e} \geq \bar{e}_{min}$ **then** | |
| 7        $r = $ PageRank($G(t_i)$) | |
| 8        add $r$ to $S$ | |
| 9 **return** $\mathrm{argmax}_{r \in S} \mathrm{RQH}(r)$ | // Eq. (2) |

---

The first step of the algorithm requires the computation of pairwise similarities, yielding time complexity of $\mathcal{O}(mn^2)$. Then, all the graphs $G(t_i)$ can be constructed in the total time of $\mathcal{O}(n^2)$, if we start with a fully connected graph and then incrementally remove the edges with the weights on the intervals $[t_{i-1}, t_i)$. Using the power method for PageRank and assuming that the number of iterations is upper-bounded with some constant [11], computing PageRank takes $\mathcal{O}(n^2)$ steps. Thus, the total number of steps in the algorithm is $\mathcal{O}(m \cdot n^2 + I \cdot n^2)$. Note that the most time-consuming step (similarity computation) can be easily parallelized, and that computing PageRank demands only vectorizable matrix-vector multiplication and vector-vector addition.

## 4    Experimental Setup

In this section, we describe the experimental procedure that we employ to investigate the following questions: i) How does FRANe compare to state-of-the-art methods for unsupervised feature ranking, and ii) What is the influence of the different parameters or FRANe on its performance?

We first give a brief description of the data sets used, continue with the evaluation procedure and finish with the parametrization of the methods. Note that the code that allows for **replicating our experiments** (including the computation of training and testing splits) is freely available at https://github. com/FRANe-team/FRANe-dev.

We obtained the data from the Scikit-feature repository [10]. We wanted to use all the datasets, but had to exclude three data sets from the study (orlraws10P, lung-small and warpAR10P) to meet the independence assumptions of the statistical tests. Table 1 gives a more detailed description of the data, including their domains. When evaluating the feature ranking algorithms, we follow the approach of [7]. Here, an algorithm is evaluated via 10-fold cross-validation. For a given partition of a data set into test part (one of the folds) and train part (the remaining 9 folds), feature ranking is computed on the train part. Then, the $n'$ top-ranked features are selected and the 5 nearest neighbor (5NN) model that uses only these features for predicting the values of all the features is trained (on the train part of the data). Finally, the performance of the feature ranking algorithm is measured in terms of the predictive performance of the 5NN model on the test set. As evaluation measure, we use the average relative mean absolute error $\overline{\text{RMAE}} = \frac{1}{n} \sum_{i=1}^{n} \frac{1}{m_{\text{TEST}}} \sum_{j=1}^{m_{\text{TEST}}} \frac{|\hat{x}_{ij} - x_{ij}|}{\sigma(f_i)}$, where $m_{\text{TEST}}$ is the number of examples in the test set, $\hat{x}_{ij}$ is the 5NN's prediction for $x_{ij}$, and $\sigma(f_i) = \sqrt{\text{Var}(f_i)}$ is the standard deviation of the feature $f_i$. A low

**Table 1.** Number of features $(n)$, examples $(m)$ and the domain of the used benchmarks.

| | $n$ | $m$ | Domain | | $n$ | $m$ | Domain |
|---|---|---|---|---|---|---|---|
| gli-85 | 22283 | 85 | biology | glioma | 4434 | 50 | biology |
| smk-can-187 | 19993 | 187 | biology | relathe | 4322 | 1427 | text data |
| cll-sub-111 | 11340 | 111 | biology | lymphoma | 4026 | 96 | biology |
| arcene | 10000 | 200 | mass spectrometry | lung | 3312 | 203 | biology |
| pixraw10p | 10000 | 100 | face image | pcmac | 3289 | 1943 | text data |
| nci9 | 9712 | 60 | biology | warppie10p | 2420 | 210 | face image |
| carcinom | 9182 | 174 | biology | colon | 2000 | 62 | biology |
| allaml | 7129 | 72 | biology | coil20 | 1024 | 1440 | face image |
| leukemia | 7070 | 72 | biology | orl | 1024 | 400 | face image |
| prostate-ge | 5966 | 102 | biology | yale | 1024 | 165 | face image |
| tox-171 | 5748 | 171 | biology | isolet | 617 | 1560 | speech recognition |
| gisette | 5000 | 7000 | digit recognition | madelon | 500 | 2600 | artificial |
| baseshock | 4862 | 1993 | text data | usps | 256 | 9298 | drawings |

value of $\overline{\text{RMAE}}$ means that the subset of $n'$ chosen features can well reconstruct all the feature values.

The obtained $\overline{\text{RMAE}}$ values are averaged over the 10 folds. To see how the predictive performance of 5NN changes as more and more top-ranked features are considered, one can build a series of 5NN models that use $n' \in \{1, 2, \ldots, 2^k\} \cup \{n\}$, where $2^k \leq n < 2^{k+1}$ features, as shown in Fig. 2. This may be more informative, but is harder to analyze when comparing different algorithms through statistical tests. For such comparisons, performance at $n' = 16$ is chosen. The hierarchical Bayesian t-test considered in this work is discussed in more detail in [2]. The test approximates the posterior probability of the difference in performance between a pair of classifiers. The posterior plot can be visualized as a simplex, where each point represents a sample from the posterior distribution. By counting such samples in different parts of the simplex, the probability of one classifier outperforming the other is estimated.

The number of iterations in FRANe was set to $I = 100$ and the threshold for the average number of edges was set to $\bar{e}_{\min} = 1$. For the decay factor $\delta$ in PageRank, the recommended value of $\delta = 0.85$ was used. For other algorithms, we used the recommended parameter values. Additionally, the number of clusters for the methods MCFS and NDFST was set to the number of classes in the datasets at hand. This was possible since we used classification datasets from the Scikit-feature repository. The classes were otherwise ignored.

## 5    Results

In this section, we first report the results of the comparison between FRANe and its competitors. We then focus on different parts of FRANe and consider alternative design choices.

The $\overline{\text{RMAE}}$ values for the different feature ranking methods, i.e., the corresponding 5NN models, are shown in Table 2. We can see that FRANe outperforms its competitors. First of all, it has the best average rank (1.88) among the considered algorithms. The second best algorithm (in terms of the average rank) is Laplace with an average rank of 2.54. The difference between FRANe and the other algorithms is even more visible when we compare the numbers of wins: FRANe is the best performing algorithm in 12 cases (46% win rate). The second highest number of wins (5) is achieved by NDFS.

To also show some statistical evidence for the quality of the FRANe rankings, we employ the Bayesian hierarchical t-test [2], since it directly answers which of the two compared algorithms is better. The other popular option – frequentist non-parametric tests such as Friedman and Bonferroni-Dunn [6] – allow for comparison of more than one algorithm, but these tests are typically too weak (as follows from their definitions [6]), and are harder to interpret.

The Bayesian comparison indicates that FRANe dominates its closest competitor (Laplace), in 26% of the cases, whereas the Laplace method is better in only 2% of the cases. In the other cases, the difference in performance is smaller than 0.001 and is considered practically insignificant. This is consistent with the

**Table 2.** The performance (measured in terms of $\overline{\text{RMAE}}$) of 5NN models that use the $n' = 16$ top-ranked features from a given feature ranking. The last two rows of the table additionally give the average rank of each algorithm and its number of wins, i.e., the number of times it is ranked first. The best result in each row is shown in bold.

| | FRANe | Laplace | NDFS | Agnos-S | MCFS | SPEC |
|---|---|---|---|---|---|---|
| gli-85 | 0.745 | **0.736** | 0.775 | 0.774 | 0.747 | 0.797 |
| smk-can-187 | 0.610 | 0.612 | 0.62 | 0.656 | 0.626 | **0.597** |
| cll-sub-111 | **0.716** | 0.738 | 0.736 | 0.763 | 0.77 | 0.777 |
| arcene | 0.759 | **0.457** | **0.457** | 0.734 | **0.457** | 0.733 |
| pixraw10p | **0.348** | 0.412 | 0.412 | 0.352 | 0.412 | 0.377 |
| nci9 | **0.763** | 0.771 | 0.771 | 0.839 | 0.771 | 0.807 |
| carcinom | 0.719 | 0.739 | 0.751 | 0.743 | **0.717** | 0.743 |
| allaml | **0.711** | 0.726 | 0.747 | 0.775 | 0.744 | 0.749 |
| leukemia | **0.824** | 0.833 | 0.833 | 0.857 | 0.833 | 0.836 |
| prostate-ge | 0.485 | 0.503 | **0.482** | 0.552 | 0.509 | 0.649 |
| tox-171 | **0.725** | 0.77 | 0.785 | 0.734 | 0.776 | 0.781 |
| gisette | **0.440** | 0.481 | 0.481 | 0.509 | 0.481 | 0.533 |
| baseshock | 0.174 | 0.188 | **0.163** | 0.182 | 0.191 | 0.197 |
| glioma | **0.609** | 0.643 | 0.636 | 0.716 | 0.615 | 0.685 |
| relathe | 0.182 | **0.174** | 0.183 | 0.284 | 0.187 | 0.218 |
| lymphoma | **0.774** | 0.873 | 0.873 | 0.804 | 0.873 | 0.873 |
| lung | **0.700** | 0.708 | 0.734 | 0.749 | 0.701 | 0.780 |
| pcmac | 0.156 | **0.147** | 0.160 | 0.168 | **0.147** | 0.163 |
| warppie10p | 0.370 | 0.526 | 0.526 | **0.316** | 0.526 | 0.526 |
| colon | **0.652** | 0.661 | 0.661 | 0.666 | 0.661 | 0.661 |
| coil20 | 0.234 | 0.364 | **0.205** | 0.407 | 0.786 | 0.528 |
| orl | 0.572 | 0.703 | 0.703 | **0.479** | 0.703 | 0.703 |
| yale | 0.608 | 0.749 | 0.749 | **0.572** | 0.749 | 0.749 |
| isolet | 0.567 | 0.548 | 0.562 | **0.523** | 0.619 | 0.643 |
| madelon | **0.853** | 0.856 | 0.856 | 0.86 | 0.856 | 0.856 |
| usps | 0.371 | 0.338 | **0.283** | 0.337 | 0.422 | 0.394 |
| average rank | **1.88** | 2.54 | 2.88 | 4.04 | 3.19 | 4.42 |
| number of wins | **12** | 4 | 5 | 4 | 3 | 1 |

results in Table 2: the overall win-rate of FRANe is notably higher (12 against 4), even though these two algorithms differ by less than one in average rank values. A detailed (and more global) comparison of the rankings (where the number of chosen features varies from 1 to $n$) on Gisette data set is given in Fig. 2. It is clear that the FRANe rankings are the best as its corresponding curve is below the curves of all other rankings.

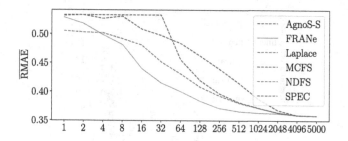

**Fig. 2.** Error curves for the different rankings on Gisette dataset.

## 5.1 Alternative Design Choices

After we have proved that FRANe offers state-of-the-art performance, we now investigate the sensitivity of its performance to varying its key components. Due to space constraints, we only vary the similarity measure used in the computation of the matrix $W$, the threshold progression that defines the list of edge-weight thresholds $T$, and the ranking quality heuristic RQH, while the node-centrality measure is left fixed (PageRank), and left for further work. We first give a brief description of the considered threshold progressions and similarity measures between *different* features $W' = \{w_{j,k} \mid j \neq k\}$, with $m' = \min W'$, and $M' = \max W'$.

*Similarity Measures.* Let $\boldsymbol{f}_j = [x_{1,j}, \ldots, x_{m,j}], \boldsymbol{f}_k = [x_{1,k}, \ldots, x_{m,k}] \in \mathbb{R}^m$ be two feature vectors. Besides correlation, other similarity measures can be used. They are all based on different distance measures $d(\boldsymbol{f}_j, \boldsymbol{f}_k)$: i) **Canberra** $\left(\sum_{i=1}^{m} \frac{|x_{i,k} - x_{i,j}|}{|x_{i,j}| + |x_{i,k}|}\right)$, ii) **Chebyshev** $\left(\max_{i=1}^{m} |x_{i,k} - x_{i,j}|\right)$, iii) **Manhattan** $\left(\sum_{i=1}^{m} |x_{i,k} - x_{i,j}|\right)$, and iv) **Euclidean** $\left(\left(\sum_{i=1}^{m} |x_{i,k} - x_{i,j}|^2\right)^{1/2}\right)$. The corresponding similarity measures are defined as $\operatorname{sim}(\boldsymbol{f}_j \boldsymbol{f}_k) = M' - d(\boldsymbol{f}_j \boldsymbol{f}_k)$.

*Threshold Functions.* The definition of the thresholds $t_i$ from Eq. (1) originally follows the **geometric** progression. The alternatives are: i) **Linear**$(m', M')$, ii) **Linear**$(\operatorname{mean}(W'), M')$, iii) **Linear**$(\operatorname{median}(W'), M')$, and iv) **Quantile**, where $t_i = i$-th $I$-quantile of $W'$'s for the latter, and $t_i = \frac{b-a}{I-1}(i-1) + a$ for **Linear**$(a, b)$. The motivation for using linear progression that starts at the mean (or its more stable analogue the median) of the $W'$ values is that, intuitively, larger thresholds are more interesting to analyze, since the corresponding graphs are sparser.

The results (see Fig. 3) show that FRANe is quite robust with respect to the chosen threshold progression and to the chosen similarity measure. Except for the correlation similarity (works best for 10/26 data sets), and the geometric threshold progression (works best in 9/26 cases), all the similarity measures and threshold progressions perform approximately equally well. Still, no fixed (progression, similarity) pair has more than 3 wins. The detailed results are available at https://github.com/FRANe-team/FRANe). They also include the experiments with RQH, where we show that RQH outperforms random search (in 22/26 cases), which is often considered a strong baseline in optimization [3].

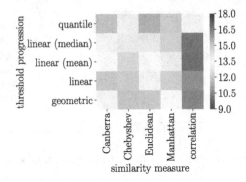

**Fig. 3.** Average ranks (over datasets) of different combinations similarity metric - threshold progression. The legend denotes the average rank of a given metric-progression combination (the lower, the better).

## 6  Conclusion

In this work we have presented FRANe, an algorithm for network-based unsupervised feature ranking. In contrast to existing approaches, FRANe attempts to *reconstruct* a representative network of *features*. By ranking *nodes* in this network via the efficient PageRank approach, we achieve state-of-the-art results for the task of unsupervised feature ranking.

The results indicate that the proposed unsupervised ranking algorithm is indeed a strong competitor to the existing approaches. Theoretical analysis indicates the $\mathcal{O}(n^2)$ complexity of the distance computation as one of the main bottlenecks. The current implementation of FRANe, however, exploits highly optimized compiled routines and scales seamlessly for each of the considered data sets. An extension which would reduce the quadratic complexity could include random subspace sampling (where the probability of choosing a feature depends on its variance).

The proposed methodology is suitable from the interpretability point of view, as the key nodes (features) and their, e.g., correlation-based neighborhoods are easily inspected. This can potentially offer novel insights into key parts of the feature space governing a given data set's structure.

Given that the main spatial bottleneck is related directly to computation of PageRank scores (maintaining the graph in the memory), we believe that an option for further scalability could potentially include distributed storage-based matrix operations [4,15], which would facilitate ranking of attributes when considering very large data sets.

As further work, we believe that distances between features could be also computed in latent space, where embeddings of features would be first obtained (via the transposed feature matrix), potentially speeding up the correlation computation, as well as providing more robust rankings. Furthermore, the body of work related to *metric learning* could similarly prove useful when determining the most suitable similarity score.

# References

1. Backstrom, L., Leskovec, J.: Supervised random walks: predicting and recommending links in social networks. In: Proceedings of the Fourth ACM International Conference on Web Search and Data Mining, WSDM 2011, pp. 635–644. Association for Computing Machinery, New York (2011). https://doi.org/10.1145/1935826.1935914
2. Benavoli, A., Corani, G., Demšar, J., Zaffalon, M.: Time for a change: a tutorial for comparing multiple classifiers through Bayesian analysis. J. Mach. Learn. Res. **18**(1), 2653–2688 (2017)
3. Bergstra, J., Bengio, Y.: Random search for hyper-parameter optimization. J. Mach. Learn. Res. **13**, 281–305 (2012)
4. Bin Abdullah, I.: Incremental PageRank for Twitter data using hadoop. Master's thesis, School of Informatics, University of Edinburgh, Scotland (2010)
5. Chiquet, J., Robin, S., Mariadassou, M.: Variational inference for sparse network reconstruction from count data. In: International Conference on Machine Learning, pp. 1162–1171. PMLR (2019)
6. Demšar, J.: Statistical comparisons of classifiers over multiple data sets. J. Mach. Learn. Res. **7**, 1–30 (2006)
7. Doquet, G., Sebag, M.: Agnostic feature selection. In: Brefeld, U., Fromont, E., Hotho, A., Knobbe, A., Maathuis, M., Robardet, C. (eds.) ECML PKDD 2019. LNCS (LNAI), vol. 11906, pp. 343–358. Springer, Cham (2020). https://doi.org/10.1007/978-3-030-46150-8_21
8. He, X., Cai, D., Niyogi, P.: Laplacian score for feature selection. In: Proceedings of the 18th International Conference on Neural Information Processing Systems, NIPS 2005, pp. 507–514. MIT Press, Cambridge (2005)
9. Langfelder, P., Horvath, S.: WGCNA: an R package for weighted correlation network analysis. BMC Bioinform. **9**(1), 559 (2008)
10. Li, J., et al.: Feature selection: a data perspective. ACM Comput. Surv. (CSUR) **50**(6), 94 (2018)
11. Page, L., Brin, S., Motwani, R., Winograd, T.: The PageRank citation ranking: bringing order to the web. Technical report 1999-66, Stanford InfoLab (November 1999)
12. Sanguinetti, G., et al.: Gene regulatory network inference: an introductory survey. In: Sanguinetti, G., Huynh-Thu, V. (eds.) Gene Regulatory Networks, pp. 1–23. Springer, New York (2019). https://doi.org/10.1007/978-1-4939-8882-2_1
13. Solorio-Fernández, S., Carrasco-Ochoa, J.A., Martínez-Trinidad, J.F.: A review of unsupervised feature selection methods. Artif. Intell. Rev. **53**(2), 907–948 (2019). https://doi.org/10.1007/s10462-019-09682-y
14. Stańczyk, U., Jain, L.C. (eds.): Feature Selection for Data and Pattern Recognition. SCI, vol. 584. Springer, Heidelberg (2015). https://doi.org/10.1007/978-3-662-45620-0
15. Wagle, N., Jasani, S., Gawand, S., Tilekar, S., Patil, P.: Twitter UserRank using hadoop MapReduce. In: Proceedings of the ACM Symposium on Women in Research 2016, WIR 2016, pp. 150–153, Association for Computing Machinery, New York (2016). https://doi.org/10.1145/2909067.2909095
16. Zhu, Z., Peng, Q., Guan, X.: Personalized PageRank based feature selection for high-dimension data. In: 2019 11th International Conference on Knowledge and Systems Engineering (KSE), pp. 1–6 (2019)

# Responsible Artificial Intelligence

# Deriving a Single Interpretable Model by Merging Tree-Based Classifiers

Valerio Bonsignori, Riccardo Guidotti, and Anna Monreale$^{(\boxtimes)}$

University of Pisa, Pisa, Italy
v.bonsignori@studenti.unipi.it,
{riccardo.guidotti,anna.monreale}@unipi.it

**Abstract.** Decision tree classifiers have been proved to be among the most interpretable models due to their intuitive structure that illustrates decision processes in form of logical rules. Unfortunately, more complex tree-based classifiers such as oblique trees and random forests overcome the accuracy of decision trees at the cost of becoming non interpretable. In this paper, we propose a method that takes as input any tree-based classifier and returns a single decision tree able to approximate its behavior. Our proposal merges tree-based classifiers by an intensional and extensional approach and applies a post-hoc explanation strategy. Our experiments shows that the retrieved single decision tree is at least as accurate as the original tree-based model, faithful, and more interpretable.

**Keywords:** Interpretable machine learning · Decision tree · Oblique tree · Model transparency · Merging decision trees

## 1 Introduction

Decision tree (DT) classifiers are very popular models still widely adopted in various business domains because their tree-like representation of knowledge is intuitive and because generally makes the decision logic employed interpretable by humans. The drawback of DTs is that their greedy training procedure returns models which are not remarkably accurate, especially for complex classification problems. To address this issue, DTs have been "empowered" either by using ensembles such for Random Forests [8] or by adopting multivariate and nonlinear splitting conditions such as in Oblique Trees [18]. Such models can reach higher levels of accuracy than regular DTs. Unfortunately, the high accuracy of these complex tree-based classifiers is paid by giving up interpretability.

In the literature, we can find two research lines to deal with the lack of interpretability of these complex tree-based classifiers. The first one relates to *tree merging procedures* [1,5,7] and the idea is to merge a set of DTs into a single one "summarizing" their behavior. Strategies for merging trees are different: joining DTs learned in parallel from disjoint subsets of data [7]; inducing a DT from the intersection of decision tables, each one representing a tree [1]; applying a

© Springer Nature Switzerland AG 2021
C. Soares and L. Torgo (Eds.): DS 2021, LNAI 12986, pp. 347–357, 2021.
https://doi.org/10.1007/978-3-030-88942-5_27

recursive *lossless* merging procedure that makes the order of the merging not relevant [5]. The second research line is related to *eXplainable Artificial Intelligence* (XAI) approaches [6]. Starting from [2], in the literature we can find a set of works aiming at approximating the behavior of a classifier with a single DT for explaining the classification reasoning. To reach this goal different strategies have been proposed: inducing a DT from a set of selected "prototypes" [10]; using genetic programming to evolve DTs to mimic the behavior of a neural network [9]; building several ensembles on synthetically augmented data and then, learning a single DT on the enriched data [3]; constructing a DT by using the set of rule conjunctions that represent the original decision forest [13]; interpreting tree ensembles by finding tree prototypes in the tree space[16].

In this paper we combine these two research lines. We propose a Single-tree Approximation MEthod (SAME) that exploits a procedure for merging decision trees, a post-hoc explanation strategy, and a combination of them to turn any tree-based classifier into a single and interpretable DT.

The implementation of SAME required to adapt existing procedures for merging traditional decision trees to oblique trees by moving from an intensional approach to an extensional one. Our experiments on eight tabular datasets show that SAME is efficient and that the retrieved single decision tree is at least as accurate as the original non interpretable tree-based model.

## 2 Setting the Stage

We address the *single tree approximation of tree-based black box classifiers* [2,6]. Consider a classification dataset $X, Y$ consisting of a set $X = \{x_1, x_2, \ldots, x_n\}$ of instances with $l$ labels (or classes) assigned to an instance in the vector $Y \in \mathbb{N}^n$. An instance of $x \in \mathbb{R}^m$ consist in a set of $m$ pairs of attribute-value $(a_i, v_i)$, where $a_i$ and $v_i$ is a value from the domain of $a_i$. We define a classifier as a function $f : \mathcal{X}^{(m)} \to \mathcal{Y}$ which maps data instances $x$ from a feature space $\mathcal{X}^{(m)}$ with $m$ input features to a decision $y$ in a label space $\mathcal{Y}$, where $y$ can take $l$ different labels. We write $f(x) = y$ to denote the decision $y$ given by $f$ on $x$. We assume that any classifier can be queried at will.

Given a not interpretable tree-based classifier $b$, such as Random Forests [8] or Oblique Trees [18], our aim is to define a function taking as input $b$, $X$, and $Y$ and returns a single DT classifier $d$ which should guarantee the following properties. First, $d$ must be able to mime the behavior of $b$, i.e., $d(x) = b(x)$ for as many instances $x \in X$ as possible. Second, the accuracy of $d$ on unseen instances should be comparable with the accuracy of $b$. Third, $d$ should not be a complex and deep tree to guarantee high levels of interpretability. The single decision tree $d$ is intrinsically transparent because it is humanly possible to understand the reasons for the decision process of every instance $d(x) = y$.

In the following we summarize some key concepts important for our proposal.

**Merging Decision Trees.** Merging DT approaches are accurately described in [14]. Four phases are identified to merge a set of trees $T_1, T_2, \ldots, T_k$ trained on various subsets of a given dataset into a unique decision tree $T$.

In the *first phase*, a decision tree $T_i$ is *transformed* into a set of rules or rule tables (also named decision regions or decision tables[1]). In the *second phase*, the decision regions of two models $T_1, T_2$ are *merged* using a specific approach. The most intuitive idea to merge two rule tables is to compute the *intersection* of all the combinations of the rules from each region and use the results as merged table model. If the regions of the rules that are being intersected are dis-joined, the resulting rule will be empty. The intersection of two overlapping regions is added to the final table model. The class label associated with it is obvious when the two initial regions have the same outcome, otherwise it is necessary to solve the class conflict by employing a specific strategy such as using the class of the rule the highest confidence or probability [1]; or associating to each region a weight and selecting the class with highest weight [15]. The approach presented in [5] allows for simultaneously merging the decision regions of every tree. It uses the notion of *condition tree*. Given a tree $T$ and a condition $C$, let $S_j$ denote the condition set of node $j$ in $T$, which is composed of conditions from root to node $j$, then a condition tree $T^{(C)}$ is composed of those nodes in $T$ such that all the conditions in $S_j$ satisfies $C$. Hence, if an inner node in $T$ is not included in $T^{(C)}$, then all its branches are not included in $T^{(C)}$. Once that two models are merged, the *third phase*, named "pruning", tries to reduce the number of regions. In [1] the regions with the lowest relative number of training samples are filtered out, while in [7] redundant rules are removed during the resolution of class conflicts. Another strategy joins adjacent regions with the same predicted class [1,15]. In [5] the final tree $T$ is pruned by removing inner nodes having as leaves the same class. Finally, the *fourth phase* consists in *growing the final DT* from the decision regions. In [1,15] the final DT is obtained by using the same procedure used to create the initial trees on the values in the regions in the final decision table. In [5] the final DT is directly obtained from the merging procedure.

We adopt the recursive merging procedure described in [5] because *(i)* it is more efficient and requires less memory than others, *(ii)* it does not require to re-train a DT at the end of the computation, *(iii)* it produces a DT with multi-way splits that is theoretically less deep than a binary DT, *(iv)* the merging method is *lossless* as it maintains for every instance the class label assigned by the tree ensembles with the same majority voting.

**Impact of Attribute Types on Tree-based Classifiers.** The aforementioned procedure for merging DTs suffers in presence of many attributes, and also in presence of large (potentially infinite) domains for each attribute. Indeed, in [5] is shown that the size of $T$ merged from the trees learned from data with categorical attributes are much smaller than the trees learned from data with numerical attributes. Therefore, in [5] numerical attributes are discretized using the *Recursive Minimal Entropy Partitioning* (RMEP) method described in [4]. RMEP recursively divides the numerical values minimizing the entropy of the target class. The obtained splits are used to define regions represented by a single representative value. In [5] Fan et al. show that there is a negligible effect on

---

[1] We use $T_i$ for DT, rule tables, decision tables, and decision regions.

the classification accuracy when using discretization w.r.t. not using it. Finally, we turn categorical attributes into numbers through *target encoding* [11].

**Post-hoc Explanation Strategy.** Research on XAI has flourished over the last few years [6,12]. Explanation methods can be categorized as: *intrinsic* or *post-hoc*, depending on whether the machine learning approach is transparent or if the explanation is retrieved by querying the model after the training; and *local* or *global*, depending on whether the explainer reveals the reasons for the prediction of a specific instance, or the overall logic of obscure model. We mention this categorization because in our work we rely on *global post-hoc* explanation. Thus, given a black box classifier $b$ trained on a dataset $X, Y$, a global post-hoc explainer $f$ applied on $b$ and $X$ aims at finding an interpretable classifier $c$, i.e., $c = f(b, X)$, such that the behavior of $c$ on $X$ is adherent with the behavior of $b$ on $X$, i.e., $b(X) \sim c(X)$. For instance, in [2] a particular DT $c$ is trained on $X, \hat{Y} = b(X)$ and the global interpretable model $c$ is returned as explanation.

## 3   Single-Tree Approximation Method

Our proposal to tackle the problem formulated in Sect. 2 consists of reducing any tree-based classifier to a single tree approximating its behavior. We name it SAME, standing for Single-tree Approximation MEthod, and we illustrate its pseudo-code in Algorithm 1. The main idea of SAME is to exploit procedures for merging DTs, a post-hoc explanation strategy, and a combination of them to turn any tree-based classifier into a single interpretable DT.

SAME takes as input a known dataset $X$, a tree-based classifier $b$, a flag $\mu$ indicating if oblique trees have to be merged, and a flag $\nu$ indicating if the post-hoc explanation approach is applied separately to each oblique tree of the forest. The algorithm returns a single DT classifier $T$. We assume that $X$ has statistical properties similar to the training set used by $b$. It works in different ways depending on the type of $b$.

- *Case 1.* If $b$ is a single DT it directly returns it (lines 9–10).
- *Case 2.* If $b$ is a forest of DT (lines 11–12), then SAME runs the *forest2single* function (lines 1–4) that exploits the *mergeTrees* procedure described in [5].
- *Case 3.* If $b$ is a single oblique tree, SAME runs the *b2forest* function to derive a random forest from $b$ and then, from the forest it merges the various trees with *forest2single* like in Case 2 (lines 13–15). The *b2forest* function (lines 5–8) classifies $X$ using the single oblique tree and then trains on $X, \hat{Y}$ a random forest classifier, i.e., it approximates the behavior of an oblique tree with a random forest.
- *Case 4.* If $b$ is a forest of oblique trees and $\mu$ is false, SAME applies the same procedure described for *Case 3*, i.e., it runs the *b2forest* function that in this case derives a random forest mimicking the forest of oblique trees and from it merges the various trees with *forest2single* (lines 16–18).
- *Case 5.* If $b$ is a forest of oblique trees and $\mu$ is true, SAME first runs the *oforest2osingle*, that as described in following subsection derives an oblique

---

**Algorithm 1:** SAME

**Input**  : $X$ - known data, $b$ - tree-based classifier, $\mu$ - merge oblique trees flag,
          $\nu$ - disjoint post-hoc explanation flag
**Output:** $T$ - single decision tree

1  **function** forest2single($b$):
2  |   $T = \{T_1, \ldots, T_k\} \leftarrow getTrees(b)$;
3  |   **return** $mergeTrees(T)$;

4  **function** b2forest($b, X$):
5  |   $\hat{Y} \leftarrow b(X)$;
6  |   **return** $trainRandomForest(X, \hat{Y})$;

7  **if** $b$ *is Single Decision Tree* **then**
8  |   $T \leftarrow b$;
9  **else if** $b$ *is Forest of Decision Trees* **then**
10 |   $T \leftarrow forest2single(b)$;
11 **else if** $b$ *is Single Oblique Trees* **then**
12 |   $RF \leftarrow b2forest(b, X)$;
13 |   $T \leftarrow forest2single(RF)$;
14 **else if** $b$ *is Forest Oblique Trees* $\wedge \neg\mu$ **then**
15 |   $RF \leftarrow b2forest(b, X)$;
16 |   $T \leftarrow forest2single(RF)$;
17 **else if** $b$ *is Forest of Oblique Trees* $\wedge \mu \wedge \neg\nu$ **then**
18 |   $OT \leftarrow oforest2osingle(b, X)$;
19 |   $RF \leftarrow b2forest(OT, X)$;
20 |   $T \leftarrow forest2single(RF)$;
21 **else if** $b$ *is Forest Oblique Trees* $\wedge \mu \wedge \nu$ **then**
22 |   $T \leftarrow \emptyset$;
23 |   **for** $OT_i \in b$ **do**
24 |   |   $RF_i \leftarrow b2forest(OT_i, X)$;
25 |   |   $T \leftarrow T \cup getTrees(RF_i)$;
26 |   $T \leftarrow forest2single(T)$;
27 **return** $T$;

---

trees from a forest of oblique trees and, then it turns the oblique tree $OT$ into a single DT as in *Case 3* (lines 19–22).

- *Case 6.* If $b$ is a forest of oblique trees $\mu$ is true and $\nu$ is true, SAME turns each oblique tree of the oblique forest into a forest of traditional DTs repetitively applying *b2forest*. Finally, it runs the *forest2single* on the union of the derived forests of DTs (lines 23–28).

SAME reduces any approximation problem with another one for which a solution is known in a sort of "cascade of approximations" making possible in this way to turn any classifier based on traditional or oblique trees into a single DT. The flags $\mu$ and $\nu$ controls the different type of approximation when the tree-based classifier is a forest of oblique trees: if $\mu$ is false, the post-hoc explanation strategy is directly employed for approximating the oblique forest; when $\mu$ is true and $\nu$ is false, the forest of oblique trees is approximated directly with the function *oforest2osingle* described in the following; when both are true, the post-hoc explanation approach is applied separately for each oblique tree.

**Merging Oblique Trees.** We define the *oforest2osingle* function used to merge a forest of oblique trees into a single oblique tree as an extension of the algorithm presented in [5]. The needs of adaptation comes from the higher complexity of the test in the nodes of oblique trees that can take the form of a multivariate test, and each multivariate test constitutes itself a meta-feature. A partition of the space using this higher level test changes the shape of the regions to be merged by the merging tree algorithm [5]. Thus, we define a different construction of the condition tree, and a more complex procedure for selecting the features for the final merge. We employ an available dataset $X$ to model the relationship between two conditions exploiting the records in $X$ satisfying those conditions. In other words, we turn the *relationship between two conditions* definition described in [5] from intentional to extensional. In [5] the relationship between two conditions is formally defined as:

**Definition 1 (Relationships of Two Conditions).** *Given two conditions $C_1$, $C_2$, where $C_1$ is a condition $s_i \in I_1$, and $C_2$ is a condition $s_j \in I_2$, with $s_i, s_j$ being a split attribute, and $I_1, I_2$ being real value intervals. If $i = j$ and $I_1 \cap I_2 = \emptyset$, then $C_1 \cap C_2 = \emptyset$, in all the other cases $C_1 \cap C_2 \neq \emptyset$.*

That is, two conditions $C_1$ and $C_2$ have a relationship ($C_1 \cap C_2 \neq \emptyset$) if they identify a common region of the data. We define the *data-driven relationship between two multivariate conditions* as follows, exploiting the notion of *coverage* of a multivariate condition defined as the set of records in $X$ satisfying (or covered by) the multivariate condition $MC$, i.e., $cov_X(MC) = \{x_i | \forall x_i \in X \text{ s.t. } MC(x_i)\}$, where $MC(x_i)$ is true if the record $x_i$ satisfies the multivariate condition $MC$.

**Definition 2 (Data-Driven Relationships of Two Multivariate Conditions).** *Given a dataset $X$ and two multivariate conditions $MC_1$, $MC_2$, we have that if $cov_X(MC_1) \cap cov_X(MC_1) = \emptyset$ then $MC_1 \cap MC_2 = \emptyset$, in all the other cases $MC_1 \cap MC_2 \neq \emptyset$.*

$MC$ indicates a multivariate condition of a given oblique tree node, that can also involve a single variable. We define an oblique condition tree as follows:

**Definition 3 (Oblique Condition Tree).** *Given an oblique decision tree $T$, a multivariate condition $MC$, and a dataset $X$, let $S_j$ denote the multivariate condition set of node $j$ in $T$ which is composed of the multivariate conditions from root to node $j$. An oblique condition tree $T^{(MC)}$ is composed of the nodes in the branch $S_j$ satisfying $\{\forall\, MC' \in S_j, MC' \cap MC \neq \emptyset\}$. If an inner node in $T$ is not included in $T^{(MC)}$, then all its branches are not included in $T^{(MC)}$.*

Given an oblique decision tree $T$ and a multivariate condition $MC$, a simple algorithm for computing $T^{(MC)}$ is to traverse $T$ depth-first from the root. For each branch of multivariate condition $MC'$ of inner node $j$, there are two cases: *(i)* if $MC'$ satisfies $MC_1 \cap MC_2 \neq \emptyset$ keep the root of that branch and search that branch recursively; *(ii)* if $MC'$ satisfies $MC_1 \cap MC_2 = \emptyset$ then the whole branch is not included in $T^{(MC)}$. The definition of pruned condition trees is directly applied to pruned oblique decision tree. Indeed, the inner node $j$ is kept in the oblique condition tree if there are records in $X$ satisfying $MC$ in both partitions after the split determined by $MC'$ in node $j$, i.e., $|cov_X(MC \wedge MC')| > 0$ and $|cov_X(MC \wedge \neg MC')| > 0$. If this is not the case and $|cov_X(MC \wedge MC')| = 0$ or $|cov_X(MC \wedge \neg MC')| = 0$, then the oblique condition tree maintains only the sub-branch covering at least one instance. If both $|cov_X(MC \wedge MC')| = 0$ and $|cov_X(MC \wedge \neg MC')| = 0$, then no sub-branches must be added to the tree.

**Table 1.** Datasets details (left). Tree-based classifiers performance (right).

| Dataset | $n$ | $m$ | $l$ | DT | | RF | | OT | | OF | |
|---|---|---|---|---|---|---|---|---|---|---|---|
| | | | | acc | F1 | acc | F1 | acc | F1 | acc | F1 |
| iris | 150 | 4 | 3 | .933 | .933 | .933 | .933 | .933 | .933 | .933 | .933 |
| cancer | 569 | 30 | 2 | .921 | .918 | .930 | .926 | .921 | .916 | .956 | .953 |
| armchair | 1000 | 2 | 3 | .920 | .922 | .902 | .902 | .920 | .922 | .922 | .924 |
| german | 1000 | 19 | 2 | .720 | .678 | .660 | .534 | .735 | .677 | .755 | .704 |
| employee | 1470 | 29 | 2 | .816 | .551 | .854 | .554 | .871 | .676 | .850 | .566 |
| compas | 7214 | 9 | 3 | .628 | .535 | .631 | .538 | .634 | .538 | .636 | .531 |
| fico | 10459 | 23 | 2 | .712 | .710 | .717 | .717 | .707 | .706 | .730 | .728 |
| adult | 32561 | 12 | 2 | .853 | .778 | .854 | .767 | .851 | .772 | .850 | .770 |

**Table 2.** Fidelity in approximating RF, OT, RF. Best values are underlined.

| Dataset | RF | | OT | | OF | | | |
|---|---|---|---|---|---|---|---|---|
| | SAME | PHDT | SAME | PHDT | SAME$_{\neg\mu}$ | SAME$_{\mu\neg\nu}$ | SAME$_{\mu\nu}$ | PHDT |
| iris | 1.00 | 1.00 | .933 | 1.00 | .733 | .333 | 1.00 | 1.00 |
| cancer | .991 | .947 | .860 | .947 | .912 | .912 | .932 | .930 |
| armchair | .892 | 1.00 | .918 | 1.00 | .980 | .838 | .972 | 1.00 |
| german | .975 | .975 | .945 | .925 | .810 | .820 | .785 | .880 |
| employee | 1.00 | .959 | .969 | .956 | .898 | .969 | .963 | .966 |
| compas | .897 | .979 | .880 | .996 | .916 | .859 | .858 | .994 |
| fico | .978 | .962 | .908 | .962 | .911 | .894 | .911 | .900 |
| adult | .995 | .994 | .988 | .951 | .964 | .992 | .970 | .988 |

Therefore, at a high level, the function *oforest2osingle* can be implemented as in [5] but updating the definition of condition tree with the definition of oblique condition tree. However, practically it is worth to mention another important difference from [5]. *Step 1* of the recursive merging procedure described in [5] determines the split attribute to use for the root of $T$ by selecting the *most frequent* split attribute: when dealing with multivariate conditions of oblique trees is not trivial to determine the most frequent attribute. Thus, we defined the following policies: *(i)* Aiming at interpretability, we prefer univariate splits, acting on a unique variable, to multivariate, splits[2]. Among traditional univariate conditions we select the most frequent one. *(ii)* Among multivariate conditions we prefer those leading to the highest information gain during the training of the oblique tree that generated that split. *(iii)* In case of multivariate conditions with the same number of splits and with the same gain, we randomly select one of them.

---

[2] We highlight that also oblique trees can adopt as best split a traditional split.

# 4    Experiments

In this section we show the effectiveness of SAME when approximating different types of tree-based classifiers on various datasets[3].

We experimented SAME on eight datasets[4]. Details are in Table 1 (left). We split each dataset into three partitions: $X_{tr}$ used for training tree-based classifiers (56%), $X_{ap}$ used by SAME for the post-hoc approximation strategies (24%), $X_{ts}$ used to measure the performance of the resultant single trees (20%).

We trained and approximated with a single decision tree the following tree-based classifiers: Decision Tree (DT) and Random Forest (RF) as implemented by *scikit-learn*; Oblique Decision Tree (ODT) and Oblique Forest (OF) as implemented in [17][5]. We select the best parameter setting for DTs and OTs using a randomized search with cross-validation on $X_{tr}$ analyzing different max depths and class weights[6]. For RFs and OFs we used ensembles composed by 20 estimators and with max depth equals to 4. For OTs we adopted the House Holder CART version [18]. Regarding the parameters of SAME, for *Case 3, 4, and 5* we adopt a RF with 20 base estimators and a 20 max depth [4, 5, 6, 7], while for *Case 6* we adopt a RF with 20 base estimators and a 20 max depth [4, 5, 6]. These parameters are the result of an a randomized search to find the best settings[7].

The classification performance are reported in Table 1 (right) in terms of accuracy and F1-score. We immediately notice that the OFs ha generally the best performance among the various tree-based classifiers. However, there is a small but statistically significant discrepancy among the accuracy scores (and F1-score) of the classifiers (non-parametric Friedman test p-value < 0.1).

To the best of our knowledge the problem treated is somewhat novel and in the literature there are not competitors explicitly designed for this task. Concerning post-hoc explanations, in line with Trepan [2], we compare SAME with PHDT, a post-hoc decision tree approximating any tree-based classifier with a DT. When the tree-based classifier is an OF, we adopt the notation $\text{SAME}_{\neg\mu}$, $\text{SAME}_{\mu\neg\nu}$, $\text{SAME}_{\mu\nu}$ to indicate *Cases 4, 5, and 6* (Sect. 3), respectively.

All the tree-based classifiers are trained on $X_{tr}$, SAME and PHDT exploit the $X_{ap}$ partition while the evaluation measures are computed on $X_{ts}$.

**Evaluation Measures.** We evaluate the performances under different perspectives on the partition $X_{ts}$. First, we check to which extent the single tree $T$ is able to accurately mime the behavior of $b$. We define the *fidelity* as $fid(Y_b, Y_T) = eval(Y_b, Y_T)$ where $Y_b = b(X_{ts})$, $Y_T = T(X_{ts})$, and *eval* can be

---

[3] Python code and datasets available at: https://github.com/valevalerio/SAME. Experiments run on Ubuntu 20.04 LTS, 252 GB RAM, 3.30GHz x 36 Intel Core i9.

[4] The datasets are available on SAME Github, on the UCI repository https://archive.ics.uci.edu/ml/index.php, and on Kaggle https://www.kaggle.com/datasets.

[5] scikit-learn: https://scikit-learn.org/, [17] Github repository https://github.com/TorshaMajumder/Ensembles_of_Oblique_Decision_Trees.

[6] max depth $\in \{5, 6, 7, 8, 10, 12, 16, unbounded\}$, class weight $\in \{normal, balanced\}$.

[7] Details of the parameters tested can be found in SAME repository.

the *accuracy* or the *F1-score*. Second, we test if $T$ can replace $b$, i.e., how much is accurate $T$ if compared with the $b$ on unseen instances $X_{ts}$. We define the *accuracy deviation* $\Delta$ as $\Delta = eval(Y, Y_T) - eval(Y, Y_b)$ where $Y$ being the vector of real class for the partition $X_{ts}$, $Y_b = b(X_{ts})$, $Y_T = T(X_{ts})$ and *eval* can be the *accuracy* or the *F1-score*. $\Delta$ is positive if $T$ is better than $b$ on unseen data, it is negative otherwise, it is zero if they have exactly the same performance[8]. Third, we measure characteristics describing a decision tree $T$ such as: *number of leaves, number of nodes, tree depth,* and *average path length*. We aim at obtaining low values since a simple model is generally more interpretable.

**Results.** We present the results obtained by approximating single trees with SAME and PHDT from DT, RF, OT, and OF with the available variants.

Table 2 reports the fidelity using the accuracy as *eval* (similar results are recorded for F1-score). We observe that both SAME and PHDT have good performance in approximating the behavior of the various tree-based classifiers. Indeed, they are even in terms of times which overcomes the other method. For SAME the best approximations are those performed using *Case 2* on the RF.

**Table 3.** Accuracy deviation on test set for RF, OT, OF. Best values are underlined.

| Dataset | RF | | OT | | OF | | | |
|---|---|---|---|---|---|---|---|---|
| | SAME | PHDT | SAME | PHDT | SAME$_{\neg\mu}$ | SAME$_{\mu\neg\nu}$ | SAME$_{\mu\nu}$ | PHDT |
| iris | 0.000 | 0.000 | 0.067 | 0.000 | -0.200 | −0.600 | 0.000 | 0.000 |
| cancer | −0.009 | 0.000 | −0.035 | −0.018 | −0.053 | −0.070 | −0.035 | −0.035 |
| armchair | −0.065 | 0.000 | −0.047 | 0.000 | −0.010 | −0.105 | −0.012 | 0.000 |
| german | −0.015 | −0.015 | −0.025 | −0.035 | −0.050 | −0.060 | −0.025 | -0.030 |
| employee | 0.000 | −0.014 | −0.010 | −0.017 | −0.061 | −0.001 | −0.010 | −0.007 |
| compas | −0.023 | 0.003 | −0.034 | −0.001 | −0.003 | −0.024 | −0.026 | −0.001 |
| fico | −0.002 | −0.003 | 0.006 | 0.001 | −0.022 | −0.009 | −0.021 | −0.016 |
| adult | −0.001 | −0.002 | −0.003 | −0.010 | −0.016 | 0.000 | −0.008 | −0.002 |

Table 3 and Table 4 shows respectively *(i)* the accuracy deviation using the accuracy as *eval* (similar results for F1-score), and *(ii)* the accuracy of the decision trees obtained from the approximation of tree-based models trained on $X_{tr}$ compared with DTs directly trained on $X_{tr}$ and tested on $X_{ts}$. In Table 3 we observe that the deviation accuracy is only limitedly smaller than zero. This indicates that the approximated trees have a predictive power comparable to the original tree-based classifiers. SAME leads to a decision tree which is more accurate than the original model four times more than PHDT does. Table 4 highlights that in five cases out of eight, the decision tree approximated by SAME is a better model than a decision tree directly trained on the training data. Table 5

---

[8] We highlight that even though they can have the same performance there is no guarantee that the mis-classification errors are the same.

**Table 4.** Accuracy on test set for single trees approximating RF, OT, OF compared with the accuracy of the DT. Best values are underlined.

| Dataset | DT | RF | | OT | | OF | | | |
|---|---|---|---|---|---|---|---|---|---|
| | | SAME | PHDT | SAME | PHDT | $\text{SAME}_{\neg\mu}$ | $\text{SAME}_{\mu\neg\nu}$ | $\text{SAME}_{\mu\nu}$ | PHDT |
| iris | .933 | .933 | .933 | <u>1.00</u> | .933 | .733 | .333 | .933 | .933 |
| cancer | <u>.921</u> | .912 | .921 | .895 | .912 | .868 | .851 | .921 | .921 |
| armchair | .920 | .855 | .920 | .855 | .902 | .910 | .815 | .910 | <u>.922</u> |
| german | .720 | .705 | .705 | .635 | .625 | .685 | .675 | <u>.730</u> | .725 |
| employee | .816 | .816 | .802 | .844 | .837 | .810 | <u>.870</u> | .840 | .843 |
| compas | .628 | .605 | .631 | .597 | .630 | .631 | .610 | .610 | <u>.635</u> |
| fico | .712 | .710 | .709 | .723 | .718 | .685 | <u>.798</u> | .709 | .714 |
| adult | .843 | <u>.852</u> | .851 | .851 | .844 | .835 | .851 | .842 | .848 |

reports the tree depth. We omit the other characteristics describing decision trees for space reasons. We observe that in general the trees returned by PHDT are more compact than those returned by SAME. However in both cases they are nearly always deeper than the original DTs.

**Table 5.** Decision trees depth. Bests values are underlined.

| Dataset | DT | RF | | OT | | OF | | | |
|---|---|---|---|---|---|---|---|---|---|
| | | SAME | PHDT | SAME | PHDT | $\text{SAME}_{\neg\mu}$ | $\text{SAME}_{\mu\neg\nu}$ | $\text{SAME}_{\mu\nu}$ | PHDT |
| iris | 3 | 4 | 3 | 5 | <u>3</u> | 4 | <u>1</u> | 4 | 3 |
| cancer | 6 | 19 | 6 | <u>5</u> | 6 | <u>1</u> | 13 | 10 | 4 |
| armchair | 7 | 4 | 6 | <u>3</u> | 6 | <u>4</u> | <u>4</u> | 5 | 5 |
| german | 4 | <u>3</u> | <u>3</u> | 21 | <u>8</u> | 13 | 13 | 14 | <u>7</u> |
| employee | 3 | 28 | <u>5</u> | 9 | <u>7</u> | 4 | 10 | 17 | <u>4</u> |
| compas | 8 | 13 | 10 | 12 | <u>10</u> | <u>5</u> | 10 | 10 | 11 |
| fico | 7 | 25 | <u>12</u> | 16 | 9 | 32 | 24 | 14 | <u>8</u> |
| adult | 8 | 15 | 10 | <u>12</u> | 13 | 13 | 10 | <u>8</u> | 11 |

# 5    Conclusion

We have presented SAME, a single-tree approximation method designed to effectively and efficiently turn every tree-based classifier into a single DT. Experimentation on various datasets reveals that SAME is competitive with baseline approaches or overcomes them. Moreover, the approximated tree can replace the original model as it can have better performance. Possible future research directions are: extending SAME to any type of tree-based and rule-based classifier and using SAME as post-hoc global explanation method for any black box.

**Acknowledgment.** Work partially supported by the European Community H2020 programme under the funding schemes: G.A. 871042 *SoBigData++*, G.A. 761758 *Humane AI*, G.A. 952215 *TAILOR* and the ERC-2018-ADG G.A. 834756 "XAI: Science and technology for the eXplanation of AI decision making".

# References

1. Andrzejak, A., et al.: Interpretable models from distributed data via merging of decision trees. In: CIDM, pp. 1–9. IEEE (2013)
2. Craven, M.W., et al.: Extracting tree-structured representations of trained networks, pp. 24–30 (1995)
3. Domingos, P.M.: Knowledge discovery via multiple models. Intell. Data Anal. **2**(1–4), 187–202 (1998)
4. Dougherty, J., et al.: Supervised and unsupervised discretization of continuous features. In: ICML, pp. 194–202. Morgan Kaufmann (1995)
5. Fan, C., et al.: Classification acceleration via merging decision trees. In: FODS, pp. 13–22. ACM (2020)
6. Guidotti, R., et al.: A survey of methods for explaining black box models. ACM Comput. Surv. **51**(5), 93:1–93:42 (2019)
7. Hall, L.O., et al.: Combining decision trees learned in parallel. In: Working Notes of the KDD-97 Workshop on Distributed Data Mining, pp. 10–15 (1998)
8. Ho, T.K.: The random subspace method for constructing decision forests. IEEE Trans. Pattern Anal. Mach. Intell. **20**(8), 832–844 (1998)
9. Johansson, U., et al.: Evolving decision trees using oracle guides. In: CIDM, pp. 238–244. IEEE (2009)
10. Krishnan, R., et al.: Extracting decision trees from trained neural networks. Pattern Recognit. **32**(12), 1999–2009 (1999)
11. Micci, D.: A preprocessing scheme for high-cardinality categorical attributes in classification and prediction problems. SIGKDD Explor. **3**(1), 27–32 (2001)
12. Miller, T.: Explanation in artificial intelligence: Insights from the social sciences. Artif. Intell. **267**, 1–38 (2019)
13. Sagi, O., Rokach, L.: Explainable decision forest: transforming a decision forest into an interpretable tree. Inf. Fusion **61**, 124–138 (2020)
14. Strecht, P.: A survey of merging decision trees data mining approaches. In Proceedings of10th Doctoral Symposium in Informatics Engineering, pp. 36–47 (2015)
15. Strecht, P., Mendes-Moreira, J., Soares, C.: Merging decision trees: a case study in predicting student performance. In: Luo, X., Yu, J.X., Li, Z. (eds.) ADMA 2014. LNCS (LNAI), vol. 8933, pp. 535–548. Springer, Cham (2014). https://doi.org/10.1007/978-3-319-14717-8_42
16. Tan, S., et al.: Tree space prototypes: another look at making tree ensembles interpretable. In: FODS, pp. 23–34. ACM (2020)
17. Torsha, M.: Ensembles of oblique decision trees (2020)
18. Wickramarachchi, D.C., et al.: HHCART: an oblique decision tree. Comput. Stat. Data Anal. **96**, 12–23 (2016)

# Ensemble of Counterfactual Explainers

Riccardo Guidotti$^{(\boxtimes)}$ and Salvatore Ruggieri

University of Pisa, Pisa, Italy
{Riccardo.Guidotti,Salvatore.Ruggieri}@unipi.it

**Abstract.** In eXplainable Artificial Intelligence (XAI), several counterfactual explainers have been proposed, each focusing on some desirable properties of counterfactual instances: minimality, actionability, stability, diversity, plausibility, discriminative power. We propose an ensemble of counterfactual explainers that boosts weak explainers, which provide only a subset of such properties, to a powerful method covering all of them. The ensemble runs weak explainers on a sample of instances and of features, and it combines their results by exploiting a diversity-driven selection function. The method is model-agnostic and, through a wrapping approach based on autoencoders, it is also data-agnostic.

## 1 Introduction

In eXplainable AI (XAI), several counterfactual explainers have been proposed, each focusing on some desirable properties of counterfactual instances. Consider an instance $x$ for which a black box decision $b(x)$ has to be explained. It should be possible to find various counterfactual instances $c$ (*availability*) which are *valid* (change the decision outcome, i.e., $b(c) \neq b(x)$), *minimal* (the number of features changed in $c$ w.r.t. $x$ should be as small as possible), *actionable* (the feature values in $c$ that differ from $x$ should be controllable) and *plausible* (the feature values in $c$ should be coherent with the reference population). The counterfactuals found should be similar to $x$ (*proximity*), but also different among each other (*diversity*). Also, they should exhibit a *discriminative power* to characterize the black box decision boundary in the feature space close to $x$. Counterfactual explanation methods should return similar counterfactuals for similar instances to explain (*stability*). Finally, they must be fast enough (*efficiency*) to allow for interactive usage.

In the literature, these desiderata for counterfactuals are typically modeled through an optimization problem [12], which, on the negative side, favors only a subset of the properties above. We propose here an *ensemble of counterfactual explainers* (ECE) that, as in the case of ensemble of classifiers, boosts weak explainers to a powerful method covering all of the above desiderata. The ensemble runs *base counterfactual explainers* (BCE) on a sample of instances and of features, and it combines their results by exploiting a diversity-driven selection function. The method is model-agnostic and, through a wrapping approach based on encoder/decoder functions, it is also data-agnostic. We will be able to

© Springer Nature Switzerland AG 2021
C. Soares and L. Torgo (Eds.): DS 2021, LNAI 12986, pp. 358–368, 2021.
https://doi.org/10.1007/978-3-030-88942-5_28

reason uniformly on counterfactuals for tabular data, images, and time series. An extensive experimentation is presented to validate the approach. We compare with state-of-the-art explanation methods on several metrics from the literature.

## 2    Related Work

Research on XAI has flourished over the last few years [5]. Explanation methods can be categorized as: *(i) intrinsic* vs *post-hoc*, depending on whether the AI model is directly interpretable, or if the explanation is computed for a given black box model; *(ii) model-specific* vs *model-agnostic*, depending on whether the approach requires access to the internals of the black box model; *(iii) local* or *global*, depending on whether the explanation regards a specific instance, or the overall logic of the black box. Furthermore, explanation methods can be categorized w.r.t. the type of explanation they return (factual or counterfactual) and w.r.t. the type of data they work with. We restrict to local and post-hoc methods returning counterfactual explanations, which is the focus of our proposal.

A recent survey of counterfactual explainers is [15]. Most of the systems are data-specific and generate synthetic (*exogenous*) counterfactuals. Some approaches search *endogenous* counterfactuals in a given dataset [9] of instances belonging to the reference population. Exogenous counterfactuals may instead break known relations between features, producing unrealistic instances. Early approaches generated exogenous counterfactuals by solving an optimization problem [12]. In our proposal, we do not rely on this family of methods as they are typically computationally expensive. Another family of approaches are closer to instance-based classification, and rely on a distance function among instances [9,10]. E.g., [10] grows a sphere around the instance to explain, stopping at the decision boundary of the black box. They are simple but effective, and the idea will be at the core of our base explainers. Some approaches deal with high dimensionality of data through autoencoders [3], which map instances into a smaller latent feature space. Search for counterfactuals is performed in the latent space, and then instances are decoded back to the original space. We rely on this idea to achieve a data-agnostic approach.

## 3    Problem Setting

A *classifier b* is a function mapping an instance $x$ from a reference population in a feature space to a nominal value $y$ also called class value or decision, i.e., $b(x) = y$. The classifier $b$ is a *black box* when its internals are either unknown to the observer or they are known but uninterpretable by humans. Examples include neural networks, SVMs, ensemble classifiers [5].

A *counterfactual* of $x$ is an instance $c$ for which the decision of the black box differs from the one of $x$, i.e., such that $b(c) \neq b(x)$. A counterfactual is *actionable* if it belongs to the reference population. Since one may not have a complete specification of the reference population, a relaxed definition of actionability is to require the counterfactual to satisfy given constraints on its feature values.

---

**Algorithm 1:** ECE

**Input** : $x$ - instance to explain, $b$ - black box, $X$ - known instances,
   $k$ - number of counterfactuals, $A$ - actionable features, $E$ - base explainers
**Output:** $C$ - $k$-counterfactual set

1 $C \leftarrow \emptyset$;                   // init. result set
2 **for** $f_k \in E$ **do**            // for each base explainer
3   $X' \leftarrow \mathcal{I}(X)$;             // sample instances
4   $A' \leftarrow \mathcal{F}(A)$;             // sample features
5   $C \leftarrow C \cup f_k(x, b, X', A')$;      // call base explainer
6 $C \leftarrow \mathcal{S}(x, C, k)$;       // select top $k$-counterfactuals
7 **return** $C$;

---

We restrict to simple constraints $a_A(c, x)$ that hold iff $c$ and $x$ have the same values over for a set $A$ of *actionable features*. Non-actionable features (such as age, gender, race) cannot be changed when searching for a counterfactual.

A $k$-*counterfactual explainer* is a function $f_k$ returning a set $C = \{c_1, \ldots, c_h\}$ of $h \leq k$ actionable counterfactuals for a given instance of interest $x$, a black box $b$, a set $X$ of known instances from the reference population, and a set $A$ of actionable features, i.e., $f_k(x, b, X, A) = C$. For endogenous approaches, $C \subseteq X$. A counterfactual explainer is model-agnostic (resp., data-agnostic) if the definition of $f_k$ does not depend on the internals of $b$ (resp., on the data type of $x$). We consider the following data types: tabular data, time series and images. For *tabular data*, an instance $x = \{(a_1, v_1), \ldots, (a_m, v_m)\}$ is a tuple of $m$ attribute-value pairs $(a_i, v_i)$, where $a_i$ is a feature (or attribute) and $v_i$ is a value from the domain of $a_i$. For example, $x = \{(age, 22), (sex, male), (income, 800)\}$. The domain of a feature can be continuous (*age, income*), or categorical (*sex*). For (univariate) *time series*, an instance $x = \langle v_1, \ldots, v_m \rangle$ is an ordered sequence of continuous values (e.g., the body temperature registered at hourly rate). For *images*, $x$ is a matrix in $\mathbb{R}^{m \times m}$ representing the intensity of the image pixels.

*Problem Statement.* We consider the problem of designing a $k$-counterfactual explainer satisfying a broad range of properties: availability, validity, actionability, plausibility, similarity, diversity, discriminative power, stability, efficiency.

## 4 Ensemble of Explainers

Our proposal to the stated problem consists of an ensemble of base explainers named ECE (Ensamble of Counterfactual Explainers). Ensemble classifiers boost the performance of weak learner base classifiers by increasing the predictive power, or by reducing bias or variance. Similarly, we aim at improving base $k$-counterfactual explainers by combining them into an ensemble of explainers.

The pseudo-code[1] of ECE is shown in Algorithm 1. It takes as input an instance to explain $x$, the black box to explain $b$, a set of known instances $X$, the number of required counterfactuals $k$, the set of actionable features $A$, a set of base $k$-counterfactual explainers $E$, and it returns (at most) $k$ counterfactuals

---

[1] Implementation and full set of parameters at https://github.com/riccotti/ECE.

$C$. Base explainers are invoked on a sample without replacement $X'$ of instances from $X$ (line 3), and on a random subset $A'$ of the actionable features $A$ (line 4), as in Random Forests. All counterfactuals produced by the base explainers are collected in a set $C$ (line 5), from which $k$ counterfactuals are selected (line 6). Actionability of counterfactuals is guaranteed by the base explainers (or by filtering out non-actionable ones from their output). Diversity is enforced by randomization (instance and feature sampling) as well as by tailored selection strategies. Stability is a result of combining multiple base explainers, analogously to the smaller variance of ensemble classification w.r.t. the base classifiers. Moreover, if all base explainers are model-agnostic, this also holds for ECE.

### 4.1   Base Explainers

All BCE's presented are parametric to a distance function $d()$ over the feature space. In the experiments, we adopt: for tabular data, a mixed distance weighting Euclidean distance for continuous features and the Jaccard dissimilarity for categorical ones; for images and times series, the Euclidean distance.

**Brute Force Explainer** (BCE-B). A brute force approach considers all subsets $\mathcal{A}$ of actionable features $A$ with cardinality at most $n$. Also, for each actionable feature, an equal-width binning into $r$ bins is computed, and for each bin the center value will be used as representative of the bin. The binning scheme considers only the known instances $X$ with black box decision different from $x$. The brute force approach consists of generating all the possible variations of $x$ with respect to any of the subset in $\mathcal{A}$ by replacing an actionable feature value in $x$ with any representative value of a bin of the feature. Variations are ranked according to their distance from $x$. For each such variation $c$, a *refine* procedure implements a bisecting strategy of the features in $c$ which are different from $x$ while maintaining $b(c) \neq b(x)$. The procedure returns either a singleton with a counterfactual or an empty set (in case $b(c) = b(x)$). The aim of *refine* is to improve similarity of the counterfactual with $x$. The procedure stops when $k$ counterfactuals have been found or there is no further candidate. The greater are $n$ and $r$, the larger number of counterfactuals to choose from, but also the higher the computational complexity of the approach, which is $O(\binom{|A|}{n} \cdot n \cdot r)$. BCE-B tackles minimization of changes and similarity, but not diversity.

**Tree-Based Explainer** (BCE-T). This proposal starts from a (surrogate/shadow [7]) decision tree $T$ trained on $X$ to mime the black box behavior. Leaves in $T$ leading to predictions different from $b(x)$ can be exploited for building counterfactuals. Basically, the splits on the path from the root to one such leaf represent conditions satisfied by counterfactuals. To ensure actionability, only splits involving actionable constraints are considered. To tackle minimality, the filtered paths are sorted w.r.t. the number of conditions not already satisfied by $x$. For each such path, we choose one instance $c$ from $X$ reaching the leaf and minimizing distance to $x$. Even though the path has been checked for actionable splits, the instance $c$ may still include changes w.r.t. $x$ that are not actionable. For this, we overwrite non-actionable features. Since not all instances at a leaf have the same class as the one predicted at the leaf, we also have to check for validity before including $c$ in the result set. The search over different paths of the decision tree allows for some diversity in the results, even though this cannot be explicitly controlled for. The computational complexity requires both a decision tree construction and a number of distance calculations.

**Generative Sphere-Based Explainer** (BCE-S). The last base counterfactual explainer relies on a generative approach growing a *sphere* of synthetic instances around $x$ [10]. Instance are generated in all directions of the feature space until the decision boundary of the black box $b$ is crossed and the closest counterfactual to $x$ is retrieved. The sphere radius is initialized to a large value, and then it is decreased until the boundary is crossed. Next, a lower bound radius and an upper bound radius are determined such that the boundary of $b$ crosses the area of the sphere between the lower bound and the upper bound radii. In its original version, the growing spheres algorithm generates instances following a uniform distribution. BCE-S adopts instead a *Gaussian-Matched* generation [1]. To ensure actionability, non-actionable features of generated instances are set as in $x$. Finally, BCE-S selects from the instances in the final ring the ones which are closest to $x$ and are valid. The complexity of the approach depends on the distance of the decision boundary from $x$, which in turn determines the number of iterations needed to compute the final ring.

### 4.2 Counterfactual Selection

The selection function $\mathcal{S}$ at line 5 of Algorithm 1 selects $k$-counterfactuals from those returned by the base explainers. This problem can be formulated as maximizing an objective function over $k$-subsets of valid counterfactuals $C$. We adopt a *density-based* objective function:

$$\underset{S \subseteq C \wedge |S| \leq k}{\arg\max} \quad |\bigcup_{c \in S} knn_C(c)| - \lambda \sum_{c \in S} d(c, x)$$

It aims at maximizing the difference between the size of neighborhood instances of the counterfactuals (a measure of diversity) and the total distance from $x$ (a measure of similarity) regularized by a parameter $\lambda$. $knn_C(c)$ returns the $h$ most similar counterfactuals to $c$ among those in $C$. We adopt the Cost Scaled Greedy (CSG) algorithm [4] for the above maximization problem.

### 4.3 Counterfactuals for Other Data Types

We enable ECE to work on data types other than tabular data by wrapping it around two functions. An *encoder* $\zeta : \mathbb{D} \rightarrow \mathbb{R}^q$ that maps an instance from its actual domain $\mathbb{D}$ to a latent space of continuous features, and a *decoder* $\eta : \mathbb{R}^q \rightarrow \mathbb{D}$ that maps an instance of the latent space back to the actual domain. Using such functions, any explainer $f_k(x, b, X, A)$ can be extended to the domain $\mathbb{D}$ by invoking $\eta(f_k(\zeta(x), b', \zeta(X), A'))$ where the black box in the latent space is $b'(x) = b(\eta(x))$. The definition of the actionable features in the latent space $A'$ depends on the actual encoder and decoder.

Let us consider the image data type (for time series, the reasoning is analogous). A natural instantiation of the wrapping that achieves dimensionality reduction with a controlled loss of information consists in the usage of *autoencoders* (AE) [8]. An AE is a neural network composed by an encoder and a decoder which are trained simultaneously for learning a representation that reduces the dimensionality while minimizing the reconstruction loss. A drawback of this approach is that we cannot easily map actionable feature in the actual domain to features in the latent space (this is a challenging research topic on its own). For this, we set $A'$ to be the whole set of latent features and hence, we are not able to deal with actionability constraints.

**Table 1.** Datasets description and black box accuracy. $n$ is the no. of instances. $m$ is the no. of features. $m_{con}$ and $m_{cat}$ are the no. of continuous and categorical features respectively. $m_{act}$ is the no. of actionable features. $m_{1h}$ is the total no. of features after one-hot encoding. Rightmost columns report classification accuracy: NN stands for DNN for tabular data, and for CNN for images and time series.

| Dataset | | $n$ | $m$ | $m_{con}$ | $m_{cat}$ | $m_{act}$ | $m_{1h}$ | $l$ | RF | NN |
|---------|--------|--------|----------------|-----|-----|-----|-----|----|-----|-----|
| tabular | adult | 32,561 | 12 | 4 | 8 | 5 | 103 | 2 | .85 | .84 |
| | compas | 7,214 | 10 | 7 | 3 | 7 | 17 | 3 | .56 | .61 |
| | fico | 10,459 | 23 | 23 | 0 | 22 | – | 2 | .68 | .67 |
| | german | 1,000 | 20 | 7 | 13 | 13 | 61 | 2 | .76 | .81 |
| img | mnist | 60k | $28 \times 28$ | all | 0 | all | – | 10 | – | .99 |
| | fashion | 60k | $28 \times 28$ | all | 0 | all | – | 10 | – | .97 |
| ts | gunpoint | 250 | 150 | all | 0 | all | – | 2 | – | .72 |
| | power | 1,096 | 24 | all | 0 | all | – | 2 | – | .98 |
| | ecg200 | 200 | 96 | all | 0 | all | – | 2 | – | .76 |

## 5  Experiments

**Experimental Settings.** We consider a few datasets widely adopted as benchmarks in the literature (see Table 1). There are three time series datasets, two image datasets, and four tabular datasets. For each tabular dataset, we have selected the set $A$ of actionable features, as follows. adult: age, education, marital status, relationship, race, sex, native country; compas: age, sex, race; fico: external risk estimate; german: age, people under maintenance, credit history, purpose, sex, housing, foreign worker.

For each dataset, we trained and explained the following black box classifiers: Random Forest (RF) as implemented by *scikit-learn*, and Deep Neural Networks (DNN) implemented by *keras* for tabular datasets, and Convolutional Neural

Networks (CNNs) implemented with *keras* for images and time series. We split tabular datasets into a 70% partition used for the training and 30% used for the test, while image and time series datasets are already released in partitioned files. For each black-box and for each dataset, we performed on the training set a random search with a 5-fold cross-validation for finding the best parameter setting. The classification accuracy on the test set is shown in Table 1 (right).

We compare our proposal against competitors from the state-of-the-art offering a software library that is updated and easy to use. DICE [12] handles categorical features, actionability, and allows for specifying the number $k$ of counterfactuals to return. However, it is not model-agnostic as it only deals with differentiable models such as DNNs. The *FAT* [13] library implements a brute force (BF) counterfactual approach. It handles categorical data but not the number $k$ of desired counterfactuals nor actionability. The *ALIBI* library implements the counterfactual explainers CEM [3,11], CEGP [14] and WACH [16]. All of them are designed to explain DNNs, do not handle categorical features and return a single counterfactual, but it is possible to enforce actionability by specifying the admissible feature ranges. Finally, CEML [2] is a model-agnostic toolbox for computing counterfactuals based on optimization that does not handle categorical features and returns a single counterfactual. We also re-implemented the case-based counterfactual explainer (CBCE) from [9]. For each tool, we use the default settings offered by the library or suggested in the reference paper. For each dataset, we explain 100 instances $x$ from the test set. The set $X$ of known instances in input to the explainers is the training set of the black box. We report aggregated results as means over the 100 instances, datasets and black boxes.

**Evaluation Metrics.** We evaluate the performances of counterfactual explainers under various perspectives [12]. The measures reported in the following are stated for a single instance $x$ to be explained, and considering the returned $k$-counterfactual set $C = f_k(x, b, X, A)$. The metrics are obtained as the mean value of the measures over all $x$'s to explain.

*Size.* The number of counterfactuals $|C|$ can be lower than $k$. We define *size* $= |C|/k$. The higher the better. Recall that by definition of a $k$-counterfactual explainer, any $c \in C$ is valid, i.e., $b(c) \neq b(x)$.

*Actionability.* It accounts for the counterfactuals in $C$ that can be realized: $act = |\{c \in C \mid a_A(c, x)\}|/k$. The higher the better.

*Implausibility.* It accounts for how close are counterfactuals to the reference population. It is the average distance of $c \in C$ from the closest instance in the known set $X$. The lower the better.

$$impl = \frac{1}{|C|} \sum_{c \in C} \min_{x \in X} d(c, x)$$

*Dissimilarity.* It measures the proximity between $x$ and the counterfactuals in $C$. The lower the better. We measure it in two fashions. The first one, named

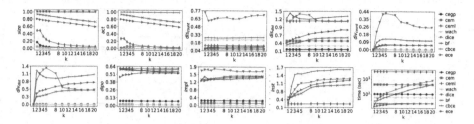

**Fig. 1.** Aggregate metrics on tabular datasets by varying $k$.

$dis_{dist}$, is the average distance between $x$ and the counterfactuals in $C$. The second one, $dis_{count}$, quantifies the average number of features changed between a counterfactual $c$ and $x$. Let $m$ be

$$dis_{dist} = \frac{1}{|C|} \sum_{c \in C} d(x, c) \qquad dis_{count} = \frac{1}{|C|m} \sum_{c \in C} \sum_{i=1}^{m} \mathbb{1}_{c_i \neq x_i}$$

*Diversity.* It accounts for a diverse set of counterfactuals, where different actions can be taken to recourse the decision of the black box. The higher the better. We denote by $div_{dist}$ the average distance between the counterfactuals in $C$, and by $div_{count}$ the average number of different features between the counterfactuals.

$$div_{dist} = \frac{1}{|C|^2} \sum_{c \in C} \sum_{c' \in C} d(c, c') \qquad div_{count} = \frac{1}{|C|^2 m} \sum_{c \in C} \sum_{c' \in C} \sum_{i=1}^{m} \mathbb{1}_{c_i \neq c'_i}$$

*Discriminative Power.* It measures the ability to distinguish through a naive approach between two different classes only using the counterfactuals in $C$. In line with [12], we implement it as follows. The sets $X_= \subset X$ and $X_{\neq} \subset X$ such that $b(X_=) = b(x)$ and $b(X_{\neq}) \neq b(x)$ are selected such that the instances in $X_=, X_{\neq}$ are the $k$ closest to $x$. Then we train a simple 1-Nearest Neighbor (1NN) classifier using $C \cup \{x\}$ as training set, and $d$ as distance function. The choice of 1NN is due to its simplicity and connection to human decision making starting from examples. We classify the instances in $X_= \cup X_{\neq}$ and we use the accuracy of the 1NN as *discriminative power* (*dipo*).

*Instability.* It measures to which extent the counterfactuals $C$ are close to the ones obtained for the closest instance to $x$ in $X$ with the same black box decision. The rationale is that similar instances should obtain similar explanations [6]. The lower the better.

$$inst = \frac{1}{1 + d(x, x')} \frac{1}{|C||C'|} \sum_{c \in C} \sum_{c' \in C'} d(c, c')$$

with $x' = argmin_{x_1 \in X \setminus \{x\}, b(x_1) = b(x)} d(x, x_1)$ and $C' = f_k(x', b, X, A)$.

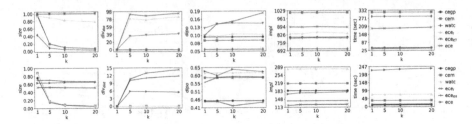

**Fig. 2.** Aggregate metrics on images ($1^{st}$ row) and time series ($2^{nd}$ row) by varying $k$.

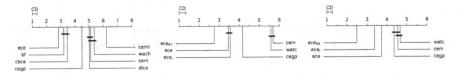

**Fig. 3.** Critical Difference (CD) diagrams for the post-hoc Nemenyi test at 95% confidence level: tabular (left), images (center), and time series (right) datasets.

*Runtime.* It measures the elapsed time required by the explainer to compute the counterfactuals. The lower the better. Experiments were performed on Ubuntu 20.04 LTS, 252 GB RAM, 3.30GHz x 36 Intel Core i9.

In line with [12,16], in the above evaluation measures, we adopt as distance $d$ the following mixed distance:

$$d(a,b) = \frac{1}{m_{con}} \sum_{i \in con} \frac{|a_i - b_i|}{MAD_i} + \frac{1}{m_{cat}} \sum_{i \in cat} \mathbb{1}_{a_i \neq b_i}$$

where *con* (resp., *cat*) is the set of continuous (resp., categorical) feature positions. Such a distance is not necessarily the one used by the compared explainers. In particular, it substantially differs from the one used by ECE.

**Parameter Tuning.** From an experimental analysis (not reported here) of the impact of the components of ECE, we set: for BCE-B, $r = 10$ and $n = 1$; and for ECE, $|E| = 10$ base explainers chosen uniformly random.

**Quantitative Evaluation.** Figure 1 shows the performance of the compared explainers on tabular data when varying $k$. From the first plot, we notice that only ECE, DICE, CBCE and BF are able to return at least 80% of the required counterfactuals. Most of the other methods only return a single one. From the second plot, we conclude that only ECE, BF and CBCE return a notable fraction of actionable counterfactuals (*act*). From the plots on dissimilarity ($dis_{count}$ and $dis_{dist}$) and diversity ($div_{count}$ and $div_{dist}$), it turns out that CBCE (and also DICE) has good values of diversity, but performs poorly w.r.t. dissimilarity. BF wins over ECE w.r.t. the $dis_{dist}$ measure, loses w.r.t. the $div_{dist}$ measure, and is substantially equivalent w.r.t. the other two measures. As for discriminative power *dipo*, ECE performs slightly lower than DICE, CBCE, BF and CEML.

Regarding plausibility (*impl*), ECE is the best performer if we exclude methods that return a single counterfactual (i.e., CEM, CEGP and WACH). Indeed, ECE *impl* is constantly smaller that DICE and BF and in line with CBCE, which is the only endogenous methods compared. Intuitively, counterfactuals returned by ECE resemble instances from the reference population. Concerning instability *inst*, ECE is slightly worse than BF and slightly better than DICE. CEML is the most stable, and CBCE the most unstable. CEM, CEGP and WACH are not shown in the instability plot because, in many cases, they do not return counterfactuals for both of the two similar instances. Finally, all the explainers, with the exception of BF and ECE, require on average a runtime of more than one minute. We summarize the performances of the approaches by the CD diagram in Fig. 3 (left), which shows the mean rank position of each method over all experimental runs (datasets × black boxes × metrics × *k*). Overall, ECE performs better than all competitors, and the difference is statistically significant.

Figure 2 shows the performance on images (first row) and time series (second row) datasets. We consider also the ECE with the identity encoder/decoder (named $ECE_I$), and with the kernel encoder/decoder ($ECE_{K7}$ for kernel of size $7 \times 7$ and $ECE_{K4}$ for kernel of size $4 \times 4$). For images, CEM, CEGP and WACH return only a single counterfactual, while ECE provides more alternatives and with the best diversity. WACH returns the least implausible counterfactuals, the variants of ECE stand in the middle, while CEM returns less realistic counterfactuals. Regarding running time, CEGP is the most efficient together with $ECE_I$ and $ECE_{K4}$. The usage of the autoencoder in ECE increases the runtime. CEM and WACH are the slowest approaches. Similar results are observed for time series, with few differences. The CD diagrams in Fig. 3 (center, right) confirm that ECE and its variants are the best performing methods.

**Acknowledgment.** Work partially supported by the European Community H2020-EU.2.1.1 programme under the G.A. 952215 *Tailor*.

# References

1. Agustsson, E., et al.: Optimal transport maps for distribution preserving operations on latent spaces of generative models. In: ICLR (2019). OpenReview.net
2. Artelt, A.: Ceml: Counterfactuals for explaining machine learning models - a Python toolbox. https://www.github.com/andreArtelt/ceml (2019–2021)
3. Dhurandhar, A., et al.: Explanations based on the missing: towards contrastive explanations with pertinent negatives. In: NeurIPS, pp. 592–603 (2018)
4. Ene, A., Nikolakaki, S.M., Terzi, E.: Team formation: Striking a balance between coverage and cost. arXiv:2002.07782 (2020)
5. Guidotti, R., Monreale, A., Ruggieri, S., Turini, F., Giannotti, F., Pedreschi, D.: A survey of methods for explaining black box models. CSUR **51**(5), 1–42 (2018)
6. Guidotti, R., Ruggieri, S.: On the stability of interpretable models. In: IJCNN, pp. 1–8. IEEE (2019)
7. Guidotti, R., et al.: Factual and counterfactual explanations for black box decision making. IEEE Intell. Syst. **34**(6), 14–23 (2019)

8. Hinton, G.E., Salakhutdinov, R.R.: Reducing the dimensionality of data with neural networks. Science **313**(5786), 504–507 (2006)
9. Keane, M.T., Smyth, B.: Good counterfactuals and where to find them: a case-based technique for generating counterfactuals for explainable AI (XAI). In: Watson, I., Weber, R. (eds.) ICCBR 2020. LNCS (LNAI), vol. 12311, pp. 163–178. Springer, Cham (2020). https://doi.org/10.1007/978-3-030-58342-2_11
10. Laugel, T., Lesot, M.-J., Marsala, C., Renard, X., Detyniecki, M.: Comparison-based inverse classification for interpretability in machine learning. In: Medina, J., et al. (eds.) IPMU 2018. CCIS, vol. 853, pp. 100–111. Springer, Cham (2018). https://doi.org/10.1007/978-3-319-91473-2_9
11. Luss, R., et al.: Generating contrastive explanations with monotonic attribute functions. arXiv:1905.12698 (2019)
12. Mothilal, R.K., Sharma, A., Tan, C.: Explaining machine learning classifiers through diverse counterfactual explanations. In: FAT*, pp. 607–617 (2020)
13. Sokol, K., et al.: FAT forensics: a python toolbox for implementing and deploying fairness, accountability and transparency algorithms in predictive systems. J. Open Source Softw. **5**(49), 1904 (2020)
14. Van Looveren, A., et al.: Interpretable counterfactual explanations guided by prototypes. arXiv:1907.02584 (2019)
15. Verma, S., Dickerson, J.P., Hines, K.: Counterfactual explanations for machine learning: A review. arXiv:2010.10596 (2020)
16. Wachter, S., et al.: Counterfactual explanations without opening the black box: automated decisions and the GDPR. Harv. JL Tech. **31**, 841–887 (2017)

# Learning Time Series Counterfactuals via Latent Space Representations

Zhendong Wang[1]([⊠]), Isak Samsten[1], Rami Mochaourab[2],
and Panagiotis Papapetrou[1]

[1] Stockholm University, Stockholm, Sweden
{zhendong.wang,samsten,panagiotis}@dsv.su.se
[2] RISE Research Institutes of Sweden, Stockholm, Sweden
rami.mochaourab@ri.se

**Abstract.** Counterfactual explanations can provide sample-based explanations of features required to modify from the original sample to change the classification result from an undesired state to a desired state; hence it provides interpretability of the model. Previous work of LatentCF presents an algorithm for image data that employs autoencoder models to directly transform original samples into counterfactuals in a latent space representation. In our paper, we adapt the approach to time series classification and propose an improved algorithm named LatentCF++ which introduces additional constraints in the counterfactual generation process. We conduct an extensive experiment on a total of 40 datasets from the UCR archive, comparing to current state-of-the-art methods. Based on our evaluation metrics, we show that the LatentCF++ framework can with high probability generate valid counterfactuals and achieve comparable explanations to current state-of-the-art. Our proposed approach can also generate counterfactuals that are considerably closer to the decision boundary in terms of margin difference.

**Keywords:** Time series classification · Interpretability · Counterfactual explanations · Deep learning

## 1 Introduction

Machine learning (ML) is developing rapidly to address real-world classification problems and automate decisions in different fields. Especially, time series classification (TSC) has gained popularity in many critical applications, such as Electrocardiogram (ECG) signal classification [9], sensor signal classification [19], and stream monitoring [16]. Nevertheless, most ML methods remain opaque, although model interpretability is crucial to gaining trust from practitioners. Recent governmental regulations, such as the EU General Data Protection Regulation (GDPR), indicate that the public is entitled to receive *"meaningful information"* from automated decision-making processes [18]. Towards that direction, Wachter et al. [18] suggested *counterfactual explanations* as a solution to provide

© Springer Nature Switzerland AG 2021
C. Soares and L. Torgo (Eds.): DS 2021, LNAI 12986, pp. 369–384, 2021.
https://doi.org/10.1007/978-3-030-88942-5_29

sample-based explanations, aligning with the data protection regulations from GDPR. Counterfactuals provide information on which features of the original test example are required to be modified and how to change the classification result from an undesired state (e.g., "abnormal" ECG) to a desired state (e.g., "normal" ECG), without opening the "black-box" classifier.

Several earlier approaches towards time series counterfactuals have appeared in the literature, with the main highlights including two techniques presented by Karlsson et al. [11]. The two techniques are model specific and they are defined for the random shapelet forest (RSF) classifier [10] as well as the k-NN classifier. Nonetheless, both techniques focus on specific evaluation metrics, i.e., *compactness*, referring to the fraction of time series points that need to alter in order to create a counterfactual, and *cost*, referring to the distance of the original time series to its counterfactual. Despite their promising performance on a large set of collection of time series datasets, both techniques are hampered by the chosen evaluation metrics as they mostly focus on minimizing the two selected metrics, which fail to assess whether the generated counterfactuals are compliant with the data distribution and they fall within the dataset manifold.

A recent approach that attempts to address some of the above limitations for image data has been proposed by Balasubramanian et al. [3], where the `LatentCF` framework was proposed to generate counterfactuals by means of a representative latent space using auto-encoders. The framework established a simple baseline for counterfactual explanations using latent spaces representations. However, the authors solely evaluated their method on image data, while we observed the limitation of ineffective counterfactual generation when we replicated their experiments. In this paper, we re-formulate this problem for the time series domain, and present an adaptation of the original approach, which we refer to as `LatentCF++`, by integrating Adam optimization and additional constraints on the latent space perturbation to generate more robust counterfactuals. Additionally, we demonstrate the generalizability considering several deep learning models serving as components to construct effective `LatentCF++` instantiations.

To highlight the importance of the problem we solve in this paper, consider the example in Fig. 1, where we provide two examples of time series counterfactuals generated by `LatentCF++` using two datasets from the UCR time series repository: *TwoLeadECG* (left) and *FordA* (right). Illustrated in blue are the original time series and in red the generated counterfactuals of the opposite class. By inspecting these counterfactuals, domain experts can not only gain improved understandings of the classifier decision boundary, but also can gain insight on how these predictions can be reversed.

**Related Work.** There is a wide range of TSC algorithms proposed using different techniques in recent years, such as elastic distance measures, intervals, shapelets and ensemble methods [2]. It is different from traditional ML classification problems due to the feature dependency of ordered data. Shapelet-based methods (e.g. shapelet transformations) identify shapelets (subsequences of whole time series) used as discriminatory features to train classifiers such as SVM and random forest [2,10]. More recently, researchers have introduced sev-

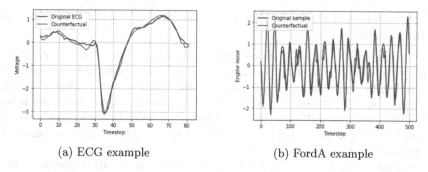

(a) ECG example                    (b) FordA example

**Fig. 1.** Examples of generated counterfactuals using `LatentCF++` on *TwoLeadECG* (left) and *FordA* (right). Illustrated in blue are the original time series and in red the generated counterfactuals of the opposite class. (Color figure online)

eral breakthrough algorithms with comparable benchmark metrics in TSC that are considered state-of-the-art, e.g., InceptionTime [7] and ROCKET [5]. Nevertheless, most of these methods are considered black-box models. As such, they lack model transparency and prediction interpretability.

Interpretability is crucial to help uncover domain findings in opaque machine learning models and has recently attained increased attention[13]. Counterfactual explanations have surged in the last few years in high-stake applications among the wide range of interpretable methods [17]. For TSC problems, Ates et al. [1] presented a counterfactual explanation solution on sample-based predictions using CORELS, focusing on multivariate time series datasets. Similarly, Karlsson et al. [11] proposed local and global solutions for counterfactual explanation problems on univariate time series data utilizing RSF and k-NN classifiers. However, both local and global approaches were proposed to provide model-specific explanations, since they cannot be applied in conjunction with other classifiers, e.g. neural networks or other state-of-the-art models in TSC.

A large number of counterfactual approaches were proposed that can provide model-agnostic explanations for any black-box classifier [17]. By utilizing a variational auto-encoder (VAE) architecture, Pawelczyk et al. [15] conducted an experiment to generate counterfactuals uniformly around spheres of the original data representation for tabular data. Joshi et al. [8] applied a VAE as the generation model, a linear softmax classifier as the classification model, to sample the set of counterfactuals with high probability paths of changes to change the outcome. The `LatentCF` approach was presented to apply gradient descent on the latent representations of input samples and transform into counterfactuals using an auto-encoder (AE) [3]. To the best of our knowledge, these counterfactual solutions using latent representations mainly focused on tabular or image data, and none of them has been generalized in the TSC domain.

**Contributions.** We propose a novel time series counterfactual generation framework, named `LatentCF++`, that ensures the robustness and closeness of the generated counterfactuals. Our specific contributions of this paper are summarized as follows:

- we formulate the time series counterfactual explanation problem for univariate time series data and provide a novel solution LatentCF++ to solve the problem by employing a latent representation for generating counterfactuals;
- we demonstrate two instantiations of our proposed solution, where we incorporate classifiers and auto-encoders based on convolutional neural networks and LSTM, together with an extension of the model based on a composite auto-encoder;
- we conduct an in-depth experiment comparing our solution and the original framework, together with current state-of-the-art RSF, k-NN and FGD as baseline models. We show that our proposed framework outperforms other methods in terms of *validity* and *margin difference*; while it achieves the comparable performance of *proximity* compared to RSF.

## 2   Problem Formulation

Similar to the definition of counterfactual time series explanation problem for multivariate time series classification [1], we define univariate time series counterfactual explanations as follows: given a black-box ML model that predicts the binary output probability from a univariate time series sample, the counterfactual method shows the modifications of the input sample that are required to change the classification result from an undesired state (e.g. negative) to a desired state (e.g. positive). We assume a given classifier in our formulation, and a pre-trained auto-encoder that can transform the time series into the latent representation and back to the original feature space. Note that we do not need to access internal structures of the classifier (e.g. weights); only the prediction function is required. Let $\mathcal{Y}$ defines the set of target class labels, and we consider a binary classification problem where $\mathcal{Y} = \{`+`,`-`\}$. The formal definition of the problem is as follows:

*Problem 1.* **Univariate time series counterfactual explanations.** Give any given classifier $f(\cdot)$ and a time series sample $x$ containing $t$ timesteps, such that the output represents as $f(x) = `-`$ with probability $\hat{y}$. In the problem, $\hat{y}$ is smaller than the decision boundary threshold $\tau$ (i.e. $\hat{y} < \tau$) as it determines negative. Our goal is to utilize an auto-encoder composed of an encode function $E(\cdot)$ and a decode function $D(\cdot)$ to find the generated counterfactual $x'$ with desired positive outcome. We want to perturb the encoded latent representation $z = E(x)$ through a gradient descent optimization iteratively to generate a new time series sample $x' = D(z)$ such that the output target $f(x') = `+`$. Finally, we aim to minimize the objective function denoting the loss between $\hat{y}$ and $\tau$.

For example, given a classifier $f(\cdot)$ and an auto-encoder with functions $E(\cdot)$ and $D(\cdot)$ trained on time series of ECG measurements, we intend to generate counterfactuals through both $E(\cdot)$ and $D(\cdot)$ functions. An exemplified counterfactual $x'$ with the desired prediction $f(x') = `+`$ (i.e. *normal*) for a time series sample $x$ with an undesired prediction $f(x) = `-`$ (i.e. *abnormal*).

# 3   The LatentCF++ Time Series Counterfactuals Framework

The LatentCF framework was presented as a simple baseline for counterfactual explanations, which employs an auto-encoder model to directly transform original samples into counterfactuals using gradient descent to search in the latent space [3]. Due to the low efficiency of gradient descent in the original implementation, it requires an extensive search for the proper learning rate to avoid getting stuck in a local optimum. As such, LatentCF often fails to generate valid counterfactuals.

Instead, we propose to improve the counterfactual generation process by integrating an adequate gradient-based optimization algorithm based on adaptive moment estimation (Adam) [12], making the counterfactual generation more robust. Adam optimization can be considered a combination of two momentum-based gradient descent algorithms AdaGrad and RMS-Prop. Employing the advantages of both allows the algorithm to deal with sparse features and non-stationary objectives and obtain faster convergence [12]. To further improve the validity of the counterfactual explanations, we add constraints (see Line 5 in Algorithm 1) to ensure that the class probability of the generated counterfactual has crossed the decision boundary threshold $\tau$. We call the improved version of the LatentCF framework, LatentCF++.

---

**Algorithm 1:** Counterfactual explanations for time series classification using LatentCF++

---

    **input**  : A time series sample $x$, threshold of decision boundary $\tau$, learning
                rate $\alpha$, loss tolerance $tol$, maximum iteration $max\_iter$
    **output**: A generated counterfactual $x'$ with desired target class $y'$

1  $z \leftarrow$ Encode$(x)$
2  $y_{pred} \leftarrow$ Predict( Decode$(z)$)
3  $loss \leftarrow$ MSE$(y_{pred} - \tau)$
4  $iter \leftarrow 0$
5  **while** $loss > tol \wedge y_{pred} < \tau \wedge iter < max\_iter$ **do**
6     |  $z \leftarrow$ AdamOptimize$(z, loss, \alpha)$
7     |  $y_{pred} \leftarrow$ Predict( Decode$(z)$)
8     |  $loss \leftarrow$ MSE( $y_{pred} - \tau$)
9     |  $iter \leftarrow iter + 1$
10 $x' \leftarrow$ Decode$(z)$
11 **return** $x'$

---

Given an input of time series sample $x$, a pre-trained encoder encodes it into latent representation $z$, followed by a decoder that reconstructs it back to the original feature space. Consecutively, a predictor function estimates the class probability of the reconstructed sample. On Line 5, the constraints are validated to guarantee that the loss iteratively decreases and that the output probability

$y_{pred}$ crosses the threshold of decision boundary $\tau$. The loss is measured using the mean of squared error between $y_{pred}$ and $\tau$. Subsequently, on Line 6 the `AdamOptimize()` function is used to update the latent representation $z$ using Adam. The output counterfactual is the decoded result $x'$ when the while loop condition breaks (i.e., either *loss* is lower than *tol*, $y_{pred}$ is larger than or equal to $\tau$, or the while loop reaches the maximum number of allowed iterations).

Finally, as we can observe from the architecture in Algorithm 1, the algorithm requires two main components: a pre-trained classifier and a pre-trained auto-encoder which can decompose into an encoder and a decoder.

## 4    Experimental Evaluation

We conduct our experiments using the UCR time series archive [4]. We mainly focus on the problem of counterfactuals for binary univariate time classification. After filtering, a subset of 40 datasets from the UCR archive is selected, containing representations from different data sources. For example, *TwoLeadECG* represents ECG measurements in the medical domain and *Wafer* exemplifies sensor data in semiconductor manufacturing. In terms of time series length, it varies from 24 (*ItalyPowerDemand*) to 2709 timesteps (*HandOutlines*) in our chosen subset. For the evaluation, we choose to apply a standard stratified split of 80% for training and the remainder for testing, for all the datasets. Moreover, to compensate for the imbalanced target classes, we apply an up-sampling technique that resamples the minority class during training.

### 4.1    Experiment Setup

There are three main deep neural network architectures that have been adopted for time series classification tasks in recent years: multi-layer perceptron (MLP), convolutional neural networks (CNN) and recurrent neural networks (RNN) [6]. In the experiment, we choose to train separate CNN and long short-term memory (LSTM, a variant of RNN) classification and auto-encoder models, as main components to apply the `LatentCF` and `LatentCF++` frameworks. Despite the fact that more recent state-of-the-art architectures have been proposed in the

**Table 1.** Summary of model components and hyperparameters for each instantiation in the experiment.

| Method | Instantiation | Auto-encoder | Classifier | Optimization | Threshold |
|---|---|---|---|---|---|
| LatentCF++ | 1dCNN | 1dCNN-AE | 1dCNN-CLF | Adam | 0.5 |
| | LSTM | LSTM-AE | LSTM-CLF | Adam | 0.5 |
| | 1dCNN-C | 1dCNN-C | | Adam | 0.5 |
| LatentCF | 1dCNN | 1dCNN-AE | 1dCNN-CLF | Vanilla GD | No |
| | LSTM | LSTM-AE | LSTM-CLF | Vanilla GD | No |

**Table 2.** Summary of architectures and hyperparameters for the deep learning models, representing different components.

| Instan. | Components | | #Layers | #Conv | #LSTM | Norm | Pooling | Output |
|---------|-----------|---------|---------|-------|-------|------|---------|--------|
| 1dCNN | 1dCNN-AE | | 5 | 5 | 0 | No | Max | Linear |
| | 1dCNN-CLF | (shallow) | 3 | 1 | 0 | Yes | Max | Sigmoid |
| | | (deep) | 4 | 3 | 0 | Yes | Max | Sigmoid |
| LSTM | LSTM-AE | | 5 | 0 | 4 | No | Max | Sigmoid |
| | LSTM-CLF | (shallow) | 2 | 0 | 1 | Yes | None | Sigmoid |
| | | (deep) | 3 | 0 | 2 | Yes | None | Sigmoid |
| 1dCNN-C | 1dCNN-C | | 8 | 6 | 0 | No | Max | Lin.+Sig. |

literature, we opt to use these simpler architectures to highlight the explainable power of latent space representations.

Thus, we construct two instantiations for both LatentCF and LatentCF++ in our experiment: 1dCNN and LSTM. We show a detailed comparison of different components and hyperparameters for each instantiation in Table 1. For illustration, the 1dCNN instantiation comprises a 1dCNN-AE auto-encoder and a 1dCNN-CLF classifier, together with an Adam optimization and the probability threshold of 0.5. Additionally, LatentCF++ is extended with one composite auto-encoder structure (1dCNN-C) instead of utilizing the two components.

More specifically, to evaluate LatentCF and LatentCF++ for each dataset, we first train five deep learning models representing different components in the framework: CNN auto-encoder (1dCNN-AE), CNN classifier (1dCNN-CLF), LSTM auto-encoder (LSTM-AE), LSTM classifier (LSTM-CLF) and CNN composite auto-encoder (1dCNN-C). Since our goal is not to assess the performance of classifiers or auto-encoders, we apply a one-size-fits-all plan to utilize a standard set of model architectures for all the datasets. Table 2 shows architectures and hyperparameters for all different deep learning models. From the table, we can see that each instantiation comprises two components - an auto-encoder and a classifier, e.g., 1dCNN consists of 1dCNN-AE and 1dCNN-CLF. Besides, the model extension 1dCNN-C contains only one component of composite auto-encoder. For each instantiation, we apply a parameter search for the learning rate between 0.001 and 0.0001, and then report the model metrics with the best validity.

Note that we have two slightly different structures (shallow and deep) for classifiers 1dCNN-CLF and LSTM-CLF, due to different timestep sizes and varied amounts of available training data. During the training, either a shallow or a deep classifier is trained for both CNN and LSTM instantiations. We then evaluate on each specific dataset using the model with the best performance. Empirically, this strategy generalizes well for all datasets in our experiment.

**CNN Models.** For the detailed architecture of 1dCNN-AE, the network contains four convolutional layers with 64, 32, 32 and 64 filters with kernel size 3 and activated using ReLU. The network's output consists of a final convolutional layer, which is corresponding to the reconstructed output. The first two layers are followed by a max pooling operation; while the next two layers are followed by

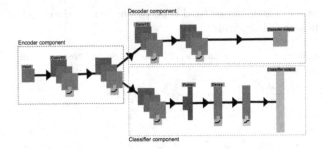

**Fig. 2.** Illustration of 1dCNN composite model architecture.

up-sampling transformations of size 2. The *deep* 1dCNN-CLF model is composed of three convolutional layers followed by a final dense layer with sigmoid activation. For each convolutional layer, the number of filters is fixed to 64 and the kernel size is set to 3, with ReLU activation and batch normalization. Moreover, the *shallow* model contains only one convolutional layer compared to three in the deep model and finally a 128-unit dense layer after the input.

**LSTM Models.** LSTM-AE consists of four consecutive LSTM layers with respectively 64, 32, 32 and 64 units with tanh activation functions. The final layer is a dense layer with a sigmoid activation function. The *shallow* model for LSTM-CLF contains one 32-unit LSTM layer with tanh activation; while the *deep* model comprises two continuous LSTM layers with 64 and 16 units, respectively. Each LSTM layer is followed by a batch normalization operation. Finally, a dense layer with sigmoid activation function is fully connected to the previous layer, where the output represents the probability of the target prediction.

**LatentCF++ Extension: Composite Model.** In addition to the two instantiations, we intend to evaluate a model extension of LatentCF++ with a CNN composite auto-encoder 1dCNN-C. The model has a unique architecture compared to the previously described models, which is shown in Fig. 2. It has three elements: an *encoder* for encoding the input into the latent space, followed by a *decoder* and a *classifier*. Accordingly, it contains two different output layers in the architecture: one convolutional layer as the decoder and a sigmoid function for the classifier. The encoding and decoding components share the same setup of layers as 1dCNN-AE: the encoding component has two convolutional layers with 64 and 32 filters followed by max pooling; the decoder comprises three convolutional layers with 32, 64 and 1 filters, respectively. While for the component of classifier, a convolutional layer (16 filters with kernel size 3) followed by a 50-unit dense layer are connected after the shared latent space; a final sigmoid dense layer is fully connected with the previous layer for output probability of the target class. In the implementation of LatentCF++, we directly apply the Predict() function on latent representation $z$ to adjust for the 1dCNN-C model, instead of using the Decode() function.

**Implementation Details.** All deep learning models are implemented in Keras[1]. For 1dCNN-AE, LSTM-AE, and 1dCNN-C, we set the training epochs to 50; while for classification models, the training epochs are set to 150 for both 1dCNN-CLF and LSTM-CLF. To reduce overfitting and improve the generalizability of our networks, we employ early stopping during model training. Adam optimizer is employed for all networks with learning rates ranging from 0.0001 and 0.001 for different models. The batch size is set to 32 for all of them.

## 4.2 Baseline Models

We adopt the first baseline model from the original paper, FGD [3], which applies the LatentCF method with only a classifier 1dCNN-CLF to perturb samples in the original feature space directly. In addition, we apply two extra baseline methods from local and global time series tweaking approaches [11] - random shapelet forest (RSF) and the k-NN counterfactual method (k-NN). Similar to the evaluation of LatentCF and LatentCF++, we apply the same parameter setup across all datasets for these baseline models. For RSF, we set the number of estimators to 50 and max depth to 5; while the other parameters are kept at their default values[2]. For k-NN, we first train a k-NN classifier with $k$ equals to 5 and the distance metric set to Euclidean; then the trained classifier is utilized to find counterfactual samples for further evaluation.

## 4.3 Evaluation Metrics

To evaluate the performance of our proposed approach in terms of explainability, we present three metrics: validity, proximity and closeness. *Validity* is defined to measure whether the generated counterfactual explanations lead to a valid transformation of the desired target class [14,17]. More specifically, it reports the fraction of counterfactuals predicted as the opposite class (i.e. have crossed the decision boundary $\tau$) according to the standalone classifier. It is defined as:

$$validity(y_{cf}, \tau) = \frac{\#(y_i \geq \tau, y_i \in y_{cf})}{\#y_{cf}}, \tag{1}$$

where $y_{cf}$ is the output probability of all the generated counterfactual samples, and $\tau$ is a user-defined threshold of the decision boundary. In our evaluation, we set the threshold $\tau$ to be 0.5.

*Proximity* is applied to measure the feature-wise distance between generated counterfactuals and corresponding original samples [14]. Karlsson et al. [11] reported a similar metric named *cost* in the evaluation of local and global time series tweaking approaches. In our case, we define proximity as Euclidean distance between transformed and original time series samples:

$$proximity(x, x') = d(x, x'), \tag{2}$$

---

[1] https://keras.io.
[2] See https://github.com/isaksamsten/wildboar.

where $d(\cdot)$ is the Euclidean distance and $x$ and $x'$ are the original time series and generated counterfactual sample respectively. We report the average value of the proximity scores for each dataset.

To measure how close is the predicted probability of a generated counter-factual compared to the decision boundary, we propose a new metric which we denote as *margin difference*. The margin difference captures the amount of information that has been altered from the original class and is defined as:

$$margin\_diff(y_{cf}, \tau) = y_{cf} - \tau, \tag{3}$$

where $y_{cf}$ is the output probability of counterfactual $x'$, and $\tau$ is the target decision boundary. Note that this metric can be either positive or negative, indicating whether the counterfactual has crossed the decision boundary or not. We record both the mean and standard deviation of margin differences for all generated counterfactuals as metrics for each dataset.

In addition, we show the classification performance of all models report as the **balanced accuracy** in the results. Moreover, we report the reconstruction loss of the auto-encoder models.

## 4.4    Results

In this section, we first compare the validity of our presented explainable models from LatentCF and LatentCF++, as well as FGD, RSF and k-NN counterfactual methods. For a detailed comparison, we choose to report metrics from a subset of 20 datasets with the sample size larger than 500. Then we report a subset average for different explainable models; together, we also present the average score (denoted as **Total avg.**) across all 40 datasets in the experiment[3].

Table 3 shows the *validity*, which we considered as the main metric for the evaluation of interpretability. Again, we report the results with the value of decision boundary $\tau = 0.5$ in Eq. 1. Across three different groups, we found that the RSF method achieved the best metric of subset average of 1.0000; in contrast, 1dCNN from the LatentCF++ method obtained the highest validity (0.9920) in terms of the total average, which indicates that an average of 99.20% of the generated counterfactuals is considered valid. In comparison, 1dCNN from the LatentCF method and FGD baseline both received validity that was lower than 0.2. This evidence suggests that our proposed LatentCF++ method can ensure a high fraction of valid counterfactual transformations similar to RSF.

In Table 4, we show a comparison for individual mean scores of *margin difference* for the subset of 20 datasets, together with the average score at the bottom. In addition, we report the average of standard deviations for both the subset (**Subset std.**) and the total (**Total std.**). From the table, we observed that 1dCNN from LatentCF++ achieved the best metric to both subset average (0.0333) and total average (0.0168), which indicates that generated counterfactuals are considerably closer to the decision boundary compared to other methods. In terms of the total average, LSTM models from LatentCF and LatentCF++

---

**Table 3.** Summary of validity for a subset of 20 different datasets in our experiment. The best score for each dataset is highlighted in bold.

| Dataset | LatentCF++ | | | LatentCF | | Baseline | | |
|---|---|---|---|---|---|---|---|---|
| | 1dCNN | LSTM | 1dCNN-C | 1dCNN | LSTM | FGD | k-NN | RSF |
| Yoga | 0.9973 | **1.0000** | 0.9912 | 0.4347 | **1.0000** | 0.4507 | **1.0000** | **1.0000** |
| TwoLeadECG | 0.9914 | 0.9052 | 0.9835 | 0.3966 | **1.0000** | 0.0000 | **1.0000** | **1.0000** |
| ItalyPower | **1.0000** | **1.0000** | 0.9912 | 0.0000 | 0.2456 | 0.0000 | **1.0000** | **1.0000** |
| MoteStrain | **1.0000** | **1.0000** | 0.9918 | 0.2017 | 0.0455 | 0.0000 | **1.0000** | **1.0000** |
| Wafer | **1.0000** | 0.8625 | 0.8117 | 0.2353 | 0.1875 | 0.0588 | **1.0000** | **1.0000** |
| FreezerRegular | **1.0000** | **1.0000** | 0.9666 | 0.5633 | **1.0000** | 0.4967 | **1.0000** | **1.0000** |
| PhalangesOutlines | **1.0000** | **1.0000** | **1.0000** | 0.3833 | 0.7048 | 0.2333 | 0.8629 | **1.0000** |
| FreezerSmall | 0.9791 | **1.0000** | 0.9790 | 0.6794 | **1.0000** | 0.5122 | **1.0000** | **1.0000** |
| HandOutlines | 0.9091 | 0.9661 | **1.0000** | 0.2727 | 0.0000 | 0.0000 | 0.9359 | **1.0000** |
| FordA | **1.0000** | 0.0000 | 0.9838 | 0.0055 | 0.0000 | 0.0000 | 0.8903 | **1.0000** |
| FordB | **1.0000** | **1.0000** | **1.0000** | 0.0027 | **1.0000** | 0.0000 | 0.9324 | **1.0000** |
| SonyAIBORobot2 | 0.9916 | 0.8560 | 0.9917 | 0.0168 | **1.0000** | 0.0000 | **1.0000** | **1.0000** |
| SemgHandGender | 0.8358 | **1.0000** | 0.8310 | 0.5970 | **1.0000** | 0.6567 | 0.6528 | **1.0000** |
| MiddlePhalanx | **1.0000** | **1.0000** | **1.0000** | 0.1111 | **1.0000** | 0.0000 | 0.8431 | **1.0000** |
| ProximalPhalanx | **1.0000** | **1.0000** | **1.0000** | 0.0000 | 0.1282 | 0.0000 | 0.6842 | **1.0000** |
| ECGFiveDays | 0.9773 | 0.8864 | 0.9773 | 0.5114 | **1.0000** | 0.0000 | **1.0000** | **1.0000** |
| DistalPhalanx | **1.0000** | **1.0000** | **1.0000** | 0.0769 | 0.3036 | 0.0000 | **1.0000** | **1.0000** |
| SonyAIBORobot1 | **1.0000** | **1.0000** | 0.9818 | 0.0182 | **1.0000** | 0.0000 | **1.0000** | **1.0000** |
| Computers | **1.0000** | 0.3824 | **1.0000** | 0.4839 | 0.5000 | 0.3548 | 0.5593 | **1.0000** |
| Strawberry | **1.0000** | **1.0000** | 0.9600 | 0.4146 | 0.0000 | 0.3333 | **1.0000** | **1.0000** |
| **Subset avg.** | 0.9841 | 0.8929 | 0.9720 | 0.2703 | 0.6058 | 0.1548 | 0.9180 | **1.0000** |
| **Total avg.** | **0.9920** | 0.8256 | 0.9615 | 0.1676 | 0.5779 | 0.0774 | 0.9496 | 0.9802 |

achieved the second and third best margin difference scores of 0.0520 and 0.0580, respectively. In comparison, RSF model received a less favourable metric of margin difference (0.0614) since it does not optimize towards the threshold. Instead, it tries to minimize the difference between the original samples and counterfactuals. 1dCNN-C, FGD and 1dCNN from LatentCF received negative scores according to total average, which means that they could not constantly guarantee counterfactuals to cross the decision boundary. Nonetheless, 1dCNN-C outperformed all the other models in 10 out of the 20 individual datasets.

In addition, we observed that the standard deviation for 1dCNN from the LatentCF++ framework resulted in the lowest of 0.0175 while k-NN achieved the maximum of 0.1218 in Table 4. Thus, this evidence shows that LatentCF++ can generate counterfactuals that are more stable in terms of margin difference. In other words, our proposed approach can better guarantee to produce counterfactuals that can cross the decision boundary and obtain more compact transformations.

From the previous comparison of validity, we found that several models received scores lower than 0.2, which means that these models cannot guarantee the fraction of valid counterfactuals. As our primary evaluation was based on the validity, we intended to compare the rest of the five best-performed models for

**Table 4.** Summary of margin difference for a subset of 20 different datasets in our experiment. The best score for each dataset is highlighted in bold.

| Dataset | LatentCF++ | | | LatentCF | | Baseline | | |
|---|---|---|---|---|---|---|---|---|
| | 1dCNN | LSTM | 1dCNN-C | 1dCNN | LSTM | FGD | k-NN | RSF |
| Yoga | **0.0005** | 0.1810 | −0.0003 | −0.0383 | 0.3115 | −0.0531 | 0.4096 | 0.0199 |
| TwoLeadECG | 0.0005 | **0.0002** | 0.0006 | −0.0347 | 0.4941 | −0.1854 | 0.4914 | 0.1936 |
| ItalyPower | **0.0005** | 0.0005 | 0.0006 | −0.3020 | −0.0854 | −0.4144 | 0.5000 | 0.2077 |
| MoteStrain | 0.0005 | 0.0005 | **0.0004** | −0.0343 | -0.0261 | −0.3766 | 0.4704 | 0.0382 |
| Wafer | **0.0005** | 0.0034 | −0.0486 | −0.2831 | −0.3181 | −0.4545 | 0.5000 | 0.0334 |
| FreezerRegular | **0.0005** | 0.4995 | −0.0028 | 0.0614 | 0.4999 | 0.0016 | 0.3916 | 0.1278 |
| PhalangesOutlines | **0.0005** | 0.0005 | 0.0005 | −0.0004 | −0.0002 | −0.0044 | 0.2783 | 0.0573 |
| FreezerSmall | 0.4774 | 0.4636 | −0.0081 | 0.1878 | 0.4766 | **0.0156** | 0.3683 | 0.1605 |
| HandOutlines | −0.0220 | 0.0061 | **0.0005** | −0.2385 | −0.5000 | −0.3531 | 0.4051 | 0.0379 |
| FordA | 0.0005 | −0.0382 | **0.0002** | −0.1302 | −0.0568 | −0.1816 | 0.2363 | 0.0196 |
| FordB | **0.0005** | 0.1443 | 0.0005 | −0.2152 | 0.2182 | −0.2709 | 0.2551 | 0.0198 |
| SonyAIBORobot2 | **0.0005** | 0.0056 | 0.0005 | −0.1871 | 0.5000 | −0.4431 | 0.3413 | 0.0311 |
| SemgHandGender | 0.2016 | 0.1822 | −0.0416 | 0.0863 | 0.1908 | 0.1640 | 0.1556 | **0.0238** |
| MiddlePhalanx | 0.0005 | 0.0005 | **0.0001** | −0.0084 | 0.0030 | −0.0088 | 0.2255 | 0.1015 |
| ProximalPhalanx | 0.0005 | **0.0004** | 0.0005 | −0.2328 | −0.0026 | −0.0963 | 0.2088 | 0.1905 |
| ECGFiveDays | 0.0005 | **0.0003** | 0.0004 | 0.0018 | 0.4985 | −0.4627 | 0.4934 | 0.0786 |
| DistalPhalanx | 0.0006 | 0.0005 | **0.0004** | −0.1098 | −0.0009 | −0.1899 | 0.3151 | 0.0932 |
| SonyAIBORobot1 | **0.0004** | 0.0005 | 0.0004 | −0.2647 | 0.4795 | −0.4233 | 0.4690 | 0.0425 |
| Computers | 0.0005 | 0.0203 | **0.0001** | −0.0082 | 0.1415 | −0.0772 | 0.0864 | 0.0202 |
| Strawberry | **0.0005** | 0.0005 | −0.0028 | −0.1076 | −0.4322 | −0.1311 | 0.4587 | 0.1245 |
| Subset avg. | **0.0333** | 0.0736 | −0.0049 | −0.0929 | 0.1196 | −0.1973 | 0.3530 | 0.0811 |
| Subset std. | 0.0348 | **0.0246** | 0.0287 | 0.2072 | 0.0453 | 0.1848 | 0.1600 | 0.0623 |
| Total avg. | **0.0168** | 0.0580 | −0.0120 | −0.0896 | 0.0520 | −0.1435 | 0.3608 | 0.0614 |
| Total std. | **0.0175** | 0.0234 | 0.0310 | 0.1193 | 0.0240 | 0.1081 | 0.1218 | 0.0576 |

further investigation. Namely, we chose to exclude 1dCNN, LSTM from LatentCF and FGD from baseline models in our comparison of proximity. Similar to the previous evaluation, we reported metrics from the subset of 20 datasets with respective subset and total average scores, as in Table 5.

In Table 5, we observed that 1dCNN from our proposed LatentCF++ framework achieved comparable proximity scores compared to the state-of-the-art method RSF, with a subset average of 0.3891 in comparison with 0.2873. This evidence indicates that the generated counterfactuals from 1dCNN are relatively closer to the original samples. In contrast, LSTM received the worst average proximity of 2.5409 among all. When we checked the individual results, we found that both 1dCNN and 1dCNN-C from LatentCF++ outperformed RSF in 9 out of 20 datasets, while RSF outperformed the others in 10 datasets. One of our key observations was that the proximity score was strongly related to the corresponding auto-encoder performance. Usually, if the 1dCNN-AE model from 1dCNN converged with a low reconstruction loss, then 1dCNN would outperform the other methods. However, it was challenging to ensure the performance of the auto-encoder since we applied a unified structure of 1dCNN-AE for all datasets.

**Table 5.** Summary of proximity for a subset of 20 different datasets in our experiment. The best score for each dataset is highlighted in bold (three methods are excluded from the comparison due to the low validity[†]).

| Dataset | LatentCF++ | | | LatentCF | | Baseline | | |
|---|---|---|---|---|---|---|---|---|
| | 1dCNN | LSTM | 1dCNN-C | 1dCNN[†] | LSTM[†] | FGD[†] | k-NN | RSF |
| Yoga | **0.2049** | 3.6447 | 0.7221 | 0.1759 | 3.6347 | 0.0277 | 1.0950 | 0.5281 |
| TwoLeadECG | **0.1447** | 1.4290 | 0.1839 | 0.1659 | 0.5681 | 0.0270 | 0.2655 | 0.1793 |
| ItalyPower | 0.4785 | 0.3986 | 0.2588 | 0.1609 | 0.3373 | 0.0066 | 0.3633 | **0.2513** |
| MoteStrain | 0.3167 | 0.4798 | 0.3884 | 0.2652 | 0.4798 | 0.0094 | 0.4673 | **0.1503** |
| Wafer | **0.2157** | 0.4416 | 0.2325 | 0.1062 | 0.3642 | 0.0069 | 0.5790 | 0.2547 |
| FreezerRegular | 0.2211 | 1.0497 | 0.1158 | 0.1812 | 0.6700 | 0.0409 | 0.0808 | **0.0569** |
| PhalangesOutlines | **0.1449** | 0.4592 | 0.2109 | 0.1060 | 0.5168 | 0.0190 | 0.2272 | 0.2238 |
| FreezerSmall | 1.9710 | 0.6919 | 0.1929 | 0.1978 | 0.5990 | 0.0319 | 0.0887 | **0.0598** |
| HandOutlines | **0.4359** | 23.9995 | 3.9276 | 0.2575 | 7.9567 | 0.0087 | 1.9754 | 1.1485 |
| FordA | 0.5670 | 2.3575 | **0.4368** | 0.2463 | 2.3087 | 0.0159 | 2.0811 | 0.4820 |
| FordB | 0.6099 | 2.2574 | 0.4261 | 0.2176 | 2.2092 | 0.0124 | 2.1105 | **0.3934** |
| SonyAIBORobot2 | 0.3234 | 0.9515 | 0.3263 | 0.2293 | 0.9466 | 0.0054 | 0.6357 | **0.2862** |
| SemgHandGender | 0.4159 | 0.8741 | 0.2637 | 0.2676 | 0.8786 | 0.0351 | 0.2999 | **0.1006** |
| MiddlePhalanx | **0.1749** | 0.6519 | 0.3936 | 0.1496 | 0.6786 | 0.0000 | 0.2646 | 0.2568 |
| ProximalPhalanx | 0.2633 | 0.5222 | 0.2933 | 0.1359 | 0.5314 | 0.0145 | **0.1577** | 0.2539 |
| ECGFiveDays | 0.1654 | 0.9480 | 0.2829 | 0.1538 | 0.8912 | 0.0057 | 0.3459 | **0.1083** |
| DistalPhalanx | 0.2258 | 0.7235 | **0.1850** | 0.1450 | 0.7282 | 0.0133 | 0.3674 | 0.2724 |
| SonyAIBORobot1 | 0.2719 | 0.5017 | 0.2440 | 0.1882 | 0.4951 | 0.0078 | 0.5752 | **0.2260** |
| Computers | 0.3759 | 1.3727 | 0.7971 | 0.3538 | 1.3243 | 0.0236 | 0.8305 | **0.1774** |
| Strawberry | **0.2545** | 1.1272 | 0.2947 | 0.2242 | 0.7843 | 0.0255 | 0.3734 | 0.3370 |
| Subset avg. | 0.3891 | 2.2441 | 0.5088 | 0.1963 | 1.3451 | 0.0169 | 0.6592 | **0.2873** |
| Total avg. | 0.8926 | 2.5409 | 0.9179 | 0.4415 | 1.9841 | 0.0087 | 1.4613 | **0.5241** |

For a detailed comparison, we investigated ECG examples of generated counterfactuals by 1dCNN, LSTM from LatentCF++, and RSF (see Fig. 3). We observed that 1dCNN and RSF generated similar counterfactuals for the ECG sample, although the counterfactual from 1dCNN appeared slightly smoother in this case. In contrast, LSTM's counterfactual poorly aligned with the original series and diverged in many timesteps. Moreover, we found that different classification models performed similarly in balanced accuracy. As to the autoencoders, LSTM-AE achieved the highest reconstruction loss over most datasets while 1dCNN-AE received the lowest reconstruction loss. This evidence explains why LSTM from LatentCF and LatentCF++ attained the worst performance when comparing proximity. It appeared that the LSTM auto-encoder could not learn a representative latent space compared to other auto-encoders in the time series domain.

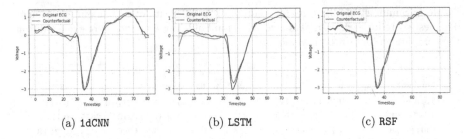

(a) 1dCNN     (b) LSTM          (c) RSF

**Fig. 3.** Examples of generated counterfactuals by 1dCNN and LSTM instantiations from LatentCF++, together with RSF in comparison. Illustrated in blue are the original time series and in red the generated counterfactuals of the opposite class. (Color figure online)

## 5   Conclusions

We presented a new counterfactual explanation framework named LatentCF++ for time series counterfactual generation. Our experimental results on the UCR archive focusing on binary classification showed that LatentCF++ substantially outperforms instantiations of its predecessor, LatentCF, and other baseline models. The results also suggest that LatentCF++ can provide robust counterfactuals that firmly guarantee validity and are considerably closer margin difference to the decision boundary. Additionally, our proposed approach achieved comparable proximity compared to the state-of-the-art time series tweaking approach RSF. Furthermore, we found that it was challenging to leverage the power of deep learning models (both classifiers and auto-encoders) for datasets with the sample size of less than 500. Hence our one-size-fits-all plan to utilize unified structures of deep learning models as components for the framework did not address some specific datasets in the experiment. However, we still showed the generalizability of our proposed framework using a unified set of model components. For future work, we plan to extend our work to generalize LatentCF++ to broader counterfactual problems using other types of data, e.g. multivariate time series, textual or tabular data. Also, we intend to conduct a qualitative analysis from domain experts to validate that the produced counterfactuals are meaningful in different application fields, such as ECG measurements in healthcare or sensor data from signal processing.

**Reproducibility.** All our code to reproduce the experiments is publicly available at our supporting website[4].

**Acknowledgments.** This work was supported by the EXTREMUM collaborative project (https://datascience.dsv.su.se/projects/extremum.html) funded by Digital Futures.

---

[4] https://github.com/zhendong3wang/learning-time-series-counterfactuals.

# References

1. Ates, E., Aksar, B., Leung, V.J., Coskun, A.K.: Counterfactual Explanations for Machine Learning on Multivariate Time Series Data. arXiv:2008.10781 [cs, stat] (August 2020)
2. Bagnall, A., Lines, J., Bostrom, A., Large, J., Keogh, E.: The great time series classification bake off: a review and experimental evaluation of recent algorithmic advances. Data Min. Knowl. Disc. **31**(3), 606–660 (2016)
3. Balasubramanian, R., Sharpe, S., Barr, B., Wittenbach, J., Bruss, C.B.: Latent-CF: a simple baseline for Reverse Counterfactual Explanations. In: NeurIPS 2020 Workshop on Fair AI in Finance (December 2020)
4. Dau, H.A., et al.: Hexagon-ML: The ucr time series classification archive (October 2018). https://www.cs.ucr.edu/~eamonn/time_series_data_2018/
5. Dempster, A., Petitjean, F., Webb, G.I.: ROCKET: exceptionally fast and accurate time series classification using random convolutional kernels. Data Min. Knowl. Disc. **34**(5), 1454–1495 (2020)
6. Ismail Fawaz, H., Forestier, G., Weber, J., Idoumghar, L., Muller, P.-A.: Deep learning for time series classification: a review. Data Min. Knowl. Disc. **33**(4), 917–963 (2019)
7. smail Fawaz, H., et al.: InceptionTime: finding AlexNet for time series classification. Data Min. Knowl. Disc. **34**(6), 1936–1962 (2020)
8. Joshi, S., Koyejo, O., Vijitbenjaronk, W., Kim, B., Ghosh, J.: Towards Realistic Individual Recourse and Actionable Explanations in Black-Box Decision Making Systems. arXiv: 1907.09615 (July 2019)
9. Kampouraki, A., Manis, G., Nikou, C.: Heartbeat time series classification with support vector machines. IEEE Trans. Inf Technol. Biomed. **13**(4), 512–518 (2009)
10. Karlsson, I., Papapetrou, P., Boström, H.: Generalized random shapelet forests. Data Min. Knowl. Disc. **30**(5), 1053–1085 (2016)
11. Karlsson, I., Rebane, J., Papapetrou, P., Gionis, A.: Locally and globally explainable time series tweaking. Knowl. Inf. Syst. **62**(5), 1671–1700 (2019)
12. Kingma, D.P., Ba, J.: Adam: a method for stochastic optimization. In: Proceedings of the 3rd International Conference on Learning Representations (ICLR 2015) (January 2015)
13. Molnar, C.: Interpretable Machine Learning - A Guide for Making Black Box Models Explainable (2019)
14. Mothilal, R.K., Sharma, A., Tan, C.: Explaining machine learning classifiers through diverse counterfactual explanations. In: Proceedings of the 2020 Conference on Fairness, Accountability, and Transparency, pp. 607–617 (January 2020)
15. Pawelczyk, M., Haug, J., Broelemann, K., Kasneci, G.: Learning model-agnostic counterfactual explanations for tabular data. In: Proceedings of The Web Conference, vol. 2020, pp. 3126–3132 (2020)
16. Rebbapragada, U., Protopapas, P., Brodley, C.E., Alcock, C.: Finding anomalous periodic time series. Mach. Learn **74**(3), 281–313 (2009)
17. Stepin, I., Alonso, J.M., Catala, A., Pereira-Fariña, M.: A survey of contrastive and counterfactual explanation generation methods for explainable artificial intelligence. IEEE Access **9**, 11974–12001 (2021)

18. Wachter, S., Mittelstadt, B., Russell, C.: Counterfactual explanations without opening the black box: automated decisions and the GDPR. SSRN Electron. J. (2017)

19. Yao, S., Hu, S., Zhao, Y., Zhang, A., Abdelzaher, T.: DeepSense: a unified Deep Learning Framework for Time-Series Mobile Sensing Data Processing. In: Proceedings of the 26th International Conference on World Wide Web. pp. 351–360 (April 2017)

# Leveraging Grad-CAM to Improve the Accuracy of Network Intrusion Detection Systems

Francesco Paolo Caforio[1]([✉]) [iD], Giuseppina Andresini[1] [iD], Gennaro Vessio[1] [iD], Annalisa Appice[1,2] [iD], and Donato Malerba[1,2] [iD]

[1] Department of Computer Science, University of Bari "Aldo Moro", Bari, Italy
{francescopaolo.caforio,giuseppina.andresini,gennaro.vessio,
annalisa.appice,donato.malerba}@uniba.it
[2] CINI - Consorzio Interuniversitario Nazionale per l'Informatica, Bari, Italy

**Abstract.** As network cyber attacks continue to evolve, traditional intrusion detection systems are no longer able to detect new attacks with unexpected patterns. Deep learning is currently addressing this problem by enabling unprecedented breakthroughs to properly detect unexpected network cyber attacks. However, the lack of decomposability of deep neural networks into intuitive and understandable components makes deep learning decisions difficult to interpret. In this paper, we propose a method for leveraging the visual explanations of deep learning-based intrusion detection models by making them more transparent and accurate. In particular, we consider a CNN trained on a 2D representation of historical network traffic data to distinguish between attack and normal flows. Then, we use the Grad-CAM method to produce coarse localization maps that highlight the most important regions of the traffic data representation to predict the cyber attack. Since decisions made on samples belonging to the same class are expected to be explained with similar localization maps, we base the final classification of a new network flow on the class of the nearest-neighbour historical localization map. Experiments with various benchmark datasets demonstrate the effectiveness of the proposed method compared to several state-of-the-art methods.

**Keywords:** Cyber-security · Network intrusion detection · Deep learning · Explainability · Grad-CAM.

## 1 Introduction

Intrusion Detection Systems (IDSs) play a crucial role in improving the security of the modern network environment by inspecting network traffic for signs of potential vulnerabilities. In today's IDSs, deep learning (DL) plays a vital role thanks to the ability to process historical network traffic data in order to learn accurate predictive models that can distinguish between attacks and

© Springer Nature Switzerland AG 2021
C. Soares and L. Torgo (Eds.): DS 2021, LNAI 12986, pp. 385–400, 2021.
https://doi.org/10.1007/978-3-030-88942-5_30

normal network flows. At present, various DL neural network architectures—e.g., autoencoders [2,25], Recurrent Neural Networks [11], Generative Adversarial Networks [27,28], Triplet Networks [5] and Convolutional Neural Networks (CNNs) [3,4,6,12]—have already enabled unprecedented breakthroughs in various network intrusion detection tasks. However, their lack of decomposability into intuitive components makes DL models difficult to interpret [15]. In particular, the difficulty in interpreting DL models undermines their actual use in production as the reasons behind their decisions are unknown [22]. Hence, in order to build confidence in the implementation of DL techniques for network IDSs, as well as to move towards their meaningful integration into the commercial network line of defence, we need to build transparent DL models that are able to explain why they predict what they predict.

In this paper, we follow this research direction and focus on dealing with the transparency issue when training CNNs for network intrusion detection. This architecture is considered here as various studies [3,4] have recently shown that it can achieve amazing results in network intrusion detection tasks once an appropriate 2D representation of network flows is adopted. In particular, we illustrate a CNN-based intrusion detection methodology, called GRACE (GRad-CAM-enhAnced Convolution neural nEtwork), which injects transparency into the CNN-based classification pipeline. To this end, we use the Gradient-weighted Class Activation Mapping (Grad-CAM) technique [18] that produces visual explanations for CNN decisions. The Grad-CAM technique exploits the gradients of the "attack" concept flowing into the final convolutional layer of the classification network, in order to produce a coarse localization map that highlights the important regions in the image for the prediction of the concept.

In particular, the main innovation of this study is the combination of the Grad-CAM explanations with the nearest-neighbour search to improve the accuracy of the CNN decisions in a *post-hoc* way. In particular, we perform a clustering step to group together Grad-CAM explanations. Then, we consider the cluster centres as explanation of the normal and attack behaviour in the network traffic. In addition, we use the cluster centres in the search for the nearest-neighbour to limit overfitting phenomena. We investigate the effectiveness of the presented methodology by exploring the feasibility of the proposed learning components on various benchmark datasets, as well as the ability of our methodology to increase the accuracy of decisions compared to competitive approaches adopted by recent literature on network intrusion detection.

The rest of this paper is organized as follows. Related works are presented in Sect. 2, while the proposed methodology is described in Sect. 3. The results of the evaluation are discussed in Sect. 4. Finally, Sect. 5 refocuses on the purpose of the research, draws conclusions and illustrates future developments.

## 2   Related Work

In recent years, DL has been recognized as an effective tool for cyber-security. However, DL techniques designed for cyber threat detection (e.g., intrusion detection, malware discovery, etc.) suffer from some limitations. In particular, neural networks are difficult to interpret and their decisions are opaque to practitioners. Even simple tasks, such as determining which features of an input contribute to a prediction, are challenging to solve with neural networks. The difficulty in interpreting DL models is nowadays a major limitation to overcome for the effective use of DL in cyber-security, sice black box DL models can be difficult to control and protect against attacks [23].

Currently, the machine learning community is dedicating increasing efforts to develop eXplainable Artificial Intelligence (XAI) techniques for interpreting DL models [7,24]. Various XAI techniques have been tested especially in computer vision to improve the reliability, transparency and fairness of DL models (see [10] for a recent survey). These techniques can be mainly categorized into two explanation strategies: *black-box* and *white-box* explanations. Black-box explanation approaches assume no knowledge on the neural network architecture and its parameters; white-box explanations techniques require that all parameters of the neural network are known in advance and can be used to extract an explanation. In addition, XAI techniques can be classified according to their explanation scope as *local* or *global*. Local-level explanation techniques (e.g., Activation Maximization, Gradient-based Saliency Map, Local Interpretable Model-Agnostic Explanations) explain the decisions taken on individual data instances. Global-level explanation techniques (e.g., Global Attribution Mapping, Neural Additive Model, Concept Activation Vectors) seek to understand the model as a whole based on groups of data instances. Finally, an explainable technique can be incorporated into the neural network model or applied as an external algorithm for explanation. In particular, various *model-agnostic post-hoc* explanation techniques have been recently tested, in which the XAI technique does not depend on the model architecture and can be applied to already trained networks [10]. For example, Gradient-weighted Class Activation Mapping (Grad-CAM) [17] is a post-hoc, model independent, gradient-based explainable method that has recently emerged in computer vision [18].

Interestingly a recent study [23] has initiated the investigation into XAI techniques to provide explanations for the decisions of a neural network in malware detection and vulnerability discovery. This compares several XAI techniques with respect to the accuracy of explanations, as well as security focused aspects, such as completeness, efficiency, and robustness. Although this study draws the conclusion that gradient-based methods can generally be recommended for explaining predictions in security systems, it also highlights that further investigations are needed in order to fully exploit explanations in cyber-security. Explanations based on locally faithful decision boundaries are used to complement an intrusion detection system in [8] and explain decisions. Local and global explanations are also studied in [22] to improve the transparency of an intrusion detection system. In our work we also proceed in the research direction started in [23] by coupling a XAI technique to a DL-based intrusion detection method. In fact, we use a gradient-based XAI white-box strategy applied in a post-hoc way to provide local explanations of neural network decisions. However, our study advances

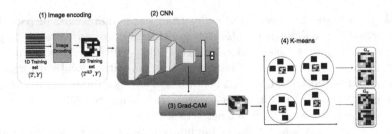

**Fig. 1.** The training stage of GRACE. (1) The training set $(\mathbf{T}, \mathbf{Y})$ is converted to $(\mathbf{T^{2D}}, \mathbf{Y})$. (2) The new training set $(\mathbf{T^{2D}}, \mathbf{Y})$ is processed to learn a 2D CNN model. (3) The last convolutional layer of the 2D CNN is used to obtain Grad-CAM heatmaps. (4) The heatmaps are used to determine the cluster centres and learn $\mathbf{G}_\oplus$—the set of the $k$ Grad-CAM centres determined on the normal traffic—and $\mathbf{G}_\ominus$—the set of the $k$ Grad-CAM centres determined on the attacks.

previous studies on XAI in cyber-security as we do not just explain the features that help recognize network attacks, but we leverage the explanatory information to improve the final accuracy and robustness of the security system.

## 3 Proposed Methodology

In this Section, we describe our novel network intrusion detection methodology, called GRACE, which achieves a combination of CNN, Grad-CAM, $k$-means and nearest-neighbour search, in order to learn a reliable predictive model that can accurately detect new signs of malicious activity in the network traffic. Since the focus of this study is on a binary classification task, all attack classes are assigned the same label regardless of the type of attack. In addition, GRACE is formulated to process flow-based characteristics (e.g., the duration of a flow) of network traffic, which aggregate information about all the packets in the network. The methodology fulfilled by GRACE consists of a training stage and a predictive stage. Both stages are described in more detail below.

### 3.1 Training Stage

The training stage block diagram of GRACE is shown in Fig. 1. During this stage, GRACE is fed with a training set $(\mathbf{T}, \mathbf{Y})$ that collects the $N$ historical network flows (training samples) that are spanned on $M$ features of a 1D feature vector $\mathbf{X^{1D}}$ and labelled with a binary target $Y$ labelled "normal" and "attack". Each feature $X_i$ of $\mathbf{X^{1D}}$ describes a characteristic of a network flow. $\mathbf{T}$ is the $N \times M$ matrix representing the training samples on the rows and the features of $\mathbf{X^{1D}}$ on the columns. $\mathbf{Y}$ is the $N \times 1$ vector that collects the labels of the training samples. GRACE learns the intrusion detection function $\mathbf{X^{1D}} \mapsto Y$ from $(\mathbf{T}, \mathbf{Y})$ with a three-step methodology. In particular, this function consists of a CNN and a collection of CNN Grad-CAMs selected as representative of normal training samples and attacks, respectively.

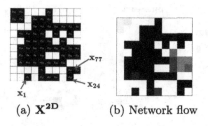

(a) $\mathbf{X^{2D}}$          (b) Network flow

**Fig. 2.** An example of a 2D grid that assigns traffic characteristics (e.g., Destination Port $(X_1)$, Idle Max $(X_{77})$ and Fwd IAT Max $(X_{24})$) to the pixel frames of the 2D grid (Fig. 2a), and the image form of a network flow showing the characteristic value in the assigned pixel (Fig. 2b).

In the first step, GRACE trains a CNN by replicating the CNN-based intrusion detection pipeline described in [4]. This pipeline is here selected as it able to train an intrusion detection model by outperforming various, recent, state-of-art intrusion detection algorithms. It is based on the idea of capturing patterns of spatial continuity among traffic characteristics and exploiting these patterns to derive an image form of the network flows. This allows us to approach the intrusion detection task as an image classification problem. In particular, the image encoding step transforms each training sample from the 1D feature vector form $\mathbf{X^{1D}}$ with size $1 \times M$ to the 2D image form $\mathbf{X^{2D}}$ with size $m \times m$ (with $M \leq m^2$). This transformation is done by assigning each feature of $\mathbf{X^{1D}}$ to a pixel frame of $\mathbf{X^{2D}}$ (see Fig. 2). A detailed description of how the feature assignment is performed is reported in [4]. Mathematically, the image transformation step transforms the training data matrix $\mathbf{T}$ of size $N \times M$ into the training data hypercube $\mathbf{T^{2D}}$ of size $N \times m \times m$. We train a CNN on $(\mathbf{T^{2D}}, \mathbf{Y})$ exploiting convolutions to better discriminate attacks from normal network flows.

In the second step, once the CNN has been trained, its last convolutional layer is used to obtain heatmaps of class activations on the input images, i.e. 2D grids of scores, calculated for each pixel in an input image, which indicates how important each pixel is with respect to a specific output class. To do this, we use the Grad-CAM method illustrated in [18]. Given an input image obtained as described above, this technique consists of extracting the output feature maps of the last convolutional layer and weighting each channel based on the gradient of the class with respect to that channel. More formally, let $y^c$ be the predicted class, and $A^k \in \mathbb{R}^{u \times v}$ be the $k$-th feature map of the last convolutional layer, where $u$ and $v$ are its width and height, obtained as the typical down-sampling of a CNN when going deeper. A "summary" of the overall feature maps can be obtained as a linear combination followed by a $ReLU$ non-linearity:

$$\text{Grad-CAM} = ReLU \left( \sum_k \alpha_k^c A^k \right),$$

where $ReLU$ is used because the pixels of interest are the ones that have a *positive* influence on the predicted class, while *negative* pixels are more likely to

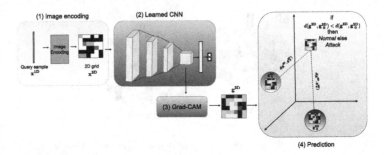

**Fig. 3.** The predictive stage of GRACE. An example query $\mathbf{x}^{1D}$ is mapped as $\mathbf{x}^{2D}$ form (1) and the learned 2D CNN is used to classify the sample (2) and to build the Grad-CAM heatmap of $\mathbf{g}^{2D}$ (3). Finally, the distance is computed between $\mathbf{g}^{2D}_{\oplus}$ and $\mathbf{g}^{2D}_{\ominus}$: these distances are processed to predict the class of $\mathbf{x}$ (4).

belong to the other class. Since some feature maps may be more important than others for the final prediction $y^c$, the average pooling of the gradient of $y^c$ with respect to the $k$-th feature map is used to weight the feature map:

$$\alpha_k^c = \frac{1}{uv} \sum_{i=1}^{u} \sum_{j=1}^{v} \frac{\partial y^c}{\partial A_{i,j}^k},$$

where $\frac{\partial y^c}{\partial A_{i,j}^k}$ measures the effect of pixel $i,j$ in the $k$-th feature map on the final prediction $y^c$. We recall that the pixel $i,j$ denotes a traffic characteristic of $X \in \mathbf{X}^{1D}$. Therefore, $\frac{\partial y^c}{\partial A_{i,j}^k}$ explains the effect of the characteristic $X$ on the final prediction $y^c$. The up-sampling of Grad-CAM to the size of the input image allows us to highlight the regions of the input image that contributed most to the final classification. In other words, this is a kind of "visual explanation".

In the third step, we perform the clustering process with the $k$-means algorithm run on the Grad-CAMs of the normal training network flows and the Grad-CAMs of the training attacks, separately. The clustering step helps to limit overfitting as already shown in [3]. As for each clustering run we process the Grad-CAMs labelled with the same class in the training set, we learn $\mathbf{G}_{\oplus}$—the set of the $k$ Grad-CAM cluster centres of the normal traffic—and $\mathbf{G}_{\ominus}$—the set of the $k$ Grad-CAM cluster centres of the attack traffic. We note that $\mathbf{G}_{\oplus}$ encloses the visual explanation of how to recognize normal traffic, while $\mathbf{G}_{\ominus}$ encloses the visual explanation of how to recognize an attack pattern, respectively.

### 3.2 Predictive Stage

The block diagram of the predictive stage of GRACE is described in Fig. 3. Let us consider a query sample $\mathbf{x}^{1D}$ as defined on the 1D feature vector $\mathbf{X}^{1D}$. First, we determine $\mathbf{x}^{2D}$, which is the image form of $\mathbf{x}^{1D}$ derived by arranging the 1D values of $\mathbf{x}^{1D}$ on the 2D grid $\mathbf{X}^{2D}$ as it has been determined in the training stage.

We classify $\mathbf{x^{2D}}$ with the CNN and visualize the CNN decision using the Grad-CAM. Let $\mathbf{g^{2D}}$ be the Grad-CAM of $\mathbf{x^{2D}}$, we determine both the nearest normal Grad-CAM cluster centroid neighbour—$\mathbf{g^{2D}_{\oplus}}$—and the nearest attacking Grad-CAM cluster centroid neighbour—$\mathbf{g^{2D}_{\ominus}}$—of $\mathbf{g^{2D}}$. The neighbourhood relation is evaluated by computing the Euclidean distance. Formally, we compute

$$\mathbf{g^{2D}_{\oplus}} = \arg\max_{\mathbf{g^{2D}_{\oplus}} \in G_{\oplus}} d(\mathbf{g^{2D}}, \mathbf{g^{2D}_{\oplus}}) \ and \ \mathbf{g^{2D}_{\ominus}} = \arg\max_{\mathbf{g^{2D}_{\ominus}} \in G_{\ominus}} d(\mathbf{g^{2D}}, \mathbf{g^{2D}_{\ominus}}), \tag{1}$$

where $d$ is the Euclidean distance, i.e., $d(\mathbf{g^{2D}}, \mathbf{g^{2D}_{\oplus}}) = \sum_{i=1}^{n}\sum_{j=1}^{m}(\mathbf{g^{2D}}[i,j] - \mathbf{g^{2D}_{\oplus}}[i,j])^2$ and $d(\mathbf{g^{2D}}, \mathbf{g^{2D}_{\ominus}}) = \sum_{i=1}^{n}\sum_{j=1}^{m}(\mathbf{g^{2D}}[i,j] - \mathbf{g^{2D}_{\ominus}}[i,j])^2$. If $d(\mathbf{g^{2D}}, \mathbf{g^{2D}_{\oplus}}) < d(\mathbf{g^{2D}}, \mathbf{g^{2D}_{\ominus}})$ then $\mathbf{x^{1D}}$ is classified as normal. Otherwise $\mathbf{x^{1D}}$ is classified as attack.

Note that the main intuition of our proposed method is that the knowledge gathered by the CNN during the supervised learning phase can be used to do an *ex-post* identification of the most salient traffic features that characterize the network flow. In fact, the existence of a one-to-one mapping between the original feature vectors and their transformation into images allows us to use Grad-CAM as a way to go up the hierarchy of CNN-induced transformations and identify the most important features. Therefore, since it can be assumed that new data traffic samples would be explained with similar explanation maps, their Grad-CAMs can be compared with those of the training (clustered) samples, to perform a kind of *post-hoc* classification, in which the knowledge of the most important features could be useful in helping GRACE to focus on the most relevant traffic patterns and thus to obtain higher accuracy than the CNN model alone.

## 4   Empirical Evaluation

We used three benchmark datasets in our evaluation, described in Sect. 4.1. The implementation details of the version of GRACE[1] used in the evaluation are illustrated in Sect. 4.2. The evaluation metrics are described in Sect. 4.3, while the experimental results are discussed in Sect. 4.4.

### 4.1   Dataset Description

We considered three benchmark intrusion detection datasets, namely KDD-CUP99,[2] NSL-KDD[3] and UNSWNB15.[4] These datasets have been recently used

---

[1] https://github.com/fpcaforio/grace/.
[2] http://kdd.ics.uci.edu//databases//kddcup99//kddcup99.html.
[3] https://www.unb.ca/cic/datasets/nsl.html.
[4] https://www.unsw.adfa.edu.au/unsw-canberra-cyber/cybersecurity/ADFA-NB15-Datasets.

**Table 1.** Dataset description. For each dataset we report: the number of attributes, the total number of network flow samples collected in the dataset, the number of normal network flows, as well as the number of attacking flows.

| | | Dataset | | |
|---|---|---|---|---|
| | | KDDCUP99 | NSL-KDD | UNSWNB15 |
| **Attributes** | Total | **42** | **42** | **43** |
| | Binary | 6 | 6 | 2 |
| | Categorical | 3 | 3 | 3 |
| | Numerical | 32 | 32 | 37 |
| | Class | 1 | 1 | 1 |
| **Training set** | Total | **494,021** | **125,973** | **82,332** |
| | Normal flows | 97,278 (19.7%) | 67,343 (53.5%) | 37,000 (44.9%) |
| | Attacking flows | 396,743 (80.3%) | 58,630 (46.5%) | 45,332 (55.1%) |
| **Testing set** | Total | **311,029** | **22,544** | **175,341** |
| | Normal flows | 60,593 (19.5%) | 9,711 (43.1%) | 56,000 (31.9%) |
| | Attacking flows | 250,436 (80.5%) | 12,833 (56.9%) | 119,341 (68.1%) |

in the evaluation of various state-of-the-art competitors also analyzed in Sect. 4.4. Each dataset includes both a labelled training set and a test set. In our evaluation study, we processed 10%KDDCUP99Train for the learning stage, while we used the entire test set, referred to as KDDCUP99Test, for the evaluation stage. This experimental scenario is commonly used in the literature. NSL-KDD was introduced in [19] as a revised version of KDDCUP99, which was obtained by removing duplicate samples from KDDCUP99. Finally, UNSWNB15 was created by the IXIA PerfectStorm tool in the Cyber Range Lab of the Australian Centre for Cyber Security (ACCS) for generating a hybrid of real modern normal activities and synthetic contemporary attack behaviours. A summary of the characteristics of these datasets is presented in Table 1.

## 4.2  Implementation Details

The proposed method was implemented in Python (version 3.7.10), using the Keras library (version 2.4.3) with TensorFlow as back-end, and the Scikit-learn library (version 0.22.2). In the pre-processing step, the categorical input features were mapped into numerical features using the one-hot-encoder strategy and then the numerical features have been scaled using the min-max normalization. The CNN architecture was designed according to the implementation described in [4] performing automatic optimization of batch size, learning rate, dropout and kernel size. This hyper-parameter optimization was conducted using the tree-structured Parzen estimation algorithm and considering 20% of the entire training data as a validation set to automatically choose the hyper-parameter configuration that achieves the best validation loss. To decide the $m \times m$ size of the network flow image transformation, we processed the image transformations built with the size $m \in [\sqrt{M}, \sqrt{M} + 1, \sqrt{M} + 2]$ (where $M$ is the number

**Table 2.** Ablation study configurations.

|                | CNN | Grad-CAM | k-means | Nearest-Neighbour search |
|----------------|-----|----------|---------|--------------------------|
| CNN            | ×   |          |         |                          |
| CNN+Grad-CAM+NN| ×   | ×        |         | ×                        |
| GRACE          | ×   | ×        | ×       | ×                        |

of features generated in the pre-processing step). For the Grad-CAM step, we automatically selected the image transformation that achieved the highest validation accuracy. The image transformation size selected with this procedure is $13 \times 13$ for KDDCUP99, $12 \times 12$ for NSL-KDD and $15 \times 15$ for UNSWNB15. The Grad-CAM algorithm has been implemented to work on the final convolutional layer of the 2D CNN. Finally, the $k$-means algorithm was run with the default parameter configuration (mini-batch size equals to 100 and number of iterations equals to 100). We set the number of clusters to 1000 in the baseline configuration.

### 4.3 Evaluation Metrics

The overall accuracy performance of the proposed methodology has been measured by analyzing the F1-score – F1 – (i.e., the harmonic mean of precision and recall) and accuracy – A – (i.e., the ratio of correctly labelled flows across all tested flows) of the learned intrusion detection models. The efficiency performance has been evaluated with the computation time (in minutes) – T – spent training the intrusion detection models. The experiments were run on Google Colab taking advantage of the GPU runtime.

### 4.4 Results

**Ablation Study.** The experimental study begun by performing an ablation study to investigate how the proposed methodology could benefit from coupling the decision made via the CNN to the nearest-neighbour search performed on the visual explanations of the CNN decisions constructed with Grad-CAM. In addition, we explored the effect of $k$-means in the nearest-neighbour step. To do this, we compared the performance of three configurations defined in Table 2. For the accuracy performance analysis, we report both F1 (Figs. 4a–4c) and A (Figs. 4d–4f) measured on predictions made on the test samples. For the efficiency analysis, we report the computation T spent completing the training stage (Figs. 4g–4i). The results show that GRACE is more accurate than its baselines, although the gain in accuracy comes at the expense of more time spent completing the training stage. GRACE also takes advantage of the clustering step, especially in NLS-KDD and UNSWNB15. This confirms that clustering mitigates overfitting when explaining CNN decisions and improves the accuracy of the final predictions made on the basis of these explanations.

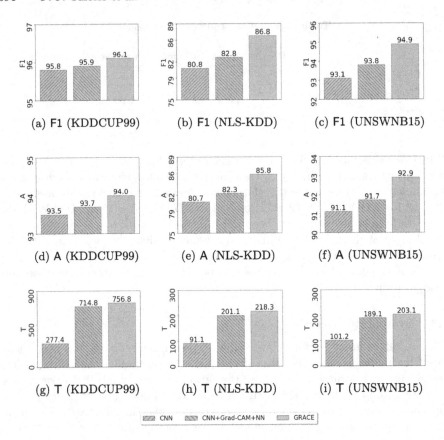

(a) F1 (KDDCUP99)    (b) F1 (NLS-KDD)    (c) F1 (UNSWNB15)

(d) A (KDDCUP99)    (e) A (NLS-KDD)    (f) A (UNSWNB15)

(g) T (KDDCUP99)    (h) T (NLS-KDD)    (i) T (UNSWNB15)

CNN    CNN+Grad-CAM+NN    GRACE

**Fig. 4.** Ablation analysis: F1 (Figs. 4a–4c), A (Figs. 4d–4f) and T (Figs. 4g–4i) of CNN, CNN+Grad-CAM+NN and GRACE on KDDCUP99, NSL-KDD and UNSWNB15.

**Sensitivity Study.** We proceeded with the analysis by studying how the performance of GRACE depends on the number of clusters discovered in the $k$-means step. We varied $k$ between 500, 1000, 5000 and 10000. As a baseline for this investigation, we also considered the configuration of GRACE without the clustering step (CNN+Grad-CAM+NN), where the nearest-neighbour search is performed considering all the training explanation samples as candidate neighbours. Again, for the accuracy performance analysis, we report both F1 (see Figs. 4a–4c) and A (Figs. 4d–4f) measured on the predictions made on the test samples. For the efficiency analysis, we report the computation T spent completing the training stage (Figs. 4g–4i). The results shown provide further empirical evidence that clustering does not decrease the accuracy performance; on the contrary, it achieves an increase in accuracy in all datasets mitigating possible overfitting in the nearest-neighbour search. Also, the smaller the number of clusters, the less the time it takes to complete the training stage. In all datasets, the default configuration

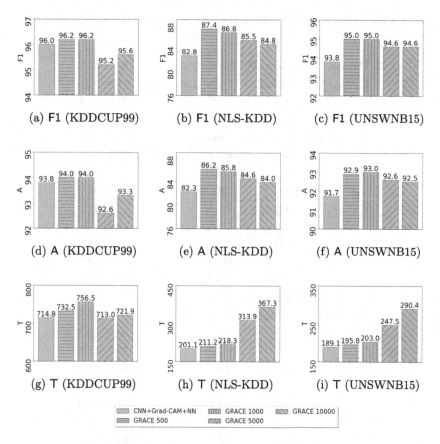

**Fig. 5.** Sensitivity analysis: F1 (Figs. 5a–5c), A (Figs. 5d–5f) and T (Figs. 5g–5i) of GRACE varying $k$ between 500, 1000, 5000 and 10000 on KDDCUP99, NSL-KDD and UNSWNB15. The baseline is CNN+Grad-CAM+NN.

with $k = 1000$ achieves the highest accuracy (or the runner-up accuracy as in NSL-KDD where the top-ranked accuracy is obtained with $k = 500$).

**Comparison with State-of-the-Art.** We also compared the accuracy performance of GRACE with that of several DL-based competitors selected from recent literature. For all methods compared, we collected the accuracy A and F1-score F1, as these metrics are commonly provided in the reference studies. The results, reported in Table 3 for all datasets, show that GRACE generally outperforms its competitors which set aside any explanation of the decisions. A few exceptions occur. The method described in [5] performs better than GRACE in NSL-KDD, but worse in KDDCUP99 and UNSWNB15. The methods described in [3] and [6] perform better than GRACE in UNSWNB15, but worse in both KDDCUP99 and NSL-KDD. Overall, the explanations can actually help build a more robust intrusion detection model.

**Table 3.** Competitor analysis. "-" denotes that no value is reported in the reference paper.

| Dataset | Algorithm | Description | A | F1 |
|---------|-----------|-------------|------|------|
| KDDCUP99 | GRACE | CNN+Grad-CAM | 94.0 | 96.1 |
| | [1] | CNN | 92.9 | 95.4 |
| | [3] | CNN | 93.5 | 95.9 |
| | [5] | Triplet | 93.5 | 95.8 |
| | [6] | CNN | 92.4 | 95.1 |
| | [9] | GAN | – | 90.0 |
| | [11] | LSTM | – | 93.2 |
| | [11] | RNN | – | 91.8 |
| | [12] | CNN | 92.4 | – |
| | [21] | DNN | 93.0 | 95.5 |
| | [27] | GAN | | 93.7 |
| | [28] | GAN | – | 88.6 |
| | [28] | GAN | – | 95.0 |
| NLS-KDD | GRACE | CNN+Grad-CAM | 85.7 | 86.8 |
| | [1] | CNN | 80.9 | 80.5 |
| | [3] | CNN | 78.7 | 77.5 |
| | [5] | Triplet | 86.6 | 87.0 |
| | [6] | CNN | 79.6 | 79.0 |
| | [14] | LSTM | 82.7 | 83.3 |
| | [20] | CNN | 80.7 | – |
| | [26] | RNN | 83.2 | – |
| UNSWNB15 | GRACE | CNN+Grad-CAM | 92.9 | 94.9 |
| | [1] | CNN | 92.4 | 94.7 |
| | [3] | CNN | 93.5 | 95.4 |
| | [5] | Triplet | 89.2 | 91.4 |
| | [6] | CNN | 93.4 | 95.2 |
| | [13] | CNN | 80.0 | 84.0 |
| | [13] | CNN | 83.0 | 86.5 |
| | [16] | CNN | 89.8 | 91.3 |
| | [16] | DNN | 86.6 | 88.9 |
| | [21] | DNN | 76.5 | 90.1 |
| | [25] | DNN | 91.2 | – |
| | [25] | MLP | 86.7 | – |
| | [25] | Autoencoder | 88.2 | – |

**Grad-CAM Analysis.** At the end of this study, we describe some explanations produced by GRACE on NSL-KDD. Figure 6 shows separately the Grad-CAMs associated with the cluster centres computed on the Grad-CAMs of the normal training samples and the Grad-CAMs of the attack training samples. For this analysis, we consider the clusters discovered in the baseline configuration ($k = 1000$). We rank the clusters in descending order based on their cardinality and we analyze the top-2 clusters for each class. Figures 6a and 6b show the Grad-CAMs computed as the centres of the top-2 attacking clusters, while Figures 6c and 6d show the Grad-CAMs computed as the centres of the top-2 normal clusters. In each centre, we highlight the top-3 characteristics that the Grad-CAM explains as the most relevant for the CNN classification (i.e., the traffic characteristics for which the highest gradient is measured). Note that this analysis reveals that the traffic characteristics most relevant to recognizing an attack are different from the characteristics most relevant to recognizing a normal sample. Interestingly, the explanation for the attack class indicates that Dst_host_rerror_rate (% of connections to current host and specified service having an S0 error) and Same_srv_rate (% of connections to the same service) are very relevant for both the top-2 attack clusters. On the other hand, Dst_host_same_src_port_rate (% of connections to current host with same src port) appears very relevant in both the top-2 normal clusters.

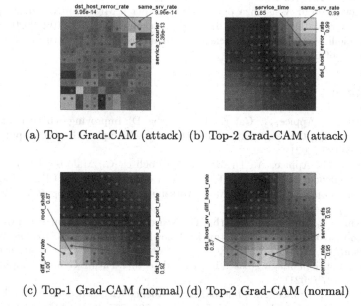

(a) Top-1 Grad-CAM (attack)  (b) Top-2 Grad-CAM (attack)

(c) Top-1 Grad-CAM (normal) (d) Top-2 Grad-CAM (normal)

**Fig. 6.** Grad-CAM centres of the top-2 clusters computed by $k$-means ($k = 1000$) on the Grad-CAMs built on the normal and attacking training traffic of NSL-KDD. The characteristics with the highest gradient are marked in the plot.

## 5 Conclusion

The machine learning community has recently seen a growing interest and concern in making the decisions made by DL models more explainable. This is true in several critical domains, but especially in cyber-security, as lack of transparency can impact security systems and make them more vulnerable to ever-changing attacks. To contribute to this research effort, in this paper we have proposed a novel IDS based on DL and XAI. The system relies on visual explanations provided by a CNN trained on historical normal and attack traffic data to better focus on the most important traffic characteristics that discriminate between the two classes. This knowledge is leveraged to perform a *post-hoc* classification that proves more accurate versus using a single "unexplained" learning model. Hence, XAI techniques are not only used to explain decisions to human experts, but to guide the system towards learning a better model. Although our work has shown that we can achieve a more robust classification of attacks by leveraging explanations in DL-based intrusion detection models, some open issues require further research. For example, a research direction is studying a strategy to take advantage of the explanatory information to identify the features that are most robust to adversarial attacks. Another direction is to explore the use of explanations to classify the attack category (e.g., Dos, DDoS, Port Scan).

**Acknowledgment.** We acknowledge the support of MUR through the project "TAL-IsMan - Tecnologie di Assistenza personALizzata per il Miglioramento della quAlitá della vitA" (Grant ID: ARS01_01116), funding scheme PON RI 2014–2020 and the project "Modelli e tecniche di data science per la analisi di dati strutturati" funded by the University of Bari "Aldo Moro".

## References

1. Andresini, G., Appice, A., Caforio, F., Malerba, D.: Improving cyber-threat detection by moving the boundary around the normal samples. Stud. Comput. Intell. **919**, 105–127 (2021)
2. Andresini, G., Appice, A., Di Mauro, N., Loglisci, C., Malerba, D.: Exploiting the auto-encoder residual error for intrusion detection. In: 2019 IEEE European Symposium on Security and Privacy Workshops (EuroS PW), pp. 281–290. IEEE (2019)
3. Andresini, G., Appice, A., Malerba, D.: Nearest cluster-based intrusion detection through convolutional neural networks. Knowl.-Based Syst. **216**, 106798 (2021)
4. Andresini, G., Appice, A., De Rose, L., Malerba, D.: Gan augmentation to deal with imbalance in imaging-based intrusion detection. Future Gener. Comput. Syst. **123**, 108–127 (2021)
5. Andresini, G., Appice, A., Malerba, D.: Autoencoder-based deep metric learning for network intrusion detection. Inf. Sci. **569**, 706–727 (2021). https://doi.org/10.1016/j.ins.2021.05.016
6. Andresini, G., Appice, A., Mauro, N.D., Loglisci, C., Malerba, D.: Multi-channel deep feature learning for intrusion detection. IEEE Access **8**, 53346–53359 (2020)
7. Arrieta, A.B., et al.: Explainable artificial intelligence (xai): Concepts, taxonomies, opportunities and challenges toward responsible ai. Inf. Fusion **58**, 82–115 (2020)

8. Burkart, N., Franz, M., Huber, M.F.: Explanation framework for intrusion detection. In: Beyerer J., Maier A., Niggemann O. (eds.) Machine Learning for Cyber Physical Systems, vol. 13, pp. 83–91. Springer, Berlin (2021). https://doi.org/10.1007/978-3-662-62746-4_9

9. Dan, L., Dacheng, C., Baihong, J., Lei, S., Jonathan, G., See-Kiong, N.: Mad-gan: Multivariate anomaly detection for time series data with generative adversarial networks. In: Artificial Neural Networks and Machine Learning, pp. 703–716 (2019)

10. Das, A., Rad, P.: Opportunities and challenges in explainable artificial intelligence (XAI): A survey. arXiv preprint arXiv:2006.11371 (2020)

11. Elsherif, A.: Automatic intrusion detection system using deep recurrent neural network paradigm. J. Inf. Secur. Cybercrimes Res. 1(1), 21–31 (2018)

12. He, Y.: Identification and processing of network abnormal events based on network intrusion detection algorithm. I. J. Netw. Secur. 21, 153–159 (2019)

13. Kim, T., Suh, S.C., Kim, H., Kim, J., Kim, J.: An encoding technique for cnn-based network anomaly detection. In: 2018 IEEE International Conference on Big Data (Big Data), pp. 2960–2965. IEEE (2018)

14. Li, Z., Rios, A.L.G., Xu, G., Trajković, L.: Machine learning techniques for classifying network anomalies and intrusions. In: 2019 IEEE International Symposium on Circuits and Systems (ISCAS), pp. 1–5. IEEE (2019)

15. Lipton, Z.C.: The mythos of model interpretability. Commun. ACM 61(10), 36–43 (2018)

16. Lopez-Martin, M., Carro, B., Sanchez-Esguevillas, A., Lloret, J.: Shallow neural network with kernel approximation for prediction problems in highly demanding data networks. Exp. Syst. Appl. 124, 196–208 (2019)

17. Selvaraju, R.R., Cogswell, M., Das, A., Vedantam, R., Parikh, D., Batra, D.: Grad-cam: Visual explanations from deep networks via gradient-based localization. In: 2017 IEEE International Conference on Computer Vision (ICCV), pp. 618–626 (2017)

18. Selvaraju, R.R., Cogswell, M., Das, A., Vedantam, R., Parikh, D., Batra, D.: Grad-cam: visual explanations from deep networks via gradient-based localization. Int. J. Comput. Vis. 128(2), 336–359 (2020)

19. Tavallaee, M., Bagheri, E., Lu, W., Ghorbani, A.A.: A detailed analysis of the KDD CUP 99 data set. In: CISDA, pp. 1–6 (2009)

20. Teyou, D., Kamdem, G., Ziazet, J.: Convolutional neural network for intrusion detection system in cyber physical systems. arXiv preprint arXiv:1905.03168 (2019)

21. Vinayakumar, R., Alazab, M., Soman, K.P., Poornachandran, P., Al-Nemrat, A., Venkatraman, S.: Deep learning approach for intelligent intrusion detection system. IEEE Access 7, 41525–41550 (2019)

22. Wang, M., Zheng, K., Yang, Y., Wang, X.: An explainable machine learning framework for intrusion detection systems. IEEE Access 8, 73127–73141 (2020)

23. Warnecke, A., Arp, D., Wressnegger, C., Rieck, K.: Evaluating explanation methods for deep learning in security. In: 2020 IEEE European Symposium on Security and Privacy (EuroS&P), pp. 158–174. IEEE (2020)

24. Xie, N., Ras, G., van Gerven, M., Doran, D.: Explainable deep learning: A field guide for the uninitiated. arXiv preprint arXiv:2004.14545 (2020)

25. Yan, J., Jin, D., Lee, C.W., Liu, P.: A comparative study of off-line deep learning based network intrusion detection. In: 10th International Conference on Ubiquitous and Future Networks, pp. 299–304 (2018)

26. Yin, C., Zhu, Y., Fei, J., He, X.: A deep learning approach for intrusion detection using recurrent neural networks. IEEE Access 5, 21954–21961 (2017)

27. Zenati, H., Foo, C.S., Lecouat, B., Manek, G., Chandrasekhar, V.R.: Efficient gan-based anomaly detection. CoRR abs/1802.06222, pp. 1–13 (2018)
28. Zenati, H., Romain, M., Foo, C.S., Lecouat, B., Chandrasekhar, V.R.: Adversarially learned anomaly detection. 2018 IEEE International Conference on Data Mining (ICDM), pp. 727–736 (2018)

# Local Interpretable Classifier Explanations with Self-generated Semantic Features

Fabrizio Angiulli$^{(\boxtimes)}$ ⓘ, Fabio Fassetti ⓘ, and Simona Nisticò ⓘ

DIMES Department, University of Calabria, Rende, Italy
{fabrizio.angiulli,fabio.fassetti,simona.nistico}@dimes.unical.it

**Abstract.** Explaining predictions of classifiers is a fundamental problem in eXplainable Artificial Intelligence (XAI). LIME (for Local Interpretable Model-agnostic Explanations) is a recently proposed XAI technique able to explain any classifier by providing an interpretable model which approximates the black-box locally to the instance under consideration. In order to build interpretable local models, LIME requires the user to explicitly define a space of interpretable components, also called artefacts, associated with the input instance. To reconstruct local black-box behaviour, the instance neighbourhood is explored by generating instance neighbours as random subsets of the provided artefacts. In this work we note that the above depicted strategy has two main flaws: first, it requires user intervention in the definition of the interpretable space and, second, the local explanation is limited to be expressed in terms the user-provided artefacts. To overcome these two limitations, in this work we propose $\mathcal{S}$-LIME, a variant of the basic LIME method exploiting unsupervised learning to replace user-provided interpretable components with self-generated semantic features. This characteristics enables our approach to sample instance neighbours in a more semantic-driven fashion and to greatly reduce the bias associated with explanations. We demonstrate the applicability and effectiveness of our proposal in the text classification domain. Comparison with the baseline highlights superior quality of the explanations provided adopting our strategy.

**Keywords:** Explainable machine learning · Local interpretable explanations · Adversarial autoencoders

## 1 Introduction

Explaining predictions of classifiers is a fundamental problem in eXplainable Artificial Intelligence (XAI). Approaches to address this problem can be summarized in different families [6]. *Gradient based methods* are characterized by the use of gradient information of the model to explain. *Layer-wise relevance propagation methods* exploit the model knowledge to explain its outcome, these methods apply only to neural networks since exploit explicitly their structure.

© Springer Nature Switzerland AG 2021
C. Soares and L. Torgo (Eds.): DS 2021, LNAI 12986, pp. 401–410, 2021.
https://doi.org/10.1007/978-3-030-88942-5_31

On the other side, there are two families that need no knowledge of the model, which can be regarded as a black-box. *Occlusion analysis methods* exploit the effect obtained from small perturbations of the sample to explain. Finally, *interpretable local surrogates methods* return an explainable decision function that intends to mimic the decision of the black-box locally [1,5].

Among them LIME (for Local Interpretable Model-agnostic Explanations), is a recently proposed XAI technique able to explain any classifier by providing an interpretable model which approximates the black-box locally to the instance under consideration. LIME has some limitations and drawbacks and for this reason in literature there are some works that propose variants of this algorithm in order to solve some of its issues or to extend it [9,11,12].

In this work we note that the above depicted strategy has two main flaws: first, it requires user intervention in the definition of the interpretable space and this task is not always easy, second, the local explanation is limited to be expressed in terms of the artefacts associated with the instance $x$ to be explained and this could not be able to capture complex patterns.

As a matter of fact, any interpretable representation returned by LIME is always a subset of the artefacts extracted from $x$. As an example, consider the textual domain, since artefacts are represented by terms occurring into the phrase $x$, this means that the explanation is always limited to an excerpt from the sentence $x$.

This means that synonyms or terms not occurring into the phrase cannot be part of the explanation, this makes this method unable to capture some types of pattern during the explanation of the black-box model. Moreover, the same consideration holds for the terms that if they occurred in the sentence would change its meaning as they are combined with other terms already present.

To overcome these two limitations, in this work we propose $S$-LIME, a variant of the basic LIME method exploiting unsupervised learning to replace user-provided interpretable components with self-generated semantic features.

This characteristic enables our approach to sample instance neighbours in a more semantic-driven fashion and to greatly reduce the bias associated with explanations.

We demonstrate the applicability and effectiveness of our proposal in the text classification domain. Comparison with the baseline highlights superior quality of the explanations provided adopting our strategy.

The rest of the work is organized as follows. Section 2 introduces preliminary notions, Sect. 3 describes the technique $S$-LIME and presents its innovative points, Sect. 4 depicts the experiments conducted to show the behavior of the algorithm and the comparison with LIME.

## 2   Preliminaries

Let $X \subset \mathbb{R}^d$ denote an *instance domain*. A *model* $f$ is either a binary indicator, that is a function mapping each instance $x \in X$ to one of two pre-defined classes identified as 0 (the negative class) and 1 (the positive class), or a function returning the probability for $x$ to belong to the positive class (while $1 - f(x)$ is the probability to belong to the negative class), we refer to binary case for sake of simplicity. $X$ is also referred to as the *original space* or *original domain* and each $x \in X$ is an instance in its *original representation*.

Since $f$ uses in general complex and incomprehensible features, a very important problem is to *explain* $f$, namely to make its output humanly understandable. More specifically, the aim here is to *explain why the model $f$ classifies a given instance $x$ as $f(x)$*. To accomplish this a possible approach, proposed in [4] and here employed, is to learn a humanly understandable model which can approximate the behaviour of the model $f$ in the proximity of $x$.

Thus, the *explanation* can be defined as a model $g \in G$, where $G$ is the set of interpretable models, like *linear models* or *decision trees*.

Given two instances $x \in X$ and $y \in X$, a *proximity function* $\pi(x, y) : X^2 \to \mathbb{R}$ is a function measuring the proximity between $x$ and $y$, used to define the around of an instance $x$.

Given, an instance $x$, a model $f$, an explanation $g$ and a proximity function $\pi$, a *fidelity function* $\mathcal{F}$ is a function returning a measure of how good is $g$ in approximating $f$ in the neighbourhood of $x$. Thus, formally given an instance $x$ and a model $f$, LIME aims to find an explanation $g$ such that the following loss function is minimized

$$\xi(x) = \mathcal{F}(x, f, g, \pi) + \Omega(g), \tag{1}$$

where $\Omega(g) : G \to \mathbb{R}$ is a function returning the complexity of the model $g$, acting as a regularization term guiding the search towards simpler models.

There are, then, two main problems to be addressed, that are the choice of the model $g$ and the way to compute the neighbourhood of $x$, which also depends on the data domain.

Authors of [4] propose to tackle these problems as follows. As for the model $g$, they propose to make it working on a different space $X'$, also called *space* of the *interpretable components* or *interpretable space*, and, in particular, to use *linear regression* as interpretable model.

The proposed fidelity function, used in LIME objective function reported in the equation, (1), is

$$\mathcal{F}(x, f, g, \pi) = \sum_{y \in X} \pi(x, y) \cdot (f(y) - g(y'))^2, \tag{2}$$

where $y'$ is the mapping of $y$ to $X'$. Thus, the model $g$ should be such that, the probability to belong to a certain class assigned by $f$ to an instance $y$ is approximately the same that $g$ assigns to the image $y'$ of $y$ in the interpretable space.

As for the neighbourhood of $x$, the proposal of [4] is to identify a set of $d'$ artefacts and consider the space $X' = \{0, 1\}^{d'}$, where each component $i$ asserts if a certain instance contains the $i$-th artefact. Then, a neighbour $y$ of $x$ is generated by mapping $x$ to an instance $x' \in X'$ and by changing some random components of $x$ from 1 to 0. As for the function $\pi$, an exponential kernel with weight $\sigma$ equipped with a suited distance $D$ can be used, namely

$$\pi(x, y) = e^{\frac{-D^2(x,y)}{\sigma^2}}. \tag{3}$$

## 3   The $\mathcal{S}$-LIME Algorithm

In this section, the technique Semantic-LIME, or $\mathcal{S}$-LIME for short, is presented. In the rest of the paper, for the sake of simplicity, we assume as reference domain the textual one, though the proposed approach is generally applicable to other domains.

LIME maps the instance $x$ to be explained to the space $X' = \{0, 1\}^{d'}$ of the artefacts (where each feature identifies the presence or the absence of a given artefact if it evaluates to 1 or 0, respectively), e.g., the terms occurring in the sentence $x$. Specifically the image $x'$ of $x$ contains all the artefacts of $X'$ and the neighbours $y'$ of $x'$ are then generated as subsets of $x'$.

It is clear from the above strategy that these neighbours $y'$ can be in general not realistic and, more importantly, that they are constrained to be excerpt from the reference statement. This may affect the quality of the provided local explanation due to the reduced expressiveness of the considered neighborhood.

Our approach replaces the above syntactic neighborhood of $x$ with a more expressive semantic neighborhood, a task which is accomplished by mapping each original instance $x$ of $X$ to a and instance $x'$ belonging to a space of features semantically related to the classification problem at hand.

With this aim we exploit unsupervised learning in order to learn both the mapping between the original space and a latent feature space and the corresponding reverse mapping. Intuitively, since the semantic of features in the original space can be hidden and, then, the instance $x$ is not explainable based on these features, the idea is to map $x$ in a "semantic" latent space $Z$ where features carry information about the classification of $x$ performed by $f$. This strategy has the advantage to allow to sample more realistic and richer neighbours of the instance to be explained.

From the operational point of view, the idea is pursued by keeping the neighbourhood of $x$ as detailed next:

1. mapping the instance $x$ to an instance $z$ of semantic latent space $Z \subset \mathbb{R}^\ell$ characterized by $\ell$ relevant features,
2. computing the neighbours of $z$ in $Z$ by randomly sampling points in the hyper-sphere centered in $z$ and having radius $\rho$,
3. considering the neighbours of $x$ obtained by mapping back the sampled points of $Z$ to $X$.

The latent space $Z$ is obtained through a task-dependent learning phase so to alleviate the user from the choice of semantic features. In terms of artefacts, the neighbours of $x$ can now be characterized both by having some artefacts that are not in $x$ and by not having some artefacts that are in $x$.

So, assume that there exist two mapping functions, *encode* : $X \to Z$ which maps an instance $x$ to its latent representation $z$, and *decode* : $Z \to X$

which maps a point in the latent space $z$ to an instance $x$, such that $x \approx decode(encode(x))$.

**Mapping Functions.** Due to the characteristics of the latent space needed by the technique, we use a *Denoising Adversarial Autoencoder* (DAAE) [8], that is an extension of Adversarial Autoencoders (AAEs) [3] that, differently from Variational Autoencoders (VAEs) [2], maintains a strong coupling between the encoder and the decoder, ensuring that the decoder does not ignore the sentence representation produced by the encoder. This problem, also know in the literature as *posterior collapse*, is indeed a very frequent problem in textual data. Furthermore, this architecture encourages to map similar sentences to similar latent representation and obtain a good trade-off between generation and reconstruction quality.

**Explanation Model.** According to the technique proposed in [4], the explanation is a humanly understandable model $g$. In particular, in this work we adopt a *decision tree*. Note that a rule-based classifier is suited and easy to understand since, in the considered domain, the instances constituting the local around of the instance to explain are encoded in a space whose features, representing the presence/absence of terms belonging to a *common dictionary*, can be exploited as *interpretable components*. To guarantee interpretability, the maximum depth of the tree is limited to 3.

As for the fidelity function, used to compute objective function (1), its expression is adapted to work in the original space as

$$\mathcal{F}(x, f, g, \pi) = \sum_{y \in X} \pi(x, y) \cdot (f(y) - g(y))^2, \qquad (4)$$

since $g$ is able to classify instances of the original space. Concerning the function $\pi$ depicted in Eq. (3), as distance $D$, the measure

$$D(x, y) = \sum_{x_i = 1} |x_i - y_i|.$$

is employed. Note that this is an asymmetric distance that highlights the agreement of $x$ and $y$ on the features characterizing $x$, namely where $x$ has value 1.

In order to highlight peculiarities of $S$-LIME, consider the sentence $x$: "*the grounds are beautiful as well*" which is classified as *positive* in the context of a review data set collected from YELP (please refer to the experimental section for the details).

As for the output of LIME, it returns that words important for classification are "beautiful", "well", "grounds" and "are". Despite LIME returns these four words as relevant, we can note that "beautiful" and "well" should have a positive impact, while "are" and "grounds" do not seem to be relevant for the classification.

As far as the output of $S$-LIME is concerned, it is reported in Fig. 1. $S$-LIME returns the words "beautiful", "well" and "not", with "beautiful" and "well" belonging to the input sentence $x$ and "not" not belonging to $x$. Thus, the two

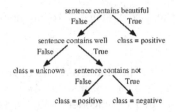

Fig. 1. $S$-LIME output example.

words semantically significant returned by LIME are returned also by $S$-LIME, while the words not relevant for the black-box model are ignored by $S$-LIME. Valuably, $S$-LIME returns also the word "not" which enriches the knowledge associated with the explanation. Indeed, Fig. 1 asserts, for explaining $x$, that

```
if the sentence contains ''beautiful'' then it is positive;
if the sentence does not contain ''beautiful'' and contains
''well'' is positive unless contains also ''not''.
```

Note that according to the above explanation, it emerges that the sentences containing "beautiful" are classified as positive by the black-box $f$ independently from the word "not". This is due to the hidden characteristics of the model $f$. Indeed, the local explanation produced by $S$-LIME suggests that in the dataset there are no sentences containing both "beautiful" and "not". To validate this suspicion, we obtained the prediction of $f$ on the artificial sentence "the grounds are not beautiful". The suspicion was real; $f$ assigned to the above sentence an high probability, namely about 0.825, of belonging to the positive class! This provides an example of the superior expressiveness of explanations provided by $S$-LIME, which allowed us to highlight a vulnerability of the black-box model.

**Algorithm.** $S$-LIME receives as input the instance $x$ to explain, the black-box model $f$, parameters $n$ and $\rho$ needed to sample neighbours, a trained auto-encoder model $\phi$ providing encoding and decoding functions, the kernel width $\sigma$. Initially, the instance $x$ is mapped to the latent space $Z$ associated with the auto-encoder $\phi$ and such mapped point is called $z^{(0)}$. Neighbours of $x$ are then provided through sampling $n$ points $z^{(i)}$ in the hyper-sphere centered in $z^{(0)}$ and having radius $\rho$ using a uniform distribution.

Then, each of these points $z^{(i)}$ is mapped to the point $y^{(i)}$ in the original space and the model $f$ is used to get the corresponding classification $\lambda_i$ The neighbours of $x$ are weighted through an exponential kernel having width $\sigma$, Using points in $y^{(i)}$, labels $\lambda_i$ and weights $w_i$, a decision tree as model, locally approximating $f$ is built.

**Interpretable Representations.** LIME provides an explanation also in terms of an *interpretable representation*, that intuitively is composed by the $K$ *interpretable components* characterizing mostly $x$. These are computed by exploiting regression as interpretable model and specifically $K$-Lasso [10] is used to obtain the $K$ features associated with the $K$ largest weights and then linear regression is perform to obtain feature weights.

To provide a similar information and also to compare the quality of our explanations with that returned by LIME, we also define our interpretable representation. Specifically, it is given by the features employed by the decision tree to split the neighbours. Moreover, we associate with each interpretable component two attributes: an *instance-attribute* and a *class-attribute*. The *instance-attribute* of an interpretable component is said to be *positive* (*negative*, resp.) if the term occurs in $x$ (does not occur in $x$, resp.), The *class-attribute* of an interpretable component is said to be *positive* (*negative*, resp.) if the term is mostly associated with the positive (negative, resp.) class, that is to say if the majority of the neighbours of $x$ having that term belongs to the positive (negative, resp.) class.

We can, hence, better characterize the interpretable components by means of the above pair *instance-attribute–class-attribute*. We have thus four types of intepretable components: *positive–positive* (*pp* for short), *positive–negative* (*pn* for short) *negative–positive* (*np* for short), and *negative–negative* (*nn* for short).

## 4   Experiments

In this section, the experiments conducted with $\mathcal{S}$-LIME are presented.

Two families of experiments have been considered. Firstly, the significance of the explanations found by the $\mathcal{S}$-LIME method and by LIME has been separately computed. Secondly, a semantic comparison between explanations is presented.

The dataset we consider in this work, named YELP, is taken from [7] and contains a subset of the *Yelp Dataset*, a collection of review taken from the homonym site, samples selected to form the subset of data are sentences with less then 16 words. A further subset of 2000 sentences is used for classification, the class of each sample represents the sentences sentiment, classes have a perfect balance so we have 1000 samples for each class. The black-box model employed here is a three-layered dense neural network.

The value of the parameters have been determined in the following way: $\rho$ has been set in order to ensure the presence of an adequate number of samples of the opposite class ($\rho = 25$); $\sigma$ has been set to obtain a good trade-off between adherence, that is the fidelity in mimicking the black-box model's behavior, and stability in explanation results in different runs ($\sigma = 5$); the number $n$ of samples generated in order to produce the explanation has been set to $n = 400$ as a trade-off between approximation quality and computational cost.

To understand the quality of the terms returned by the methods, we introduce here a score intending to measure the usefulness of the term in discriminating between the two classes. Let $n$ denote the number of examples (sentences) composing the training set which contain the term $t$ and let $k$ be the number of sentences of the most frequent class associated with these $n$ sentences (thus $k \geq n/2$). A trivial significance score would be represented by the ratio $k/n$, representing the relative frequency of the term in its majority class. However, since the statistical validity of this ratio is related to the number $n$ of sentences, we use the following corrected version of the ratio $n/k$, also referred to as *significance* (where we assume that $n$ is increased by 2 and $k$ is increased by 1 to apply a correction dealing with the case $k = n$):

$$significance(t) = \begin{cases} 0, & \text{if } \frac{k}{n} - s \le \frac{n-k}{n} + s \text{ with } s = \sqrt{\frac{k}{n^2}\left(1 - \frac{k}{n}\right)}; \\ 2\left(\frac{k}{n} - s - \frac{1}{2}\right), & \text{otherwise.} \end{cases} \quad (5)$$

In the formula above, the value $p = k/n$ corresponds to the (relative) expected value $E[B]/n = np/n$ of a binomial random variable $B$ having probability of success $p$, while the value $s = \sqrt{np(1-p)}/n$ corresponds to the (relative) standard deviation of $B$. Hence, if the probability of success $p$ is within two standard deviations from the probability of failure $1 - p$ we assume the ability of the term to discriminate between the two classes is low and set the significance to 0. Otherwise, we take the corrected value $k/n - s$, which is greater than $1/2$ by construction, and linearly scale it between 0 and 1.

**Significance of Explanations.** In order to evaluate the significance of the explanations, by fixing a number $n$ of features, the set of $n$ most relevant features returned by LIME and the set of features used by a decision tree forced to use no more than $n$ nodes are considered. Since for textual data set features are words, the significance of a set of features is related to the significance of a word. In order to measure the significance of the word, its capability in discriminating between classes is considered. In particular, given a word $w$, consider the set $S_w$ of data set sentences containing $w$ and partition this set in $S_w^{c_0}$ and $S_w^{c_1}$, where $S_w^{c_i}$ is the subset containing sentences belonging to class $c_i$. Thus, the significance of $w$ is that defined in Eq. (5) and given a set of words representing an explanation $e$, the significance of $e$ is the mean significance of the words in $e$.

Figure 2(a) reports the significance of the words selected for explanations. In particular, for each method $m$ and for each sentence $s$ the mean significance of words selected by $m$ to explain $s$ is considered. Such plot shows how the proposed method is able to detect more significant words with respect to LIME.

In order to compare the differences between sets of words provided as explanations by the two methods, for each sentence $s$ let $e_s^l$ be the set of words associated with the explanation provided by LIME and let $e_s^{\mathcal{L}}$ be the set of words associated with the explanation provided by $\mathcal{S}$-LIME, the set $e_s^l \setminus e_s^{\mathcal{L}}$ of words in $e_s^l$ and not in $e_s^{\mathcal{L}}$ and the set $e_s^{\mathcal{L}} \setminus e_s^l$ of words in $e_s^{\mathcal{L}}$ and not in $e_s^l$ are considered and the significance of the words there contained is taken into account.

Figure 2(b) reports, separately, the mean significance of the words selected for explaining sentences belonging to the three different sets, $e_s^l \setminus e_s^{\mathcal{L}}$, $e_s^l \cap e_s^{\mathcal{L}}$ and $e_s^{\mathcal{L}} \setminus e_s^l$. Such plot highlights how the words detected by LIME and not detected by the proposed technique are very less important than those detected by LIME and also by the proposed technique, while there is a large set of words detected by the proposed technique and not detected by LIME whose significance is comparable to that of words detected by both methods.

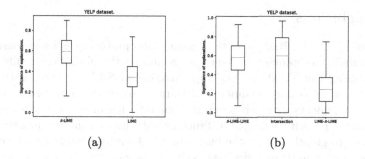

(a)                                    (b)

**Fig. 2.** Mean significance of explanations and comparison.

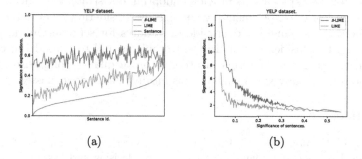

(a)                                    (b)

**Fig. 3.** Importance of single explanations and comparison.

**Differences Between Explanations of a Sentence.** Fig. 3 reports a more detailed view about performance of the two methods on each sentence of the set. In particular, for each sentence $s$ the significance of $s$ is computed as the mean significance of its words. Figure 3(a) reports for each sentence the mean significance of the words in the explanations provided by the two methods and the significance of the sentence. Note that in all cases the set detected by $S$-LIME is important independently from the significance of the sentence words. This does not hold for LIME whose explanations strictly depend on the significance of the sentence. Moreover, the explanations provided by $S$-LIME are meanly more important than that detected by LIME and the two methods become comparable only if the sentence contains a large number of important words.

Figure 3(b) reports for each sentence $s$ on the $x$ axis the significance $is$ of $s$ and on the $y$ axis the values $\frac{ie^{\mathcal{L}}}{is}$ and $\frac{ie^{l}}{is}$ where $ie^{\mathcal{L}}$ is the significance of the explanation of $s$ provided by $S$-LIME and $ie^{l}$ is the significance of the explanation of $s$ provided by LIME. The plot highlights how the proposed method is able to select a set of important words even if the sentence has few relevant words, while this is the case for LIME which meanly select subsets of words slightly more important than the set of words composing the sentence.

# 5    Conclusions

In this work, the classifier prediction explanation problem has been considered. We presented a technique based on approximating the black-box locally to the instance under consideration. The approach, called S-LIME, extends the basic LIME method exploiting unsupervised learning to replace user-provided interpretable components with self-generated semantic features. This characteristics enables our approach to sample instance neighbours in a more semantic-driven fashion and to greatly reduce the bias associated with explanations. We demonstrate the applicability and effectiveness of our proposal in the text classification domain. Comparison with the baseline on Yelp textual dataset, highlights superior quality of the explanations provided adopting our strategy. As future work, we intend enlarge the experimental campaign by considering other datasets and also different data domains, to provide procedures for semi-automatic parameters setting, to consider also multi-class problems and to extend the concept of significance to group of terms in order improve quality comparison among explanations, especially when single terms hardly characterize classes.

# References

1. Guidotti, R., Monreale, A., Ruggieri, S., Pedreschi, D., Turini, F., Giannotti, F.: Local rule-based explanations of black box decision systems. arXiv preprint arXiv:1805.10820 (2018)
2. Kingma, D.P., Welling, M.: Auto-encoding variational bayes (2013)
3. Makhzani, A., Shlens, J., Jaitly, N., Goodfellow, I., Frey, B.: Adversarial autoencoders. arXiv preprint arXiv:1511.05644 (2015)
4. Ribeiro, M.T., Singh, S., Guestrin, C.: why should i trust you? explaining the predictions of any classifier. In: Proceedings of the 22nd ACM SIGKDD International Conference on Knowledge Discovery and Data Mining, pp. 1135–1144 (2016)
5. Ribeiro, M.T., Singh, S., Guestrin, C.: Anchors: high-precision model-agnostic explanations. In: Proceedings of the AAAI Conference on Artificial Intelligence, vol. 32 (2018)
6. Samek, W., Montavon, G., Lapuschkin, S., Anders, C.J., Müller, K.R.: Explaining deep neural networks and beyond: a review of methods and applications. Proc. IEEE 109(3), 247–278 (2021)
7. Shen, T., Lei, T., Barzilay, R., Jaakkola, T.: Style transfer from non-parallel text by cross-alignment. arXiv preprint arXiv:1705.09655 (2017)
8. Shen, T., Mueller, J., Barzilay, R., Jaakkola, T.: Educating text autoencoders: latent representation guidance via denoising. In: International Conference on Machine Learning, pp. 8719–8729. PMLR (2020)
9. Sokol, K., Hepburn, A., Santos-Rodriguez, R., Flach, P.: blimey: surrogate prediction explanations beyond lime. arXiv preprint arXiv:1910.13016 (2019)
10. Tibshirani, R.: Regression shrinkage and selection via the lasso. J. Royal Stat. Soc. Ser. B (Methodological) 58(1), 267–288 (1996)
11. Visani, G., Bagli, E., Chesani, F.: Optilime: Optimized lime explanations for diagnostic computer algorithms. arXiv preprint arXiv:2006.05714 (2020)
12. Zafar, M.R., Khan, N.M.: Dlime: a deterministic local interpretable model-agnostic explanations approach for computer-aided diagnosis systems. arXiv preprint arXiv:1906.10263 (2019)

# Privacy Risk Assessment of Individual Psychometric Profiles

Giacomo Mariani[1], Anna Monreale[1(✉)], and Francesca Naretto[2]

[1] University of Pisa, Pisa, Italy
g.mariani16@studenti.unipi.it, anna.monreale@di.unipi.it
[2] Scuola Normale Superiore, Pisa, Italy
francesca.naretto@sns.it

**Abstract.** In the modern Internet era the usage of social media such as Twitter and Facebook is constantly increasing. These social media are accumulating a lot of textual data, because individuals often use them for sharing their experiences and personal facts writing text messages. These data hide individual psychological aspects that might represent a valuable alternative source with respect to the classical clinical texts. In many studies, text messages are used to extract individuals psychometric profiles that help in analysing the psychological behaviour of users. Unfortunately, both text messages and psychometric profiles may reveal personal and sensitive information about users, leading to privacy violations. Therefore, in this paper, we propose a study of privacy risk for psychometric profiles: we empirically analyse the privacy risk of different aspects of the psychometric profiles, identifying which psychological facts expose users to an identity disclosure.

**Keywords:** Privacy risk assessment · Textual data · Psychometric profile

## 1 Introduction

In the digital era textual data is the main component of human digital communications. People write in different social media, such as Facebook and Twitter, for sharing (personal) information, emotions; write emails for business; write posts in online platforms assuring anonymity, such as Reddit, an online platform for sharing personal thinking and their experience and for helping other people.

The great availability of this type of data is important for many researchers because allows for interesting studies on opinion diffusion [15], on fake news diffusion [21] on sentiment analysis [5], on the relationship between mental health and language [4,20,23]. These studies, especially those aiming at capturing sentiments and emotions [5] and the psychological behaviour of users, base their study on the opportunity to extract a psychometric profile from users text messages by using tools like LIWC (Language Inquiry and Word Count)[13].

However, textual data may carry sensitive personal information and the psychometric profile might capture very specific cognitive and emotional aspects

© Springer Nature Switzerland AG 2021
C. Soares and L. Torgo (Eds.): DS 2021, LNAI 12986, pp. 411–421, 2021.
https://doi.org/10.1007/978-3-030-88942-5_32

about the writer to make her identifiable. This re-identification may lead to the disclosure of very sensitive information and the violation of fundamental right to privacy: many research works proved that the psychometric profiles can help in detecting mental disorders [4,20,23] and it is clear that an inference like that on individuals may mean an invasion of their private sphere. Therefore, before using these profiles it would be opportune to empirically assess the real privacy risk and identify the suitable privacy mitigation. In recent years, privacy has been studied in several contexts, from location based services [24] to GPS trajectories [1], from mobile phone data [16] to retail data [10], from social networks [14] to text data [2]. Privacy risks in text data have been studied under different perspectives. Some techniques aims at protecting private information of the individuals mentioned in text documents [2,19]; while others aim at protecting the authorship disclosure [9]. Some methods focus on detecting (quasi-)identifying information, suppress disclosive items or replace them with general name entity values (like "person", "date"). An approach typically used for text protection is generalization [3,6,19].

In this paper we are not interested in protecting text data but we focus on quantitatively measuring the privacy risk of the user's psychometric profile. To the best of our knowledge, no work in the literature has studied this aspect. The most important works about privacy risk assessment is the Linddun methodology [7], a privacy-aware framework, useful for modeling privacy threats in software-based systems, and PRUDEnce [18] a framework enabling a systematic evaluation of privacy risk in different contexts (e.g., mobility, purchases) and for data with different nature (e.g., tabular data, sequential data, spatio-temporal data).

In this paper, starting from PRUDEnce [18], we define a privacy attack on psychometric profiles and we study how empirical privacy risk distributes over the population. Simulating the privacy attack on the emails of Enron corpus [8] we found out that different aspects of the psychometric profile lead to diverse privacy risk distribution. In particular, we found that some features describing the affective and social aspects lead to lower privacy risk with respect to linguistic and cognitive dimensions.

## 2    Individual Psychometric Profile

Given the set of text messages written by a user $M^u = m_1^u, \ldots, m_t^u$, where each $m_i^u$ might be an email, a Facebook post, etc., we extract her psychometric profile using LIWC [13], a dictionary-based approach able to capture the users' language habits. LIWC, is widely used in computational linguistics as a source of features for psychological and psycholinguistic analysis.

LIWC Dictionary contains both style words (function words, articles, pronouns, auxiliary verbs, etc.) and content words (conveying the content of communication). These two types of words have a different role, but both are interesting from a psychological point of view because it is important not only what people say but also how their communication is structured. The built-in dictionary was created from 100, 000 text files containing over 250 million words, that result into a dictionary of almost $6,400$ words, word stems and emoticons. LIWC processes the text in input, word by word, to obtain the final vector representation $P$, composed of 93 variables, i.e., $P = p_1, \ldots, p_{93}$. The first feature $p_1$ represents the *word count*, i.e., the number of words in a text message. Features $p_2, \ldots p_8$ represent 7 variables summarizing aspects such as the emotional tone, the authenticity, etc. Features $p_9, \ldots p_{81}$ represent different aspects: Linguistic Dimensions and Grammar aspects, Psychological Processes (e.g., Affective, Social, Perceptual, Cognitive process, etc.). These variables have a percentage value associated, i.e., the percentage of words in the text that belong to the corresponding LIWC category. The last set of features $p_{82}, \ldots, p_{93}$ corresponds to variables counting the punctuation marks in the text. Given $M^u$, i.e., the list of text messages of a user $u$, we extract from each message $m_i^u \in M^u$ the LIWC features composing the psychometric profile derived from $m_i^u$, $P^{m_i^u}$. Thus, given a user $u$ and other messages $M^u$ we derive a set of $|M_u|$ psychometric profiles. i.e., $P^u = \{P^{(m_1^u)}, P^{(m_2^u)}, \ldots, P^{(m_t^u)}\}$. In the following we denote with $p_j^{(m_i^u)}$ the $j$-$th$ LIWC feature we extracted from the text message $m_i^u$ of written by the user $u$.

## 3  Privacy Risk Assessment

In this paper, we consider the work proposed in [18] which presents PRUDEnce , a framework able to assess the privacy risk of human data. In this setting, a Service Developer (SD) asks a Data Provider (DP) to deliver data to perform an analysis or develop a service. The DP must assess the privacy risk of the individuals who generated the data being analysed before the data sharing. This task is mandatory to guarantee the right to privacy of individuals. Once assessed the privacy risk, the DP can choose how to protect the data before sharing them, selecting the most appropriate privacy-preserving technology. PRUDEnce considers the privacy risk of an individual as her maximum probability of re-identification in a dataset with respect to a set of attacks. An attack assumes that an adversary gets access to a dataset, then, using some previously obtained *background knowledge* (bk ), (i.e. the knowledge of a portion of an individual's data) the adversary tries to re-identify all the records in the dataset regarding that individual. An attack is defined by a matching function, which represents the process with which an adversary exploits the bk to find the corresponding individual in the data. For the attack definition, PRUDEnce exploits the notions of background knowledge category, configuration and instance. The first one denotes the type of information known by the adversary about a specific set of dimensions of an individual's data, e.g. the time in which she wrote an email (temporal dimension), an idiom of that person (linguistic dimension). The number of the elements known by the adversary is called background knowledge configuration $BK = \{BK_1, BK_2, \ldots, BK_h\}$ in which each subset contains the background knowledge of a specific length. If the background knowledge configuration is a psychometric information about the user, $BK_2$ contains all the possible background knowledge with 2 psychometric information about the individual under analysis. Finally, an instance of background knowledge $b \in BK_h$

is defined as the specific information known by the adversary, such as the exact value of a feature. Consider a text written by a user $u$. On it, we can aggregate and extract several information about the psychometric profile of $u$, obtaining a record $P^{(m^u)} = \{p_1, p_2, p_3, p_4\}$, a list of variables. On $P^{(m^u)}$ the DP can generate all the possible instances of background knowledge an adversary may use to re-identify the whole record $P^{(m^u)}$. If the adversary knows 2 variables' values, we have $BK_2 = \{(p_1, p_2), (p_2, p_3), (p_3, p_4), (p_1, p_3), (p_1, p_4), (p_2, p_4)\}$. The adversary may know $b = (p_1, p_3) \in BK_2$ and tries to detect all the other values' variables of $u$ to reconstruct the whole $P^{(m^u)}$. Let $\mathcal{D}$ be a database, $D$ a dataset derived from $\mathcal{D}$ (e.g., an aggregated data structure depending on psychometric variables), the probability of re-identification is defined as follows.

**Definition 1.** *Given an attack, a function matching$(d, b)$ indicating whether or not a record $d \in D$ matches the background knowledge $b$, and a function $M(D, b) = \{d \in D | matching(d, b) = True\}$, we define the probability of re-identification of an individual $u$ in dataset $D$ as: $PR_D(d = u|b) = \frac{|M(D_u, b)|}{|M(D, b)|}$. It is the probability to associate record $d \in D$ to individual $u$, given background knowledge $b$. Here, $D_u$ denotes the set of records that represent the data of the user $u \in D$.*

Note that $PR_D(d = u|b) = 0$ if the user $u$ is not in $D$. Since each background knowledge $b$ has its own probability of re-identification, we define the risk of re-identification of an individual as the maximum probability of re-identification over the set of possible background knowledge:

**Definition 2.** *The risk of re-identification (or privacy risk) of an individual $u$ given a set of background knowledge $BK_h$ is her maximum probability of re-identification $Risk_h(u, D) = \max PR_D(d = u|b)$ for $b \in BK_h$. It has the lower bound $\frac{|D_u|}{|D|}$ (a random choice in $D$), and $Risk_h(u, D) = 0$ if $u \notin D$.*

An individual is hence associated with several privacy risks, each for every background knowledge of an attack.

### 3.1    Privacy Risk Assessment for Psychometric Profiles

PRUDEnce is a general framework, in which, depending on the human data under analysis, the privacy attack to simulate varies. PRUDEnce has been used for studying the privacy risk evaluation on mobility datasets [11,11,18], on purchase datasets [10,12] and also on multi-dimensional data [17]. To the best of our knowledge, there is no study analysing the privacy risk of the psychometric profiles derivable from message text written by users. In the literature there are many data-driven studies based on textual data where documents are preprocessed to extract psychometric features useful for data mining and ML tools. The analysis of how privacy risk distributes over the population in this context is important because often these profiles are good proxy of sensitive inferences as mental disorders [4,20,23]. We define a privacy attack on psychometric features extracted from textual data. This attack exploits apriori knowledge on these features to understand whether they can lead to a re-identification of users.

We consider a DP that maintains a database $\mathcal{D}$ constituted of unstructured textual data and a SP that wants to use (a subset of) these data to deliver a data-driven service based on some features derived from $\mathcal{D}$ . The database $\mathcal{D}$ taken into account is composed of a set of users $U = \{u_1, \ldots, u_n\}$. Each user $u \in U$ owns a *corpus* of $t$ unstructured text documents $M^u = \{m_1^u, \ldots, m_t^u\}$, therefore we have that $\mathcal{D} = \bigcup_{u \in U} M^u$. Given a text document $m_i^u \in \mathcal{D}$, the DP extracts a set of features describing some property of the text and the user style. We call these set of features *document features*, and we denote it as $P^{m_i^u} = \left\{ p_1^{(m_i^u)}, \ldots, p_h^{(m_i^u)} \right\}$. Note that the features can be numerical or not, depending on the context and the type of use of these data. Hence, for each user $u \in U$ we have a set of document features, one for each document in the corpus of that user; mathematically: $P^u = \{P^{(m_1^u)}, P^{(m_2^u)}, \ldots P^{(m_n^u)}\}$. We denote by $D = \bigcup_{u \in U} P^u$ the dataset of psychometric profiles of users $U$.

We consider the scenario in which the attacker knows that an unknown user $\hat{u} \in U$ is associated to $l$ feature values $b = \{p_{b_1}, \ldots, p_{b_l}\}$ with $l \leq h$, where $b$ belongs to a psychometric profile extracted from a text message $m_i^{\hat{u}}$ . The attacker wants to identify $\hat{u}$ exploiting the records in $D$ that have features corresponding to the ones in $b$. We subdivide all the possible background knowledge configurations with respect to the number of attributes: $BK_j$ is the set of all the possible feature sets in $D$ with $j$ attributes, and we denote the set of all the background knowledge configurations $BK = \{BK_1, \ldots, BK_h\}$. In order to model the probability of re-identification (Definition 1) for this attack we define the *matching* function for $d = P^{m_k^u}$ and $b = \{p_{b_1}, \ldots, p_{b_h}\} \in BK_h$ in a dataset $D$ as:

$$matching(d, b) = \begin{cases} True, & p_1^{(m_k^u)} = p_{b_1} \wedge p_2^{(m_k^u)} = p_{b_2} \wedge \ldots \wedge p_h^{(m_k^u)} = p_{b_h} \\ False, & \text{otherwise} \end{cases} \quad (1)$$

We recall that, based on this function definition, we can compute the probability of re-identification (Definition 1) as well as the privacy risk of each user (Definition 2). Intuitively, a certain user $u$ is more easily re-identified if in $D$ there are a low number of profiles compatible with the background knowledge $b$.

## 4    Experiments

In this section we present the results of the simulation of the attack presented in Sect. 3[1]. We analyse two settings. (1) We perform the simulation assuming

---

[1] The implementation of these attacks, written in Python 3.7, is available on Github https://github.com/karjudev/text-privacy. For conducting the experiments we used a server with 16x Intel(R) Xeon(R) Gold 5120 CPU @ 2.20 GHz (64 bits), 63 gb RAM.

**Table 1.** List of LIWC features divided into 13 categories

| Feature category | Category name | Features |
|---|---|---|
| Affective processes | affective | *affect, posemo, negemo, anx, anger, sad* |
| Biological process | biological | *bio, body, health, sexual, ingest* |
| Cognitive processes | cognitive | *cogproc, insight, cause, discrep, tentat, certain, differ* |
| Drives | drives | *drives, affiliation, achieve, power, reward, risk* |
| Perceptual processes | perceptual | *percept, see, hear, feel* |
| Social processes | social | *social, family, friend, female, male* |
| Language | language | *WC, Analytic, Clout, Authentic, Tone, Sixltr, Dic, WPS* |
| Time | time | *focuspresent, focuspast, focusfuture* |
| Grammar | grammar | *verb, adj, compare, interrog, number, quant* |
| Personal concerns | personal | *work, leisure, home, money, relig, death* |
| Informal | informal | *informal, swear, netspeak, assent, nonflu, filler* |
| Linguistic | linguistic | *function, pronoun, ppron, i, we, you, shehe, they, ipron, article, prep, auxverb, adverb, conj, negate* |
| Punctuation | punctuation | *AllPunc, Period, Comma, Colon, SemiC, QMark, Exclam, Dash, Quote, Apostro, Parenth, OtherP* |

an adversary that may know any combination of $h$ LIWC features belonging to a user profile. In this case we consider 10 sets of background knowledge configuration $BK_h$, ranging from $h = [1, 10]$. (2) We perform an analysis aiming to understand how the risk distribution varies over the categories of LIWC features. In this case we identify 13 categories of features to be analyzed. For each category we simulate the attack assuming an adversary that knows a combination of LIWC features in that category. We varied the background knowledge configuration size $h$ from 1 to the total number of features in that category. We report the 13 LIWC categories in Table 1.

*Dataset and Pre-processing.* The data used to test the privacy risk assessment tool is a pre-processed version the Enron corpus [8][2] The original version was subject to a massive pre-processing, consisting of:

1. *Removal of the email header*: Enron corpus contains plain text email messages, with email addresses. We extract only the unstructured body from the original texts.
2. *Redundant string removal*: emails can be original messages, replies to other emails or forwards of a message. In the two latter cases, a string separates an optional new text from the original message, that we remove.
3. *Removal of non-contributing messages*: We restrict our analysis only to messages longer than 2 characters[3]. We drop the duplicate messages[4]. We restrict the dataset only to e-mail addresses in the **enron.com** domain[5]. We exclude the records with a value of Word Count (WC) less than 1[6].

---

[2] Composed of $517,401$ messages from 158 different authors.
[3] 482,117 messages from 20,192 authors.
[4] 230,571 messages from 20,192 authors.
[5] 176,243 messages from 6,410 authors.
[6] 176,207 messages from 6,410 authors.

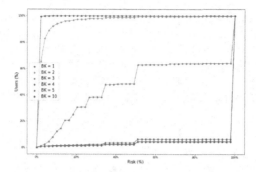

**Fig. 1.** Cumulative distribution of risk in the dataset. For each risk value (on the $x$-axis) is reported the number of users that have at least such risk (on the $y$-axis). Note that we report only $BK = [1, 5]$ and $BK = 10$ as from $BK = 5$ onwards the behaviour is similar: increasing the knowledge of the attacker, we reach a plateau (visible in $BK = 5, 10$).

After this pre-processing, the resulting dataset $\mathcal{D}$ consists in $176, 207$ email messages from $6, 410$ unique email addresses. On $\mathcal{D}$ we use LIWC to extract the psychometric profile from each email to get the dataset $D$. This profile consists of 93 features that are discretized into $k = 18$ equal frequency buckets[7]. After the computation of the psychometric profiles, we perform a correlation analysis on these features, using the Pearson correlation coefficient, observing that the features `Tone`, `affect`, `pron`, `focuspresent` are highly correlated (more than 80%) with other features, so we removed them.

*Experimental Results.* The first experiment simulates the attack where an adversary knows any combinations of $h$ features of a psychometric profile (ranging from $h = [1, 10]$). Figure 1 shows the results of this simulation, depicting the cumulative distribution, where for each risk value (on $x$-axis) we reported how many users have *at least* such risk (on $y$-axis).

We observe that the risk distribution for the users behaves as already observed for retail [12] and mobility [11,18] data: low risk values for a small background knowledge size, increasing until a stabilisation point, where the (high) risk for background knowledge sizes from 4 to 10 is approximately the same.

After finding typical privacy risk behaviour in this setting, we deepen our analysis by studying the risk among the different categories of LWIC features. In the second experiment we considered 13 categories of LIWC features, reported in Table 1. In this case, for the simulation of the attacks we assume that the adversary only knows some values' features belonging to a specific category. Hence, for each category we compute the attack detailed in Sect. 3, considering a background knowledge size $h$ from 1 to the total number of features for that category. Interestingly, varying the category considered, the privacy attack may be more or less effective. In Fig. 2 we observe how some kind of psychometric

---

[7] The number is obtained using the Sturges formula [22].

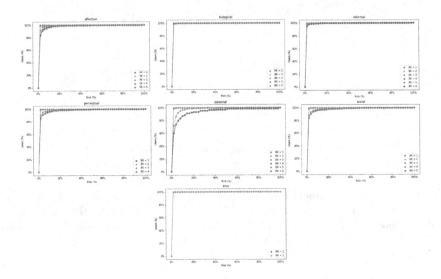

**Fig. 2.** LIWC categories exposing the user to low risk

features do not expose the user to a high risk. It is interesting to note that we have very different categories in this group. We find `social` and `informal` that focus mainly on the presence of colloquial terms or names[8]. For them, we were expecting a very high risk of privacy as we are analysing a dataset of business emails, where this kind of terms should have a low probability of use for every user. However, since the population's behavior is very similar in the dataset, the number of re-identified users is very low. We have a similar effect also for the `personal concerns` category, that considers the usage of references to work, money, etc. In this case we have very common words in this type of data and hence the users have similar values. Instead, it is peculiar the case of `affective`, `biological` and `perceptual`. Some of these are part of the psychological analysis of the text, such as `perceptual`, which analyses the presence of phrases related to what one feels and understands, or `biological`, which analyses references about health. Thus, despite referring to information that may appear very personal, these categories do not expose the user to a high privacy risk. Different, instead, is the case of `time`: there are only two features, so the `bk` of the adversary is very limited. In this case, we can say that the privacy risk is lower for dimensionality reasons.

In Fig. 3, we report other categories of psychometric features that expose the user to a higher risk. The categories `language` and `linguistic` are the most dangerous ones in our context as they expose the users to a higher privacy risk: with a `bk` of 5, almost the entirety of the users can be re-identified. These categories analyse the way in which a user writes: the language used, such as the emotional tone, the authoritarian tone; number of words used; and the linguistic

---

[8] Colloquial terms are: "mom" or "dad" and "mate" or "buddy".

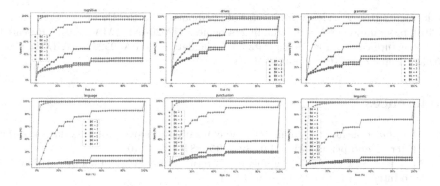

**Fig. 3.** LIWC categories exposing the user to high risk

style, such as the use of pronouns, articles and prepositions. Thus, it is noticeable that the way in which a user writes characterises her even in a professional context like that under analysis. Another interesting aspect regards `cognitive` and `drives`: they are the only psychological category among the ones at high risk. `cognitive` analyses the presence of words about insight, causation and discrepancy[9]. `drives`, instead, is more focused on the concepts of achievement, reward and risk. We observe that, even if the risk overall the dataset is not high, as the case of `language` and `linguistic`, for these psychological categories we can see that starting from a `bk` of length 4 the privacy risk increases considerably, meaning that they tend to distinguish better different users.

## 5   Conclusions

We defined and analysed a privacy risk attack for psychometric profiles. Psychometric profiles are used in many data mining and ML applications, hence it is crucially important to evaluate the real privacy risk the authors of such data are facing. We defined a privacy risk attack considering a corpus of documents for each user. For each document, we extracted a psychometric profile. The attack assumes that there is an adversary who obtains some psychometric information about an individual. Given the dataset containing the all psychometric profiles, the adversary tries to re-identify all the records in the dataset belonging to that individual. The experimental results show that the privacy risk in this context depends on the kind of psychometric features the adversary knows: there are categories that expose users to an extremely high privacy risk, namely `grammar`, `language`, `punctuation`, `linguistic`, `drives` and `cognitive`.

Our future research agenda includes the simulation of this privacy risk on textual data with different characterisation and the application of some privacy protection strategies.

---

[9] Words like "think", "know", "always", "never" and "should".

**Acknowledgment.** This work is partially supported by the European Community H2020 programme under the funding schemes: H2020-INFRAIA-2019-1: Research Infrastructure G.A. 871042 SoBigData++ (sobigdata.eu), G.A. 952215 TAILOR, G.A. 952026 Humane AI NET (humane-ai.eu).

# References

1. Abul, O., Bonchi, F., Nanni, M.: Anonymization of moving objects databases by clustering and perturbation. Inf. Syst. **35**, 884–910 (2010)
2. Anandan, B., Clifton, C.: Significance of term relationships on anonymization. In: Web Intelligence/IAT Workshops (2011)
3. Chakaravarthy, V.T., Gupta, H., Roy, P., Mohania, M.K.: Efficient techniques for document sanitization. In: CIKM (2008)
4. Choudhury, M., Counts, S., Horvitz, E.: Predicting postpartum changes in emotion and behavior via social media. In: Conference on Human Factors in Computing Systems - Proceedings (2013)
5. Crossley, S., Kyle, K., McNamara, D.: Sentiment analysis and social cognition engine (seance): an automatic tool for sentiment, social cognition, and social-order analysis. Behav. Res. Methods **49**, 803–821 (2017)
6. Cumby, C.M., Ghani, R.: A machine learning based system for semi-automatically redacting documents. In: IAAI (2011)
7. Deng, M., et al.: A privacy threat analysis framework: supporting the elicitation and fulfillment of privacy requirements. Requir. Eng. **16**, 3–32 (2011)
8. Klimt, B., Yang, Y.: Introducing the enron corpus. In: CEAS (2004)
9. Li, Y., Baldwin, T., Cohn, T.: Towards robust and privacy-preserving text representations. In: ACL, no. 2 (2018)
10. Pellungrini, R., Monreale, A., Guidotti, R.: Privacy risk for individual basket patterns. In: MIDAS/PAP@PKDD/ECML (2018)
11. Pellungrini, R., Pappalardo, L., Pratesi, F., Monreale, A.: Analyzing privacy risk in human mobility data. In: STAF Workshops (2018)
12. Pellungrini, R., Pratesi, F., Pappalardo, L.: Assessing privacy risk in retail data. In: PAP@PKDD/ECML (2017)
13. Pennebaker, J.W., Boyd, R.L., Jordan, K., Blackburn, K.: The development and psychometric properties of liwc2015. Technical report (2015)
14. Pensa, R.G., di Blasi, G.: A semi-supervised approach to measuring user privacy in online social networks. In: DS (2016)
15. del Pilar Salas-Zárate, M., et al.: A study on LIWC categories for opinion mining in spanish reviews. J. Inf. Sci. **40**, 749–760 (2014)
16. Pratesi, F., Gabrielli, L., Cintia, P., Monreale, A., Giannotti, F.: PRIMULE: privacy risk mitigation for user profiles. Data Knowl. Eng. **125**, 101786 (2020)
17. Pratesi, F., Monreale, A., Giannotti, F., Pedreschi, D.: Privacy preserving multidimensional profiling. In: GOODTECHS (2017)
18. Pratesi, F., et al.: Prudence: a system for assessing privacy risk vs utility in data sharing ecosystems. Trans. Data Priv. (2018)
19. Sánchez, D., Batet, M.: Toward sensitive document release with privacy guarantees. Eng. Appl. Artif. Intell. **59**, 23–34 (2017)
20. Shen, J.H., Rudzicz, F.: Detecting anxiety through reddit. In: Proceedings of the Fourth Workshop on Computer Linguistics and Clinical Psychology-From Linguistic Signal to Clinical Reality (2017)

21. Shrestha, A., Spezzano, F., Joy, A.: Detecting fake news spreaders in social networks via linguistic and personality features. In: CLEF (Working Notes) (2020)
22. Sturges, H.A.: The choice of a class interval. J. Am. Stat. Assoc. **21**, 65–66 (1926)
23. Tadesse, M.M., Lin, H., Xu, B., Yang, L.: Detection of depression-related posts in reddit social media forum. IEEE Access **7**, 44883–44893 (2019)
24. Xiao, Y., Xiong, L.: Protecting locations with differential privacy under temporal correlations. In: CCS (2015)

# The Case for Latent Variable Vs Deep Learning Methods in Misinformation Detection: An Application to COVID-19

Caitlin Moroney[1], Evan Crothers[2], Sudip Mittal[3], Anupam Joshi[4],
Tülay Adalı[4], Christine Mallinson[4], Nathalie Japkowicz[1],
and Zois Boukouvalas[1(✉)]

[1] American University, Washington, D.C. 20016, USA
cm0246b@student.american.edu, {japkowic,boukouva}@american.edu
[2] University of Ottawa, Ottawa, ON, Canada
ecrot027@uottawa.ca
[3] Mississippi State University, Mississippi State, Starkville, MS 39762, USA
mittal@cse.msstate.edu
[4] University of Maryland, Baltimore County, Baltimore, MD 21250, USA
{joshi,adali,mallinson}@umbc.edu

**Abstract.** The detection and removal of misinformation from social media during high impact events, e.g., COVID-19 pandemic, is a sensitive application since the agency in charge of this process must ensure that no unwarranted actions are taken. This suggests that any automated system used for this process must display both high prediction accuracy as well as high explainability. Although Deep Learning methods have shown remarkable prediction accuracy, accessing the contextual information that Deep Learning-based representations carry is a significant challenge. In this paper, we propose a data-driven solution that is based on a popular latent variable model called Independent Component Analysis (ICA), where a slight loss in accuracy with respect to a BERT model is compensated by interpretable contextual representations. Our proposed solution provides direct interpretability without affecting the computational complexity of the model and without designing a separate system. We carry this study on a novel labeled COVID-19 Twitter dataset that is based on socio-linguistic criteria and show that our model's explanations highly correlate with humans' reasoning.

**Keywords:** Misinformation detection · Knowledge discovery · Independent Component Analysis · Explainability

## 1 Introduction

With the evolution of social media, there has been a fundamental change in how misinformation is propagated especially during high impact events. A recent example of a high impact event is the COVID-19 disease where misinformation is

© Springer Nature Switzerland AG 2021
C. Soares and L. Torgo (Eds.): DS 2021, LNAI 12986, pp. 422–432, 2021.
https://doi.org/10.1007/978-3-030-88942-5_33

dangerously spreading and includes conspiracy theories, harmful health advises, racism, among many others.

Recent machine learning advances have shown significant promise for the detection of misinformation. Examples include approaches based on hand-crafted features and approaches based on deep learning. Although approaches based on hand-crafted features provide at a certain level interpretable results [9], in most cases the selection of the features is tied to the particular application affecting the generalization ability of the model. On the other hand, approaches based on deep learning effectively learn the latent representations and have shown great promise in terms of prediction performance [12]. However, the connections between high level features and low representation space are usually accessed by using or designing a separate system resulting in high computational or construction overhead [10]. In this study, we present a computationally efficient data-driven solution that is based on a latent variable model called independent component analysis (ICA) such that detection of misinformation and knowledge discovery can be achieved jointly. Our method achieves a prediction performance close to that of deep learning while at the same time offering the kind of interpretability that deep learning, even with the help of a separate explainability system, cannot achieve.

This work makes several contributions. First, it proposes a new method for misinformation detection based on ICA. Second, it demonstrates how to highlight the connections between the low dimensional representation space and the high level features. Finally, it makes available a new labeled and annotated COVID-19 Twitter dataset[1] as well as a set of rules for label generation based on socio-linguistic criteria.

## 2  Development of Labeled Twitter COVID-19 Dataset

In constructing our labeled Twitter dataset we initially randomly collected a sample of 282,201 Twitter users from Canada[2] by using the Conditional Independence Coupling (CIC) method [20]. CIC matches the prior distribution of the population, in this case the Canadian general population, ensuring that the sample is balanced for gender, race and age. All tweets posted by these 282,201 people from January 1, 2020 to March 13, 2020 were collected and a random subset of 1,600 tweets was further analyzed to create a manageable and balanced dataset of both real tweets and tweets that contain misinformation. Note here that we follow current literature that defines misinformation as *an umbrella term to include all false or inaccurate information that is spread in social media*. This is a useful heuristic because, on a social media platform where any user can publish anything, it is otherwise difficult to determine whether a piece of misinformation is deliberately created or not [21]. To eliminate data bias two subject

---

[1] Dataset is available at https://zoisboukouvalas.github.io/Code.html.
[2] We thank Dr. Kenton White, Chief Scientist at Advanced Symbolics Inc, for providing the initial Twitter dataset.

**Table 1.** 17 linguistic characteristics identified on the 560 Twitter dataset

| Linguistic attribute | Example from dataset |
|---|---|
| Hyperbolic, intensified, superlative, or emphatic language [2,16] | e.g., 'blame', 'accuse', 'refuse', 'catastrophe', 'chaos', 'evil' |
| Greater use of punctuation and/or special characters [2,15] | e.g., 'YA THINK!!?!!?!', 'Can we PLEASE stop spreading the lie that Coronavirus is super super contagious? It's not. It has a contagious rating of TWO' |
| Strongly emotional or subjective language [2,16] | e.g., 'fight', 'danger', 'hysteria', 'panic', 'paranoia', 'laugh', 'stupidity' or other words indicating fear, surprise, alarm, anger, and so forth |
| Greater use of verbs of perception and/or opinion [15] | e.g., 'hear', 'see', 'feel', 'suppose', 'perceive', 'look', 'appear', 'suggest', 'believe', 'pretend' |
| Language related to death and/or war [8] | e.g., 'martial law', 'kill', 'die', 'weapon', 'weaponizing' |
| Greater use of proper nouns [11] | e.g., 'USSR lied about Chernobyl. Japan lied about Fukushima. China has lied about Coronavirus. Countries lie. Ego, global' |
| Shorter and/or simpler, language [11] | e.g., '#Iran just killed 57 of our citizens. The #coronavirus is spreading for Canadians Our economy needs support.' |
| Hate speech [8] and/or use of racist or stereotypical language | e.g., 'foreigners', 'Wuhan virus', reference to Chinese people eating cats and dogs |
| First and second person pronouns [15,16] | e.g., 'I', 'me', 'my', 'mine', 'you', 'your', 'we', 'our' |
| Direct falsity claim and/or a truth claim [2] | e.g., 'propaganda', 'fake news', 'conspiracy', 'claim', 'misleading', 'hoax' |
| Direct health claim | e.g., 'cure', 'breakthrough', posting infection statistics |
| Repetitive words or phrases [11] | e.g., 'Communist China is lying about true extent of Coronavirus outbreak - If Communist China doesn't come clean' |
| Mild or strong expletives, curses, slurs, or other offensive terms | e.g., 'bitch', 'WTF', 'dogbreath', 'Zombie homeless junkies', 'hell', 'screwed' |
| Language related to religion | e.g., 'secular', 'Bible' |
| Politically biased terms | e.g., 'MAGA', 'MAGAt', 'Chinese regime', 'deep state', 'Communist China' |
| Language related to financial or economic impact | e.g., 'THE STOCK MARKET ISN'T REAL THE ECONOMY ISN'T REAL THE CORONAVIRUS ISN'T REAL FAKE NEWS REEEEEEEEEEEEEEEE' |
| Language related to the Trump presidential election, campaign, impeachment, base, and rallies | e.g., 'What you are watching with the CoronaVirus has been planned and orchestrated. ' |

matter experts from our group, Dr. Boukouvalas and Dr. Mallinson, independently reviewed each tweet to determine whether or not a post should be labeled as real or misinformation.

Tweets were labeled as misinformation if they included content that promotes *political bias*, *conspiracy*, *propaganda*, *anger*, or *racism* and thus such tweets could affect decision making and create social and political unrest during COVID-19. Tweets that were not labeled as misinformation by both experts were kept for a second review and finally marked as misinformation if both experts agreed on their decision. To obtain a balanced dataset we randomly down-sampled the real class and also manually checked this class for consistency and validity with respect to reliability. The final dataset consists of 280 real and 280 misinformation tweets.

The set of tweets that contains misinformation was further analyzed for the presence of linguistic attributes that might indicate unreliability and provide a set of linguistic rules of potential use to label further data sets and to assess the interpretation ability of our model. This was done by reviewing each tweet for, first, the presence of linguistic characteristics previously identified in the literature as being indicative of or associated with misinformation, bias, and/or less reliable sources in news media; and second, for the presence of any additional distinguishing linguistic characteristics that appeared to be indicative of misinformation in this dataset. A list of 17 linguistic characteristics was developed and is presented in Table 1 along with instances of each characteristics drawn from the dataset.

# 3   Tweet Representations Generation

## 3.1   Transformer Language Models

For comparison to state-of-the-art deep learning methods, we compare our results against a suite of Transformer language models. Specifically, we evaluate "base" and "large" variants of Bidirectional Encoder Representations from Transformers (BERT) [7], Robustly Optimized BERT Pretraining Approach (RoBERTA) [14], and Efficiently Learning an Encoder that Classifies Token Replacements Accurately (ELECTRA) [6].

The bidirectional aspect of BERT comes from using a masked language model (MLM) pre-training objective, which allows the model to incorporate information from both the left and right contexts [7]. RoBERTA and ELECTRA represent subsequent improvements of these state-of-the-art results through further investigation of improved pre-training methods [6,14]. To facilitate reproducibility, we use the pre-trained HuggingFace PyTorch implementation of each of these models. The base version of each model contains 12 encoder layers, 768 hidden units, and 12 attention heads (for a total of over 110M parameters), while the large version of each model contains 24 encoder layers, 1024 hidden units, and 16 attention heads (for a total of over 330M parameters).

The classification model takes as input the Transformer output encoding; for each tweet, we obtain a vector of length equal to the dimension of the final hidden layer of the model (i.e., 768 for base models and 1024 for large models). Aggregate representations for the input sequence (in our experiments, a tweet) can be generated by taking the output vector of a [CLS] token prepended to the input sequence. However, we found that taking the mean across the vectors for all of a segment's tokens provided a better classification performance, mirroring results from other research into BERT-based sequence representations [17]. As such, our sequence representations are generated through taking the mean across all of a segment's token vectors.

## 3.2   Independent Component Analysis

We formulate the problem of tweet representation generation in the following way. Let $\mathbf{X} \in \mathbb{R}^{d \times V}$ be the observation matrix which denotes the word-word co-occurrence matrix and incorporates contextual information from the raw tweet text data. The model is given by $\mathbf{X} = \mathbf{AS}$, where $\mathbf{A} \in \mathbb{R}^{d \times N}$ is the mixing matrix and $\mathbf{S} \in \mathbb{R}^{N \times V}$ are the latent variables. Since in our study $d > N$, we reduce to the case where $d = N$ using principal component analysis (PCA). The most popular way to estimate the latent variables is by using ICA [1]. The assumption of independence of the sources not only enables a unique decomposition under minimal model assumptions but also results in interpretable contextual representations through their linear mixing coefficients. For more information about the mathematical formulation of ICA we refer the reader to [5].

The estimated columns of the mixing matrix $\mathbf{A}$ denote the weight features that will be used for the construction of the tweet representations. An estimate of $\mathbf{A}$ is computed as $\hat{\mathbf{A}} = (\mathbf{F})^\dagger (\mathbf{W})^{-1}$, where $(\mathbf{F})^\dagger$ denotes the pseudo-inverse of the matrix that is formed by the eigen-vectors with the first $N$ highest eigenvalues of $\mathbf{X}$ and $\mathbf{W}$ is the estimated demixing matrix resulted from ICA. In this work, we used the entropy bound minimization (ICA-EBM) algorithm [13], due to the fact that it has shown superior performance in a wide range of applications. To construct the individual tweet representations, we average over the estimated rows of $\mathbf{A}$ for the words in each tweet to obtain a single $N$-dimensional vector representation for each tweet. For our study we have selected $N = 250$.

## 4   Results and Discussion

### 4.1   Prediction Performance

We evaluate the same classification algorithm, Support Vector Machines (SVM) using a linear kernel[3]. In addition to the BERT versions and ICA tweet representations, we consider three other popular latent variable methods: Nonnegative matrix factorization (NMF) [3], Dictionary Learning (DL) [19], and Latent Dirichlet allocation (LDA) [4]. To construct tweet representations using NMF, DL, and LDA we followed a similar procedure as we did with ICA. To measure performance, we employed the standard suite of evaluation metrics, i.e., accuracy, F1 score, precision, and recall. We report the macro-averaged versions of these scores. For all of the experiments, hyper-parameter optimization and model training and testing is done using a nested five fold cross validation scheme.

From Table 2, we see that prediction accuracy using RoBERTa-Large word representations performs the best in terms of accuracy and F1 score. However, the ICA method is able to achieve very similar performance to that of RoBERTa-Large and in some cases better than other BERT versions such as ELECTRA-Base and ELECTRA-Large. Performance using tweet representations derived

---

[3] It is worth mentioning that for all methods similar results were obtained with the sigmoid and and the rbf kernel.

**Table 2.** Prediction performance

| Method | Accuracy | Recall | Precision | F1 |
|---|---|---|---|---|
| LDA | 0.754 | 0.868 | 0.712 | 0.777 |
| ICA | **0.862** | **0.931** | **0.820** | **0.871** |
| DL | 0.779 | 0.740 | 0.799 | 0.767 |
| NMF | 0.832 | 0.872 | 0.810 | 0.838 |
| BERT-Base-Uncased | 0.868 | 0.883 | 0.858 | 0.869 |
| BERT-Base-Cased | 0.866 | 0.889 | 0.855 | 0.870 |
| BERT-Large-Uncased | 0.880 | 0.888 | 0.877 | 0.881 |
| BERT-Large-Cased | 0.875 | **0.895** | 0.863 | 0.876 |
| ELECTRA-Base | 0.848 | 0.837 | 0.861 | 0.847 |
| ELECTRA-Large | 0.832 | 0.838 | 0.828 | 0.832 |
| RoBERTa-Base | 0.873 | 0.862 | **0.888** | 0.873 |
| RoBERTa-Large | **0.886** | 0.891 | 0.883 | **0.886** |

from DL and LDA was significantly lower than that of ICA and BERT across all metrics.

## 4.2 Explainability

ICA has the advantage over Deep Learning techniques of being able to provide contextual interpretations through the estimated mixing matrix $\hat{A}$. It does so by first ordering the ICA features by magnitude for a given tweet vector representation. For each tweet, the most important features, which may be considered as topics, are then extracted. From the matrix $\hat{A}$, we then select the columns corresponding to the most important topics for the chosen tweet, and for each column we sort the rows, corresponding to vocabulary words, by magnitude. This allows us to obtain the most important words in each topic for the most important topics in each tweet.

The results of this extraction are shown at the bottom of Figs. 1 and 2. Feature 1 and Feature 2 represent the dominant words belonging to the highest two features extracted by ICA on 2 real and 2 misinformation tweets that were all classified correctly by the ICA-based method. From Fig. 1 and Fig. 2 we see that the words listed in the main features of the two real cases (Cases 1 and 2) do not match the rules extracted in Table 1, except for one: hyperbolic language ("apocalyptic"). This suggests that the two main features extracted in each of these two real cases support the classifier's decision, since the words most strongly associated with the features that caused that decision do not trigger many rules believed to represent language used in misinformation. On the other hand, in the two misinformation cases (Cases 3 and 4), many rules are triggered including hate speech, hyperbolic language, and strongly emotional language. This suggests that the two main features extracted in each of these misinformation cases support the classifier's decision, since the words most strongly associated with the features that caused that decision do trigger many rules believed to represent language used in misinformation. Furthermore, the fact that the two closely related misinformation tweets of Cases 3 and 4 triggered the same two features shows the consistency of our approach. Figure 3, (Case 5), shows the explainability results for the case where the ICA-based method did not classify the tweet correctly. In this case, we see that the second main feature is the same as the one that was picked by the two misinformation tweets enabling a user to understand why the ICA-based method predicted this tweet as misinformation.

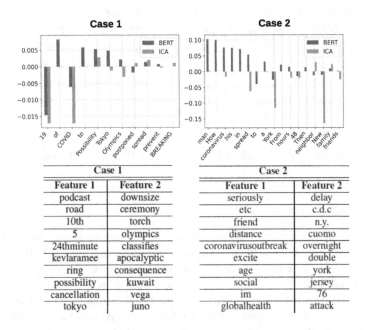

| Case 1 | |
|---|---|
| Feature 1 | Feature 2 |
| podcast | downsize |
| road | ceremony |
| 10th | torch |
| 5 | olympics |
| 24thminute | classifies |
| kevlaramee | apocalyptic |
| ring | consequence |
| possibility | kuwait |
| cancellation | vega |
| tokyo | juno |

| Case 2 | |
|---|---|
| Feature 1 | Feature 2 |
| seriously | delay |
| etc | c.d.c |
| friend | n.y. |
| distance | cuomo |
| coronavirusoutbreak | overnight |
| excite | double |
| age | york |
| social | jersey |
| im | 76 |
| globalhealth | attack |

**Fig. 1.** Top: Local explanations by LIME for both BERT and ICA. Orange bars correspond to the ICA-based approach and the blue bars correspond to the BERT-based approach. Furthermore, 0 is neutral whereas positive values are misinformation and negative values real; Bottom: Feature 1 and Feature 2 represent the dominant words belonging to the highest two features extracted by ICA on two real tweets that ICA-based method correctly classified; Tweets: **Case 1 (predicted as real by ICA and by BERT)**: BREAKING: Possibility Tokyo Olympics postponed to prevent spread of COVID-19; **Case 2 (predicted as real by ICA and as misinformation by BERT)**: From a man to his family. Then to a neighbor. Then to friends. How coronavirus spread in New York in 48 h. (Color figure online)

While, as just discussed, the context in which a decision is made can easily be extracted from the ICA-based method, in recent years, efforts have been made to extract information from opaque classifiers. One such effort is the popular local interpretable model-agnostic explanations (LIME) system [18] which produces local explanations for classifier decisions. This technique, however, comes at a cost since, for example, LIME took, on average, 6,400.1 s to process a single tweet explanation for the BERT and SVM pipeline and 70.3 s for the ICA and SVM pipeline whereas the extraction of ICA's main features was instantaneous. In addition to the cost, we argue that LIME does not consistently outfit the BERT-based method (or the ICA-based method for that matter) with a satisfying explainability. In particular, looking at the top graphs in Fig. 1 and Fig. 2, we notice inconsistencies in LIME's explanations. In these graphs, the orange bars correspond to the ICA-based approach and the blue bars correspond to the BERT-based approach. Furthermore, 0 is neutral whereas positive values are misinformation and negative values real. In Case 1, BERT issued the correct classification. However, LIME's explanation for this classification is that Covid and 19 were reliable words, whereas irrelevant stop words such as "Of" and "to" gave the system indication that it was misinformation. In case 2 that BERT wrongly classified as misinformation, that unreliability is given by the irrelevant

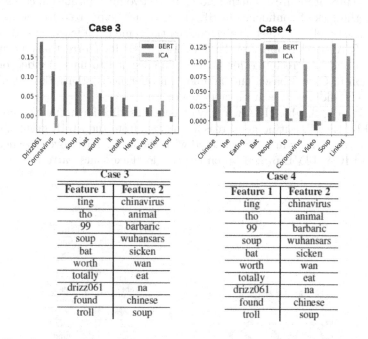

| Case 3 | |
| --- | --- |
| Feature 1 | Feature 2 |
| ting | chinavirus |
| tho | animal |
| 99 | barbaric |
| soup | wuhansars |
| bat | sicken |
| worth | wan |
| totally | eat |
| drizz061 | na |
| found | chinese |
| troll | soup |

| Case 4 | |
| --- | --- |
| Feature 1 | Feature 2 |
| ting | chinavirus |
| tho | animal |
| 99 | barbaric |
| soup | wuhansars |
| bat | sicken |
| worth | wan |
| totally | eat |
| drizz061 | na |
| found | chinese |
| troll | soup |

**Fig. 2.** Top: Local explanations by LIME for both BERT and ICA. Orange bars correspond to the ICA-based approach and the blue bars correspond to the BERT-based approach. Furthermore, 0 is neutral whereas positive values are misinformation and negative values real; Bottom: Feature 1 and Feature 2 represent the dominant words belonging to the highest two features extracted by ICA on two misinformation tweets that ICA-based method correctly classified; Tweets: **Case 3 (predicted as misinformation by ICA and by BERT):** @Drizz061 Have you even tried bat soup? Coronavirus is totally worth it; **Case 4 (predicted as misinformation by ICA and as real by BERT):** Chinese People Eating Bat Soup Linked to the Coronavirus-Video. (Color figure online)

words "man", "Coronavirus", "spread" etc. This does not inspire confidence in LIME the way ICA's features did for the ICA-based explainability method. Furthermore, unlike the features extracted by ICA, LIME does not associate the words of a tweet with words associated in other tweets (local explanations). This makes knowledge discovery and explainability a real challenge, since there is no direct way to associate the linguistic attributes of Table 1 with the words contained in a given misinformation tweet. On the other hand, ICA can simultaneously use the context for classification and make it explicit in its explanations. To illustrate this idea, let's take the example of the expression "bat soup". If the expression was used only in the context of sentences such as "Did you know that in some countries bat soup is a delicacy?" in the corpus, then the context, would yield a classification of "real" as well as ICA features that do not trigger any misinformation rules from Table 1, whereas in the context of this corpus, the expression "bat soup" triggers many of these rules as mentioned when discussing the ICA results of Figure 2. Furthermore, in the two misinformation cases (Cases 3 and 4), the lack of confidence in BERT is, in fact, supported by the fact that the two closely related tweets which elicited the same main features and classification by ICA received opposite classifications by BERT (correct misinformation for Case 3 and incorrect real for Case 4) and seemed to have decided on misinformation in Case 3 based on the unknown reference: "@Drizz061", which, in passing, ICA did not give much credence to.

Case 5 in Fig. 3, which ICA wrongly classified as misinformation was correctly classified by BERT, but no good reason emerges from the LIME graphs, except for the fact that the values it associated with the words hover over 0, whereas in the case of ICA, LIME picked up on the words "Fake" and "arrested" which is

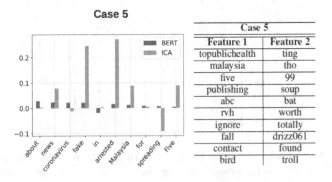

**Fig. 3.** Left: Local explanations by LIME for both BERT and ICA. Orange bars correspond to the ICA-based approach and the blue bars correspond to the BERT-based approach. Furthermore, 0 is neutral whereas positive values are misinformation and negative values real; Right: Feature 1 and Feature 2 represent the dominant words belonging to the highest two features extracted by ICA on one real tweet that ICA classified incorrectly; Tweet: **Case 5 (predicted as misinformation by ICA and as real by BERT)**: Five arrested in Malaysia for spreading fake news about coronavirus. (Color figure online)

different from the explanation given by the feature and, in some way, does make sense according to Table 1 rules: Direct falsity claim ("Fake") and Language related to Death and/or War ("arrested"), which suggests that ICA features together with LIME give a relatively full picture of ICA's mechanism, but that LIME, which is the only category of tools that can be used in Deep learning settings does not provide the user with a clear window into its decision making approach.

## 5 Conclusion

Although Deep Learning recently became the approach of choice for practical NLP tasks such as the detection and removal of misinformation from social media, this study argues that latent variable decomposition methods can be quite competitive and come with added advantages: *simplicity, efficiency,* and most importantly, *built-in explainability.* After presenting a new Covid-19 misinformation data set, we demonstrate that an ICA- based classification approach is almost as accurate as a BERT-based approaches while efficiently extracting features bearing more resemblance to the socio-linguistic rules used to build the data set than the information extracted by LIME, a state-of-the-art explainability tool.

## References

1. Adalı, T., Anderson, M., Fu, G.S.: Diversity in independent component and vector analyses: identifiability, algorithms, and applications in medical imaging. IEEE Signal Process. Mag. **31**(3), 18–33 (2014)
2. Baly, R., Karadzhov, G., Alexandrov, D., Glass, J., Nakov, P.: Predicting factuality of reporting and bias of news media sources. arXiv preprint arXiv:1810.01765 (2018)
3. Berry, M.W., Browne, M., Langville, A.N., Pauca, V.P., Plemmons, R.J.: Algorithms and applications for approximate nonnegative matrix factorization. Comput. Stat. Data Anal. **52**(1), 155–173 (2007)
4. Blei, D.M., Ng, A.Y., Jordan, M.I.: Latent dirichlet allocation. J. Mach. Learn. Res. **3**, 993–1022 (2003)
5. Boukouvalas, Z., Levin-Schwartz, Y., Mowakeaa, R., Fu, G.S., Adalı, T.: Independent component analysis using semi-parametric density estimation via entropy maximization. In: 2018 IEEE Statistical Signal Processing Workshop (SSP), pp. 403–407. IEEE (2018)
6. Clark, K., Luong, M.T., Le, Q.V., Manning, C.D.: Electra: pre-training text encoders as discriminators rather than generators. arXiv preprint arXiv:2003.10555 (2020)
7. Devlin, J., Chang, M., Lee, K., Toutanova, K.: BERT: pre-training of deep bidirectional transformers for language understanding. CoRR abs/1810.04805 (2018). http://arxiv.org/abs/1810.04805
8. Asr, F.T.: The language gives it away: How an algorithm can help us detect fake news (2019). https://theconversation.com/the-language-gives-it-away-howan-algorithm-can-help-us-detectfake-news-120199, online

9. Gupta, A., Kumaraguru, P.: Credibility ranking of tweets during high impact events. In: Proceedings of the 1st Workshop on Privacy and Security in Online Social Media, p. 2. ACM (2012)
10. Hansen, L.K., Rieger, L.: Interpretability in intelligent systems – a new concept? In: Samek, W., Montavon, G., Vedaldi, A., Hansen, L.K., Müller, K.-R. (eds.) Explainable AI: Interpreting, Explaining and Visualizing Deep Learning. LNCS (LNAI), vol. 11700, pp. 41–49. Springer, Cham (2019). https://doi.org/10.1007/978-3-030-28954-6_3
11. Horne, B.D., Adali, S.: This just. In: fake news packs a lot in title, uses simpler, repetitive content in text body, more similar to satire than real news. In: Eleventh International AAAI Conference on Web and Social Media (2017)
12. Islam, M.R., Liu, S., Wang, X., Xu, G.: Deep learning for misinformation detection on online social networks: a survey and new perspectives. Social Netw. Anal. Min. **10**(1), 1–20 (2020). https://doi.org/10.1007/s13278-020-00696-x
13. Li, X., Adali, T.: Independent component analysis by entropy bound minimization. IEEE Trans. Signal Process. **58**(10), 5151–5164 (2010)
14. Liu, Y., et al.: Roberta: a robustly optimized bert pretraining approach. arXiv preprint arXiv:1907.11692 (2019)
15. Perez-Rosas, V., Kleinberg, B., Lefevre, A., Mihalcea, R.: Automatic detection of fake news. arXiv preprint arXiv:1708.07104 (2017)
16. Rashkin, H., Choi, E., Jang, J.Y., Volkova, S., Choi, Y.: Truth of varying shades: Analyzing language in fake news and political fact-checking. In: Proceedings of the 2017 Conference on Empirical Methods in Natural Language Processing, pp. 2931–2937 (2017)
17. Reimers, N., Gurevych, I.: Sentence-bert: sentence embeddings using siamese bert-networks. arXiv preprint arXiv:1908.10084 (2019)
18. Ribeiro, M.T., Singh, S., Guestrin, C.: "Why should i trust you?": explaining the predictions of any classifier (2016). http://arxiv.org/abs/1602.04938
19. Tošić, I., Frossard, P.: Dictionary learning. IEEE Signal Process. Mag. **28**(2), 27–38 (2011)
20. White, K., Li, G., Japkowicz, N.: Sampling online social networks using coupling from the past. In: 2012 IEEE 12th International Conference on Data Mining Workshops, pp. 266–272. IEEE (2012)
21. Wu, L., Morstatter, F., Carley, K.M., Liu, H.: Misinformation in social media: definition, manipulation, and detection. ACM SIGKDD Explor. Newsl **21**(2), 80–90 (2019)

# Spatial, Temporal and Spatiotemporal Data

# Local Exceptionality Detection in Time Series Using Subgroup Discovery: An Approach Exemplified on Team Interaction Data

Dan Hudson[1,2(✉)] [iD], Travis J. Wiltshire[2] [iD], and Martin Atzmueller[1] [iD]

[1] Semantic Information Systems Group, Osnabrück University, Osnabrück, Germany
{daniel.dominic.hudson,martin.atzmueller}@uni-osnabrueck.de
[2] Department of Cognitive Science and AI,
Tilburg University, Tilburg, The Netherlands
t.j.wiltshire@tilburguniversity.edu

**Abstract.** In this paper, we present a novel approach for local exceptionality detection on time series data. This method provides the ability to discover interpretable patterns in the data, which can be used to understand and predict the progression of a time series. As an exploratory approach, the results can be used to generate hypotheses about the relationships between the variables describing a specific process and its dynamics. We detail our approach in a concrete instantiation and exemplary implementation, specifically in the field of teamwork research. Using a real-world dataset of team interactions we discuss the results and showcase the presented novel analysis options. In addition, we outline possible implications of the results in terms of understanding teamwork.

**Keywords:** Subgroup discovery · Exceptional model mining · Time series · Teamwork research · Multimodal analysis.

## 1 Introduction

Methods for local exceptionality detection such as subgroup discovery [2] and its variant exceptional model mining (EMM) [8] are established knowledge discovery techniques for finding interpretable patterns. Basically, they identify patterns relating different attributes of a dataset that are interesting according to some target model, thus providing explicit and *interpretable* rules to associate descriptive properties found in the data instances. Considering time series and/or event data, the investigation of subgroup discovery has been limited, mainly focusing on aggregating/averaging time overall [14] or by considering aggregates on sets of discrete-valued events [17], compared to continuous-valued time series which we consider in this work.

© Springer Nature Switzerland AG 2021
C. Soares and L. Torgo (Eds.): DS 2021, LNAI 12986, pp. 435–445, 2021.
https://doi.org/10.1007/978-3-030-88942-5_34

In this paper, we present a novel approach for performing subgroup discovery and EMM on time series. We propose an extensible approach, in particular relating to feature and target construction on dynamic time series data.

Time series exceptionality analysis is a vast field, e. g., [1,12]. Here, for example, methods for change detection [1], anomaly detection also including symbolic representations [5,9,15] and time series classification are relevant, where typically global approaches are addressed, in contrast to local exceptionality detection which we focus on in this work. An approach based for compressing event logs based on the minimum description length (MDL) principle was presented by [11], making it possible to detect local patterns in temporal data, however without focusing on exceptionality. Compared to this work, which focuses on event sequences as inputs, our approach aims to find meaningful representations of (potentially complex) continuous-valued time series, and assesses the discovered patterns by reference to a target variable rather than compression.

In order to demonstrate our approach, we exemplify its application on a case study conducted in the area of social sensing, wherein team interactions are examined through a multimodal, sensor-based approach. In general, the study of teamwork looks at how groups of multiple individuals work toward a common goal [19] through collaborative team processes [10]. Recent work [18,20] has emphasised the need to understand dynamics within team processes, by embracing methodologies that record teams over time. For example, body movement, along with dynamics of speaking and turn-taking, are well understood to be important social signals used in cooperation and teamwork [16,25]. Although they can be quantified from video and audio recordings, it is difficult to establish the important relationships between these social signals in an empirical way when using multiple time-varying modalities.

Since it is not obvious how to start analysing these time-varying signals, the application of exploratory analysis methods such as subgroup discovery is well suited for such an analysis, to provide first insights and to support hypothesis generation. Our approach leads to interpretable rules which are plausible due to the use of expert knowledge in feature selection (described further in Sect. 3). We therefore choose this case study to showcase our method and to discuss the respective results, using a real-world dataset of 27 video and audio recordings of teams performing a collaborative task, taken from the ELEA corpus [21].

Our contributions are summarised as follows:

1. We present a novel methodological approach as an iterative human-guided process that makes it possible to use subgroup discovery on time series data. We discuss according feature extraction and target construction, e. g., using different time lags, for making the results predictive at different timescales.
2. We showcase our approach through a study in the context of team research. For this analysis, we also introduce a new quality function, adapting the concept of dynamic complexity from team research. In addition, we present a novel subgroup visualisation for multi-dimensional parameter analysis, i. e., the *subgroup radar plot*, also enabling user-guided subgroup assessment.

3. In our case study, we search for relationships between multimodal data, i. e., body movement and speech in time series. As evaluated by a domain specialist, this gives rise to several meaningful hypotheses that can be investigated in future work in the field of teamwork study.

# 2  Method

Below, we present our process model for local exceptionality detection in time series, and then we discuss the individual steps of our approach in detail.

*Overview.* Our proposed approach is visualised in Fig. 1 as a linear workflow, which can be executed in multiple iterations: First, the time series is split into slices, i.e., non-overlapping subsequences of equal length, so that it is possible to investigate moments when a time-varying target variable reaches an extreme value. For each slice, we extract a set of descriptive features. These are (optionally) discretised, e. g., the value of each feature can be converted into 'low', 'medium' and 'high' based on tercile boundaries across the slices. The choice of the appropriate length for a slice should be driven by the application, i. e., to include enough time points to allow a variety of features to be extracted, such as frequency components and estimates of entropy; also, it should be small enough that dynamics of the time series are not likely to change multiple times within a slice. Then, the target variable is prepared. We also propose to investigate the relationship between attributes of the time slices to the target variable at different lags, which necessitates performing the analysis with multiple copies of the dataset (with the lag applied). With a lag of zero, our process discovers subgroups that are informative about how various attributes covary with the target, e. g., for investigating how a system-level process is reflected in multiple variables. With higher lags, the process has a more predictive focus, relating to especially high/low target value at a later point.

**Fig. 1.** The workflow of our methodological process to perform subgroup discovery on time series data. With a human-in-the-loop, the workflow can be iteratively applied.

*Time Series Feature Extraction.* We convert time series into 'slices' which can then individually be summarised with static rather than time-varying attributes. To obtain the necessary features, we use the TSFresh package in

Python [6], which computes a large number of features specifically to summarise time series. Examples of features computed by TSFresh are, e. g.,: (a) mean value, (b) absolute energy, (c) autocorrelation at different lags, (d) Fourier coefficients, (e) binned entropy, (f) sample entropy, (g) root mean square, etc. This approach makes it possible to perform subgroup discovery on the features extracted for each slice, while still retaining some of the variation that is observed as the original time series progresses (since different slices will correspond to different points in time in the original series).

**Subgroup Discovery for Local Exceptionality Detection.** Subgroup discovery aims at finding a combination of *selectors* or *selection expressions*, in a form similar to rules (e. g., `PropertyA = True` or `PropertyB > 1.5`), which function as membership criteria for a subgroup: any data points that satisfy the criteria are part of the subgroup. A *subgroup description* (or *pattern*) combines selectors into a Boolean formula. For a typical conjunctive description language, a pattern $P = \{sel_1, \ldots, sel_k\}$ is defined out of a set $S$ of selectors $sel_j \in S$, which are interpreted as a conjunction, i.e. $p = sel_1 \wedge \ldots \wedge sel_k$. A *subgroup* corresponding to a pattern then contains all instances $d$ of a database $D$, i. e., $d \in D$ for which the respective formula for the pattern evaluates to true. Specifying subgroups in this way is useful because the rules are easy to interpret and relate directly to known properties of the data points – also called *instances*.

The key question is determining which subgroups are interesting, e. g., because they have a particularly high average target value compared to the population mean, as observed for the whole dataset. The interestingness of a pattern is determined by a quality function $q : 2^S \rightarrow \mathbb{R}$. It maps every pattern in the search space to a real number that reflects the interestingness of a pattern (or the extension of the pattern, respectively). Many quality functions for a single target feature, e. g., in the binary or numerical case, trade off the size $n = |ext(P)|$ of a subgroup and the deviation $t_P - t_0$, where $t_P$ is the average value in the subgroup identified by the pattern $P$ and $t_0$ the average value of the target feature in the general population. Thus, standard quality functions are of the form

$$q_a(P) = n^a \cdot (t_P - t_0), \ a \in [0; 1].$$

For binary target concepts, this includes, e. g., a *simplified binomial* function $q_{0.5}$ for $a = 0.5$, or the *Piatetsky-Shapiro* quality function $q_1$ with $a = 1$, cf. [2]. Recently, [4] described the use of quality functions which also include a term to quantify dispersion of the target feature within the subgroup, which increases the consistency within subgroups with respect to the target feature.

# 3    Results: Case Study on Team Interaction Data

Below, we discuss a case study applying our approach in the context of interactive team cognition [7]. We investigate the 'dynamic complexity' of speech amongst team members (described in detail in Sect. 3), which is a method to quantify interaction dynamics that is also sensitive to moments of transition where the dynamics are changing. These moments are potentially to the benefit or the detriment of the team (we provide further discussion in [24]). Subgroup discovery provides interpretable patterns of interesting situations/events, where the dynamic complexity shows *exceptional* local deviations, indicating interesting points in the respective team interaction.

**Dataset.** The data used in this case study comes from the Emergent Leadership (ELEA) corpus [21], which contains recordings of groups of individuals who have been tasked to work together to rank a list of items given a disaster scenario. In particular, the task was to rank the importance of items such as 'chocolate' and 'newspapers' for the situation in which the group has been stranded in a freezing-cold winter environment. The corpus includes audio recordings from a single microphone in the centre of the room, and video recordings from webcams facing the participants. Both types of recording are available for 27 groups, each consisting of 3–4 participants. Via the video recordings, we quantify body movement during the team task. See [13] for a detailed discussion on how to quantify body movement in this context. Intuitively, we relate the body movement modality to the modality of speech, using the audio recordings to quantify speech dynamics. As a target we use dynamic complexity, cf. Sect. 3 (below), to estimate speech dynamics.

**Feature Selection.** TSFresh is not domain-specific, and therefore extracts generic features from time series. An important step in our process is to identify which features are potentially relevant and interpretable for the application being considered. We selected a subset of 91 from the 300+ features extracted by TSFresh. At a high level, these features can be categorised as follows:

(a) Descriptive statistics (mean, variance, quantiles, standard deviations); (b) Average and variance of the changes between successive time points; (c) Measures of complexity, such as Lempel-Ziv complexity, as well as multiple forms of entropy; (d) 'Matrix profile' statistics, which can be informative about repetitive or anomalous sub-sequences of the time slices; (e) Measures based on the number of peaks or extreme points; (f) Strength of different frequency components in the signal; (g) Measures based on autocorrelation at different lags; (h) Measures based on how well the data fits a certain (e. g., linear) model.

Features were selected based on potential relevance to body movement in social interactions, through discussion with an expert in interactive team cognition. For example, when considering coefficients of the Fourier transform, we included the magnitude since a large degree of movement at a specific frequency is something that can be visibly interpreted from the video recordings of body movement, but excluded the angle (or phase at the start of the time slice) since this is hard to visually comprehend (without performing further analysis to look for, e.g., synchrony) and therefore seems unlikely to be a usable social signal.

**Target Modeling – Dynamic Complexity.** As the target variable for subgroup discovery, we focused on the dynamic complexity of the speech recordings. The dynamic complexity measure is used to quantify how complex the behaviour of a system is, and provides us with a way to characterise the dynamics of speech, in a manner which could potentially be useful for (e.g.) detecting moments when a phase transition between patterns of behaviour is likely [22]. It is calculated over a moving window by combining two components, called the *fluctuation* and *distribution*, which respectively correspond to: the degree to which frequent and large oscillations are observed in the window, and, the degree to which the observed values refuse to favour a particular region of the measurement scale (instead being distributed equally across the possible range of values). The product of *fluctuation* and *distribution* gives the *dynamic complexity* of the window. For a detailed discussion we refer to [22,24]. We convert the audio data, which is sampled 40,000 Hz, to a more coarse-grained dataset by computing the energy of each second of audio, which allows us to calculate dynamic complexity on a scale appropriate to interaction behaviour (a timescale of seconds to minutes).

We evaluated two variations of dynamic complexity, captured as a target attribute, which we constructed in the spirit of EMM. First, we model the dynamic complexity as a Gaussian distribution of values, and use the z-score normalised mean as the target variable to determine which subgroups are most interesting. Second, we perform a linear regression of the dynamic complexity against time as a target model, and use the resulting slope as the target attribute. In both cases, the quality function we use to rank subgroups is the simple binomial quality function ($q_{0.5}$, see above), which tends to favour smaller subgroups with a more extreme target value. To balance this, we set the minimum subgroup size to 20 so that they do not become too small to be meaningful.

*Results.* As stated earlier, we are able to perform subgroup discovery at different time lags, making the task predictive when using lags greater than zero, or an exploration of the relationships between variables at lag zero. First, we discuss the 0-lag results, with slices of 1 min. A selection of five subgroups is presented in Table 1. The subgroups are also visualised as subgroup radar plots in Figs. 2(a), indicating the most important quality parameters, and 3(a), additionally showing how the subgroups differ according to 5 key selector variables. In general, these presented novel subgroup visualisations (Figs. 2, 3) allow a seamless overview–zoom–detail cycle, according to the *Information Seeking Mantra* by Shneiderman [23]: Overview first (macroscopic view), browsing and zooming

**Table 1.** A selection of subgroups discovered using the SD-Map algorithm [3] at a lag of 0 min: subgroup pattern, a textual description, size ($S$) and mean z-score ($\varnothing$).

| | Pattern | Description | $|S|$ | $\varnothing$ |
|---|---|---|---|---|
| 1 | mean_change_quantiles_f_agg_"mean"_<br>_isabs_False_qh_0.6_ql_0.2=low, AND<br>mean_longest_strike_below_mean=high,<br>AND mean_quantile_q_0.8=low | The value of changes around the mean (after values have been restricted to remain between the 0.2 and 0.6 quantiles) is low across the team. Team members tend to have at least one long sequence of values below the mean. The 0.8 quantile also tends to be low. | 21 | 1.137 |
| 2 | mean_lempel_ziv_complexity_<br>_bins_100=low, AND<br>mean_longest_strike_below_mean=high,<br>AND mean_quantile_q_0.8=low | The Lempel-Ziv measure of complexity is neither high nor low across the team. Team members tend to have at least one long sequence of values below the mean. The 0.8 quantile also tends to be low. | 25 | 1.026 |
| 3 | mean_longest_strike_below_mean=high,<br>AND mean_mean=low | Team members tend to have at least one long sequence of values below the mean. The average value of movement is generally low across the team. | 38 | 0.81 |

(mesoscopic analysis), and details on demand (microscopic focus) – from basic quality parameters to subgroup description and its combination. This allows a human-in-the-loop approach, which proved beneficial for our domain specialist when inspecting the results of our knowledge discovery methodology.

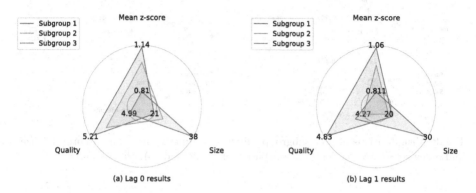

(a) Lag 0 results                    (b) Lag 1 results

**Fig. 2.** Visualisations of how subgroups differ according to quality, mean z-score, and size. The results are shown at: (a) a lag of 0 min, and (b) a lag of 1 min.

Overall, our results show that it is possible to discover relatively small subgroups (size 20–40 compared to a population size of 327) whose dynamic complexity has on average a z-score of around 1, as the mean, averaged across members of the subgroup. Since the outputs are interpretable, it is possible to speculate about what they mean in the context of body movement and speech dynamics. Many of the subgroups, such as those shown in Table 1, suggest that low amounts of movement might be indicative of complex speech dynamics, particularly if there is a long sequence of low values. This perhaps suggests that while speech dynamics are becoming chaotic, the team members become more still – for example moving less as they focus more on the discussion. This is a hypothesis generated from the data which further work could seek to verify. In order to consider the impact of the window size, we also performed subgroup discovery using 30-second slices of the time series. This uncovered subgroups which often used the same rules, e.g., stating that the change around the mean between the 0.2 and 0.6 quantiles be low across the team, and, that the mean and various quantiles should also be low. There were some differences, especially that these subgroups incorporated more rules concerning variability between team members with respect to their number of values below the mean/above the mean and their Lempel-Ziv complexity, suggesting that certain types of imbalance in the teams may also help to identify moments of high dynamic complexity in speech. Overall, changing the window size in this way did not have a large impact on the discovered subgroups.

Next, we discuss results when applying a lag of 1 min, discovering ways to predict high complexity in speech dynamics from the body movement signals a minute earlier. A selection of subgroups are presented in Table 2, Fig. 2(b),

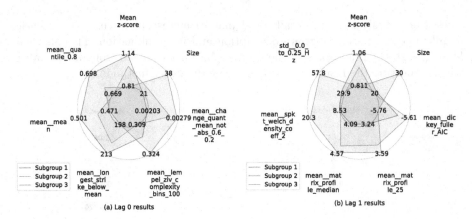

**Fig. 3.** Visualisations of how subgroups differ according to the mean z-score, the size, and 5 key selector variables – with (a) a lag of 0 min, and (b) a lag of 1 min.

and Fig. 3(b), to give an idea of how exceptional the subgroups are compared to the population overall. Like with the 0-lag results, it appears to be possible to discover subgroups of around 20–40 members which have an average z-score of close to 1. The features used to define subgroups, however, are different. Some of the subgroups, such as the first two listed in Table 2, suggest that the team members might have similar, low values for the low-frequency movement components during the minute preceding a period of high dynamic complexity. Features based on 'matrix profile' statistics are used to define many subgroups. Looking at these features, it seems that high dynamic complexity can be expected following a period when the body movement signal does not have clear repetitions in its structure. This could indicate that complexity and a lack of pattern in body movement is predictive of chaotic speech dynamics shortly thereafter. This is another example of a data-driven hypothesis that future confirmatory work could verify.

Furthermore, besides the mean dynamic complexity of speech, we performed the analysis with two related target concepts. Specifically, we considered (1) the slope when conducting a linear regression of the dynamic complexity against time, and (2) the change between successive windows. The results in this case would be informative about periods when the complexity of speech dynamics is increasing. Many of these subgroups included rules that complexity of body movement (measured through Lempel-Ziv, Fourier entropy and binned entropy) is generally high across the team, and some subgroups also indicated that a large number of peaks in the slices of body movement was related to increasing dynamic complexity in speech. This could indicate that complexity in speech increases fastest while movement is already complex. From this analysis, the subgroups also suggested a relationship between increasing speech complexity and a medium-strength frequency component in body movement at 0.5–0.75 Hz, and also medium values for the mean and various quantiles. Results when using

**Table 2.** A selection of subgroups discovered using the SD-Map algorithm [3] at a lag of 1 min: subgroup pattern, a textual description, size ($S$) and mean z-score ($\varnothing$).

| | Pattern | Description | $|S|$ | $\varnothing$ |
|---|---|---|---|---|
| 1 | mean_augmented_dickey_fuller_ _attr."teststat"_autolag_"AIC"=medium, AND mean_spkt_welch_density_coeff_2 =low, AND std_0.0_to_0.25_Hz=low | The signal is neither relatively well-modelled nor relatively poorly-modelled by a process with a unit root, according to the Augmented Dickey-Fuller test. The strength of the frequency component at 0.234 Hz is low, and there is low variability of the strength of frequency components between 0.0 Hz and 0.25 Hz among team members. | 21 | 1.062 |
| 2 | mean_matrix_profile_feature_"25"_ _threshold_0.98=high, AND mean_spkt_welch_density_coeff_2=low, AND std_root_mean_square=low | The 0.25 quantile of the similarity of subwindows within the signal to other subwindows within the signal is low, suggesting that a reasonable proportion of subsequences (of the time series) are unusual (not repeating). The strength of the frequency component at 0.234 Hz is low. How well the time slices can be modelled by a linear progression is not varied across the team members. | 20 | 0.964 |
| 3 | mean_matrix_profile_feature_"median"_ _threshold_0.98=high, AND std_change_quantiles_f_agg_"var"_ _isabs_True_qh_0.6_ql_0.4=low, AND std_quantile_q_0.9=low | The median similarity of subwindows within the signal to other subwindows within the signal is low, suggesting that subsequences (of the time series) tend to be relatively unusual (not repeating). The absolute value of changes around the mean (after values have been restricted to remain between the 0.4 and 0.6 quantiles) is consistent across the team. The 0.9 quantile also has low variability. | 30 | 0.811 |

the change between successive as a target variable also suggest that the frequency components at 0.0–0.25 Hz should be low. The reliability and significance of these frequency components as indicators of increasing speech complexity could be investigated through future work.

## 4  Conclusions

In conclusion, we have presented a novel approach for local exceptionality detection in time series using subgroup discovery – as a workflow that can be applied in multiple iterations for modeling features and the target, respectively. With these, it is possible to identify features which are strongly associated with or highly predictive of a target variable. We demonstrated the approach via a case study of analysing team interaction data, matching body movement information to a measure of the dynamics of speech. Among other things, this showcased the hypothesis-generating capabilities of our approach. Future research could consider possible refinements, e. g., by considering to extract features from overlapping windows at different offsets as an alternative to non-overlapping windows.

**Acknowledgements.** The research leading to this work has received funding by the Dutch Research Council (NWO), project KIEM ICT ODYN (ENPPS.KIEM.019.016), and by the German Research Foundation (DFG), project "MODUS" (grant AT 88/4-1).

# References

1. Aminikhanghahi, S., Cook, D.J.: A survey of methods for time series change point detection. Knowl. Inf. Syst. **51**(2), 339–367 (2016). https://doi.org/10.1007/s10115-016-0987-z
2. Atzmueller, M.: Subgroup discovery. WIREs DMKD **5**(1), 35–49 (2015)
3. Atzmueller, M., Puppe, F.: SD-map – a fast algorithm for exhaustive subgroup discovery. In: Fürnkranz, J., Scheffer, T., Spiliopoulou, M. (eds.) PKDD 2006. LNCS (LNAI), vol. 4213, pp. 6–17. Springer, Heidelberg (2006). https://doi.org/10.1007/11871637_6
4. Boley, M., Goldsmith, B.R., Ghiringhelli, L.M., Vreeken, J.: Identifying consistent statements about numerical data with dispersion-corrected subgroup discovery. Data Min. Knowl. Disc. **31**(5), 1391–1418 (2017). https://doi.org/10.1007/s10618-017-0520-3
5. Chandola, V., Banerjee, A., Kumar, V.: Anomaly detection for discrete sequences: a survey. IEEE TKDE **24**(5), 823–839 (2010)
6. Christ, M., Braun, N., Neuffer, J., Kempa-Liehr, A.W.: Time series feature extraction on basis of scalable hypothesis tests (tsfresh-a python package). Neurocomputing **307**, 72–77 (2018)
7. Cooke, N.J., Gorman, J.C., Myers, C.W., Duran, J.L.: Interactive team cognition. Cogn. Sci. **37**(2), 255–285 (2013)
8. Duivesteijn, W., Feelders, A.J., Knobbe, A.: Exceptional model mining. Data Min. Knowl. Disc. **30**(1), 47–98 (2015). https://doi.org/10.1007/s10618-015-0403-4
9. Ramirez, E., Wimmer, M., Atzmueller, M.: A computational framework for interpretable anomaly detection and classification of multivariate time series with application to human gait data analysis. In: Marcos, M., et al. (eds.) KR4HC/TEAAM-2019. LNCS (LNAI), vol. 11979, pp. 132–147. Springer, Cham (2019). https://doi.org/10.1007/978-3-030-37446-4_11
10. Fiore, S.M., Smith-Jentsch, K.A., Salas, E., Warner, N., Letsky, M.: Towards an understanding of macrocognition in teams: developing and defining complex collaborative processes and products. Theor. Iss. Ergon. Sci. **11**(4), 250–271 (2010)
11. Galbrun, E., Cellier, P., Tatti, N., Termier, A., Crémilleux, B.: Mining periodic patterns with a MDL criterion. In: Berlingerio, M., Bonchi, F., Gärtner, T., Hurley, N., Ifrim, G. (eds.) ECML PKDD 2018. LNCS (LNAI), vol. 11052, pp. 535–551. Springer, Cham (2019). https://doi.org/10.1007/978-3-030-10928-8_32
12. Hamilton, J.D.: Time Series Analysis. Princeton University Press, Princeton (2020)
13. Hudson, D., Wiltshire, T.J., Atzmueller, M.: Multisyncpy: a python package for assessing multivariate coordination dynamics. PsyArXiv (2021)
14. Jin, N., Flach, P., Wilcox, T., Sellman, R., Thumim, J., Knobbe, A.: Subgroup discovery in smart electricity meter data. IEEE TII **10**(2), 1327–1336 (2014)
15. Keogh, E., Lin, J., Fu, A.: Hot Sax: efficiently finding the most unusual time series subsequence. In: Proceedings ICDM. IEEE (2005)
16. Kim, T., Chang, A., Holland, L., Pentland, A.S.: Meeting mediator: enhancing group collaboration using sociometric feedback. In: Proceedings Conference on Computer supported cooperative work, pp. 457–466. ACM (2008)

17. Knobbe, A., Orie, J., Hofman, N., van der Burgh, B., Cachucho, R.: Sports analytics for professional speed skating. Data Min. Knowl. Disc. **31**(6), 1872–1902 (2017). https://doi.org/10.1007/s10618-017-0512-3
18. Kozlowski, S.W., Chao, G.T.: Unpacking team process dynamics and emergent phenomena: challenges, conceptual advances, and innovative methods. Am. Psychol. **73**(4), 576 (2018)
19. Salas, E., Dickinson, T.L., Converse, S.A., Tannenbaum, S.I.: Toward an understanding of team performance and training. In: Swezey, R.W., Salas, E. (eds.) Teams: Their Training and Performance, pp. 3–29. Ablex Publishing, Norwood (1992)
20. Salas, E., Reyes, D.L., McDaniel, S.H.: The science of teamwork: progress, reflections, and the road ahead. Am. Psychol. **73**(4), 593 (2018)
21. Sanchez-Cortes, D., Aran, O., Mast, M.S., Gatica-Perez, D.: A nonverbal behavior approach to identify emergent leaders in small groups. IEEE Trans. Multimedia **14**(3), 816–832 (2011)
22. Schiepek, G., Strunk, G.: The identification of critical fluctuations and phase transitions in short term and coarse-grained time series–a method for the real-time monitoring of human change processes. Biol. Cybern. **102**(3), 197–207 (2010)
23. Shneiderman, B.: The eyes have it: a task by data type taxonomy for information visualizations. In: Proceedings IEEE Symposium on Visual Languages, pp. 336–343 (1996)
24. Wiltshire, T.J., Hudson, D., Lijdsman, P., Wever, S., Atzmueller, M.: Social analytics of team interaction using dynamic complexity heat maps and network visualizations. arXiv preprint arXiv:2009.04445 (2020)
25. Wiltshire, T.J., Steffensen, S.V., Fiore, S.M.: Multiscale movement coordination dynamics in collaborative team problem solving. Appl. Ergon. **79**, 143–151 (2019)

# Neural Additive Vector Autoregression Models for Causal Discovery in Time Series

Bart Bussmann$^{(\boxtimes)}$, Jannes Nys, and Steven Latré

IDLab, University of Antwerp - imec, Antwerpen, Belgium
{bart.bussmann,jannes.nys,steven.latre}@uantwerpen.be

**Abstract.** Causal structure discovery in complex dynamical systems is an important challenge for many scientific domains. Although data from (interventional) experiments is usually limited, large amounts of observational time series data sets are usually available. Current methods that learn causal structure from time series often assume linear relationships. Hence, they may fail in realistic settings that contain nonlinear relations between the variables. We propose Neural Additive Vector Autoregression (NAVAR) models, a neural approach to causal structure learning that can discover nonlinear relationships. We train deep neural networks that extract the (additive) Granger causal influences from the time evolution in multi-variate time series. The method achieves state-of-the-art results on various benchmark data sets for causal discovery, while providing clear interpretations of the mapped causal relations.

**Keywords:** Causal discovery · Time series · Deep learning

## 1 Introduction

Discovering mechanisms and causal structures is an important challenge for many scientific domains. Randomized control trials may not always be feasible, practical or ethical, such as in the domain of climate sciences and genetics. Therefore, when no interventional data is available, we are forced to rely on observational data only.

In dynamical systems, the arrow of time simplifies the analysis of possible causal interactions in the sense that we can assume that only preceding signals are a potential cause of the current observations. A common approach is to test time-lagged causal associations in the framework of Granger causality [10]. These methods often model the time-dependence via linear causal relationships, with Vector AutoRegression (VAR) models as the most common approach.

Even though there is extensive literature on nonlinear causal discovery (e.g. [17,31]) relatively few others (e.g. [14,32]) have harnessed the power of deep learning for causal discovery in time series. These methods operate within the Granger causality framework and use deep neural networks to model the time

© Springer Nature Switzerland AG 2021
C. Soares and L. Torgo (Eds.): DS 2021, LNAI 12986, pp. 446–460, 2021.
https://doi.org/10.1007/978-3-030-88942-5_35

dependencies and interactions between the variables. In principle, deep learning approaches make it possible to model causal relationships, even when they are nonlinear. While these methods have a high degree of expressiveness, this flexibility comes at a cost: interpretation of the causal relations learned by black-box methods is hindered, while this is essentially the goal of causal structure learning. To overcome this, these methods learn to set certain input weights to zero, which they interpret as an absence of Granger Causality.

In this work, we propose the Neural Additive Vector Autoregression (NAVAR) model to resolve this problem. NAVAR assumes an additive structure, where the predictions depend linearly on independent nonlinear functions of the individual input variables. We model these nonlinear functions using neural networks. In comparison to other works using Granger causality for causal discovery in time series, our work differs in the following ways:

1. Compared to common linear methods, our method can easily capture (highly) nonlinear relations.
2. While being able to model nonlinear relations, NAVAR maintains a clear interpretation of the causal dependencies between pairs of variables. In contrast to other deep learning methods that resort to feature importance methods, NAVAR uses the interpretational power of additive models to discover Granger causal relationships.
3. By using an additive model of learned transformations of the input variables, our model allows not only for the discovery of causal relationships between pairs of time series but also inspection of the functional form of these causal dependencies. Thanks to the additive structure, we can inspect the direct contribution of every input variable to every output variable.
4. The additive structure allows us to score and rank causal relations. Since we can compute the direct contribution of each input variable to each output variable independently, the variability of these contributions can be used as evidence for the existence of a causal link.

The rest of this paper is structured as follows: Sect. 2 introduces the Granger causality framework and VAR models. In Sect. 3 we generalize this notion to the additive nonlinear case and introduce NAVAR models that can estimate Granger causality using neural networks. In Sect. 4 we evaluate the performance of NAVAR on various benchmarks and compare it to existing methods. Finally, in Sect. 5 we discuss related work and in Sect. 6 we conclude and discuss directions for future work.

## 2    Granger Causality and the VAR Model

Let $X_{1:T} = \{X_{1:T}^{(1)}, X_{1:T}^{(2)}, .., X_{1:T}^{(N)}\}$ be a multivariate time series with $N$ variables and $T$ time steps. Our goal is to discover the causal relations between this set of time series. (Pairwise) Granger causality is one of the classical frameworks to discover causal relationships between time series. In this framework, we model the time series as:

$$X_t^{(i)} = g^i(X_{<t}^{(1)}, ..., X_{<t}^{(N)}) + \eta_t^i \tag{1}$$

where $X_{<t}^{(i)} = X_{1:t-1}^{(i)}$ denotes the past of $X^{(i)}$, and $\eta_t$ is an independent noise vector. A variable $X^{(i)}$ is said to Granger cause another variable $X^{(j)}$ if the past of the set of all (input) variables $\{X_{<t}^{(1)}, ..., X_{<t}^{(i)}, ..., X_{<t}^{(N)}\}$ allows for better predictions for $X_t^{(j)}$ compared to the same set where the past of $X^{(i)}$ is not included: $\{X_{<t}^{(1)}, ..., X_{<t}^{(i-1)}, X_{<t}^{(i+1)}, ..., X_{<t}^{(N)}\}$. Granger causality approaches assume causal sufficiency. We refer to the directed graph with the variables $X^{(i)}$ as vertices, and links representing Granger causality between two variables as the Granger causal graph.

In the VAR framework, the time series $X_t^{(j)}$ is assumed to be a linear combination of all past values (up to some maximum lag $K$) and independent noise term. This means that every value $X_t^{(j)}$ can be modeled as:

$$X_t^{(j)} = \beta^j + \sum_{i=1}^{N} \sum_{k=1}^{K} [A_k]^{ij} X_{t-k}^{(i)} + \eta_t^j \tag{2}$$

Where $A_k$ is a $N \times N$ time-invariant matrix which identifies the interaction between the variables, $\beta$ is a $N$-dimensional bias vector, and $\eta_t$ is an independent noise vector with zero mean. A common approach to infer which pairs of variables are *not* Granger causal is to identify $i$ and $j$ for which $[A_k]^{ij} = 0$ for all time lags $k = 1, ..., K$.

## 3   NAVAR: Neural Additive Vector AutoRegression

The idea underlying the linear VAR model is simple and it can be surprisingly effective. For instance, in the NeurIPS 2019 Causality for Climate competition, the winners used four variations based on the standard linear VAR model [34]. However, a limitation of the VAR model is that it can only model linear interactions. Guided by the success and reliability of VAR models for Granger-causal discovery, in this work, we generalize the VAR model to allow for nonlinear additive relationships between variables:

$$X_t^{(j)} = \beta^j + \sum_{i=1}^{N} f^{ij}(X_{t-K:t-1}^{(i)}) + \eta_t^j \tag{3}$$

Here, $f^{ij}$ is a nonlinear function describing the relationship between the past $K$ values of $X^{(i)}$ on the current value of $X^{(j)}$. Note that the VAR model is the special case where $f^{ij}$ is linear. We can identify Granger causality in the following way: if variable $X^{(i)}$ is not a Granger cause of another variable $X^{(j)}$ then $f^{ij}$ is invariant to the values of $X_{t-K:t-1}^{(i)}$. In other words, if $f^{ij}$ is a constant function of *all* values $X_{t-K:t-1}^{(i)}$, then $X^{(i)}$ is not a Granger cause of $X^{(j)}$.

The choice of this additive model is built on the following assumption. In many practical applications, the functional dependence of a variable $X_t^{(i)}$ on the history of a variable $X_{<t}^{(i)}$ is complex, with e.g. nonlinear functions across multiple time lags. However, dependencies on multiple time series can usually

be well approximated by additive models. Therefore, we introduce an additive structure for the contributions stemming from the different variables, but do not impose an additive restriction to contributions from different time lags.

We choose to use deep neural networks (DNN) to model the nonlinear function $f^{ij}$. In our method, dubbed Neural Additive Vector AutoRegression (NAVAR), we train separate models on the past of each variable to predict its contribution to the value of all variables at the next time step. In particular, at every time step $t$, we pass the past values of a variable $X^{(i)}$ to a neural network $f$ with $N$ output nodes to compute its contribution to all other variables $X^{(j)}$:

$$c_t^{i \to j} = [f_{\theta_i}(X_{t-K:t-1}^{(i)})]^j \tag{4}$$

The function $[f_{\theta_i}]^j$ is the $j$th output of the neural network $f$ with parameters $\theta_i$. A graphical overview of the method can be found in Fig. 1. In principle, one can choose a wide variety of neural networks for $f$, e.g. Multi-Layered Perceptrons (MLP), Recurrent Neural Networks (RNN), and Convolutional Neural Networks (CNN). In our experiments, we consider MLPs and LSTMs [9] to demonstrate the concept, since the additive structure is key to its success. In the LSTM version of our model, single time steps of a variable are sequentially passed to the networks, and thus the networks predict the contributions based only on $X_{t-1}$ and its recurrent hidden states (in contrast to K inputs to the MLP). Therefore, the size of the LSTM network does not increase for larger lags and is thus particularly scalable to longer lags. Although these backbones already outperform

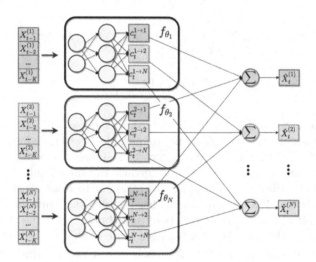

**Fig. 1.** Graphical representation of the NAVAR model with MLPs. For every time step $t$ and every variable $X^{(i)}$, we compute a nonlinear combination of its past $X_{t-K:t-1}^{(i)}$ (with a maximum time lag $K$) as the contribution to every other variable. To compute the estimate of $X_t^{(j)}$, all contributions $c_t^{i \to j}$ are summed.

the state of the art, we envision that more complex backbone architectures for $f$ could potentially further increase performance.

The resulting prediction for $X^{(j)}$ at time $t$ is the sum of all its incoming contributions:

$$\hat{X}_t^{(j)} = \beta^j + \sum_{i=1}^{N} c_t^{i \to j} \tag{5}$$

We choose this additive structure of neural networks as it is a natural extension of the VAR framework with nonlinearities (see Eq. 3) and it allows us to uncover the causal links from $X^{(i)}$ to $X^{(j)}$ by inspecting the direct contributions $c_t^{i \to j}$. Granger causality requires us to estimate the predictions for $X_t^{(j)}$ when the past of $X^{(i)}$ is not included, which in our framework can be directly obtained by ignoring the corresponding contribution $c_t^{i \to j}$ in Eq. (5). This is a key feature of our method that allows it to be scalable: we avoid the necessity to perform multiple fits of a neural network, such as a fit including and excluding the past of variable $X^{(i)}$, when testing the predictive power due to $X^{(i)}$ (see the discussion in Related Work).

The regression networks are trained using the MSE loss function. We introduce an $l_1$ penalty to the contributions $c_t^{i \to j}$ in order to promote sparsity in the resulting causal link structure. Assuming that large causal networks will have a similar number of causes per variable compared to smaller networks, we choose to penalize the sum of the absolute value of received contributions per variable instead of the mean contribution size. This results in the following loss function for the predictions at a time step $t$:

$$\mathcal{L}_t(\beta, \theta) = \frac{1}{N} \sum_{j=1}^{N} \left( \beta^j + \sum_{i=1}^{N} [f_{\theta_i}(X_{t-K:t-1}^{(i)})]^j - X_t^{(j)} \right)^2 \\ + \frac{\lambda}{N} \sum_{i,j=1}^{N} \left| [f_{\theta_i}(X_{t-K:t-1}^{(i)})]^j \right| \tag{6}$$

Furthermore, we add a weight decay term to the loss with coefficient $\mu$.

In order to make the contributions comparable, every individual time series is normalized such that it has mean zero and standard deviation one before training. After training the networks, we deduce the causal links from the variability of the contributions in Eq. (4). The rationale to reconstruct the Granger causal graph is that if a certain variable has a large causal influence on another variable, then it will send a large variety of contributions over the course of time. However, if a variable $X^{(i)}$ is not a Granger cause of another variable $X^{(j)}$ then $f^{ij}$ is a constant function, because $X^{(j)}$ is invariant to the values of $X_{t-K:t-1}^{(i)}$. To score a potential causal link $X^{(i)} \to X^{(j)}$ with the trained neural network, we therefore compute the standard deviation of the set of contributions $c_t^{i \to j}$ for all $t \in \{K+1, T\}$:

$$\text{score}(i \to j) = \sigma(\{c_{K+1}^{i \to j}, c_{K+2}^{i \to j}, ..., c_T^{i \to j}\}) \tag{7}$$

In all of our experiments, we use the ReLU activation function and the Adam optimizer [15] to train our networks. Our implementation of NAVAR and code to reproduce the experiments can be found at: https://github.com/bartbussmann/NAVAR.

## 4   Experiments

### 4.1   Interpretable Contributions

First, we investigate the ability of our model to learn interpretable nonlinear causal dependencies on a toy dataset. We construct the dataset with three variables ($N = 3$) and 4000 time steps ($T = 4000$) based on the following SCM:

$$X_t^{(1)} = \cos(X_{t-1}^{(2)}) + \tanh(X_{t-1}^{(3)}) + \eta_t^1$$
$$X_t^{(2)} = 0.35 \cdot X_{t-1}^{(2)} + X_{t-1}^{(3)} + \eta_t^2$$
$$X_t^{(3)} = \left| 0.5 \cdot X_{t-1}^{(1)} \right| + \sin(2X_{t-1}^{(2)}) + \eta_t^3$$

where $\eta_t^i \sim \mathcal{N}(0, 1)$ for $i = 1, 2, 3$.

We train a NAVAR (MLP) model on this dataset and investigate the learned contributions between pairs of variables. In Fig. 2 we find that the model has learned contributions that are similar to the ground truth causal relationship. Furthermore, we find that for the pairs of variables that are not Granger causal, the learned contribution function has very little variability. This illustrates that our rationale for using the standard deviation of the learned contributions as measure for Granger causal influence is appropriate.

Next, we investigate how to interpret the contributions when the underlying data contains nonlinear interactions across multiple time lags. To this end, we

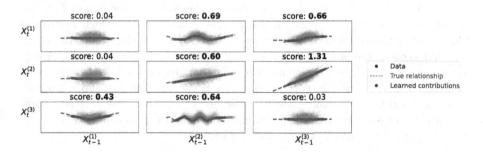

**Fig. 2.** The learned contributions between pairs of variables in our synthetic dataset. The learned contribution functions closely reflect the true causal influence, showing the power of NAVAR models in both Granger causal discovery and interpretability. The causality score from Eq. (7) is given for each potential link. The scores of true causal relationships are presented in boldface.

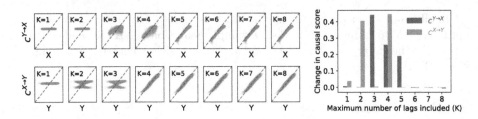

**Fig. 3.** Left Panel: NAVAR discovers coupled nonlinearities within time series across multiple lags. Contributions are shown for different $K$ to study the time lag (by masking the input of the fitted model at higher lags). The diagonal represents learned contributions that perfectly predict the target variable. At lags with true causal relationships, the standard deviation of the contribution increases and the mean squared error (distance to the diagonal) decreases. Right panel: change in causal score from Eq. (7) when including $K$ lags, with respect to the case of including $K-1$ lags. For $c^{Y \to X}$, we observe a high causal score contribution at $K = 3, 4, 5$, while for $c^{X \to Y}$ we observe high scores at $K = 2, 4$, both in agreement with the underlying SCM in Eq. (8).

construct a second synthetic dataset with two variables ($N = 2$) and 4000 time steps ($T = 4000$) based on the following structural causal model:

$$X_t = \cos(Y_{t-3} + Y_{t-4} + Y_{t-5}) + \eta_t^1$$
$$Y_t = X_{t-2} \cdot X_{t-4} + \eta_t^2 \tag{8}$$

where $\eta_t^i \sim \mathcal{N}(0, 0.1)$ for $i = 1, 2$

We train a NAVAR (MLP) model with a maximum lag $K = 8$. Although we do not enforce interpretable additive contributions of individual time lags and thus cannot extract the isolated causal influence of individual time lags, we can still investigate the effect of leaving time lags out. Therefore, we mask the input of the fitted model from a certain maximum time lag. In Fig. 3 three observations can be made: (1) after adding a lag with a true causal link the standard deviation of the contribution increases significantly, which motivated the use of our score function; (2) for time lags with a true causal link the mean squared error decreases; (3) for time lags without a true causal relationship neither of these change significantly, showing that the model did not pick up on spurious contributions (i.e. correlations). We point out that the above analysis is made feasible due to the additive structure which allows us to study pairs of variables in isolation from other contributions, and the sparsity penalty that forces the model to consider mostly direct causes.

## 4.2    CauseMe - Earth Sciences Data

We evaluate our algorithm on various datasets on the CauseMe platform [18]. The CauseMe platform provides benchmark datasets to assess and compare the performance of causal discovery methods. The available benchmarks contain both datasets generated from synthetic models mimicking real challenges, as well as real-world data sets in the earth sciences where the causal structure is known with high confidence. The datasets vary in dimensionality, complexity, and sophistication, and come with various challenges that are common in real datasets, such as autocorrelation, nonlinearities, chaotic dynamics, extreme events, nonstationarity, and measurement errors [28]. On the platform, users have registered over 80 methods for Granger cause discovery.

We compare our methods with four baseline methods implemented by the platform, namely: VAR [30], Adaptive LASSO [36], PCMCI [27], and FullCI [29]. The VAR and Adaptive Lasso methods are both linear regression methods, where the latter consists of computing several Lasso regressions with iterative feature re-weighting. PCMCI and FullCI are constraint-based methods and perform conditional independence tests. Both of these algorithms come with three different independence tests, namely the linear ParCorr test and the nonlinear GPDC and CMI tests. For these methods, we report the results of the best scoring independence test. Furthermore, we compare NAVAR with SLARAC and SELVAR [34], the two algorithms that won the NeurIPS 2019 Causality for Climate competition. SLARAC fits a VAR model on bootstrap samples of the data, each time choosing a random number of lags to include, whereas SELVAR selects edges employing a hill-climbing procedure based on the leave-one-out residual sum of squares of a VAR model.

Every experiment (e.g. Climate, with $N = 40$, $T = 250$) consists of 200 datasets. For every experiment, we tune our hyperparameters (hidden units, batch size, learning rate, contribution penalty coefficient $\lambda$, and weight decay $\mu$) on the first five datasets, of which we use the first 80% for training and the final 20% for validation. The optimal hyperparameters are tabulated in Appendix A[1]. We set the maximum lag parameter $K$ based on information provided by

**Table 1.** Average AUROC on various datasets of the CauseMe platform. Performance of the baseline methods Adaptive LASSO, PCMCI, and FullCI are not available for the hybrid and real-world datasets. For each dataset, we provide the total number of time steps $T$ and the number of variables $N$. Datasets with purely linear dynamics are indicated by an asterisk. Models with the highest AUROC are indicated in boldface.

|  | Nonlinear VAR | | | | Climate* | Weather | River |
|---|---|---|---|---|---|---|---|
|  | N = 3 T = 300 | N = 5 T = 300 | N = 10 T = 300 | N = 20 T = 300 | N = 40 T = 250 | N = 10 T = 2000 | N = 12 T = 4600 |
| NAVAR (MLP) | 0.86 | **0.86** | **0.89** | **0.89** | 0.80 | 0.89 | **0.94** |
| NAVAR (LSTM) | 0.85 | 0.84 | 0.84 | 0.81 | 0.80 | 0.89 | **0.94** |
| SELVAR | **0.88** | **0.86** | 0.86 | 0.85 | 0.81 | 0.90 | 0.87 |
| SLARAC | 0.74 | 0.76 | 0.78 | 0.78 | **0.95** | **0.95** | 0.93 |
| VAR | 0.72 | 0.69 | 0.68 | 0.66 | 0.80 | 0.79 | 0.71 |
| Ad. LASSO | 0.82 | 0.79 | 0.79 | 0.78 | – | – | – |
| PCMCI | 0.85 | 0.82 | 0.83 | 0.82 | – | – | – |
| FullCI | 0.83 | 0.81 | 0.81 | 0.82 | – | – | – |

---

[1] Appendices and code can be found at https://github.com/bartbussmann/NAVAR.

CauseMe, and train on every dataset for 5000 epochs. The AUROC scores are calculated by the CauseMe platform, where self-links are ignored.

We run our method on the synthetic nonlinear VAR dataset, the hybrid climate and weather dataset, and the real-world river run-off dataset. The results in Table 1 show that NAVAR (MLP) models outperform the other methods on most of the nonlinear VAR datasets. Interestingly, where the performance of most methods declines as the number of variables $N$ increases, the performance of NAVAR (MLP) does not decrease. Noting the relative poor performance of SLARAC on the nonlinear VAR dataset compared to its performance on the linear climate dataset, we conclude that this algorithm is very well suited for discovering *exactly linear* relationships. Although NAVAR models might be slightly too flexible for linear datasets, it outperforms the other methods on the real-world river run-off dataset. This strengthens our intuition that many real-world processes can be modeled by an additive combination of nonlinear functions.

## 4.3   DREAM3 - Gene Expression Data

Next, we evaluate our algorithm on the DREAM3 dataset, a simulated gene expression dataset [26]. The benchmark consists of five different datasets of E.Coli and yeast gene networks, each consisting of $N = 100$ variables. For every dataset, 46 time series are available, but every time series consists of only $T = 21$ time steps. We compare NAVAR to other neural approaches to Granger causality, namely componentwise-MLP (cMLP) and componentwise-LSTM (cLSTM) [32], Temporal Causal Discovery Framework (TCDF) [19], and (economy) Statistical Recurrent Units ((e)SRU) [14] (see Related Work).

Similar to the models in [14], we assume a maximum lag of 2 for the MLP models and use 10 hidden units per layer. We calculate the AUROC by increasing a threshold over the causal score, where self-links are ignored in the calculation.

Table 2. Average AUROC on the DREAM3 gene expression dataset. Neural methods are indicated with an asterisk, and their scores are obtained from [14]. Models with the highest AUROC are indicated in boldface.

| Model | E.Coli 1 | E.Coli 2 | Yeast 1 | Yeast 2 | Yeast 3 |
|---|---|---|---|---|---|
| NAVAR (MLP)* | 0.696 | 0.649 | 0.681 | **0.601** | 0.594 |
| NAVAR (LSTM)* | **0.715** | **0.682** | **0.695** | 0.599 | **0.597** |
| cMLP* | 0.644 | 0.568 | 0.585 | 0.506 | 0.528 |
| cLSTM* | 0.629 | 0.609 | 0.579 | 0.519 | 0.555 |
| TCDF* | 0.614 | 0.647 | 0.581 | 0.556 | 0.557 |
| SRU* | 0.657 | 0.666 | 0.617 | 0.575 | 0.550 |
| eSRU* | 0.660 | 0.629 | 0.627 | 0.557 | 0.550 |
| SELVAR | 0.551 | 0.536 | 0.556 | 0.516 | 0.534 |
| SLARAC | 0.580 | 0.509 | 0.526 | 0.503 | 0.494 |

The hyperparameters are tuned using a 80/20% training/validation split, where we train on the first 80% of timesteps, and select the hyperparameters with lowest mean squared error on the final 20% time steps. The selected hyperparameters are reported in Appendix A. The hyperparameters of the other neural models are tuned in tantamount manner and can be found in [14, Appendix G]. We report the average AUROC over 100 different runs of the NAVAR model.

The results in Table 2 show that using deep learning to extract causal structure in time series is a non-trivial task. Our method, however, obtains the best result on all datasets. Since both the MLP and LSTM backbone outperform the other methods, we believe this is due to the imposed structure of our architecture, where the direct contributions of a variable form a more reliable indicator for causality than the methods that rely entirely on induced sparseness in the weight matrices, such as in cMLP, cLSTM, and (e)SRU. Furthermore, using permutation importance with neural networks, as in the TCDF model, is known to generate misleading conclusions [11]. The large difference in performance between NAVAR (MLP) and NAVAR (LSTM) on the E.Coli datasets, demonstrate the benefits of exploring different backbones for different applications.

The linear methods are consistently outperformed by all neural methods on this dataset, which clearly indicates the importance of nonlinearity in causal structure discovery. On top of that, we also immediately obtain interpretable predictions, as shown in Fig. 4, where we show an example of the learned causal contributions in the E.Coli 1 gene network. The model captures that the mNRA levels of gene 0 are mostly influenced by the past mNRA levels of this gene itself. However, at the end of the time series, as the levels of gene 0 go down, the influence from gene 1 and 14 pushes the gene 0 levels further down.

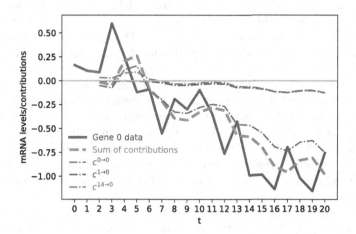

**Fig. 4.** Example of three learned contributions to gene 0 of E.Coli 1 of the DREAM3 dataset. The original data (blue) are normalized. The final prediction is computed by summing the contributions from all genes. (Color figure online)

# 5    Related Work

## 5.1    Neural Methods to Causal Structure Learning

Recently, there has been a rise in interest in applying deep learning to causal structure learning, especially within the framework of Structural Causal Models (SCM) [21,24]. Research in larger graphs was limited due to a combinatorially intractable search space of possible causal graphs. A key ingredient to the solution was presented by Zheng et al. [35], who formulated structure learning as a continuous optimization problem. One of the key advantages of using neural networks is that one can combine the structure learning objective and the prediction objective into a single optimization problem. Other methods that explore this avenue are [16], which extends the [35] method to nonlinear functions modeled by neural networks, while still imposing acyclicity in the causal network. Here, causal links are approximated by neural network paths. Bengio et al. [4] and Ke et al. [13] use a meta-learning transfer objective to identify causal structures from interventional data. The structural learning objective is optimized by varying mask variables that represent the presence/missing of a causal link. Kalainathan et al. [12] explore the use of generative models and adversarial learning to reconstruct the causal graph.

## 5.2    Causal Structure Learning for Time Series Data

Since there is a direct connection between differential equations and structural causal models [5], the functioning of many complex dynamical systems can be understood in terms of causal relationships. Therefore, there has been considerable research devoted to discovering causal relationships in time series. Discovering causal relationships in these temporal settings is more straightforward than in iid data, in the sense that we can use the time-order to establish the directionality of a causal relationship. Approaches that leverage this assumption exist in many variations, such as non-parametric [3,7], model-based [17,22], constraint-based [29], and information theoretic [20] approaches.

Despite the broad range of research in Granger causality in time series, only limited research has applied the representational power of deep learning to this task. A possible reason for this is that the main challenge in causal structure learning is that the final product is the *interpretation* of the dependencies between the variables, which are directly related to the causal connections. However, interpretation is known to be the Achilles heel of black-box tools such as deep learning.

Other works that do use neural networks, such as [1,8,33], first focused on a brute force approach to estimate feature importance, where the Granger causal link $i \to j$ is estimated by the predictive power of a model for $X_t^{(j)}$ that includes the past of all variables, compared to a similar model where the past of the variable $X^{(i)}$ is excluded from the input. However, such an approach is not scalable when the number of variables increases.

The Temporal Causal Discovery Framework (TCDF) [19] uses a attention-based (causal) convolutional neural network. They consider attention scores and introduce permutation importance to identify causal links in an additional causal validation step. Most similar to our work, Tank et al. [32] proposed a neural Granger causal model by using sparse component-wise MLPs (cMLP) and LSTMs (cLSTM). This approach induces sparsity on the causal links by using a hierarchical group regularization. Khanna and Tan [14] use (economical) Statistical Recurrent Units to model the Granger causal dependencies, in a similar vein to the cLSTMs of [32]. Both methods use proximal gradient descent with line search to obtain interpretable results. Proximal optimization is necessary to induce *exact* zeros in the weight matrices of the first layer. Exact zeros are then interpreted as a missing Granger causal link.

In contrast, we do not limit the input features of our model, but instead, enforce interpretability directly into the architecture of our neural network by restricting the function class to produce additive features. This helps in extracting the correct causal relationships between variables, as we can directly regularize the causal summary graph instead of individual input features. Since every prediction is a sum of scalar contributions from the other variables, disentangling the effect of the different inputs becomes trivial and causal influence can be deduced intuitively.

### 5.3   Neural Networks as Generalised Additive Models

In this work, we restrict the structure of the network in order to find the Granger causes of each time series. In particular, our model can be viewed as a Generalized Additive Model (GAM). In the general case, a GAM takes the form:

$$g(E[y]) = \beta + f_1(x_1) + f_2(x_2) + .. + f_n(x_n) \tag{9}$$

One of the main advantages of using GAMs is that the models are considerably more interpretable than many black-box methods since the individual contributions are disentangled and evident. The benefit of assuming additive models was studied in [6,23], but not in the context of neural networks or time series.

The use of deep learning to represent the functions $f_i$ in Eq. (9) was first explored in [25] under the name Generalized Additive Neural Networks (GANNs). For a long time after, this avenue has not been explored further. Interestingly, however, in parallel to this work Agarwal et al. [2] explored the power of Neural Additive Models (NAM) as a predictive model for tabular data with mixed data types. Agarwal et al. [2] introduced exp-centered hidden units (ExU) to allow neural networks to easily approximate 'jumpy functions', which is necessary when considering tabular data.

## 6   Discussion

We presented a neural additive extension to the autoregression framework for (Granger) causal discovery in time series, which we call NAVAR models. The choice of this architecture was guided by the success of VAR models in this context as well as by generalised additive methods as a natural extension to linear methods. We showed that neural additive models have the power to discover nonlinear relationships between time series, while they can still provide an intuitive interpretation of the learned causal interactions. Despite the fact that NAVAR does not account for higher-order interaction terms, benchmarks over a variety of datasets show that NAVAR models are more reliable than existing methods in uncovering the causal structure.

There are many interesting directions for future research. We have shown that NAVAR models already work with MLPs and LSTMs as backbone, but we can easily imagine more complex architectures, such as (dilated) CNNs and Transformers. Furthermore, it could be interesting to investigate bayesian neural networks in order to evaluate the uncertainty of a found causal model. Finally, important future work could be improvements to the model that explicitly account for unobserved confounders, non-stationarity, and contemporaneous causes.

**Acknowledgements.** This project has received funding from the European Union's Horizon 2020 research and innovation programme under the Marie Skłodowska-Curie grant agreement No 813114.

# References

1. Abbasvandi, Z., Nasrabadi, A.M.: A self-organized recurrent neural network for estimating the effective connectivity and its application to EEG data. Comput. Biol. Med. **110**, 93–107 (2019)
2. Agarwal, R., Frosst, N., Zhang, X., Caruana, R., Hinton, G.E.: Neural additive models: Interpretable machine learning with neural nets. arXiv preprint arXiv:2004.13912 (2020)
3. Baek, E., Brock, W.: A general test for nonlinear granger causality: bivariate model. In: Iowa State University and University of Wisconsin at Madison Working Paper (1992)
4. Bengio, Y., et al.: A meta-transfer objective for learning to disentangle causal mechanisms (2019). arXiv preprint arXiv:1901.10912
5. Bongers, S., Mooij, J.M.: From random differential equations to structural causal models: the stochastic case. arXiv preprint arXiv:1803.08784 (2018)
6. Bühlmann, P., Peters, J., Ernest, J., et al.: Cam: causal additive models, high-dimensional order search and penalized regression. Ann. Stat. **42**(6), 2526–2556 (2014)
7. Chen, Y., Bressler, S.L., Ding, M.: Frequency decomposition of conditional granger causality and application to multivariate neural field potential data. J. Neurosci. Methods **150**(2), 228–237 (2006)
8. Duggento, A., Guerrisi, M., Toschi, N.: Echo state network models for nonlinear granger causality. bioRxiv, pp. 651–679 (2019)
9. Gers, F.A., Schmidhuber, J., Cummins, F.: Learning to forget: continual prediction with LSTM. Neural Comput. **12**(10), 2451–2471 (2000)
10. Granger, C.W.J.: Investigating causal relations by econometric models and cross-spectral methods. Econometrica J. Econometric Soc., pp. 424–438 (1969)

11. Hooker, G., Mentch, L.: Please stop permuting features: an explanation and alternatives. arXiv preprint arXiv:1905.03151 (2019)
12. Kalainathan, D., Goudet, O., Guyon, I., Lopez-Paz, D., Sebag, M.: Sam: structural agnostic model, causal discovery and penalized adversarial learning. arXiv preprint arXiv:1803.04929 (2018)
13. Ke, N.R., et al.: Learning neural causal models from unknown interventions. arXiv preprint arXiv:1910.01075 (2019)
14. Khanna, S., Tan, V.F.A.: Economy statistical recurrent units for inferring nonlinear granger causality. arXiv preprint arXiv:1911.09879 (2019)
15. Kingma, D.P., Ba, J.: Adam: a method for stochastic optimization. arXiv preprint arXiv:1412.6980 (2014)
16. Lachapelle, S., Brouillard, P., Deleu, T., Lacoste-Julien, S.: Gradient-based neural dag learning. arXiv preprint arXiv:1906.02226 (2019)
17. Marinazzo, D., Liao, W., Chen, H., Stramaglia, S.: Nonlinear connectivity by granger causality. Neuroimage **58**(2), 330–338 (2011)
18. Muñoz-Marí, J., Mateo, G., Runge, J., Camps-Valls, G.: Causeme: an online system for benchmarking causal discovery methods. In: Preparation (2020)
19. Nauta, M., Bucur, D., Seifert, C.: Causal discovery with attention-based convolutional neural networks. Mach. Learn. Knowl. Extract. **1**(1), 312–340 (2019)
20. Papana, A., Kyrtsou, C., Kugiumtzis, D., Diks, C.: Detecting causality in nonstationary time series using partial symbolic transfer entropy: evidence in financial data. Comput. Econ. **47**(3), 341–365 (2016)
21. Pearl, J.: Causal diagrams for empirical research. Biometrika **82**(4), 669–688 (1995)
22. Peters, J., Janzing, D., Schölkopf, B.: Causal inference on time series using restricted structural equation models. Adv. Neural Inf. Process. Syst. **26**, 154–162 (2013)
23. Peters, J., Mooij, J.M., Janzing, D., Schölkopf, B.: Causal discovery with continuous additive noise models. J. Mach. Learn. Res. **15**(1), 2009–2053 (2014)
24. Peters, J., Janzing, D., Schölkopf, B.: Elements of Causal Inference: Foundations and Learning Algorithms. MIT press, Cambridge (2017)
25. Potts, W.J.E.: Generalized additive neural networks. In: Proceedings of the Fifth ACM SIGKDD International Conference on Knowledge Discovery and Data Mining, pp. 194–200 (1999)
26. Prill, R.J., et al.: Towards a rigorous assessment of systems biology models: the dream3 challenges. PloS one **5**(2), e9202 (2010)
27. Runge, J.: Causal network reconstruction from time series: from theoretical assumptions to practical estimation. Chaos Interdisc. J. Nonlinear Sci. **28**(7), 075310 (2018)
28. Runge, J., et al.: Inferring causation from time series in earth system sciences. Nat. Commun. **10**(1), 1–13 (2019)
29. Runge, J., Nowack, P., Kretschmer, M., Flaxman, S., Sejdinovic, D.: Detecting and quantifying causal associations in large nonlinear time series datasets. Sci. Adv. **5**(11), eaau4996 (2019)
30. Seabold, S., Perktold, J.: Statsmodels: econometric and statistical modeling with python. In: Proceedings of the 9th Python in Science Conference, vol. 57, p. 61. Scipy (2010)
31. Stephan, K.E., et al.: Nonlinear dynamic causal models for fmri. Neuroimage **42**(2), 649–662 (2008)
32. Tank, A., Covert, I., Foti, N., Shojaie, A., Fox, E.: Neural granger causality for nonlinear time series. Stat **1050**, 16 (2018)

33. Wang, Y., et al.: Estimating brain connectivity with varying-length time lags using a recurrent neural network. IEEE Trans. Biomed. Eng. **65**(9), 1953–1963 (2018)
34. Weichwald, S., Jakobsen, M.E., Mogensen, P.B., Petersen, L., Thams, N., Varando, G.: Causal structure learning from time series: Large regression coefficients may predict causal links better in practice than small p-values. arXiv preprint arXiv:2002.09573 (2020)
35. Zheng, X., Aragam, B., Ravikumar, P.K., Xing, E.P.: Dags with no tears: continuous optimization for structure learning. In: Advances in Neural Information Processing Systems, pp. 9472–9483 (2018)
36. Zou, H.: The adaptive lasso and its oracle properties. J. Am. Stat. Assoc **101**(476), 1418–1429 (2006)

# Spatially-Aware Autoencoders for Detecting Contextual Anomalies in Geo-Distributed Data

Roberto Corizzo[1]($\boxtimes$)(iD), Michelangelo Ceci[2,3](iD), Gianvito Pio[2](iD),
Paolo Mignone[2](iD), and Nathalie Japkowicz[1](iD)

[1] Department of Computer Science, American University,
Washington, DC 20016, USA
{rcorizzo,japkowic}@american.edu
[2] Department of Computer Science, University of Bari Aldo Moro, Bari, Italy
{michelangelo.ceci,gianvito.pio,paolo.mignone}@uniba.it
[3] Department of Knowledge Technologies, Jožef Stefan Institute, Ljubljana, Slovenia

**Abstract.** The huge amount of data generated by sensor networks enables many potential analyses. However, one important limiting factor for the analyses of sensor data is the possible presence of anomalies, which may affect the validity of any conclusion we could draw. This aspect motivates the adoption of a preliminary anomaly detection method. Existing methods usually do not consider the spatial nature of data generated by sensor networks. Properly modeling the spatial nature of the data, by explicitly considering spatial autocorrelation phenomena, has the potential to highlight the degree of agreement or disagreement of multiple sensor measurements located in different geographical positions. The intuition is that one could improve anomaly detection performance by considering the spatial context. In this paper, we propose a spatially-aware anomaly detection method based on a stacked auto-encoder architecture. Specifically, the proposed architecture includes a specific encoding stage that models the spatial autocorrelation in data observed at different locations. Finally, a distance-based approach leverages the embedding features returned by the auto-encoder to identify possible anomalies. Our experimental evaluation on real-world geo-distributed data collected from renewable energy plants shows the effectiveness of the proposed method, also when compared to state-of-the-art anomaly detection methods.

**Keywords:** Anomaly detection · Auto-encoders · Geo-distributed data

## 1 Introduction

The increasing adoption of sensor networks leads to the generation of a large amount of data, that could fruitfully be analyzed to support decision-making processes in multiple real-world sectors. Machine learning and data mining methods

© Springer Nature Switzerland AG 2021
C. Soares and L. Torgo (Eds.): DS 2021, LNAI 12986, pp. 461–471, 2021.
https://doi.org/10.1007/978-3-030-88942-5_36

for the analysis of data generated by sensor networks have been adopted in multiple application domains. However, it is noteworthy that data collected through sensor networks are inherently affected by anomalies. This is due to the nature of the sensors, which operate in an external environment, and to the nature of the network (grid), which can be subject to communication issues. Therefore, directly using raw sensor data to solve the task at hand may result in a degraded accuracy of the models [10]. For this reason, incorporating preliminary anomaly detection phases in the data analysis workflow appears fundamental. Recently, this thread has attracted increasing interest, with emerging approaches tailored for specific representations and for the detection of specific anomalies [19].

Sensor networks also open to the possibility to collect observations for a set of properties of interest in multiple geographical locations. In the literature, statistical techniques have been investigated to analyze geo-distributed sensor data in a combined manner, trying to improve the performance of the learning models. For instance, the incorporation of statistical indicators of spatio-temporal autocorrelation in classical machine learning algorithms has been successfully investigated in [8,17]. However, this opportunity has been often disregarded by recent anomaly detection approaches, often based on deep neural network architectures [19]. The goal of this paper is to fill this gap. Specifically, we propose a method to solve *unsupervised* anomaly detection tasks, where the considered anomalies are *contextual* [9,10]. More in detail, anomalies are detected on a single geographic position on the basis of the multi-dimensional sensor data observed at that location and its neighboring locations (*diffused context* [9,10]).

Methodologically, we propose a neural network architecture, based on stacked auto-encoders, that incorporates a specific spatial encoding component to capture spatial autocorrelation phenomena. We argue that capturing the agreement (or disagreement) of the measurement of the same physical property, at the same time point, in multiple locations may boost the anomaly detection accuracy of the model. The adoption of stacked auto-encoders in our method is motivated by their ability to learn non-linear representations that effectively incorporate salient features [7]. The hidden layers of the model architecture are usually chosen to have a reduced number of neurons, compared to the input layer, representing data with a reduced dimensionality.

Auto-encoders have already been exploited to solve anomaly detection tasks in [3,20], mainly leveraging the reconstruction error. A popular approach is to train the auto-encoder on background data, which is assumed to belong to the *normal* class (i.e., without anomalies). After the training stage, new instances fed to the model are expected to exhibit a low reconstruction error if they belong to the normal class. On the contrary, anomalies are expected to show a high reconstruction error, due to the fact that they belong to a different distribution.

Although this approach appears relatively intuitive, *i)* it does not take into account the spatial dimension in the data for the identification of the anomalies and *ii)* it might be susceptible to noise introduced in the data. Both these aspects are typical of data generated by sensor networks and the analysis of such type of data requires to overcome them. As for *i)*, we explicitly consider spatial autocorrelation in the learning phase and, as for *ii)* we propose to identify anomalies in the embedding space rather than in the original feature space, to be more robust to the presence of noise in the data. For this purpose, we propose to leverage the feature extraction capability of the model and perform anomaly detection by analyzing the embedding bottleneck features of the stacked auto-encoder.

In summary, the contributions of this paper are the following: *i)* we propose a stacked auto-encoder architecture which incorporates a spatial encoding stage in its architecture, to explicitly model spatial autocorrelation in geo-distributed multi-variate sensor data; *ii)* we devise a distance-based anomaly detection technique that leverages the distance among data observations, represented according to an embedding space learned by the stacked auto-encoder; *iii)* we evaluate the proposed approach on real-world datasets related to the renewable energy field.

## 2   Background

Data anomalies can usually be classified in three categories: point, contextual, and collective anomalies [9]. In this paper we address the detection of contextual anomalies, where the context is represented by the spatial dimension of a data observation [12,16]. For instance, a contextual anomaly could be represented by an abrupt temperature value measurement at one geographical location.

In general, the identification of contextual anomalies can be carried out with supervised, semi-supervised or unsupervised machine learning approaches [9].

Although there are several machine learning based methods, unsupervised ones are better suited for domains characterized by a scarce availability (or by the total absence) of labeled data, which is the case in many real-world scenarios.

Among existing methods, it is worth mentioning One-Class SVM (OCSVM) [15], that learns a separating hyperplane in a high-dimensional space [15]. Once the model is learned, OCSVM can classify a new data observation as similar (i.e., normal) or different (i.e., anomaly) with respect to the training data distribution, according to its position within the decision boundary. In this line of research, OCSVM models have also been adopted in ensemble settings [1,18].

Isolation Forests [13] exploit a combination of tree-based models, through which calculate an isolation score for each data observation. Specifically, the score of an observation is computed as the average path length from the root of the tree to the node containing the single observation. A short path indicates that an observation is easy to isolate from the others due to significantly different attribute values compared to the training data points.

In this scenario, methods based on auto-encoders and stacked auto-encoders [20] have demonstrated superior performance. This behavior is theoretically motivated by their ability to construct representations, with a low reconstruction error, based on non-linear combinations of the input features [4].

Although auto-encoders have seen particular interest for anomaly detection from images [20], in this work we adopt such models and investigate their effectiveness for the detection of abrupt changes in multivariate time-series data. Moreover, we introduce a novel component in the architecture to explicitly model spatial autocorrelation phenomena.

# 3    Method

The method proposed in this paper is able to analyze multi-variate sensor data (related, for example, to temperature, wind speed, pressure, etc.) collected from multiple geo-distributed locations. Specifically, considering a discrete timeline and a set of locations $L$, let $x_{t,l}$ be the vector of measurements at time $t$ and location $l \in L$. The multi-variate data coming from sensors can be represented as an unbounded sequence (i.e., a stream) of sets:

$$D = \langle \{x_{1,1}, \ldots, x_{1,|L|}\}, \{x_{2,1}, \ldots, x_{2,|L|}\}, \ldots, \{x_{t,1}, \ldots, x_{t,|L|}\} \rangle$$

We learn a stacked auto-encoder using $D$ as input data representation to subsequently carry out the anomaly detection task. The adoption of stacked auto-encoders is motivated by their ability to extract layer-wise representations, at increasing levels of abstraction. In general, the first layer of a stacked auto-encoder learns simple features (e.g., edges, in the image domain), whereas deeper layers learn features at increasing levels of complexity and summarization (e.g., co-occurring edges that form corners). This characteristic allows to model complex properties of background data in the embedding space, that may in turn lead to an increased ability to discriminate between normal and anomalous instances.

In the following subsection, we describe the proposed strategy to explicitly consider spatial autocorrelation phenomena in the auto-encoder architecture.

## 3.1    Spatial Encoding Stage

The proposed auto-encoder architecture features a *spatial encoding* stage which is based on LISA (Local Indicators of Spatial Autocorrelation) [2], that simultaneously exploits data available at every location.

In order to describe how the spatial encoding stage works, we introduce how the computation of LISA is performed. The spatial neighborhood of the sensor network is expressed as a matrix, and for each location and data observation at a time point $t$, LISA is computed using such a matrix. Specifically, the first step is to define a neighborhood matrix $\Lambda \in \mathbb{R}^{|L| \times |L|}$, such that:

$$\Lambda[i,j] = 1 - dist(l_i, l_j)/maxDist \tag{1}$$

where $l_i \in L$ and $l_j \in L$ are two locations, $dist(l_i, l_j)$ is the spatial distance (in kilometers) between the two locations $l_i$ and $l_j$, and $maxDist$ is the maximum pairwise spatial distance observed in the sensor network.

A subsequent step computes the deviation of each data feature with respect to the mean, leveraging z-score normalization. Intuitively, in our approach we are interested in identifying the contribution of the neighboring locations for each feature observed at each time point. Therefore, given $D_t = \{x_{t,1}, \ldots, x_{t,|L|}\} \in D$, the subset of all data observations for all the locations at a specific time $t$, the z-scores for a location $l \in L$ are calculated as: $z_{t,l}^{(f)} = \left( x_{t,l}^{(f)} - \overline{D_t^{(f)}} \right) / \sigma_{D_t^{(f)}}$, where $f$ is a generic feature measured by a sensor (that is, a generic element

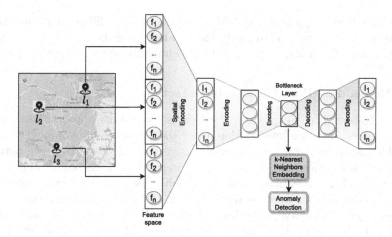

**Fig. 1.** A graphical representation of our spatially-aware auto-encoder architecture.

of the vector $x_{t,l}$); $\overline{D_t^{(f)}}$ represents the average value of the feature $f$ in all the locations; $\sigma_{D_t^{(f)}}$ represents the standard deviation of the feature $f$ in all the locations. Leveraging $z_{t,l}^{(f)}$, it is possible to compute LISA for the variable $f$ of the location $l_i$ for time $t$ (according to [2]) as follows:

$$I_{l_i,t}^{(f)} = z_{t,l_i}^{(f)} \cdot \sum_{l_j \in L, i \neq j} \Lambda[i,j] \cdot z_{t,l_j}^{(f)} \tag{2}$$

Following the aforementioned process, for each time point $t$, the spatial encoding stage extracts a new representation $S_t$ as follows:

$$S_t = \{[I_{l_1,t}^{(f_1)}, \dots, I_{l_1,t}^{(f_n)}], [I_{l_2,t}^{(f_1)}, \dots, I_{l_2,t}^{(f_n)}], \dots, [I_{|L|,t}^{(f_1)}, \dots, I_{|L|,t}^{(f_n)}]\} \tag{3}$$

### 3.2 Encoding and Decoding Stage

The subsequent encoding stages extract new representations with a lower dimensionality than the input data, similarly to the typical auto-encoder architecture. In our model, we perform two encoding stages after the spatial encoding stage (see Fig. 1), with 1/2 and 1/4 of the input features, respectively. The architecture is trained end-to-end leveraging historical data which represent normal behavior conditions. We assume that historical data contains no anomalies (or a negligible amount), and use the trained model for anomaly detection purposes.

Starting from the dataset $D$, the stacked auto-encoder aims at learning the encoding function $e : \mathcal{X} \rightarrow \mathcal{F}$ and the decoding function $d : \mathcal{F} \rightarrow \mathcal{X}$, such that:

$$\langle e(\cdot), d(\cdot) \rangle = \underset{\langle e(\cdot), d(\cdot) \rangle}{\operatorname{argmin}} \|D - d(e(D))\|^2, \tag{4}$$

where $\mathcal{X}$ is the input space of $D$, and $\mathcal{F}$ is the learned embedding space.

The functions $e(\cdot)$ and $d(\cdot)$ should be parametric and differentiable according to a distance function. Consequently, the parameters of the encoding and decoding functions defined above are optimized by minimizing the reconstruction loss.

### 3.3 Embedding-Based Anomaly Detection

To detect anomalies, we propose a $k$-Nearest Neighbors approach that leverages the encoded data representation. Once the auto-encoder is trained with the available historical data, we compute the average Euclidean distance between each data observation and its nearest $k$ observations in the embedding space. Coherently, when a new observation is available, we encode it in the embedding space, and compute its average distance w.r.t. the nearest $k$ observations. If the distance is greater than a given threshold, then the observation is considered as an anomaly. In this work, we do not adopt a manual threshold, but estimate it from the data distribution. Specifically, we use $[\bar{d} + 3 \cdot \sigma]$, where $\bar{d}$ is the average pairwise distance observed between each training data observation and its nearest observations, while $\sigma$ is the standard deviation of the observed distances.

Note that the identification of the $k$ nearest neighbors in our method is based on the Hybrid Spill Tree (HSP) [14], a distributed data structure (variant of metric trees) for high-dimensional indexing, that allows to retrieve the $k$ nearest neighbors of an observation in $O(log|D|)$.

## 4 Experiments

### 4.1 Datasets

The datasets considered in our experiment consist of weather variables (such as temperature, humidity, etc.) monitored at hourly granularity by sensors placed on renewable energy plants, located in different geographical areas. In particular, we considered the following datasets analyzed also in previous studies [11]:

- **PV Italy.** The dataset consists of data collected every 15 min (from 2:00 AM to 8.00 PM, every day) by sensors located on 17 photovoltaic power plants located in Italy. The time period spans from January 1st, 2012 to May 4th, 2014. More details about data preprocessing steps can be found in [6].
- **Wind NREL.** This dataset (www.nrel.gov/wind) was modeled using the Weather Research & Forecasting model. Five plants with the highest production have been selected, obtaining the time series of wind speed and production observed every 10 min, for a time period of two years (from January 1st, 2005 to December 31st, 2006). Hourly aggregation was performed.

For both datasets, we consider the following features: latitude and longitude of each plant; day and hour; altitude and azimuth; weather conditions, i.e., ambient temperature, irradiance, pressure, wind speed, wind bearing, humidity, dew point, cloud cover, and a descriptive weather summary. Weather conditions are either measured (training phase) or forecasted (detection phase). In particular, all the weather data were extracted from Forecast.io, except for the expected altitude and azimuth, that were extracted from SunPosition (www.susdesign. com/sunposition), and the expected irradiance (PV Italy dataset only), that was extracted from PVGIS (re.jrc.ec.europa.eu/pvg_tools/en/#MR).

For each dataset, we build the testing set by selecting all the instances (measurements at hourly granularity, observed at all the plants) belonging to 10 randomly selected days. We analyze the anomaly detection capabilities of the method, considering three different training window sizes: 30, 60 and 90 days. This means that, for each day in the testing set, we train the model using historical data belonging to 30, 60 or 90 days, respectively, preceding the considered testing day, with the goal of identifying anomalies for all the measurements belonging to the considered day of the testing set. For evaluation purposes, anomalies are artificially introduced by perturbating the correct attribute values. This is done on 25% of instances on 50% of the features.

## 4.2 Competitor Systems and Experimental Setup

In line with the discussion of existing works reported in Sect. 2, in our experiments we considered, as possible competitors, the most suitable class of approaches to address the task of interest in our study, that are mainly based one-class classification. Indeed, they offer the flexibility to learn a model from an initial (regular) data distribution and are able to flag data that significantly differ from the learned distribution. In particular, we considered three state-of-the-art competitor methods falling in this class, namely One-Class SVM (**OCSVM**) [15], **Isolation Forest** [13], and an **Auto-encoder** architecture that bases the detection of anomalies on the reconstruction error [3,20]. These approaches are widely adopted, and generally provide highly accurate detections.

Their parameters were set to the values suggested in their respective papers. In particular, for One-Class SVM, we choose a Radial Basis Function (RBF) kernel and select the best value for the $\gamma$ parameter in the set $\gamma \in \{0.1, scale, auto\})$. The *auto* configuration corresponds to $\gamma = \frac{1}{n\_features}$, whereas the *scale* configuration corresponds to $\gamma = \frac{1}{n\_features \cdot var(X)}$, where $var(X)$ represents the variance of the training data. For Isolation Forest, we set: the number of base estimators in the ensemble $n\_estimators \in \{10, 25, 50\}$; the number of features to draw at random for each base estimator equal to the number of the whole set of features. For the auto-encoder, we followed the heuristics proposed by [5]: we initially experimented with different configurations for *learning rate* (negative powers of 10, starting from a default value of 0.01) and *batch size* (powers of 2) using a 20% validation set. Preliminary results suggested that the different configurations did not affect performance metrics significantly. For this reason, the experiments were performed with the following parameters: $epochs=500, learning\ rate=0.0001, batch\ size=32$. Moreover, we experimented with two different values of its parameter $p$, i.e., $p \in \{1.5, 3\}$ (if the reconstruction error deviates more than $p \cdot \sigma$ from the one observed on the training set, the instance is marked as an anomaly).

As regards our method, we report the results with different values of $k$, namely $k \in \{50, 100, 150\}$. Finally, in order to specifically evaluate the contribution provided by the spatial embedding component, we also report the results obtained by a simplified version of the proposed architecture, that does not exploit the spatial embedding step. We call this variant **Without SE**.

All the results were collected in terms of Precision, Recall and F-Score.

## 4.3   Results and Discussion

In Table 1 we report all the results obtained in our experiments. First, we can observe that our approach generally obtains the best results among all the considered methods. Looking specifically at the results on PV Italy, we can observe that the best F-Score results are obtained with a time window of 30 days. This means that the kNN-based approach that we propose achieves optimal results even with a limited view on historical data. Looking at Precision and Recall, it is clear that our approach is sensitive to anomalies, but robust to false detections: the results in terms of precision ($\sim$98–99%) indicate that the false positive rate is around 1–2%, while the Recall results indicate a good rate of detected anomalies, i.e., around 75%. Such results are not obtained by competitor systems, that show a significantly lower Precision ($\sim$85% in the best case, obtained by the Auto-encoder, $\sigma = 1.5$), and recall, especially in the case of Isolation Forest.

Looking at the simpler version of our method (Without SE), we can observe comparable, but lower results than those achieved by the full variant of our method. This behavior confirms that the proposed architecture, based on kNN on the embedded instances, is generally effective and is further supported by the spatial encoding step that takes into account spatial autocorrelation phenomena.

A closer look at the results obtained on the Wind NREL dataset reveals a similar situation. In this case, we can only observe one case (i.e., window size = 90 days and $k = 150$) in which the best results are achieved by the variant of our method that does not exploit the spatial embedding component. However, the difference with the full version of our method is negligible, and it may be possibly due to the fact that, in this dataset, less features are correlated to the spatial dimension, with respect to the photovoltaic power plants in PV Italy (see, e.g., the irradiance feature). Nevertheless, a contribution of the spatial encoding step can still be observed when the time window is limited to 30 or 60 days.

Focusing on the best F-score results achieved by the considered methods, measured over all the values of their parameters, we can easily observe that the proposed method always outperforms all the other competitors. We can also observe a slightly higher Recall exhibited by the auto-encoder, but at the price of a significantly lower Precision. However, our method generally leads to a 7%-9% improvement in terms of F-score in all the cases with respect to the auto-encoder.

**Table 1.** Anomaly detection results obtained considering varying training sliding window sizes. Best F-Score results for each Window size configuration are marked in bold.

| PV Italy | 30 days | | | 60 days | | | 90 days | | |
|---|---|---|---|---|---|---|---|---|---|
| **Our method** | Prec | Rec | F-Score | Prec | Rec | F-Score | Prec | Rec | F-Score |
| $k=50$ | 0.9853 | 0.7433 | 0.8472 | 0.9801 | 0.7437 | 0.8444 | 0.9764 | 0.7412 | 0.8419 |
| $k=100$ | 0.9896 | 0.7452 | 0.8500 | 0.9853 | 0.7455 | 0.8478 | 0.9787 | 0.7430 | 0.8436 |
| $k=150$ | 0.9925 | 0.7464 | **0.8519** | 0.9871 | 0.7458 | **0.8488** | 0.9814 | 0.7433 | **0.8451** |
| **Without SE** | Prec | Rec | F-Score | Prec | Rec | F-Score | Prec | Rec | F-Score |
| $k=50$ | 0.9248 | 0.7729 | 0.8250 | 0.9161 | 0.7948 | 0.8344 | 0.9118 | 0.7965 | 0.8337 |
| $k=100$ | 0.9517 | 0.7528 | 0.8261 | 0.9323 | 0.7687 | 0.8260 | 0.9228 | 0.7711 | 0.8236 |
| $k=150$ | 0.9651 | 0.7427 | 0.8277 | 0.9451 | 0.7553 | 0.8246 | 0.9349 | 0.7600 | 0.8229 |
| **Auto-encoder** | Prec | Rec | F-Score | Prec | Rec | F-Score | Prec | Rec | F-Score |
| $p = 1.5$ | 0.8516 | 0.8179 | 0.7836 | 0.8502 | 0.8145 | 0.7774 | 0.8467 | 0.8132 | 0.7771 |
| $p = 3$ | 0.7880 | 0.7703 | 0.7055 | 0.7877 | 0.7698 | 0.7049 | 0.7892 | 0.7722 | 0.7087 |
| **OCSVM** | Prec | Rec | F-Score | Prec | Rec | F-Score | Prec | Rec | F-Score |
| | 0.7880 | 0.7386 | 0.7305 | 0.7880 | 0.7386 | 0.7305 | 0.7880 | 0.7386 | 0.7305 |
| **Isolation forest** | Prec | Rec | F-Score | Prec | Rec | F-Score | Prec | Rec | F-Score |
| n_estimators=10 | 0.6277 | 0.4409 | 0.4033 | 0.6277 | 0.4409 | 0.4033 | 0.6277 | 0.4409 | 0.4033 |
| n_estimators=25 | 0.6799 | 0.5092 | 0.5007 | 0.6799 | 0.5092 | 0.5007 | 0.6799 | 0.5092 | 0.5007 |
| n_estimators=50 | 0.6879 | 0.5406 | 0.5450 | 0.6879 | 0.5406 | 0.5450 | 0.6879 | 0.5406 | 0.5450 |
| *Wind NREL* | 30 days | | | 60 days | | | 90 days | | |
| **Our method** | Prec | Recall | F-Score | Prec | Recall | F-Score | Prec | Recall | F-Score |
| $k=50$ | 0.9871 | 0.7558 | 0.8523 | 0.9726 | 0.7675 | 0.8490 | 0.9636 | 0.7675 | 0.8453 |
| $k=100$ | 0.9992 | 0.7508 | **0.8569** | 0.9887 | 0.7567 | 0.8530 | 0.9837 | 0.7533 | 0.8494 |
| $k=150$ | 0.9992 | 0.7508 | **0.8569** | 0.9938 | 0.7542 | **0.8553** | 0.9915 | 0.7533 | 0.8534 |
| **Without SE** | Prec | Recall | F-Score | Prec | Recall | F-Score | Prec | Recall | F-Score |
| $k=50$ | 0.9768 | 0.7586 | 0.8478 | 0.9585 | 0.7664 | 0.8421 | 0.9590 | 0.7700 | 0.8439 |
| $k=100$ | 0.9930 | 0.7547 | 0.8548 | 0.9781 | 0.7559 | 0.8474 | 0.9772 | 0.7647 | 0.8503 |
| $k=150$ | 0.9970 | 0.7526 | 0.8564 | 0.9874 | 0.7529 | 0.8511 | 0.9884 | 0.7581 | **0.8537** |
| **Auto-encoder** | Prec | Rec | F-Score | Prec | Rec | F-Score | Prec | Rec | F-Score |
| $p = 1.5$ | 0.8371 | 0.8344 | 0.7907 | 0.8704 | 0.8408 | 0.7997 | 0.8652 | 0.8384 | 0.7971 |
| $p = 3$ | 0.7159 | 0.7880 | 0.7068 | 0.7164 | 0.7888 | 0.7085 | 0.6924 | 0.7888 | 0.7083 |
| **OCSVM** | Prec | Rec | F-Score | Prec | Rec | F-Score | Prec | Rec | F-Score |
| | 0.8341 | 0.8040 | 0.8050 | 0.8341 | 0.8040 | 0.8050 | 0.8341 | 0.8040 | 0.8050 |
| **Isolation forest** | Prec | Rec | F-Score | Prec | Rec | F-Score | Prec | Rec | F-Score |
| n_estimators=10 | 0.7045 | 0.3792 | 0.3198 | 0.7045 | 0.3792 | 0.3198 | 0.7045 | 0.3792 | 0.3198 |
| n_estimators=25 | 0.8076 | 0.4128 | 0.3585 | 0.8076 | 0.4128 | 0.3585 | 0.8076 | 0.4128 | 0.3585 |
| n_estimators=50 | 0.8181 | 0.4272 | 0.3729 | 0.8181 | 0.4272 | 0.3729 | 0.8181 | 0.4272 | 0.3729 |

# 5   Conclusion

In this paper we presented a novel anomaly detection method based on an auto-encoder architecture that features a spatial encoding stage to model spatial auto-correlation. The proposed architecture is unsupervised, and the model is trained using historical data. The anomaly detection task is carried out by comparing the projection of new observations in the embedding space to their nearest neighbors.

This strategy allows us to detect anomalies using a distance-based approach that exploits a threshold automatically estimated from training embedded data.

The experimental evaluation performed on two datasets, related to real-world sensor networks of power plants, showed significant improvements in terms of F-Score, that reaches 9.18% compared to auto-encoders based on reconstruction error. A direct comparison with a variant of the proposed method, that does not exploit the spatial encoding component, also revealed the positive contribution coming from the explicit consideration of the spatial information.

As future work we will investigate other approaches to model spatio-temporal autocorrelation, as part of the neural network architecture. Moreover, we will conduct an extensive experimental evaluation involving datasets related to other domains, and affected by different amounts and types of anomalies.

**Acknowledgement.** The authors acknowledge the support of the U.S. DARPA through the project "Lifelong Streaming Anomaly Detection" (Grant N. A19-0131-003 and A21-0113-002), and of the EU Commission through the H2020 project "IMPETUS-Intelligent Management of Processes, Ethics and Technology for Urban Safety" (Grant n. 883286). GP acknowledges the support of Ministry of Universities and Research (MUR) through the project "Big Data Analytics", AIM 1852414, activity 1, line 1. PM acknowledges the support of Apulia Region through the project "Metodi per l'ottimizzazione delle reti di distribuzione di energia e per la pianificazione di interventi manutentivi ed evolutivi" (CUP H94I20000410008, Grant n. 7EDD092A) in the context of "Research for Innovation - REFIN". We also acknowledge the support of NVIDIA through the donation of a Titan V GPU.

# References

1. Aggarwal, C.C.: Outlier ensembles: position paper. ACM SIGKDD Explor. Newsl **14**(2), 49–58 (2013)
2. Anselin, L.: Local indicators of spatial association-LISA. Geograph. Anal. **27**(2), 93–115 (1995)
3. Beggel, L., Pfeiffer, M., Bischl, B.: Robust anomaly detection in images using adversarial autoencoders. In: Brefeld, U., Fromont, E., Hotho, A., Knobbe, A., Maathuis, M., Robardet, C. (eds.) ECML PKDD 2019. LNCS (LNAI), vol. 11906, pp. 206–222. Springer, Cham (2020). https://doi.org/10.1007/978-3-030-46150-8_13
4. Bengio, Y.: Learning Deep Architectures for AI. Now Publishers Inc. (2009)
5. Bengio, Y.: Practical recommendations for gradient-based training of deep architectures. In: Montavon, G., Orr, G.B., Müller, K.-R. (eds.) Neural Networks: Tricks of the Trade. LNCS, vol. 7700, pp. 437–478. Springer, Heidelberg (2012). https://doi.org/10.1007/978-3-642-35289-8_26
6. Ceci, M., Corizzo, R., Fumarola, F., Malerba, D., Rashkovska, A.: Predictive modeling of pv energy production: How to set up the learning task for a better prediction? IEEE Trans. Ind. Inf. **13**(3), 956–966 (2016)
7. Ceci, M., Corizzo, R., Japkowicz, N., Mignone, P., Pio, G.: Echad: embedding-based change detection from multivariate time series in smart grids. IEEE Access **8**, 156053–156066 (2020)
8. Ceci, M., Corizzo, R., Malerba, D., Rashkovska, A.: Spatial autocorrelation and entropy for renewable energy forecasting. Data Min. Knowl. Disc. **33**(3), 698–729 (2019). https://doi.org/10.1007/s10618-018-0605-7

9. Chandola, V., Banerjee, A., Kumar, V.: Anomaly detection: a survey. ACM Comput. Surv. (CSUR) **41**(3), 15 (2009)
10. Corizzo, R., Ceci, M., Japkowicz, N.: Anomaly detection and repair for accurate predictions in geo-distributed big data. Big Data Res. **16**, 18–35 (2019)
11. Corizzo, R., Pio, G., Ceci, M., Malerba, D.: DENCAST: distributed density-based clustering for multi-target regression. J. Big Data **6**(1), 1–27 (2019). https://doi.org/10.1186/s40537-019-0207-2
12. Kou, Y., Lu, C.T., Chen, D.: Spatial weighted outlier detection. In: Proceedings of SIAM International Conference on Data Mining 2006, pp. 614–618. SIAM (2006)
13. Liu, F.T., Ting, K.M., Zhou, Z.H.: Isolation forest. In: 2008 Eighth IEEE International Conference on Data Mining. pp. 413–422. IEEE (2008)
14. Liu, T., Moore, A.W., Gray, A., Yang, K.: An investigation of practical approximate nearest neighbor algorithms. In: NIPS 2004, pp. 825–832. MIT Press (2004)
15. Schölkopf, B., Williamson, R.C., Smola, A.J., Shawe-Taylor, J., Platt, J.C.: Support vector method for novelty detection. In: Advances in Neural Information Processing Systems, pp. 582–588 (2000)
16. Shekhar, S., Lu, C.T., Zhang, P.: Detecting graph-based spatial outliers: algorithms and applications (a summary of results). In: ACM SIGKDD, pp. 371–376 (2001)
17. Stojanova, D., Ceci, M., Appice, A., Malerba, D., Džeroski, S.: Dealing with spatial autocorrelation when learning predictive clustering trees. Ecol. Inf. **13**, 22–39 (2013)
18. Xing, H.J., Liu, W.T.: Robust adaboost based ensemble of one-class support vector machines. Inf. Fusion **55**, 45–58 (2020)
19. Zhang, C., Song, D., Chen, Y., Feng, X., et al.: A deep neural network for unsupervised anomaly detection and diagnosis in multivariate time series data. In: AAAI 2019, vol. 33, pp. 1409–1416 (2019)
20. Zhou, C., Paffenroth, R.C.: Anomaly detection with robust deep autoencoders. In: ACM SIGKDD 2017, pp. 665–674 (2017)

# Author Index

Printed in the United States
by Baker & Taylor Publisher Services